SELECTED SOLUTIONS MANUAL

JOSEPH TOPICH

Virginia Commonwealth University

McMURRY FAY

Chemistry

FOURTH EDITION

Pearson Education, Inc.
Upper Saddle River, NJ 07458

Assistant Editor: Melanie VanBenthuysen
Executive Editor: Kent Porter Hamann
Editor-in-Chief: John Challice
Vice President of Production & Manufacturing: David W. Riccardi
Executive Managing Editor: Kathleen Schiaparelli
Assistant Managing Editor: Becca Richter
Production Editor: Dana Dunn
Supplement Cover Management/Design: Paul Gourhan
Manufacturing Buyer: Ilene Kahn
Cover Photo: Quade Paul and Jeffrey A. Scovil

© 2004 Pearson Education, Inc.
Pearson Prentice Hall
Pearson Education, Inc.
Upper Saddle River, NJ 07458

Printed in the United States of America

10 9 8 7 6 5 4 3 2

ISBN 0-13-140214-5

Pearson Education Ltd., *London*
Pearson Education Australia Pty. Ltd., *Sydney*
Pearson Education Singapore, Pte. Ltd.
Pearson Education North Asia Ltd., *Hong Kong*
Pearson Education Canada, Inc., *Toronto*
Pearson Educación de Mexico, S.A. de C.V.
Pearson Education—Japan, *Tokyo*
Pearson Education Malaysia, Pte. Ltd.
Pearson Education, *Upper Saddle River, New Jersey*

CONTENTS

eMedia Solutions
CONTENTS

Preface

Chemistry is the study of the composition and properties of matter along with the changes that matter undergoes. The principles of chemistry are learned through a combination of different experiences. You read about them in your textbook. You hear about them from your chemistry instructor. You see them first hand in the laboratory, and probably most importantly, you use chemistry principles to solve chemistry problems.

Problem solving is your key to success in chemistry! CHEMISTRY, 4/e by McMurry and Fay contains thousands of problems for you to work. Develop a problem solving strategy. Read each problem carefully. List the information contained in the problem. Understand what the problem is asking. Use your knowledge of chemistry principles to identify connections between the information in the problem and the solution that you are seeking. Set up and attempt to solve the problem. Look at your answer. Is it reasonable? Are the units correct? Then, and only then, check your answer with the *Selected Solutions Manual*.

The *Selected Solutions Manual* to accompany CHEMISTRY, 4/e by McMurry and Fay contains the solutions to all in-chapter, and even numbered understanding key concept and end-of-chapter problems.

I have worked to ensure that the solutions in this manual are as error free as possible. Solutions have been double-checked and in many cases triple-checked. Small differences in numerical answers between student results and those in the *Selected Solutions Manual* may result because of rounding and significant figure differences. It should also be noted that there is, in many cases, more than one acceptable set up for a problem.

I would like to thank John McMurry and Robert Fay for the opportunity to contribute to their CHEMISTRY package. I also want to thank them for their helpful comments as I worked on this solutions manual. I also want to acknowledge and thank David Ball (accuracy checker) and the Prentice Hall staff. Finally, I want to thank in a very special way my wife, Ruth, and our daughter, Judy, for their constant encouragement and support as I worked on this project.

Joseph Topich
Department of Chemistry
Virginia Commonwealth University

1   Chemistry: Matter and Measurement

1.1   (a) Cd   (b) Sb   (c) Am

1.2   (a) silver   (b) rhodium   (c) rhenium   (d) cesium   (e) argon   (f) arsenic

1.3   (a) Ti, metal   (b) Te, semimetal   (c) Se, nonmetal
       (d) Sc, metal   (e) At, semimetal   (f) Ar, nonmetal

1.4   The three "coinage metals" are copper (Cu), silver (Ag), and gold (Au).

1.5   (a) The decimal point must be shifted ten places to the right so the exponent is −10. The
       result is 3.72×10^{-10} m.
       (b) The decimal point must be shifted eleven places to the left so the exponent is 11. The
       result is 1.5×10^{11} m.

1.6   (a) microgram   (b) decimeter   (c) picosecond
       (d) kiloampere   (e) millimole

1.7 $°C = \dfrac{5}{9} \times (°F - 32) = \dfrac{5}{9} \times (98.6 - 32) = 37.0°C$

       $K = °C + 273.15 = 37.0 + 273.15 = 310.2$ K

1.8   (a) $K = °C + 273.15 = -78 + 273.15 = 195.15 \ K = 195 \ K$

       (b) $°F = (\dfrac{9}{5} \times °C) + 32 = (\dfrac{9}{5} \times 158) + 32 = 316.4°F = 316°F$

       (c) $°C = K - 273.15 = 375 - 273.15 = 101.85°C = 102°C$

       $°F = (\dfrac{9}{5} \times °C) + 32 = (\dfrac{9}{5} \times 101.85) + 32 = 215.33°F = 215°F$

1.9 $d = \dfrac{m}{V} = \dfrac{27.43 \ g}{12.40 \ cm^3} = 2.212 \ g/cm^3$

1.10 $volume = 9.37 \ g \times \dfrac{1 \ mL}{1.483 \ g} = 6.32 \ mL$

1.11   The actual mass of the bottle and the acetone = 38.0015 g + 0.7791 g = 38.7806 g. The
       measured values are 38.7798 g, 38.7795 g, and 38.7801 g. These values are both close to
       each other and close to the actual mass. Therefore the results are both precise and
       accurate.

1

1.12 (a) 76.600 kg has 5 significant figures because zeros at the end of a number and after the decimal point are always significant.
(b) 4.502 00 x 10³ g has 6 significant figures because zeros in the middle of a number are significant and zeros at the end of a number and after the decimal point are always significant.
(c) 3000 nm has 1, 2, 3, or 4 significant figures because zeros at the end of a number and before the decimal point may or may not be significant.
(d) 0.003 00 mL has 3 significant figures because zeros at the beginning of a number are not significant and zeros at the end of a number and after the decimal point are always significant.
(e) 18 students has an infinite number of significant figures since this is an exact number.
(f) 3 x 10⁻⁵ g has 1 significant figure.
(g) 47.60 mL has 4 significant figures because a zero at the end of a number and after the decimal point is always significant.
(h) 2070 mi has 3 or 4 significant figures because a zero in the middle of a number is significant and a zero at the end of a number and before the decimal point may or may not be significant.

1.13 (a) Since the digit to be dropped (the second 4) is less than 5, round down. The result is 3.774 L.
(b) Since the digit to be dropped (0) is less than 5, round down. The result is 255 K.
(c) Since the digit to be dropped is equal to 5 with nothing following, round down. The result is 55.26 kg.

1.14 (a)

24.567 g	This result should be expressed with 3 decimal places.
+ 0.044 78 g	Since the digit to be dropped (7) is greater than 5, round up.
24.611 78 g	The result is 24.612 g (5 significant figures).

(b) 4.6742 g / 0.003 71 L = 1259.89 g/L
0.003 71 has only 3 significant figures so the result of the division should have only 3 significant figures. Since the digit to be dropped (first 9) is greater than 5, round up. The result is 1260 g/L (3 significant figures), or 1.26 x 10³ g/L.

(c)

0.378 mL	This result should be expressed with 1 decimal place. Since
+ 42.3 mL	the digit to be dropped (9) is greater than 5, round up. The
− 1.5833 mL	result is 41.1 mL (3 significant figures).
41.0947 mL	

1.15 The level of the liquid in the thermometer is just past halfway between the 32°C and 33°C marks on the thermometer. The temperature is 32.6°C (3 significant figures).

1.16 (a) Calculation: °F = ($\frac{9}{5}$ x °C) + 32 = ($\frac{9}{5}$ x 1064) + 32 = 1947°F

Ballpark estimate: °F ≈ 2 x °C if °C is large. The melting point of gold ≈ 2000°F.

(b) r = d/2 = 3 x 10⁻⁶ m = 3 x 10⁻⁴ cm; h = 2 x 10⁻⁶ m = 2 x 10⁻⁴ cm

Calculation: volume = $\pi r^2 h$ = (3.1416)(3 x 10⁻⁴ cm)²(2 x 10⁻⁴ cm) = 6 x 10⁻¹¹ cm³

Ballpark estimate: volume = $\pi r^2 h \approx 3 r^2 h \approx$ 3(3 x 10⁻⁴ cm)²(2 x 10⁻⁴ cm) ≈ 5 x 10⁻¹¹ cm³

1.17　1 carat = 200 mg = 200 x 10⁻³ g = 0.200 g

Mass of Hope Diamond in grams = 44.4 carats x $\dfrac{0.200\ g}{1\ carat}$ = 8.88 g

1 ounce = 28.35 g

Mass of Hope Diamond in ounces = 8.88 g x $\dfrac{1\ ounce}{28.35\ g}$ = 0.313 ounces

1.18　An LD₅₀ value is the amount of a substance per kilogram of body weight that is a lethal dose for 50% of the test animals.

1.19　mass of salt = 155 lb x $\dfrac{453.6\ g}{1\ lb}$ x $\dfrac{1\ kg}{1000\ g}$ x $\dfrac{4\ g}{1\ kg}$ = 281.2 g or 300 g

1.20

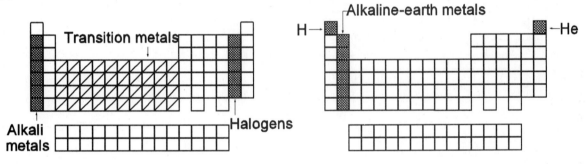

1.22　red – gas; blue – 42; green – sodium

1.24　(a) Darts are clustered together (good precision) but are away from the bullseye (poor accuracy).
　　　(b) Darts are clustered together (good precision) and hit the bullseye (good accuracy).
　　　(c) Darts are scattered (poor precision) and are away from the bullseye (poor accuracy).

1.26

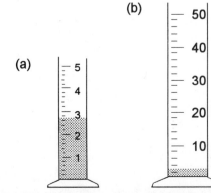

The 5 mL graduated cylinder is marked every 0.2 mL and can be read to ± 0.02 mL. The 50 mL graduated cylinder is marked every 2 mL and can only be read to ± 0.2 mL. The 5 mL graduated cylinder will give more accurate measurements.

3

Additional Problems
Elements and the Periodic Table

1.28   114 elements are presently known. About 90 elements occur naturally.

1.30   There are 18 groups in the periodic table. They are labeled as follows:
        1A, 2A, 3B, 4B, 5B, 6B, 7B, 8B (3 groups), 1B, 2B, 3A, 4A, 5A, 6A, 7A, 8A

1.32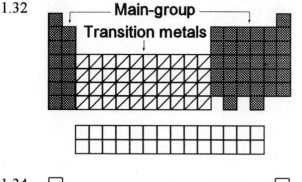

1.34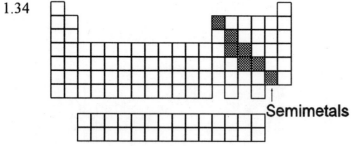

A semimetal is an element with properties that fall between those of metals and nonmetals.

1.36   Li, Na, K, Rb, and Cs

1.38   F, Cl, Br, and I

1.40   (a) gadolinium, Gd   (b) germanium, Ge   (c) technetium, Tc   (d) arsenic, As

1.42   (a) Te, tellurium   (b) Re, rhenium   (c) Be, beryllium
        (d) Ar, argon   (e) Pu, plutonium

1.44   (a) Tin is Sn: Ti is titanium.   (b) Manganese is Mn: Mg is magnesium.
        (c) Potassium is K: Po is polonium.   (d) The symbol for helium is He. The
                                                     second letter is lowercase.

Units and Significant Figures

1.46   Mass measures the amount of matter in an object, whereas weight measures the pull of
        gravity on an object by the earth or other celestial body.

1.48   (a) kilogram, kg   (b) meter, m   (c) kelvin, K   (d) cubic meter, m^3

1.50   A Celsius degree is larger than a Fahrenheit degree by a factor of $\frac{9}{5}$.

1.52   The volume of a cubic decimeter (dm^3) and a liter (L) are the same.

1.54   Only (a) is exact because it is obtained by counting. (b) and (c) are not exact because they result from measurements.

1.56   cL is centiliter (10^{-2} L)

1.58   1 mg = 1 x 10^{-3} g and 1 pg = 1 x 10^{-12} g

$$\frac{1 \times 10^{-3}\,g}{1\ mg} \times \frac{1\ pg}{1 \times 10^{-12}\,g} = 1 \times 10^9\ pg/mg$$

35 ng = 35 x 10^{-9} g
$$\frac{35 \times 10^{-9}\,g}{35\ ng} \times \frac{1\ pg}{1 \times 10^{-12}\,g} = 3.5 \times 10^4\ pg/35\ ng$$

1.60   (a)   5 pm = 5 x 10^{-12} m

$$5 \times 10^{-12}\ m \times \frac{100\ cm}{1\ m} = 5 \times 10^{-10}\ cm$$

$$5 \times 10^{-12}\ m \times \frac{1\ nm}{1 \times 10^{-9}\ m} = 5 \times 10^{-3}\ nm$$

(b) $8.5\ cm^3 \times \left(\dfrac{1\ m}{100\ cm}\right)^3 = 8.5 \times 10^{-6}\ m^3$

$8.5\ cm^3 \times \left(\dfrac{10\ mm}{1\ cm}\right)^3 = 8.5 \times 10^3\ mm^3$

(c) $65.2\ mg \times \dfrac{1 \times 10^{-3}\ g}{1\ mg} = 0.0652\ g$

$65.2\ mg \times \dfrac{1 \times 10^{-3}\ g}{1\ mg} \times \dfrac{1\ pg}{1 \times 10^{-12}\ g} = 6.52 \times 10^{10}\ pg$

1.62   (a) 35.0445 g has 6 significant figures because zeros in the middle of a number are significant.
(b) 59.0001 cm has 6 significant figures because zeros in the middle of a number are significant.
(c) 0.030 03 kg has 4 significant figures because zeros at the beginning of a number are not significant and zeros in the middle of a number are significant.
(d) 0.004 50 m has 3 significant figures because zeros at the beginning of a number are not significant and zeros at the end of a number and after the decimal point are always significant.

(e) 67,000 m² has 2, 3, 4, or 5 significant figures because zeros at the end of a number and before the decimal point may or may not be significant.

(f) 3.8200×10^3 L has 5 significant figures because zeros at the end of a number and after the decimal point are always significant.

1.64   To convert 3,666,500 m³ to scientific notation, move the decimal point 6 places to the left and include an exponent of 10^6. The result is 3.6665×10^6 m³.

1.66   (a) To convert 453.32 mg to scientific notation, move the decimal point 2 places to the left and include an exponent of 10^2. The result is 4.5332×10^2 mg.
       (b) To convert 0.000 042 1 mL to scientific notation, move the decimal point 5 places to the right and include an exponent of 10^{-5}. The result is 4.21×10^{-5} mL.
       (c) To convert 667,000 g to scientific notation, move the decimal point 5 places to the left and include an exponent of 10^5. The result is 6.67×10^5 g.

1.68   (a) Since the digit to be dropped (0) is less than 5, round down. The result is 3.567×10^4 or 35,670 m (4 significant figures).
       Since the digit to be dropped (the second 6) is greater than 5, round up. The result is 35,670.1 m (6 significant figures).
       (b) Since the digit to be dropped is 5 with nonzero digits following, round up. The result is 69 g (2 significant figures).
       Since the digit to be dropped (0) is less than 5, round down. The result is 68.5 g (3 significant figures).
       (c) Since the digit to be dropped is 5 with nothing following, round down. The result is 4.99×10^3 cm (3 significant figures).
       (d) Since the digit to be dropped is 5 with nothing following, round down. The result is 2.3098×10^{-4} kg (5 significant figures).

1.70   (a) $4.884 \times 2.05 = 10.012$
       The result should contain only 3 significant figures because 2.05 contains 3 significant figures (the smaller number of significant figures of the two). Since the digit to be dropped (1) is less than 5, round down. The result is 10.0.

       (b) $94.61 / 3.7 = 25.57$
       The result should contain only 2 significant figures because 3.7 contains 2 significant figures (the smaller number of significant figures of the two). Since the digit to be dropped (second 5) is 5 with nonzero digits following, round up. The result is 26.

       (c) $3.7 / 94.61 = 0.0391$
       The result should contain only 2 significant figures because 3.7 contains 2 significant figures (the smaller number of significant figures of the two). Since the digit to be dropped (1) is less than 5, round down. The result is 0.039.

       (d)   5502.3   This result should be expressed with no decimal places. Since the
                     24   digit to be dropped (3) is less than 5, round down. The result is
                  +   0.01   5526.
                  ─────────
                   5526.31

(e) $\begin{array}{r} 86.3 \\ +\ \ 1.42 \\ -\ \ 0.09 \\ \hline 87.63 \end{array}$ This result should be expressed with only 1 decimal place. Since the digit to be dropped (3) is less than 5, round down. The result is 87.6.

(f) $5.7 \times 2.31 = 13.167$

The result should contain only 2 significant figures because 5.7 contains 2 significant figures (the smaller number of significant figures of the two). Since the digit to be dropped (second 1) is less than 5, round down. The result is 13.

Unit Conversions

1.72 (a) $0.25\ \text{lb} \times \dfrac{453.59\ \text{g}}{1\ \text{lb}} = 113.4\ \text{g} = 110\ \text{g}$

(b) $1454\ \text{ft} \times \dfrac{12\ \text{in}}{1\ \text{ft}} \times \dfrac{2.54\ \text{cm}}{1\ \text{in}} \times \dfrac{1\ \text{m}}{100\ \text{cm}} = 443.2\ \text{m}$

(c) $2{,}941{,}526\ \text{mi}^2 \times \left(\dfrac{1.6093\ \text{km}}{1\ \text{mi}}\right)^2 \times \left(\dfrac{1000\ \text{m}}{1\ \text{km}}\right)^2 = 7.6181 \times 10^{12}\ \text{m}^2$

1.74 (a) $1\ \text{acre–ft} \times \dfrac{1\ \text{mi}^2}{640\ \text{acres}} \times \left(\dfrac{5280\ \text{ft}}{1\ \text{mi}}\right)^2 = 43{,}560\ \text{ft}^3$

(b) $116\ \text{mi}^3 \times \left(\dfrac{5280\ \text{ft}}{1\ \text{mi}}\right)^3 \times \dfrac{1\ \text{acre–ft}}{43{,}560\ \text{ft}^3} = 3.92 \times 10^8\ \text{acre–ft}$

1.76 (a) $\dfrac{200\ \text{mg}}{100\ \text{mL}} \times \dfrac{1000\ \text{mL}}{1\ \text{L}} = 2000\ \text{mg/L}$

(b) $\dfrac{200\ \text{mg}}{100\ \text{mL}} \times \dfrac{1 \times 10^{-3}\ \text{g}}{1\ \text{mg}} \times \dfrac{1\ \mu\text{g}}{1 \times 10^{-6}\ \text{g}} = 2000\ \mu\text{g/mL}$

(c) $\dfrac{200\ \text{mg}}{100\ \text{mL}} \times \dfrac{1 \times 10^{-3}\ \text{g}}{1\ \text{mg}} \times \dfrac{1000\ \text{mL}}{1\ \text{L}} = 2\ \text{g/L}$

(d) $\dfrac{200\ \text{mg}}{100\ \text{mL}} \times \dfrac{1 \times 10^{-3}\ \text{g}}{1\ \text{mg}} \times \dfrac{1000\ \text{mL}}{1\ \text{L}} \times \dfrac{1\ \text{ng}}{1 \times 10^{-9}\ \text{g}} \times \dfrac{1 \times 10^{-6}\ \text{L}}{1\ \mu\text{L}} = 2000\ \text{ng}/\mu\text{L}$

(e) $2\ \text{g/L} \times 5\ \text{L} = 10\ \text{g}$

1.78 $55\ \dfrac{\text{mi}}{\text{h}} \times \dfrac{5280\ \text{ft}}{1\ \text{mi}} \times \dfrac{12\ \text{in}}{1\ \text{ft}} \times \dfrac{2.54\ \text{cm}}{1\ \text{in}} \times \dfrac{1\ \text{h}}{3600\ \text{s}} \times \dfrac{2.5 \times 10^{-4}\ \text{s}}{1\ \text{shake}} = 0.61\ \dfrac{\text{cm}}{\text{shake}}$

Temperature

1.80 $\quad °F = (\frac{9}{5} \times °C) + 32$

$\qquad °F = (\frac{9}{5} \times 39.9°C) + 32 = 103.8°F \qquad$ (goat)

$\qquad °F = (\frac{9}{5} \times 22.2°C) + 32 = 72.0°F \qquad$ (Australian spiny anteater)

1.82 $\quad °C = \frac{5}{9} \times (°F - 32) = \frac{5}{9} \times (6192 - 32) = 3422°C$

$\qquad K = °C + 273.15 = 3422 + 273.15 = 3695.15 \text{ K or } 3695 \text{ K}$

1.84 $\quad$ Ethanol boiling point $\quad 78.5°C \qquad 173.3°F \qquad 200°E$
$\qquad$ Ethanol melting point $-117.3°C \qquad -179.1°F \qquad 0°E$

(a) $\dfrac{200°E}{[78.5°C - (-117.3°C)]} = \dfrac{200°E}{195.8°C} = 1.021 \text{ °E/°C}$

(b) $\dfrac{200°E}{[173.3°F - (-179.1°F)]} = \dfrac{200°E}{352.4°F} = 0.5675 \text{ °E/°F}$

(c) $°E = \dfrac{200}{195.8} \times (°C + 117.3)$

H_2O melting point $= 0°C; \quad °E = \dfrac{200}{195.8} \times (0 + 117.3) = 119.8°E$

H_2O boiling point $= 100°C; \quad °E = \dfrac{200}{195.8} \times (100 + 117.3) = 222.0°E$

(d) $°E = \dfrac{200}{352.4} \times (°F + 179.1) = \dfrac{200}{352.4} \times (98.6 + 179.1) = 157.6°E$

(e) $°F = \left(°E \times \dfrac{352.4}{200}\right) - 179.1 = \left(130 \times \dfrac{352.4}{200}\right) - 179.1 = 50.0°F$

Since the outside temperature is 50.0°F, I would wear a sweater or light jacket.

Density

1.86 $\quad 250 \text{ mg} \times \dfrac{1 \times 10^{-3} \text{ g}}{1 \text{ mg}} = 0.25 \text{ g}; \qquad V = 0.25 \text{ g} \times \dfrac{1 \text{ cm}^3}{1.40 \text{ g}} = 0.18 \text{ cm}^3$

$\qquad 500 \text{ lb} \times \dfrac{453.59 \text{ g}}{1 \text{ lb}} = 226{,}795 \text{ g}; \quad V = 226{,}795 \text{ g} \times \dfrac{1 \text{ cm}^3}{1.40 \text{ g}} = 161{,}996 \text{ cm}^3 = 162{,}000 \text{ cm}^3$

1.88 $\quad d = \dfrac{m}{V} = \dfrac{220.9 \text{ g}}{(0.50 \times 1.55 \times 25.00) \text{ cm}^3} = 11.4 \dfrac{g}{cm^3} = 11 \dfrac{g}{cm^3}$

1.90 $\quad d = \dfrac{m}{V} = \dfrac{8.763 \text{ g}}{(28.76 - 25.00) \text{ mL}} = \dfrac{8.763 \text{ g}}{3.76 \text{ mL}} = 2.331 \dfrac{\text{g}}{\text{cm}^3} = 2.33 \dfrac{\text{g}}{\text{cm}^3}$

General Problems

1.92 (a) selenium, Se $\qquad$ (b) rhenium, Re $\qquad$ (c) cobalt, Co $\qquad$ (d) rhodium, Rh

1.94 NaCl melting point = 1074 K

$°C = K - 273.15 = 1074 - 273.15 = 800.85°C = 801°C$

$°F = (\dfrac{9}{5} \times °C) + 32 = (\dfrac{9}{5} \times 800.85) + 32 = 1473.53°F = 1474°F$

NaCl boiling point = 1686 K

$°C = K - 273.15 = 1686 - 273.15 = 1412.85°C = 1413°C$

$°F = (\dfrac{9}{5} \times °C) + 32 = (\dfrac{9}{5} \times 1412.85) + 32 = 2575.13°F = 2575°F$

1.96 $\quad V = 112.5 \text{ g} \times \dfrac{1 \text{ mL}}{1.4832 \text{ g}} = 75.85 \text{ mL}$

1.98 $\quad V = 8.728 \times 10^{10} \text{ lb} \times \dfrac{453.59 \text{ g}}{1 \text{ lb}} \times \dfrac{1 \text{ mL}}{1.831 \text{ g}} \times \dfrac{1 \text{ L}}{1000 \text{ mL}} = 2.162 \times 10^{10} \text{ L}$

1.100 (a) $\text{density} = \dfrac{1 \text{ lb}}{1 \text{ pint}} \times \dfrac{8 \text{ pints}}{1 \text{ gal}} \times \dfrac{1 \text{ gal}}{3.7854 \text{ L}} \times \dfrac{453.59 \text{ g}}{1 \text{ lb}} \times \dfrac{1 \text{ L}}{1000 \text{ mL}} = 0.95861 \text{ g/mL}$

(b) area in m^2 =

$1 \text{ acre} \times \dfrac{1 \text{ mi}^2}{640 \text{ acres}} \times \left(\dfrac{5280 \text{ ft}}{1 \text{ mi}}\right)^2 \times \left(\dfrac{12 \text{ in}}{1 \text{ ft}}\right)^2 \times \left(\dfrac{2.54 \text{ cm}}{1 \text{ in}}\right)^2 \times \left(\dfrac{1 \text{ m}}{100 \text{ cm}}\right)^2 = 4047 \text{ m}^2$

(c) mass of wood =

$1 \text{ cord} \times \dfrac{128 \text{ ft}^3}{1 \text{ cord}} \times \left(\dfrac{12 \text{ in}}{1 \text{ ft}}\right)^3 \times \left(\dfrac{2.54 \text{ cm}}{1 \text{ in}}\right)^3 \times \dfrac{0.40 \text{ g}}{1 \text{ cm}^3} \times \dfrac{1 \text{ kg}}{1000 \text{ g}} = 1450 \text{ kg} = 1400 \text{ kg}$

(d) mass of oil =

$1 \text{ barrel} \times \dfrac{42 \text{ gal}}{1 \text{ barrel}} \times \dfrac{3.7854 \text{ L}}{1 \text{ gal}} \times \dfrac{1000 \text{ mL}}{1 \text{ L}} \times \dfrac{0.85 \text{ g}}{1 \text{ mL}} \times \dfrac{1 \text{ kg}}{1000 \text{ g}} = 135.1 \text{ kg} = 140 \text{ kg}$

(e) fat Calories =

$0.5 \text{ gal} \times \dfrac{32 \text{ servings}}{1 \text{ gal}} \times \dfrac{165 \text{ Calories}}{1 \text{ serving}} \times \dfrac{30.0 \text{ Cal from fat}}{100 \text{ Cal total}} = 792 \text{ Cal from fat}$

1.102 (a) number of Hershey's Kisses =

$2.0 \text{ lb} \times \dfrac{453.59 \text{ g}}{1 \text{ lb}} \times \dfrac{1 \text{ serving}}{41 \text{ g}} \times \dfrac{9 \text{ kisses}}{1 \text{ serving}} = 199 \text{ kisses} = 200 \text{ kisses}$

(b) Hershey's Kiss volume = $\dfrac{41 \text{ g}}{1 \text{ serving}} \times \dfrac{1 \text{ serving}}{9 \text{ kisses}} \times \dfrac{1 \text{ mL}}{1.4 \text{ g}} = 3.254 \text{ mL} = 3.3 \text{ mL}$

(c) Calories/Hershey's Kiss = $\dfrac{230\ Cal}{1\ serving} \times \dfrac{1\ serving}{9\ kisses}$ = 25.55 Cal/kiss = 26 Cal/kiss

(d) % fat Calories =

$\dfrac{13\ g\ fat}{1\ serving} \times \dfrac{9\ Cal\ from\ fat}{1\ g\ fat} \times \dfrac{1\ serving}{230\ Cal\ total} \times 100\% = 51\%$ Calories from fat

1.104 $^\circ C = \dfrac{5}{9} \times (^\circ F - 32)$;   Set $^\circ C = {}^\circ F$: $^\circ C = \dfrac{5}{9} \times (^\circ C - 32)$

Solve for $^\circ C$: $^\circ C \times \dfrac{9}{5} = {}^\circ C - 32$

$(^\circ C \times \dfrac{9}{5}) - {}^\circ C = -32$

$^\circ C \times \dfrac{4}{5} = -32$

$^\circ C = \dfrac{5}{4}(-32) = -40^\circ C$

The Celsius and Fahrenheit scales "cross" at $-40^\circ C$ ($-40^\circ F$).

1.106   Convert 8 min, 25 s to s.   8 min x $\dfrac{60\ s}{1\ min}$ + 25 s = 505 s

Convert 293.2 K to $^\circ F$ $293.2 - 273.15 = 20.05^\circ C$

$^\circ F = (\dfrac{9}{5} \times 20.05) + 32 = 68.09^\circ F$

Final temperature = 68.09$^\circ F$ + 505 s x $\dfrac{3.0^\circ F}{60\ s}$ = 93.34$^\circ F$

$^\circ C = \dfrac{5}{9} \times (93.34 - 32) = 34.1^\circ C$

1.108   Average brass density = (0.670)(8.92 g/cm³) + (0.330)(7.14 g/cm³) = 8.333 g/cm³

length = 1.62 in x $\dfrac{2.54\ cm}{1\ in}$ = 4.115 cm

diameter = 0.514 in x $\dfrac{2.54\ cm}{1\ in}$ = 1.306 cm

volume = $\pi r^2 h$ = (3.1416)[(1.306 cm)/2]²(4.115 cm) = 5.512 cm³

mass = 5.512 cm³ x $\dfrac{8.333\ g}{1\ cm^3}$ = 45.9 g

1.110   (a) Gallium is a metal.

(b) Indium, which is right under gallium in the periodic table, should have similar chemical properties.

(c) Ga density = $\dfrac{0.2133\ lb}{1\ in.^3} \times \dfrac{453.59\ g}{1\ lb} \times \dfrac{1\ in.^3}{(2.54\ cm)^3}$ = 5.904 g/cm³

(d) Ga boiling point   2204°C   1000°G
     Ga melting point   29.78°C   0°G

$$\frac{1000°G - 0°G}{2204°C - 29.78°C} = \frac{1000°G}{2174.22°C} = 0.4599 \text{ °G/°C}$$

°G = 0.4599 x (°C – 29.78)
°G = 0.4599 x (801 – 29.78) = 355°G
The melting point of sodium chloride (NaCl) on the gallium scale is 355°G.

Atoms, Molecules, and Ions

2.1 First, find the S:O ratio in each compound.
Substance A: S:O mass ratio = (6.00 g S) / (5.99 g O) = 1.00
Substance B: S:O mass ratio = (8.60 g S) / (12.88 g O) = 0.668

$$\frac{\text{S:O mass ratio in substance A}}{\text{S:O mass ratio in substance B}} = \frac{1.00}{0.668} = 1.50 = \frac{3}{2}$$

2.2 $0.0002 \text{ in} \times \dfrac{2.54 \text{ cm}}{1 \text{ in}} \times \dfrac{1 \text{ Au atom}}{2.9 \times 10^{-8} \text{ cm}} = 2 \times 10^4 \text{ Au atoms}$

2.3 $1 \times 10^{19} \text{ C atoms} \times \dfrac{1.5 \times 10^{-10} \text{ m}}{\text{C atom}} \times \dfrac{1 \text{ km}}{1000 \text{ m}} \times \dfrac{1 \text{ time}}{40,075 \text{ km}} = 37.4 \text{ times} \approx 40 \text{ times}$

2.4 $^{75}_{34}\text{Se}$ has 34 protons, 34 electrons, and (75 − 34) = 41 neutrons.

2.5 $^{35}_{17}\text{Cl}$ has (35 − 17) = 18 neutrons. $^{37}_{17}\text{Cl}$ has (37 − 17) = 20 neutrons.

2.6 The element with 47 protons is Ag. The mass number is the sum of the protons and the neutrons, 47 + 62 = 109. The isotope symbol is $^{109}_{47}\text{Ag}$.

2.7 atomic mass = (0.6917 x 62.94 amu) + (0.3083 x 64.93 amu) = 63.55 amu

2.8 $2.15 \text{ g} \times \dfrac{1 \text{ amu}}{1.6605 \times 10^{-24} \text{ g}} \times \dfrac{1 \text{ Cu}}{63.55 \text{ amu}} = 2.04 \times 10^{22} \text{ Cu atoms}$

2.9

```
        H   H
        |   |
   H — C — N — H
        |
        H
```

2.10 Figure (b) represents a collection of hydrogen peroxide (H_2O_2) molecules.

2.11 adrenaline, $C_9H_{13}NO_3$

2.12 (a) LiBr is composed of a metal (Li) and nonmetal (Br) and is ionic.
(b) $SiCl_4$ is composed of only nonmetals and is molecular.
(c) BF_3 is composed of only nonmetals and is molecular.
(d) CaO is composed of a metal (Ca) and nonmetal (O) and is ionic.

2.13   Figure (a) most likely represents an ionic compound because there are no discrete molecules, only a regular array of two different chemical species (ions). Figure (b) most likely represents a molecular compound because discrete molecules are present.

2.14   (a) HF is an acid. In water, HF dissociates to produce H^+(aq).
        (b) $Ca(OH)_2$ is a base. In water, $Ca(OH)_2$ dissociates to produce OH^-(aq).
        (c) LiOH is a base. In water, LiOH dissociates to produce OH^-(aq).
        (d) HCN is an acid. In water, HCN dissociates to produce H^+(aq).

2.15   (a) CsF, cesium fluoride   (b) K_2O, potassium oxide   (c) CuO, copper(II) oxide
        (d) BaS, barium sulfide   (e) $BeBr_2$, beryllium bromide

2.16   (a) vanadium(III) chloride, VCl_3 (b) manganese(IV) oxide, MnO_2
        (c) copper(II) sulfide, CuS   (d) aluminum oxide, Al_2O_3

2.17   red – potassium sulfide, K_2S
        green – strontium iodide, SrI_2
        blue – gallium oxide, Ga_2O_3

2.18   (a) NCl_3, nitrogen trichloride   (b) P_4O_6, tetraphosphorus hexoxide
        (c) S_2F_2, disulfur difluoride   (d) SeO_2, selenium dioxide

2.19   (a) disulfur dichloride, S_2Cl_2 (b) iodine monochloride, ICl
        (c) nitrogen triiodide, NI_3

2.20   (a) $Ca(ClO)_2$, calcium hypochlorite
        (b) $Ag_2S_2O_3$, silver(I) thiosulfate or silver thiosulfate
        (c) NaH_2PO_4, sodium dihydrogen phosphate   (d) $Sn(NO_3)_2$, tin(II) nitrate
        (e) $Pb(CH_3CO_2)_4$, lead(IV) acetate   (f) $(NH_4)_2SO_4$, ammonium sulfate

2.21   (a) lithium phosphate, Li_3PO_4 (b) magnesium hydrogen sulfate, $Mg(HSO_4)_2$
        (c) manganese(II) nitrate, $Mn(NO_3)_2$ (d) chromium(III) sulfate, $Cr_2(SO_4)_3$

2.22   Drawing 1 represents ionic compounds with one cation and two anions. Only (c) $CaCl_2$ is consistent with drawing 1.
        Drawing 2 represents ionic compounds with one cation and one anion. Both (a) LiBr and (b) $NaNO_2$ are consistent with drawing 2.

2.23   (a) HIO_4, periodic acid   (b) $HBrO_2$, bromous acid   (c) H_2CrO_4, chromic acid

2.24   A normal visual image results when light from the sun or other source reflects off an object, strikes the retina in our eye, and is converted into electrical signals that are processed by the brain. The image obtained with a scanning tunneling microscope, by contrast, is a three-dimensional, computer-generated data plot that uses tunneling current to mimic depth perception. The nature of the computer-generated image depends on the identity of the molecules or atoms on the surface, on the precision with which the probe tip is made, on how the data are manipulated, and on other experimental variables.

Understanding Key Concepts

2.26   To obey the law of mass conservation, the correct drawing must have the same number of red and yellow spheres as in drawing (a). The correct drawing is (d).

2.28.   (a) alanine, $C_3H_7NO_2$ (b) ethylene glycol, $C_2H_6O_2$ (c) acetic acid, $C_2H_4O_2$

2.30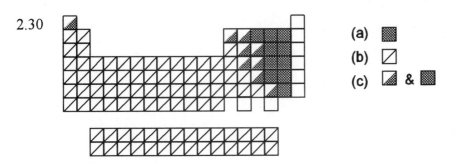

Additional Problems
Atomic Theory

2.32   The law of mass conservation in terms of Dalton's atomic theory states that chemical reactions only rearrange the way that atoms are combined; the atoms themselves are not changed.
         The law of definite proportions in terms of Dalton's atomic theory states that the chemical combination of elements to make different substances occurs when atoms join together in small, whole-number ratios.

2.34   First, find the C:H ratio in each compound.
         Benzene: C:H mass ratio = (4.61 g C) / (0.39 g H) = 12
         Ethane: C:H mass ratio (4.00 g C) / (1.00 g H) = 4.00
         Ethylene: C:H mass ratio = (4.29 g C) / (0.71 g H) = 6.0

$$\frac{\text{C:H mass ratio in benzene}}{\text{C:H mass ratio in ethane}} = \frac{12}{4.00} = \frac{3}{1}$$

$$\frac{\text{C:H mass ratio in benzene}}{\text{C:H mass ratio in ethylene}} = \frac{12}{6.0} = \frac{2}{1}$$

$$\frac{\text{C:H mass ratio in ethylene}}{\text{C:H mass ratio in ethane}} = \frac{6.0}{4.00} = \frac{3}{2}$$

2.36   (a) For benzene:

$$4.61 \text{ g} \times \frac{1 \text{ amu}}{1.6605 \times 10^{-24} \text{ g}} \times \frac{1 \text{ C atom}}{12.011 \text{ amu}} = 2.31 \times 10^{23} \text{ C atoms}$$

$$0.39 \text{ g} \times \frac{1 \text{ amu}}{1.6605 \times 10^{-24} \text{ g}} \times \frac{1 \text{ H atom}}{1.008 \text{ amu}} = 2.3 \times 10^{23} \text{ H atoms}$$

$$\frac{C}{H} = \frac{2.31 \times 10^{23} \text{ C atoms}}{2.3 \times 10^{23} \text{ H atoms}} = \frac{1 \text{ C}}{1 \text{ H}}$$

A possible formula for benzene is CH.

For ethane:

$$4.00 \text{ g} \times \frac{1 \text{ amu}}{1.6605 \times 10^{-24} \text{ g}} \times \frac{1 \text{ C atom}}{12.011 \text{ amu}} = 2.01 \times 10^{23} \text{ C atoms}$$

$$1.00 \text{ g} \times \frac{1 \text{ amu}}{1.6605 \times 10^{-24} \text{ g}} \times \frac{1 \text{ H atom}}{1.008 \text{ amu}} = 5.97 \times 10^{23} \text{ H atoms}$$

$$\frac{C}{H} = \frac{2.01 \times 10^{23} \text{ C atoms}}{5.97 \times 10^{23} \text{ H atoms}} = \frac{1 \text{ C}}{3 \text{ H}}$$

A possible formula for ethane is CH_3.

For ethylene:

$$4.29 \text{ g} \times \frac{1 \text{ amu}}{1.6605 \times 10^{-24} \text{ g}} \times \frac{1 \text{ C atom}}{12.011 \text{ amu}} = 2.15 \times 10^{23} \text{ C atoms}$$

$$0.71 \text{ g} \times \frac{1 \text{ amu}}{1.6605 \times 10^{-24} \text{ g}} \times \frac{1 \text{ H atom}}{1.008 \text{ amu}} = 4.2 \times 10^{23} \text{ H atoms}$$

$$\frac{C}{H} = \frac{2.15 \times 10^{23} \text{ C atoms}}{4.2 \times 10^{23} \text{ H atoms}} = \frac{1 \text{ C}}{2 \text{ H}}$$

A possible formula for ethylene is CH_2.

(b) The results in part (a) give the smallest whole-number ratio of C to H for benzene, ethane, and ethylene, and these ratios are consistent with their modern formulas.

2.38 (a) $(1.67 \times 10^{-24} \frac{g}{H \text{ atom}})(6.02 \times 10^{23} \text{ H atoms}) = 1.01 \text{ g}$

This result is numerically equal to the atomic mass of H in grams.

(b) $(26.558 \times 10^{-24} \frac{g}{O \text{ atom}})(6.02 \times 10^{23} \text{ O atoms}) = 16.0 \text{ g}$

This result is numerically equal to the atomic mass of O in grams.

2.40 Assume a 1.00 g sample of the binary compound of zinc and sulfur.
0.671 × 1.00 g = 0.671 g Zn;   0.329 × 1.00 g = 0.329 g S

$$0.671 \text{ g} \times \frac{1 \text{ amu}}{1.6605 \times 10^{-24} \text{ g}} \times \frac{1 \text{ Zn atom}}{65.39 \text{ amu}} = 6.18 \times 10^{21} \text{ Zn atoms}$$

$$0.329 \text{ g} \times \frac{1 \text{ amu}}{1.6605 \times 10^{-24} \text{ g}} \times \frac{1 \text{ S atom}}{32.066 \text{ amu}} = 6.18 \times 10^{21} \text{ S atoms}$$

$$\frac{Zn}{S} = \frac{6.18 \times 10^{21} \text{ Zn atoms}}{6.18 \times 10^{21} \text{ S atoms}} = \frac{1 \text{ Zn}}{1 \text{ S}}; \text{ therefore the formula is ZnS.}$$

Elements and Atoms

2.42   The atomic number is equal to the number of protons.
        The mass number is equal to the sum of the number of protons and the number of neutrons.

2.44   Atoms of the same element that have different numbers of neutrons are called isotopes.

2.46   The subscript giving the atomic number of an atom is often left off of an isotope symbol because one can readily look up the atomic number in the periodic table.

2.48   (a) carbon, C (b) argon, Ar (c) vanadium, V

2.50   (a) $^{220}_{86}Rn$ (b) $^{210}_{84}Po$ (c) $^{197}_{79}Au$

2.52   (a) $^{15}_{7}N$, 7 protons, 7 electrons, $(15 - 7) = 8$ neutrons

        (b) $^{60}_{27}Co$, 27 protons, 27 electrons, $(60 - 27) = 33$ neutrons

        (c) $^{131}_{53}I$, 53 protons, 53 electrons, $(131 - 53) = 78$ neutrons

        (d) $^{142}_{58}Ce$, 58 protons, 58 electrons, $(142 - 58) = 84$ neutrons

2.54   (a) $^{24}_{12}Mg$, magnesium   (b) $^{58}_{28}Ni$, nickel
        (c) $^{104}_{46}Pd$, palladium   (d) $^{183}_{74}W$, tungsten

2.56 $(0.199 \times 10.0129 \text{ amu}) + (0.801 \times 11.009\,31 \text{ amu}) = 10.8 \text{ amu for B}$

2.58 $24.305 \text{ amu} = (0.7899 \times 23.985 \text{ amu}) + (0.1000 \times 24.986 \text{ amu}) + (0.1101 \times Z)$
        Solve for Z. $Z = 25.982$ amu for ^{26}Mg.

Compounds and Mixtures, Molecules and Ions

2.60   (a) muddy water, heterogeneous mixture
        (b) concrete, heterogeneous mixture
        (c) house paint, homogeneous mixture
        (d) a soft drink, homogeneous mixture (heterogeneous mixture if it contains CO_2 bubbles)

2.62   An atom is the smallest particle that retains the chemical properties of an element. A molecule is matter that results when two or more atoms are joined by covalent bonds. H and O are atoms, H_2O is a water molecule.

2.64 A covalent bond results when two atoms share several (usually two) of their electrons. An ionic bond results from a complete transfer of one or more electrons from one atom to another. The C–H bonds in methane (CH_4) are covalent bonds. The bond in NaCl (Na^+Cl^-) is an ionic bond.

2.66 Element symbols are composed of one or two letters. If the element symbol is two letters, the first letter is uppercase and the second is lowercase. CO stands for carbon and oxygen in carbon monoxide.

2.68 (a) Be^{2+}, 4 protons and 2 electrons (b) Rb^+, 37 protons and 36 electrons
(c) Se^{2-}, 34 protons and 36 electrons (d) Au^{3+}, 79 protons and 76 electrons

2.70 C_3H_8O

2.72

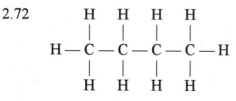

Acids and Bases

2.74 (a) HI, acid (b) CsOH, base (c) H_3PO_4, acid
(d) $Ba(OH)_2$, base (e) H_2CO_3, acid

2.76 $HI(aq) \rightarrow H^+(aq) + I^-(aq)$; the anion is I^-
$H_3PO_4(aq) \rightarrow H^+(aq) + H_2PO_4^-(aq)$; the predominant anion is $H_2PO_4^-$
$H_2CO_3(aq) \rightarrow H^+(aq) + HCO_3^-(aq)$; the predominant anion is HCO_3^-

Naming Compounds

2.78 (a) KCl (b) $SnBr_2$ (c) CaO (d) $BaCl_2$ (e) AlH_3

2.80 (a) barium ion (b) cesium ion (c) vanadium(III) ion
(d) hydrogen carbonate ion (e) ammonium ion (f) nickel(II) ion
(g) nitrite ion (h) chlorite ion (i) manganese(II) ion (j) perchlorate ion

2.82 (a) SO_3^{2-} (b) PO_4^{3-} (c) Zr^{4+} (d) CrO_4^{2-} (e) $CH_3CO_2^-$ (f) $S_2O_3^{2-}$

2.84 (a) zinc(II) cyanide (b) iron(III) nitrite (c) titanium(IV) sulfate
(d) tin(II) phosphate (e) mercury(I) sulfide (f) manganese(IV) oxide
(g) potassium periodate (h) copper(II) acetate

2.86 (a) Na^+ and SO_4^{2-}; therefore the formula is Na_2SO_4
(b) Ba^{2+} and PO_4^{3-}; therefore the formula is $Ba_3(PO_4)_2$
(c) Ga^{3+} and SO_4^{2-}; therefore the formula is $Ga_2(SO_4)_3$

_____

General Problems

2.88   atomic mass = (0.205 x 69.924 amu) + (0.274 x 71.922 amu)
                       + (0.078 x 72.923 amu) + (0.365 x 73.921 amu)
                       + (0.078 x 75.921 amu) = 72.6 amu

2.90   (a) sodium bromate   (b) phosphoric acid
       (c) phosphorous acid   (d) vanadium(V) oxide

2.92   For NH_3, $(2.34 \text{ g N}) \left(\dfrac{3 \times 1.0079 \text{ amu H}}{14.0067 \text{ amu N}} \right) = 0.505 \text{ g H}$

      For N_2H_4, $(2.34 \text{ g N}) \left(\dfrac{4 \times 1.0079 \text{ amu H}}{2 \times 14.0067 \text{ amu N}} \right) = 0.337 \text{ g H}$

2.94 TeO_4^{2-}, tellurate; TeO_3^{2-}, tellurite.
      TeO_4^{2-} and TeO_3^{2-} are analogous to SO_4^{2-} and SO_3^{2-}.

2.96   (a) I^- (b) Au^{3+} (c) Kr

2.98 $\dfrac{39.9626 \text{ amu}}{15.9994 \text{ amu}} = \dfrac{X}{16.0000 \text{ amu}};$ $X = 39.9641$ amu for ^{40}Ca prior to 1961.

2.100 (a) calcium-40, ^{40}Ca
       (b) Not enough information, several different isotopes can have 63 neutrons.
       (c) The neutral atom contains 26 electrons. The ion is iron-56, $^{56}Fe^{3+}$.
       (d) Se^{2-}

2.102 $^1H^{35}Cl$ has 18 protons, 18 neutrons, and 18 electrons.
       $^1H^{37}Cl$ has 18 protons, 20 neutrons, and 18 electrons.
       $^2H^{35}Cl$ has 18 protons, 19 neutrons, and 18 electrons.
       $^2H^{37}Cl$ has 18 protons, 21 neutrons, and 18 electrons.
       $^3H^{35}Cl$ has 18 protons, 20 neutrons, and 18 electrons.
       $^3H^{37}Cl$ has 18 protons, 22 neutrons, and 18 electrons.

2.104 (a) Mg^{2+} and Cl^-, $MgCl_2$, magnesium chloride
       (b) Ca^{2+} and O^{2-}, CaO, calcium oxide
       (c) Li^+ and N^{3-}, Li_3N, lithium nitride
       (d) Al^{3+} and O^{2-}, Al_2O_3, aluminum oxide

2.106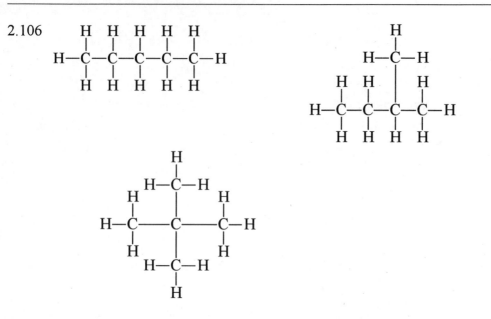

2.108   Molecular mass = (8 x 12.011 amu) + (9 x 1.0079 amu) + (1 x 14.0067 amu)
+ (2 x 15.9994 amu) = 151.165 amu

2.110   (a) Aspirin is likely a molecular compound because it is composed of only nonmetal elements.
(b) Assume a 100.0 g sample of aspirin. It would contain: 60.00 g C, 4.48 g H, and 35.52 g O.

$$60.00 \text{ g} \times \frac{1 \text{ amu}}{1.6605 \times 10^{-24} \text{ g}} \times \frac{1 \text{ C atom}}{12.011 \text{ amu}} = 3.008 \times 10^{24} \text{ C atoms}$$

$$4.48 \text{ g} \times \frac{1 \text{ amu}}{1.6605 \times 10^{-24} \text{ g}} \times \frac{1 \text{ H atom}}{1.008 \text{ amu}} = 2.68 \times 10^{24} \text{ H atoms}$$

$$35.52 \text{ g} \times \frac{1 \text{ amu}}{1.6605 \times 10^{-24} \text{ g}} \times \frac{1 \text{ O atom}}{15.999 \text{ amu}} = 1.337 \times 10^{24} \text{ O atoms}$$

The atom ratio in aspirin is:

$C_{3.008 \times 10^{24}} H_{2.68 \times 10^{24}} O_{1.337 \times 10^{24}}$, divide each subscript by 1×10^{24}

$C_{3.008} H_{2.68} O_{1.337}$, divide each subscript by the smallest, 1.337

$C_{3.008 / 1.337} H_{2.68 / 1.337} O_{1.337 / 1.337}$

$C_{2.25} H_2 O$, multiply each subscript by 4

$C_{(2.25 \times 4)} H_{(2 \times 4)} O_{(1 \times 4)}$

$C_9 H_8 O_4$

2.112   65.39 amu = (0.4863 x 63.929 amu) + (0.2790 x Z) + (0.0410 x 66.927 amu)
+ (0.1875 x 67.925 amu) + (0.0062 x 69.925 amu)
Solve for Z.
65.39 amu = 47.00 amu + (0.2790 x Z)
65.39 amu − 47.00 amu = 18.39 amu = 0.2790 x Z
18.39 amu/0.2790 = Z
Z = 65.91 amu for ^{66}Zn

3 Formulas, Equations, and Moles

3.1 $2 \text{ KClO}_3 \rightarrow 2 \text{ KCl} + 3 \text{ O}_2$

3.2   (a) $C_6H_{12}O_6 \rightarrow 2 \text{ C}_2\text{H}_6\text{O} + 2 \text{ CO}_2$
(b) $4 \text{ Fe} + 3 \text{ O}_2 \rightarrow 2 \text{ Fe}_2\text{O}_3$
(c) $4 \text{ NH}_3 + \text{Cl}_2 \rightarrow \text{N}_2\text{H}_4 + 2 \text{ NH}_4\text{Cl}$

3.3 $3 \text{ A}_2 + 2 \text{ B} \rightarrow 2 \text{ BA}_3$

3.4   (a) Fe_2O_3: $2(55.85) + 3(16.00) = 159.7$ amu
(b) H_2SO_4: $2(1.01) + 1(32.07) + 4(16.00) = 98.1$ amu
(c) $\text{C}_6\text{H}_8\text{O}_7$: $6(12.01) + 8(1.01) + 7(16.00) = 192.1$ amu
(d) $\text{C}_{16}\text{H}_{18}\text{N}_2\text{O}_4\text{S}$: $16(12.01) + 18(1.01) + 2(14.01) + 4(16.00) + 1(32.07) = 334.4$ amu

3.5 $\text{Fe}_2\text{O}_3(s) + 3 \text{ CO}(g) \rightarrow 2 \text{ Fe}(s) + 3 \text{ CO}_2(g)$

$$0.500 \text{ mol Fe}_2\text{O}_3 \times \frac{3 \text{ mol CO}}{1 \text{ mol Fe}_2\text{O}_3} = 1.50 \text{ mol CO}$$

3.6 $\text{C}_5\text{H}_{11}\text{NO}_2\text{S}$: $5(12.01) + 11(1.01) + 1(14.01) + 2(16.00) + 1(32.07) = 149.24$ amu

3.7 $\text{C}_9\text{H}_8\text{O}_4$, 180.2 amu; $500 \text{ mg} = 500 \times 10^{-3} \text{ g} = 0.500 \text{ g}$

$$0.500 \text{ g} \times \frac{1 \text{ mol}}{180.2 \text{ g}} = 2.77 \times 10^{-3} \text{ mol aspirin}$$

$$2.77 \times 10^{-3} \text{ mol} \times \frac{6.02 \times 10^{23} \text{ molecules}}{1 \text{ mol}} = 1.67 \times 10^{21} \text{ aspirin molecules}$$

3.8   salicylic acid, $\text{C}_7\text{H}_6\text{O}_3$, 138.1 amu;   acetic anhydride, $\text{C}_4\text{H}_6\text{O}_3$, 102.1 amu
aspirin, $\text{C}_9\text{H}_8\text{O}_4$, 180.2 amu;   acetic acid, $\text{C}_2\text{H}_4\text{O}_2$, 60.1 amu

$$4.50 \text{ g C}_7\text{H}_6\text{O}_3 \times \frac{1 \text{ mol C}_7\text{H}_6\text{O}_3}{138.1 \text{ g C}_7\text{H}_6\text{O}_3} \times \frac{1 \text{ mol C}_4\text{H}_6\text{O}_3}{1 \text{ mol C}_7\text{H}_6\text{O}_3} \times \frac{102.1 \text{ g C}_4\text{H}_6\text{O}_3}{1 \text{ mol C}_4\text{H}_6\text{O}_3} = 3.33 \text{ g C}_4\text{H}_6\text{O}_3$$

$$4.50 \text{ g C}_7\text{H}_6\text{O}_3 \times \frac{1 \text{ mol C}_7\text{H}_6\text{O}_3}{138.1 \text{ g C}_7\text{H}_6\text{O}_3} \times \frac{1 \text{ mol C}_9\text{H}_8\text{O}_4}{1 \text{ mol C}_7\text{H}_6\text{O}_3} \times \frac{180.2 \text{ g C}_9\text{H}_8\text{O}_4}{1 \text{ mol C}_9\text{H}_8\text{O}_4} = 5.87 \text{ g C}_9\text{H}_8\text{O}_4$$

$$4.50 \text{ g C}_7\text{H}_6\text{O}_3 \times \frac{1 \text{ mol C}_7\text{H}_6\text{O}_3}{138.1 \text{ g C}_7\text{H}_6\text{O}_3} \times \frac{1 \text{ mol C}_2\text{H}_4\text{O}_2}{1 \text{ mol C}_7\text{H}_6\text{O}_3} \times \frac{60.1 \text{ g C}_2\text{H}_4\text{O}_2}{1 \text{ mol C}_2\text{H}_4\text{O}_2} = 1.96 \text{ g C}_2\text{H}_4\text{O}_2$$

3.9 C_2H_4, 28.1 amu; C_2H_6O, 46.1 amu

$$4.6 \text{ g } C_2H_4 \times \frac{1 \text{ mol } C_2H_4}{28.1 \text{ g } C_2H_4} \times \frac{1 \text{ mol } C_2H_6O}{1 \text{ mol } C_2H_4} \times \frac{46.1 \text{ g } C_2H_6O}{1 \text{ mol } C_2H_6O} = 7.5 \text{ g } C_2H_6O \text{ (theoretical yield)}$$

$$\text{Percent yield} = \frac{\text{Actual yield}}{\text{Theoretical yield}} \times 100\% = \frac{4.7 \text{ g}}{7.5 \text{ g}} \times 100\% = 63\%$$

3.10 CH_4, 16.04 amu; CH_2Cl_2, 84.93 amu;   1.85 kg = 1850 g

$$1850 \text{ g } CH_4 \times \frac{1 \text{ mol } CH_4}{16.04 \text{ g } CH_4} \times \frac{1 \text{ mol } CH_2Cl_2}{1 \text{ mol } CH_4} \times \frac{84.93 \text{ g } CH_2Cl_2}{1 \text{ mol } CH_2Cl_2} = 9800 \text{ g } CH_2Cl_2 \text{ (theoretical yield)}$$

Actual yield = (9800 g)(0.431) = 4220 g CH_2Cl_2

3.11 Li_2O, 29.9 amu: 65 kg = 65,000 g; H_2O, 18.0 amu: 80.0 kg = 80,000 g

$$65,000 \text{ g } Li_2O \times \frac{1 \text{ mol } Li_2O}{29.9 \text{ g } Li_2O} = 2.17 \times 10^3 \text{ mol } Li_2O$$

$$80,000 \text{ g } H_2O \times \frac{1 \text{ mol } H_2O}{18.0 \text{ g } H_2O} = 4.44 \times 10^3 \text{ mol } H_2O$$

The reaction stoichiometry between Li_2O and H_2O is one to one. There are twice as many moles of H_2O as there are moles of Li_2O. Therefore, Li_2O is the limiting reactant.
$(4.44 \times 10^3 \text{ mol} - 2.17 \times 10^3 \text{ mol}) = 2.27 \times 10^3 \text{ mol } H_2O$ remaining

$$2.27 \times 10^3 \text{ mol } H_2O \times \frac{18.0 \text{ g } H_2O}{1 \text{ mol } H_2O} = 40,860 \text{ g } H_2O = 40.9 \text{ kg} = 41 \text{ kg } H_2O$$

3.12   LiOH, 23.9 amu; CO_2, 44.0 amu

$$500.0 \text{ g LiOH} \times \frac{1 \text{ mol LiOH}}{23.9 \text{ g LiOH}} \times \frac{1 \text{ mol } CO_2}{1 \text{ mol LiOH}} \times \frac{44.0 \text{ g } CO_2}{1 \text{ mol } CO_2} = 921 \text{ g } CO_2$$

3.13   (a) $A + B_2 \rightarrow AB_2$
There is a 1:1 stoichiometry between the two reactants. A is the limiting reactant because there are fewer reactant A's than there are reactant B_2's.
(b) 1.0 mol of AB_2 can be made from 1.0 mol of A and 1.0 mol of B_2.

3.14   (a) 125 mL = 0.125 L; (0.20 mol/L)(0.125 L) = 0.025 mol $NaHCO_3$
(b) 650.0 mL = 0.6500 L; (2.50 mol/L)(0.6500 L) = 1.62 mol H_2SO_4

3.15   (a) NaOH, 40.0 amu; 500.0 mL = 0.5000 L

$$1.25 \frac{\text{mol NaOH}}{L} \times 0.500 \text{ L} \times \frac{40.0 \text{ g NaOH}}{1 \text{ mol NaOH}} = 25.0 \text{ g NaOH}$$

(b) $C_6H_{12}O_6$, 180.2 amu

$$0.250 \, \frac{\text{mol } C_6H_{12}O_6}{L} \times 1.50 \, L \times \frac{180.2 \, \text{g } C_6H_{12}O_6}{1 \, \text{mol } C_6H_{12}O_6} = 67.6 \, \text{g } C_6H_{12}O_6$$

3.16 $C_6H_{12}O_6$, 180.2 amu;

$$25.0 \, \text{g } C_6H_{12}O_6 \times \frac{1 \, \text{mol } C_6H_{12}O_6}{180.2 \, \text{g } C_6H_{12}O_6} = 0.1387 \, \text{mol } C_6H_{12}O_6$$

$$0.1387 \, \text{mol} \times \frac{1 \, L}{0.20 \, \text{mol}} = 0.69 \, L; \quad 0.69 \, L = 690 \, \text{mL}$$

3.17 $C_{27}H_{46}O$, 386.7 amu; 750 mL = 0.750 L

$$0.005 \, \frac{\text{mol } C_{27}H_{46}O}{L} \times 0.750 \, L \times \frac{386.7 \, \text{g } C_{27}H_{46}O}{1 \, \text{mol } C_{27}H_{46}O} = 1 \, \text{g } C_{27}H_{46}O$$

3.18 $M_i \times V_i = M_f \times V_f; \quad M_f = \frac{M_i \times V_i}{V_f} = \frac{3.50 \, M \times 75.0 \, \text{mL}}{400.0 \, \text{mL}} = 0.656 \, M$

3.19 $M_i \times V_i = M_f \times V_f; \quad V_i = \frac{M_f \times V_f}{M_i} = \frac{0.500 \, M \times 250.0 \, \text{mL}}{18.0 \, M} = 6.94 \, \text{mL}$

Dilute 6.94 mL of 18.0 M H_2SO_4 with enough water to make 250.0 mL of solution. The resulting solution will be 0.500 M H_2SO_4.

3.20 50.0 mL = 0.0500 L; (0.100 mol/L)(0.0500 L) = 5.00 x 10^{-3} mol NaOH

$$5.00 \times 10^{-3} \, \text{mol NaOH} \times \frac{1 \, \text{mol } H_2SO_4}{2 \, \text{mol NaOH}} = 2.50 \times 10^{-3} \, \text{mol } H_2SO_4$$

$$\text{volume} = 2.50 \times 10^{-3} \, \text{mol} \times \frac{1 \, L}{0.250 \, \text{mol}} = 0.0100 \, L; \quad 0.0100 \, L = 10.0 \, \text{mL } H_2SO_4$$

3.21 $HNO_3(aq) + KOH(aq) \rightarrow KNO_3(aq) + H_2O(l)$
25.0 mL = 0.0250 L and 68.5 mL = 0.0685 L

$$0.150 \, \frac{\text{mol KOH}}{L} \times 0.0250 \, L \times \frac{1 \, \text{mol } HNO_3}{1 \, \text{mol KOH}} = 3.75 \times 10^{-3} \, \text{mol } HNO_3$$

$$HNO_3 \, \text{molarity} = \frac{3.75 \times 10^{-3} \, \text{mol}}{0.0685 \, L} = 5.47 \times 10^{-2} \, M$$

3.22 From the reaction stoichiometry, moles NaOH = moles CH_3CO_2H
(0.200 mol/L)(0.0947 L) = 0.018 94 mol NaOH = 0.018 94 mol CH_3CO_2H

$$\text{molarity} = \frac{0.018 \, 94 \, \text{mol}}{0.0250 \, L} = 0.758 \, M$$

3.23   For dimethylhydrazine, $C_2H_8N_2$, divide each subscript by 2 to obtain the empirical formula. The empirical formula is CH_4N. $C_2H_8N_2$, 60.1 amu or 60.1 g/mol

$$\% \, C = \frac{2 \times 12.0 \text{ g}}{60.1 \text{ g}} \times 100\% = 39.9\%$$

$$\% \, H = \frac{8 \times 1.01 \text{ g}}{60.1 \text{ g}} \times 100\% = 13.4\%$$

$$\% \, N = \frac{2 \times 14.0 \text{ g}}{60.1 \text{ g}} \times 100\% = 46.6\%$$

3.24   Assume a 100.0 g sample. From the percent composition data, a 100.0 g sample contains 14.25 g C, 56.93 g O, and 28.83 g Mg.

$$14.25 \text{ g C} \times \frac{1 \text{ mol C}}{12.0 \text{ g C}} = 1.19 \text{ mol C}$$

$$56.93 \text{ g O} \times \frac{1 \text{ mol O}}{16.0 \text{ g O}} = 3.56 \text{ mol O}$$

$$28.83 \text{ g Mg} \times \frac{1 \text{ mol Mg}}{24.3 \text{ g Mg}} = 1.19 \text{ mol Mg}$$

$Mg_{1.19}C_{1.19}O_{3.56}$; divide each subscript by the smallest, 1.19.
$Mg_{1.19/1.19}C_{1.19/1.19}O_{3.56/1.19}$
The empirical formula is $MgCO_3$.

3.25 $$1.161 \text{ g H}_2O \times \frac{1 \text{ mol H}_2O}{18.0 \text{ g H}_2O} \times \frac{2 \text{ mol H}}{1 \text{ mol H}_2O} = 0.129 \text{ mol H}$$

$$2.818 \text{ g CO}_2 \times \frac{1 \text{ mol CO}_2}{44.0 \text{ g CO}_2} \times \frac{1 \text{ mol C}}{1 \text{ mol CO}_2} = 0.0640 \text{ mol C}$$

$$0.129 \text{ mol H} \times \frac{1.01 \text{ g H}}{1 \text{ mol H}} = 0.130 \text{ g H}$$

$$0.0640 \text{ mol C} \times \frac{12.0 \text{ g C}}{1 \text{ mol C}} = 0.768 \text{ g C}$$

$$1.00 \text{ g total} - (0.130 \text{ g H} + 0.768 \text{ g C}) = 0.102 \text{ g O}$$

$$0.102 \text{ g O} \times \frac{1 \text{ mol O}}{16.0 \text{ g O}} = 0.006 \, 38 \text{ mol O}$$

$C_{0.0640}H_{0.129}O_{0.006\,38}$; divide each subscript by the smallest, 0.006 38.
$C_{0.0640/0.006\,38}H_{0.129/0.006\,38}O_{0.006\,38/0.006\,38}$
$C_{10.03}H_{20.22}O_1$ The empirical formula is $C_{10}H_{20}O$.

3.26   The empirical formula is CH_2O, 30 amu: molecular mass = 150 amu.

$$\frac{\text{molecular mass}}{\text{empirical formula mass}} = \frac{150 \text{ amu}}{30 \text{ amu}} = 5; \text{ therefore}$$

molecular formula = 5 × empirical formula = $C_{(5 \times 1)}H_{(5 \times 2)}O_{(5 \times 1)} = C_5H_{10}O_5$

3.27   (a) Assume a 100.0 g sample. From the percent composition data, a 100.0 g sample contains 21.86 g H and 78.14 g B.

$$21.86 \text{ g H} \times \frac{1 \text{ mol H}}{1.01 \text{ g H}} = 21.6 \text{ mol H}$$

$$78.14 \text{ g B} \times \frac{1 \text{ mol B}}{10.8 \text{ g B}} = 7.24 \text{ mol B}$$

$B_{7.24} H_{21.6}$; divide each subscript by the smaller, 7.24.
$B_{7.24/7.24} H_{21.6/7.24}$ The empirical formula is BH_3, 13.8 amu.
27.7 amu / 13.8 amu = 2; molecular formula = $B_{(2 \times 1)}H_{(2 \times 3)} = B_2H_6$.
(b) Assume a 100.0 g sample. From the percent composition data, a 100.0 g sample contains 6.71 g H, 40.00 g C, and 53.28 g O.

$$6.71 \text{ g H} \times \frac{1 \text{ mol H}}{1.01 \text{ g H}} = 6.64 \text{ mol H}$$

$$40.00 \text{ g C} \times \frac{1 \text{ mol C}}{12.0 \text{ g C}} = 3.33 \text{ mol C}$$

$$53.28 \text{ g O} \times \frac{1 \text{ mol O}}{16.0 \text{ g O}} = 3.33 \text{ mol O}$$

$C_{3.33} H_{6.64} O_{3.33}$; divide each subscript by the smallest, 3.33.
$C_{3.33/3.33} H_{6.64/3.33} O_{3.33/3.33}$ The empirical formula is CH_2O, 30.0 amu.
90.08 amu / 30.0 amu = 3; molecular formula = $C_{(3 \times 1)}H_{(3 \times 2)}O_{(3 \times 1)} = C_3H_6O_3$

3.28   Main sources of error in calculating Avogadro's number by spreading oil on a pond are:
(i)   the assumption that the oil molecules are tiny cubes
(ii) the assumption that the oil layer is one molecule thick
(iii) the assumption of a molecular mass of 200 for the oil

3.29   area of oil = $2.0 \times 10^7 \text{ cm}^2$
volume of oil = 4.9 cm^3 = area x $4\,l$ = $(2.0 \times 10^7 \text{ cm}^2) \times 4\,l$

$$l = \frac{4.9 \text{ cm}^3}{(2.0 \times 10^7 \text{ cm}^2)(4)} = 6.125 \times 10^{-8} \text{ cm}$$

area of oil = $2.0 \times 10^7 \text{ cm}^2 = l^2 \times N = (6.125 \times 10^{-8} \text{ cm})^2 \times N$

$$N = \frac{2.0 \times 10^7 \text{ cm}^2}{(6.125 \times 10^{-8} \text{ cm})^2} = 5.33 \times 10^{21} \text{ oil molecules}$$

$$\text{moles of oil} = (4.9 \text{ cm}^3) \times (0.95 \text{ g/cm}^3) \times \frac{1 \text{ mol oil}}{200 \text{ g oil}} = 0.0233 \text{ mol oil}$$

$$\text{Avogadro's number} = \frac{5.33 \times 10^{21} \text{ molecules}}{0.0233 \text{ mol}} = 2.3 \times 10^{23} \text{ molecules/mole}$$

Understanding Key Concepts

3.30   The concentration of a solution is cut in half when the volume is doubled. This is best represented by box (b).

3.32   The molecular formula for cytosine is $C_4H_5N_3O$.

$$\text{mol } CO_2 = 0.001 \text{ mol cyt } \times \frac{4\ C}{\text{cyt}} \times \frac{1\ CO_2}{C} = 0.004 \text{ mol } CO_2$$

$$\text{mol } H_2O = 0.001 \text{ mol cyt } \times \frac{5\ H}{\text{cyt}} \times \frac{1\ H_2O}{2\ H} = 0.0025 \text{ mol } H_2O$$

3.34 $C_{17}H_{18}F_3NO$ $17(12.01) + 18(1.01) + 3(19.00) + 1(14.01) + 1(16.00) = 309.36$ amu

3.36   (a) $A_2 + 3\ B_2 \rightarrow 2\ AB_3$; B_2 is the limiting reactant because it is completely consumed.
(b) For 1.0 mol of A_2, 3.0 mol of B_2 are required. Because only 1.0 mol of B_2 is available, B_2 is the limiting reactant.

$$1 \text{ mol } B_2 \times \frac{2 \text{ mol } AB_3}{3 \text{ mol } B_2} = 2/3 \text{ mol } AB_3$$

Additional Problems
Balancing Equations

3.38   Equation (b) is balanced, (a) is not balanced .

3.40   (a) $Mg + 2\ HNO_3 \rightarrow H_2 + Mg(NO_3)_2$
(b) $CaC_2 + 2\ H_2O \rightarrow Ca(OH)_2 + C_2H_2$
(c) $2\ S + 3\ O_2 \rightarrow 2\ SO_3$
(d) $UO_2 + 4\ HF \rightarrow UF_4 + 2\ H_2O$

Molecular Masses and Moles

3.42 Hg_2Cl_2: $2(200.59) + 2(35.45) = 472.1$ amu
$C_4H_8O_2$: $4(12.01) + 8(1.01) + 2(16.00) = 88.1$ amu
CF_2Cl_2: $1(12.01) + 2(19.00) + 2(35.45) = 120.9$ amu

3.44   One mole equals the atomic mass or molecular mass in grams.
(a) Ti, 47.88 g   (b) Br_2, 159.81 g   (c) Hg, 200.59 g   (d) H_2O, 18.02 g

3.46   There are 2 ions per each formula unit of NaCl. (2.5 mol)(2 mol ions/mol) = 5.0 mol ions

3.48   There are 3 ions (one Mg^{2+} and 2 Cl^-) per each formula unit of $MgCl_2$.
$MgCl_2$, 95.2 amu

$$27.5 \text{ g } MgCl_2 \times \frac{1 \text{ mol } MgCl_2}{95.2 \text{ g } MgCl_2} \times \frac{3 \text{ mol ions}}{1 \text{ mol } MgCl_2} = 0.867 \text{ mol ions}$$

3.50 $\text{Molar mass} = \dfrac{3.28 \text{ g}}{0.0275 \text{ mol}} = 119 \text{ g/mol}$;   molecular mass = 119 amu.

3.52 $FeSO_4$, 151.9 amu; 300 mg = 0.300 g

$$0.300 \text{ g FeSO}_4 \text{ x } \frac{1 \text{ mol FeSO}_4}{151.9 \text{ g FeSO}_4} = 1.97 \text{ x } 10^{-3} \text{ mol FeSO}_4$$

$$1.97 \text{ x } 10^{-3} \text{ mol FeSO}_4 \text{ x } \frac{6.02 \text{ x } 10^{23} \text{ Fe(II) atoms}}{1 \text{ mol FeSO}_4} = 1.19 \text{ x } 10^{21} \text{ Fe(II) atoms}$$

3.54 $C_8H_{10}N_4O_2$, 194.2 amu; 125 mg = 0.125 g

$$0.125 \text{ g caffeine x } \frac{1 \text{ mol caffeine}}{194.2 \text{ g caffeine}} = 6.44 \text{ x } 10^{-4} \text{ mol caffeine}$$

$$0.125 \text{ g caffeine x } \frac{1 \text{ mol caffeine}}{194.2 \text{ g caffeine}} \text{ x } \frac{6.022 \text{ x } 10^{23} \text{ molecules}}{1 \text{ mol}} = 3.88 \text{ x } 10^{20}$$
caffeine molecules

3.56 (a) $1.0 \text{ g Li x } \frac{1 \text{ mol Li}}{6.94 \text{ g Li}} = 0.14 \text{ mol Li}$

(b) $1.0 \text{ g Au x } \frac{1 \text{ mol Au}}{197.0 \text{ g Au}} = 0.0051 \text{ mol Au}$

(c) penicillin G: $C_{16}H_{17}N_2O_4SK$, 372.5 amu

$$1.0 \text{ g x } \frac{1 \text{ mol penicillin G}}{372.5 \text{ g penicillin G}} = 2.7 \text{ x } 10^{-3} \text{ mol penicillin G}$$

Stoichiometry Calculations

3.58 TiO_2, 79.88 amu; $100.0 \text{ kg Ti x } \frac{79.88 \text{ kg TiO}_2}{47.88 \text{ kg Ti}} = 166.8 \text{ kg TiO}_2$

3.60 (a) $2 \text{ Fe}_2O_3 + 3 \text{ C} \rightarrow 4 \text{ Fe} + 3 \text{ CO}_2$

(b) Fe_2O_3, 159.7 amu; $525 \text{ g Fe}_2O_3 \text{ x } \frac{1 \text{ mol Fe}_2O_3}{159.7 \text{ g Fe}_2O_3} \text{ x } \frac{3 \text{ mol C}}{2 \text{ mol Fe}_2O_3} = 4.93 \text{ mol C}$

(c) $4.93 \text{ mol C x } \frac{12.01 \text{ g C}}{1 \text{ mol C}} = 59.2 \text{ g C}$

3.62 (a) $2 \text{ Mg} + O_2 \rightarrow 2 \text{ MgO}$
(b) Mg, 24.30 amu; O_2, 32.00 amu; MgO, 40.30 amu

$$25.0 \text{ g Mg x } \frac{1 \text{ mol Mg}}{24.30 \text{ g Mg}} \text{ x } \frac{1 \text{ mol O}_2}{2 \text{ mol Mg}} \text{ x } \frac{32.00 \text{ g O}_2}{1 \text{ mol O}_2} = 16.5 \text{ g O}_2$$

$$25.0 \text{ g Mg x } \frac{1 \text{ mol Mg}}{24.30 \text{ g Mg}} \text{ x } \frac{2 \text{ mol MgO}}{2 \text{ mol Mg}} \text{ x } \frac{40.30 \text{ g MgO}}{1 \text{ mol MgO}} = 41.5 \text{ g MgO}$$

(c) $25.0 \text{ g O}_2 \times \dfrac{1 \text{ mol O}_2}{32.00 \text{ g O}_2} \times \dfrac{2 \text{ mol Mg}}{1 \text{ mol O}_2} \times \dfrac{24.30 \text{ g Mg}}{1 \text{ mol Mg}} = 38.0 \text{ g Mg}$

$25.0 \text{ g O}_2 \times \dfrac{1 \text{ mol O}_2}{32.00 \text{ g O}_2} \times \dfrac{2 \text{ mol MgO}}{1 \text{ mol O}_2} \times \dfrac{40.30 \text{ g MgO}}{1 \text{ mol MgO}} = 63.0 \text{ g MgO}$

3.64 (a) $2 \text{ HgO} \rightarrow 2 \text{ Hg} + \text{O}_2$
(b) HgO, 216.6 amu; Hg, 200.6 amu; O_2, 32.0 amu

$45.5 \text{ g HgO} \times \dfrac{1 \text{ mol HgO}}{216.6 \text{ g HgO}} \times \dfrac{2 \text{ mol Hg}}{2 \text{ mol HgO}} \times \dfrac{200.6 \text{ g Hg}}{1 \text{ mol Hg}} = 42.1 \text{ g Hg}$

$45.5 \text{ g HgO} \times \dfrac{1 \text{ mol HgO}}{216.6 \text{ g HgO}} \times \dfrac{1 \text{ mol O}_2}{2 \text{ mol HgO}} \times \dfrac{32.00 \text{ g O}_2}{1 \text{ mol O}_2} = 3.36 \text{ g O}_2$

(c) $33.3 \text{ g O}_2 \times \dfrac{1 \text{ mol O}_2}{32.00 \text{ g O}_2} \times \dfrac{2 \text{ mol HgO}}{1 \text{ mol O}_2} \times \dfrac{216.6 \text{ g HgO}}{1 \text{ mol HgO}} = 451 \text{ g HgO}$

3.66 $2.00 \text{ g Ag} \times \dfrac{1 \text{ mol Ag}}{107.9 \text{ g Ag}} = 0.0185 \text{ mol Ag}$

$0.657 \text{ g Cl} \times \dfrac{1 \text{ mol Cl}}{35.45 \text{ g Cl}} = 0.0185 \text{ mol Cl}$

$\text{Ag}_{0.0185}\text{Cl}_{0.0185}$ Divide both subscripts by 0.0185. The empirical formula is AgCl.

Limiting Reactants and Reaction Yield

3.68 $3.44 \text{ mol N}_2 \times \dfrac{3 \text{ mol H}_2}{1 \text{ mol N}_2} = 10.3 \text{ mol H}_2 \text{ required.}$

Because there is only 1.39 mol H_2, H_2 is the limiting reactant.

$1.39 \text{ mol H}_2 \times \dfrac{2 \text{ mol NH}_3}{3 \text{ mol H}_2} \times \dfrac{17.03 \text{ g NH}_3}{1 \text{ mol NH}_3} = 15.8 \text{ g NH}_3$

$1.39 \text{ mol H}_2 \times \dfrac{1 \text{ mol N}_2}{3 \text{ mol H}_2} \times \dfrac{28.01 \text{ g N}_2}{1 \text{ mol N}_2} = 13.0 \text{ g N}_2 \text{ reacted}$

$3.44 \text{ mol N}_2 \times \dfrac{28.01 \text{ g N}_2}{1 \text{ mol N}_2} = 96.3 \text{ g N}_2 \text{ initially}$

$(96.3 \text{ g} - 13.0 \text{ g}) = 83.3 \text{ g N}_2 \text{ left over}$

3.70 C_2H_4, 28.05 amu; Cl_2, 70.91 amu; $\text{C}_2\text{H}_4\text{Cl}_2$, 98.96 amu

$15.4 \text{ g C}_2\text{H}_4 \times \dfrac{1 \text{ mol C}_2\text{H}_4}{28.05 \text{ g C}_2\text{H}_4} = 0.549 \text{ mol C}_2\text{H}_4$

$3.74 \text{ g Cl}_2 \times \dfrac{1 \text{ mol Cl}_2}{70.91 \text{ g Cl}_2} = 0.0527 \text{ mol Cl}_2$

Because the reaction stoichiometry between C_2H_4 and Cl_2 is one to one, Cl_2 is the limiting reactant.

$$0.0527 \text{ mol } Cl_2 \text{ x } \frac{1 \text{ mol } C_2H_4Cl_2}{1 \text{ mol } Cl_2} \text{ x } \frac{98.96 \text{ g } C_2H_4Cl_2}{1 \text{ mol } C_2H_4Cl_2} = 5.22 \text{ g } C_2H_4Cl_2$$

3.72 $CaCO_3$, 100.1 amu; HCl, 36.46 amu

$$CaCO_3 + 2\,HCl \rightarrow CaCl_2 + H_2O + CO_2$$

$$2.35 \text{ g } CaCO_3 \text{ x } \frac{1 \text{ mol } CaCO_3}{100.1 \text{ g } CaCO_3} = 0.0235 \text{ mol } CaCO_3$$

$$2.35 \text{ g } HCl \text{ x } \frac{1 \text{ mol } HCl}{36.46 \text{ g } HCl} = 0.0645 \text{ mol } HCl$$

The reaction stoichiometry is 1 mole of $CaCO_3$ for every 2 moles of HCl. For 0.0235 mol $CaCO_3$, we only need 2(0.0235 mol) = 0.0470 mol HCl. We have 0.0645 mol HCl; therefore $CaCO_3$ is the limiting reactant.

$$0.0235 \text{ mol } CaCO_3 \text{ x } \frac{1 \text{ mol } CO_2}{1 \text{ mol } CaCO_3} \text{ x } \frac{22.4 \text{ L}}{1 \text{ mol } CO_2} = 0.526 \text{ L } CO_2$$

3.74 $CH_3CO_2H + C_5H_{12}O \rightarrow C_7H_{14}O_2 + H_2O$

CH_3CO_2H, 60.05 amu; $C_5H_{12}O$, 88.15 amu; $C_7H_{14}O_2$, 130.19 amu

$$3.58 \text{ g } CH_3CO_2H \text{ x } \frac{1 \text{ mol } CH_3CO_2H}{60.05 \text{ g } CH_3CO_2H} = 0.0596 \text{ mol } CH_3CO_2H$$

$$4.75 \text{ g } C_5H_{12}O \text{ x } \frac{1 \text{ mol } C_5H_{12}O}{88.15 \text{ g } C_5H_{12}O} = 0.0539 \text{ mol } C_5H_{12}O$$

Because the reaction stoichiometry between CH_3CO_2H and $C_5H_{12}O$ is one to one, isopentyl alcohol ($C_5H_{12}O$) is the limiting reactant.

$$0.0539 \text{ mol } C_5H_{12}O \text{ x } \frac{1 \text{ mol } C_7H_{14}O_2}{1 \text{ mol } C_5H_{12}O} \text{ x } \frac{130.19 \text{ g } C_7H_{14}O_2}{1 \text{ mol } C_7H_{14}O_2} = 7.02 \text{ g } C_7H_{14}O_2$$

7.02 g $C_7H_{14}O_2$ is the theoretical yield. Actual yield = (7.02 g)(0.45) = 3.2 g.

3.76 $CH_3CO_2H + C_5H_{12}O \rightarrow C_7H_{14}O_2 + H_2O$

CH_3CO_2H, 60.05 amu; $C_5H_{12}O$, 88.15 amu; $C_7H_{14}O_2$, 130.19 amu

$$1.87 \text{ g } CH_3CO_2H \text{ x } \frac{1 \text{ mol } CH_3CO_2H}{60.05 \text{ g } CH_3CO_2H} = 0.0311 \text{ mol } CH_3CO_2H$$

$$2.31 \text{ g } C_5H_{12}O \text{ x } \frac{1 \text{ mol } C_5H_{12}O}{88.15 \text{ g } C_5H_{12}O} = 0.0262 \text{ mol } C_5H_{12}O$$

Because the reaction stoichiometry between CH_3CO_2H and $C_5H_{12}O$ is one to one, isopentyl alcohol ($C_5H_{12}O$) is the limiting reactant.

$$0.0262 \text{ mol } C_5H_{12}O \text{ x } \frac{1 \text{ mol } C_7H_{14}O_2}{1 \text{ mol } C_5H_{12}O} \text{ x } \frac{130.19 \text{ g } C_7H_{14}O_2}{1 \text{ mol } C_7H_{14}O_2} = 3.41 \text{ g } C_7H_{14}O_2$$

3.41 g $C_7H_{14}O_2$ is the theoretical yield.

% Yield = $\dfrac{\text{Actual yield}}{\text{Theoretical yield}}$ x 100% = $\dfrac{2.96\text{ g}}{3.41\text{ g}}$ x 100% = 86.8%

Molarity, Solution Stoichiometry, Dilution, and Titration

3.78 (a) 35.0 mL = 0.0350 L; $1.200 \dfrac{\text{mol } HNO_3}{L}$ x 0.0350 L = 0.0420 mol HNO_3

    (b) 175 mL = 0.175 L; $0.67 \dfrac{\text{mol } C_6H_{12}O_6}{L}$ x 0.175 L = 0.12 mol $C_6H_{12}O_6$

3.80 $BaCl_2$, 208.2 amu

15.0 g $BaCl_2$ x $\dfrac{1\text{ mol } BaCl_2}{208.2\text{ g } BaCl_2}$ = 0.0720 mol $BaCl_2$

0.0720 mol x $\dfrac{1.0\text{ L}}{0.45\text{ mol}}$ = 0.16 L; 0.16 L = 160 mL

3.82 NaCl, 58.4 amu; 400 mg = 0.400 g; 100 mL = 0.100 L

0.400 g NaCl x $\dfrac{1\text{ mol NaCl}}{58.4\text{ g NaCl}}$ = 0.006 85 mol NaCl

molarity = $\dfrac{0.006\ 85\text{ mol}}{0.100\text{ L}}$ = 0.0685 M

3.84 NaCl, 58.4 amu; KCl, 74.6 amu; $CaCl_2$, 111.0 amu; 500 mL = 0.500 L

4.30 g NaCl x $\dfrac{1\text{ mol NaCl}}{58.4\text{ g NaCl}}$ = 0.0736 mol NaCl

0.150 g KCl x $\dfrac{1\text{ mol KCl}}{74.6\text{ g KCl}}$ = 0.002 01 mol KCl

0.165 g $CaCl_2$ x $\dfrac{1\text{ mol } CaCl_2}{111.0\text{ g } CaCl_2}$ = 0.001 49 mol $CaCl_2$

0.0736 mol + 0.002 01 mol + 2(0.001 49 mol) = 0.0786 mol Cl^-

Na^+ molarity = $\dfrac{0.0736\text{ mol}}{0.500\text{ L}}$ = 0.147 M

Ca^{2+} molarity = $\dfrac{0.001\ 49\text{ mol}}{0.500\text{ L}}$ = 0.002 98 M

K^+ molarity = $\dfrac{0.002\ 01\text{ mol}}{0.500\text{ L}}$ = 0.004 02 M

Cl^- molarity = $\dfrac{0.0786\text{ mol}}{0.500\text{ L}}$ = 0.157 M

Chapter 3 – Formulas, Equations, and Moles

3.86 $M_f \times V_f = M_i \times V_i$; $M_f = \dfrac{M_i \times V_i}{V_f} = \dfrac{12.0\,M \times 35.7\,mL}{250.0\,mL} = 1.71\,M\,HCl$

3.88 $2\,HBr(aq) + K_2CO_3(aq) \rightarrow 2\,KBr(aq) + CO_2(g) + H_2O(l)$
K_2CO_3, 138.2 amu; 450 mL = 0.450 L

$0.500\,\dfrac{mol\,HBr}{L} \times 0.450\,L = 0.225\,mol\,HBr$

$0.225\,mol\,HBr \times \dfrac{1\,mol\,K_2CO_3}{2\,mol\,HBr} \times \dfrac{138.2\,g\,K_2CO_3}{1\,mol\,K_2CO_3} = 15.5\,g\,K_2CO_3$

3.90 $H_2C_2O_4$, 90.04 amu

$3.225\,g\,H_2C_2O_4 \times \dfrac{1\,mol\,H_2C_2O_4}{90.04\,g\,H_2C_2O_4} \times \dfrac{2\,mol\,KMnO_4}{5\,mol\,H_2C_2O_4} = 0.0143\,mol\,KMnO_4$

$0.0143\,mol \times \dfrac{1\,L}{0.250\,mol} = 0.0572\,L = 57.2\,mL$

Formulas and Elemental Analysis

3.92 CH_4N_2O, 60.1 amu

$\%\,C = \dfrac{12.0\,g\,C}{60.1\,g} \times 100\% = 20.0\%$

$\%\,H = \dfrac{4 \times 1.01\,g\,H}{60.1\,g} \times 100\% = 6.72\%$

$\%\,N = \dfrac{2 \times 14.0\,g\,N}{60.1\,g} \times 100\% = 46.6\%$

$\%\,O = \dfrac{16.0\,g\,O}{60.1\,g} \times 100\% = 26.6\%$

3.94 Assume a 100.0 g sample. From the percent composition data, a 100.0 g sample contains 24.25 g F and 75.75 g Sn.

$24.25\,g\,F \times \dfrac{1\,mol\,F}{19.00\,g\,F} = 1.276\,mol\,F$

$75.75\,g\,Sn \times \dfrac{1\,mol\,Sn}{118.7\,g\,Sn} = 0.6382\,mol\,Sn$

$Sn_{0.6382}F_{1.276}$; divide each subscript by the smaller, 0.6382.
$Sn_{0.6382/0.6382}F_{1.276/0.6382}$ The empirical formula is SnF_2.

3.96 Mass of toluene sample = 45.62 mg = 0.045 62 g; mass of CO_2 = 152.5 mg = 0.1525 g; mass of H_2O = 35.67 mg = 0.035 67 g

$0.1525\,g\,CO_2 \times \dfrac{1\,mol\,CO_2}{44.01\,g\,CO_2} \times \dfrac{1\,mol\,C}{1\,mol\,CO_2} = 0.003\,465\,mol\,C$

$$\text{mass C} = 0.003\ 465 \text{ mol C} \times \frac{12.011 \text{ g C}}{1 \text{ mol C}} = 0.041\ 62 \text{ g C}$$

$$0.035\ 67 \text{ g H}_2\text{O} \times \frac{1 \text{ mol H}_2\text{O}}{18.02 \text{ g H}_2\text{O}} \times \frac{2 \text{ mol H}}{1 \text{ mol H}_2\text{O}} = 0.003\ 959 \text{ mol H}$$

$$\text{mass H} = 0.003\ 959 \text{ mol H} \times \frac{1.008 \text{ g H}}{1 \text{ mol H}} = 0.003\ 991 \text{ g H}$$

The (mass C + mass H) = 0.041 62 g + 0.003 991 g = 0.045 61 g. The calculated mass of (C + H) essentially equals the mass of the toluene sample, this means that toluene contains only C and H and no other elements.

$C_{0.003\ 465}H_{0.003\ 959}$; divide each subscript by the smaller, 0.003 465.

$C_{0.003\ 465\ /\ 0.003\ 465}H_{0.003\ 959\ /\ 0.003\ 465}$

$CH_{1.14}$; multiply each subscript by 7 to obtain integers.

The empirical formula is C_7H_8.

3.98 Let X equal the molecular mass of cytochrome c.

$$0.0043 = \frac{55.847 \text{ amu}}{X}; \quad X = \frac{55.847 \text{ amu}}{0.0043} = 13,000 \text{ amu}$$

3.100 Let X equal the molecular mass of disilane.

$$0.9028 = \frac{2 \times 28.09 \text{ amu}}{X}; \quad X = \frac{2 \times 28.09 \text{ amu}}{0.9028} = 62.23 \text{ amu}$$

62.23 amu − 2(Si atomic mass) = 62.23 amu − 2(28.09 amu) = 6.05 amu
6.05 amu is the total mass of H atoms.

$$6.05 \text{ amu} \times \frac{1 \text{ H atom}}{1.01 \text{ amu}} = 6 \text{ H atoms}; \qquad \text{Disilane is Si}_2\text{H}_6.$$

General Problems

3.102 (a) $C_6H_{12}O_6$, 180.2 amu

$$\% \text{ C} = \frac{6 \times 12.01 \text{ g C}}{180.2 \text{ g}} \times 100\% = 39.99\%$$

$$\% \text{ H} = \frac{12 \times 1.008 \text{ g H}}{180.2 \text{ g}} \times 100\% = 6.713\%$$

$$\% \text{ O} = \frac{6 \times 16.00 \text{ g O}}{180.2 \text{ g}} \times 100\% = 53.27\%$$

(b) H_2SO_4, 98.08 amu

$$\% \text{ H} = \frac{2 \times 1.008 \text{ g H}}{98.08 \text{ g}} \times 100\% = 2.055\%$$

$$\% \text{ S} = \frac{32.07 \text{ g S}}{98.08 \text{ g}} \times 100\% = 32.70\%$$

$$\% \text{ O} = \frac{4 \times 16.00 \text{ g O}}{98.08 \text{ g}} \times 100\% = 65.25\%$$

(c) $KMnO_4$, 158.0 amu

$$\% \, K = \frac{39.10 \text{ g K}}{158.0 \text{ g}} \times 100\% = 24.75\%$$

$$\% \, Mn = \frac{54.94 \text{ g Mn}}{158.0 \text{ g}} \times 100\% = 34.77\%$$

$$\% \, O = \frac{4 \times 16.00 \text{ g O}}{158.0 \text{ g}} \times 100\% = 40.51\%$$

(d) $C_7H_5NO_3S$, 183.2 amu

$$\% \, C = \frac{7 \times 12.01 \text{ g C}}{183.2 \text{ g}} \times 100\% = 45.89\%$$

$$\% \, H = \frac{5 \times 1.008 \text{ g H}}{183.2 \text{ g}} \times 100\% = 2.751\%$$

$$\% \, N = \frac{14.01 \text{ g N}}{183.2 \text{ g}} \times 100\% = 7.647\%$$

$$\% \, O = \frac{3 \times 16.00 \text{ g O}}{183.2 \text{ g}} \times 100\% = 26.20\%$$

$$\% \, S = \frac{32.07 \text{ g S}}{183.2 \text{ g}} \times 100\% = 17.51\%$$

3.104 (a) $SiCl_4 + 2 H_2O \rightarrow SiO_2 + 4 HCl$
       (b) $P_4O_{10} + 6 H_2O \rightarrow 4 H_3PO_4$
       (c) $CaCN_2 + 3 H_2O \rightarrow CaCO_3 + 2 NH_3$
       (d) $3 NO_2 + H_2O \rightarrow 2 HNO_3 + NO$

3.106 Assume a 100.0 g sample of ferrocene. From the percent composition data, a 100.0 g sample contains 5.42 g H, 64.56 g C, and 30.02 g Fe.

$$5.42 \text{ g H} \times \frac{1 \text{ mol H}}{1.01 \text{ g H}} = 5.37 \text{ mol H}$$

$$64.56 \text{ g C} \times \frac{1 \text{ mol C}}{12.01 \text{ g C}} = 5.376 \text{ mol C}$$

$$30.02 \text{ g Fe} \times \frac{1 \text{ mol Fe}}{55.85 \text{ g Fe}} = 0.5375 \text{ mol Fe}$$

$C_{5.376}H_{5.37}Fe_{0.5375}$; divide each subscript by the smallest, 0.5375.
$C_{5.376/0.5375}H_{5.37/0.5375}Fe_{0.5375/0.5375}$ The empirical formula is $C_{10}H_{10}Fe$.

3.108 Na_2SO_4, 142.04 amu; Na_3PO_4, 163.94 amu; Li_2SO_4, 109.95 amu; 100.00 mL = 0.10000 L

$$0.550 \text{ g Na}_2\text{SO}_4 \times \frac{1 \text{ mol Na}_2\text{SO}_4}{142.04 \text{ g Na}_2\text{SO}_4} = 0.003 \, 872 \text{ mol Na}_2\text{SO}_4$$

$$1.188 \text{ g Na}_3\text{PO}_4 \times \frac{1 \text{ mol Na}_3\text{PO}_4}{163.94 \text{ g Na}_3\text{PO}_4} = 0.007 \, 247 \text{ mol Na}_3\text{PO}_4$$

$$0.223 \text{ g Li}_2\text{SO}_4 \times \frac{1 \text{ mol Li}_2\text{SO}_4}{109.95 \text{ g Li}_2\text{SO}_4} = 0.002\,028 \text{ mol Li}_2\text{SO}_4$$

$$\text{Na}^+ \text{ molarity} = \frac{(2 \times 0.003\,872 \text{ mol}) + (3 \times 0.007\,247 \text{ mol})}{0.100\,00 \text{ L}} = 0.295 \text{ M}$$

$$\text{Li}^+ \text{ molarity} = \frac{2 \times 0.002\,028 \text{ mol}}{0.100\,00 \text{ L}} = 0.0406 \text{ M}$$

$$\text{SO}_4^{2-} \text{ molarity} = \frac{(1 \times 0.003\,872 \text{ mol}) + (1 \times 0.002\,028 \text{ mol})}{0.100\,00 \text{ L}} = 0.0590 \text{ M}$$

$$\text{PO}_4^{3-} \text{ molarity} = \frac{1 \times 0.007\,247 \text{ mol}}{0.100\,00 \text{ L}} = 0.0725 \text{ M}$$

3.110 High resolution mass spectrometry is capable of measuring the mass of molecules with a particular isotopic composition.

3.112 The combustion reaction is: $2 \text{ C}_8\text{H}_{18} + 25 \text{ O}_2 \rightarrow 16 \text{ CO}_2 + 18 \text{ H}_2\text{O}$
C_8H_{18}, 114.23 amu; CO_2, 44.01 amu

$$\text{pounds CO}_2 = 1.00 \text{ gal} \times \frac{3.7854 \text{ L}}{1 \text{ gal}} \times \frac{1000 \text{ mL}}{1 \text{ L}} \times \frac{0.703 \text{ g C}_8\text{H}_{18}}{1 \text{ mL}} \times \frac{1 \text{ mol C}_8\text{H}_{18}}{114.23 \text{ g C}_8\text{H}_{18}} \times$$

$$\frac{16 \text{ mol CO}_2}{2 \text{ mol C}_8\text{H}_{18}} \times \frac{44.01 \text{ g CO}_2}{1 \text{ mol CO}_2} \times \frac{1 \text{ lb}}{453.59 \text{ g}} = 18.1 \text{ pounds CO}_2$$

3.114 AgCl, 143.32 amu; CO_2, 44.01 amu; H_2O, 18.02 amu

$$\text{mol Cl in 1.00 g of X} = 1.95 \text{ g AgCl} \times \frac{1 \text{ mol AgCl}}{143.32 \text{ g AgCl}} \times \frac{1 \text{ mol Cl}}{1 \text{ mol AgCl}} = 0.0136 \text{ mol Cl}$$

$$\text{mass Cl} = 0.0136 \text{ mol Cl} \times \frac{35.453 \text{ g Cl}}{1 \text{ mol Cl}} = 0.482 \text{ g Cl}$$

$$\text{mol C in 1.00 g of X} = 0.900 \text{ g CO}_2 \times \frac{1 \text{ mol CO}_2}{44.01 \text{ g CO}_2} \times \frac{1 \text{ mol C}}{1 \text{ mol CO}_2} = 0.0204 \text{ mol C}$$

$$\text{mass C} = 0.0204 \text{ mol C} \times \frac{12.011 \text{ g C}}{1 \text{ mol C}} = 0.245 \text{ g C}$$

$$\text{mol H in 1.00 g of X} = 0.735 \text{ g H}_2\text{O} \times \frac{1 \text{ mol H}_2\text{O}}{18.02 \text{ g H}_2\text{O}} \times \frac{2 \text{ mol H}}{1 \text{ mol H}_2\text{O}} = 0.0816 \text{ mol H}$$

$$\text{mass H} = 0.0816 \text{ mol H} \times \frac{1.008 \text{ g H}}{1 \text{ mol H}} = 0.0823 \text{ g H}$$

$$\text{mass N} = 1.00 \text{ g} - \text{mass Cl} - \text{mass C} - \text{mass H} = 1.00 - 0.482 \text{ g} - 0.245 \text{ g} - 0.0823 \text{ g} = 0.19 \text{ g N}$$

$$\text{mol N in 1.00 g of X} = 0.19 \text{ g N} \times \frac{1 \text{ mol N}}{14.01 \text{ g N}} = 0.014 \text{ mol N}$$

Determine empirical formula.

$C_{0.0204}H_{0.0816}N_{0.014}Cl_{0.0136}$, divide each subscript by the smallest, 0.0136.

$C_{0.0204\,/\,0.0136}H_{0.0816\,/\,0.0136}N_{0.014\,/\,0.0136}Cl_{0.0136\,/\,0.0136}$

$C_{1.5}H_6NCl$, multiply each subscript by 2 to get integers.

The empirical formula is $C_3H_{12}N_2Cl_2$.

3.116   Let SA stand for salicylic acid.

$$\text{mol C in 1.00 g of SA} = 2.23 \text{ g CO}_2 \times \frac{1 \text{ mol CO}_2}{44.01 \text{ g CO}_2} \times \frac{1 \text{ mol C}}{1 \text{ mol CO}_2} = 0.0507 \text{ mol C}$$

$$\text{mass C} = 0.0507 \text{ mol C} \times \frac{12.011 \text{ g C}}{1 \text{ mol C}} = 0.609 \text{ g C}$$

$$\text{mol H in 1.00 g of SA} = 0.39 \text{ g H}_2O \times \frac{1 \text{ mol H}_2O}{18.02 \text{ g H}_2O} \times \frac{2 \text{ mol H}}{1 \text{ mol H}_2O} = 0.043 \text{ mol H}$$

$$\text{mass H} = 0.043 \text{ mol H} \times \frac{1.008 \text{ g H}}{1 \text{ mol H}} = 0.043 \text{ g H}$$

$$\text{mass O} = 1.00 \text{ g} - \text{mass C} - \text{mass H} = 1.00 - 0.609 \text{ g} - 0.043 \text{ g} = 0.35 \text{ g O}$$

$$\text{mol O in 1.00 g of} = 0.35 \text{ g N} \times \frac{1 \text{ mol O}}{16.00 \text{ g O}} = 0.022 \text{ mol O}$$

Determine empirical formula.

$C_{0.0507}H_{0.043}O_{0.022}$, divide each subscript by the smallest, 0.022.

$C_{0.0507\,/\,0.022}H_{0.043\,/\,0.022}O_{0.022\,/\,0.022}$

$C_{2.3}H_2O$, multiply each subscript by 3 to get integers.

The empirical formula is $C_7H_6O_3$. The empirical formula mass = 138.12 g/mol

Because salicylic acid has only one acidic hydrogen, there is a 1 to 1 mol ratio between salicylic acid and NaOH in the acid-base titration.

$$\text{mol SA in 1.00 g SA} = 72.4 \text{ mL} \times \frac{1 \text{ L}}{1000 \text{ mL}} \times \frac{0.100 \text{ mol NaOH}}{1 \text{ L}} \times \frac{1 \text{ mol SA}}{1 \text{ mol NaOH}} =$$
$$0.00724 \text{ mol SA}$$

$$\text{SA molar mass} = \frac{1.00 \text{ g}}{0.00724 \text{ mol}} = 138 \text{ g/mol}$$

Because the empirical formula mass and the molar mass are the same, the empirical formula is the molecular formula for salicylic acid.

3.118   Let X equal the mass of benzoic acid and Y the mass of gallic acid in the 1.00 g mixture. Therefore, X + Y = 1.00 g.

Because both acids contain only one acidic hydrogen, there is a 1 to 1 mol ratio between each acid and NaOH in the acid-base titration.

In the titration, mol benzoic acid + mol gallic acid = mol NaOH

$$\text{Therefore, } X \times \frac{1 \text{ mol BA}}{122 \text{ g BA}} + Y \times \frac{1 \text{ mol GA}}{170 \text{ g GA}} = \text{mol NaOH}$$

$$\text{mol NaOH} = 14.7 \text{ mL x } \frac{1 \text{ L}}{1000 \text{ mL}} \text{ x } \frac{0.500 \text{ mol NaOH}}{1 \text{ L}} = 0.00735 \text{ mol NaOH}$$

We have two unknowns, X and Y, and two equations.

X + Y = 1.00 g

$$\text{X x } \frac{1 \text{ mol BA}}{122 \text{ g BA}} + \text{Y x } \frac{1 \text{ mol GA}}{170 \text{ g GA}} = 0.00735 \text{ mol NaOH}$$

Rearrange to get X = 1.00 g – Y and then substitute it into the second equation to solve for Y.

$$(1.00 \text{ g} - \text{Y}) \text{ x } \frac{1 \text{ mol BA}}{122 \text{ g BA}} + \text{Y x } \frac{1 \text{ mol GA}}{170 \text{ g GA}} = 0.00735 \text{ mol NaOH}$$

$$\frac{1 \text{ mol}}{122} - \frac{\text{Y mol}}{122 \text{ g}} + \frac{\text{Y mol}}{170 \text{ g}} = 0.00735 \text{ mol}$$

$$-\frac{\text{Y mol}}{122 \text{ g}} + \frac{\text{Y mol}}{170 \text{ g}} = 0.00735 \text{ mol} - \frac{1 \text{ mol}}{122} = -8.47 \text{ x } 10^{-4} \text{ mol}$$

$$\frac{(-\text{Y mol})(170 \text{ g}) + (\text{Y mol})(122 \text{ g})}{(170 \text{ g})(122 \text{ g})} = -8.47 \text{ x } 10^{-4} \text{ mol}$$

$$\frac{-48 \text{ Y mol}}{20740 \text{ g}} = -8.47 \text{ x } 10^{-4} \text{ mol}; \quad \frac{48 \text{ Y}}{20740 \text{ g}} = 8.47 \text{ x } 10^{-4}$$

$$\text{Y} = \frac{(20740 \text{ g})(8.47 \text{ x } 10^{-4})}{48} = 0.366 \text{ g}$$

X = 1.00 g – 0.366 g = 0.634 g

In the 1.00 g mixture there is 0.63 g of benzoic acid and 0.37 g of gallic acid.

3.120 FeO, 71.85 amu; Fe_2O_3, 159.7 amu

Let X equal the mass of FeO and Y the mass of Fe_2O_3 in the 10.0 g mixture. Therefore, X + Y = 10.0 g.

$$\text{mol Fe} = 7.43 \text{ g x } \frac{1 \text{ mol Fe}}{55.85 \text{ g Fe}} = 0.133 \text{ mol Fe}$$

$$\text{mol FeO} + 2 \text{ x mol Fe}_2\text{O}_3 = 0.133 \text{ mol Fe}$$

$$\text{X x } \frac{1 \text{ mol FeO}}{71.85 \text{ g FeO}} + 2 \text{ x } \left(\text{Y x } \frac{1 \text{ mol Fe}_2\text{O}_3}{159.7 \text{ g Fe}_2\text{O}_3} \right) = 0.133 \text{ mol Fe}$$

Rearrange to get X = 10.0 g – Y and then substitute it into the equation above to solve for Y.

$$(10.0 \text{ g} - \text{Y}) \text{ x } \frac{1 \text{ mol FeO}}{71.85 \text{ g FeO}} + 2 \text{ x } \left(\text{Y x } \frac{1 \text{ mol Fe}_2\text{O}_3}{159.7 \text{ g Fe}_2\text{O}_3} \right) = 0.133 \text{ mol Fe}$$

$$\frac{10.0 \text{ mol}}{71.85} - \frac{\text{Y mol}}{71.85 \text{ g}} + \frac{2 \text{ Y mol}}{159.7 \text{ g}} = 0.133 \text{ mol}$$

$$-\frac{\text{Y mol}}{71.85 \text{ g}} + \frac{2 \text{ Y mol}}{159.7 \text{ g}} = 0.133 \text{ mol} - \frac{10.0 \text{ mol}}{71.85} = -0.0062 \text{ mol}$$

$$\frac{(-\text{Y mol})(159.7 \text{ g}) + (2 \text{ Y mol})(71.85 \text{ g})}{(71.85 \text{ g})(159.7 \text{ g})} = -0.0062 \text{ mol}$$

$$\frac{-16.0 \text{ Y mol}}{11474 \text{ g}} = -0.0062 \text{ mol}; \quad \frac{16.0 \text{ Y}}{11474 \text{ g}} = 0.0062$$

Y = (0.0062)(11474 g)/16.0 = 4.44 g = 4.4 g Fe_2O_3

X = 10.0 g – Y = 10.0 g – 4.4 g = 5.6 g FeO

3.122 $C_6H_{12}O_6 + 6\,O_2 \rightarrow 6\,CO_2 + 6\,H_2O$; $C_6H_{12}O_6$, 180.16 amu; CO_2, 44.01 amu

$$66.3 \text{ g } C_6H_{12}O_6 \times \frac{1 \text{ mol } C_6H_{12}O_6}{180.16 \text{ g } C_6H_{12}O_6} \times \frac{6 \text{ mol } CO_2}{1 \text{ mol } C_6H_{12}O_6} \times \frac{44.01 \text{ g } CO_2}{1 \text{ mol } CO_2} = 97.2 \text{ g } CO_2$$

$$66.3 \text{ g } C_6H_{12}O_6 \times \frac{1 \text{ mol } C_6H_{12}O_6}{180.16 \text{ g } C_6H_{12}O_6} \times \frac{6 \text{ mol } CO_2}{1 \text{ mol } C_6H_{12}O_6} \times \frac{25.4 \text{ L } CO_2}{1 \text{ mol } CO_2} = 56.1 \text{ L } CO_2$$

3.124 Mass of Cu = 2.196 g; mass of S = 2.748 g – 2.196 g = 0.552 g S

(a) %Cu = $\dfrac{2.196 \text{ g}}{2.748 \text{ g}} \times 100\% = 79.91\%$

%S = $\dfrac{0.552 \text{ g}}{2.748 \text{ g}} \times 100\% = 20.1\%$

(b) 2.196 g Cu x $\dfrac{1 \text{ mol Cu}}{63.55 \text{ g Cu}} = 0.034\,55 \text{ mol Cu}$

0.552 g S x $\dfrac{1 \text{ mol S}}{32.07 \text{ g S}} = 0.0172 \text{ mol S}$

$Cu_{0.03455}S_{0.0172}$; divide each subscript by the smaller, 0.0172.
$Cu_{0.03455 / 0.0172}S_{0.0172 / 0.0172}$ The empirical formula is Cu_2S.
(c) Cu_2S, 159.16 amu

$$\frac{5.6 \text{ g } Cu_2S}{1 \text{ cm}^3} \times \frac{1 \text{ mol } Cu_2S}{159.16 \text{ g } Cu_2S} \times \frac{2 \text{ mol } Cu^+ \text{ ions}}{1 \text{ mol } Cu_2S} \times \frac{6.022 \times 10^{23} \text{ } Cu^+ \text{ ions}}{1 \text{ mol } Cu^+ \text{ ions}}$$
$= 4.2 \times 10^{22} \text{ } Cu^+ \text{ ions/cm}^3$

3.126 PCl_3, 137.33 amu; PCl_5, 208.24 amu
Let Y = mass of PCl_3 in the mixture, and (10.00 – Y) = mass of PCl_5 in the mixture.
fraction Cl in PCl_3 = $\dfrac{(3)(35.453 \text{ g/mol})}{137.33 \text{ g/mol}} = 0.774\,48$

fraction Cl in PCl_5 = $\dfrac{(5)(35.453 \text{ g/mol})}{208.24 \text{ g/mol}} = 0.851\,25$

(mass of Cl in PCl_3) + (mass of Cl in PCl_5) = mass of Cl in the mixture
0.774 48Y + 0.851 25(10.00 g – Y) = (0.8104)(10.00 g)
Y = 5.32 g PCl_3 and 10.00 – Y = 4.68 g PCl_5

3.128　NH_4NO_3, 80.04 amu;　$(NH_4)_2HPO_4$, 132.06 amu

Assume you have a 100.0 g sample of the mixture.

Let X = grams of NH_4NO_3 and $(100.0 - X)$ = grams of $(NH_4)_2HPO_4$.

Both compounds contain 2 nitrogen atoms per formula unit.

Because the mass % N in the sample is 30.43%, the 100.0 g sample contains 30.43 g N.

$$\text{mol } NH_4NO_3 = (X) \times \frac{1 \text{ mol } NH_4NO_3}{80.04 \text{ g}}$$

$$\text{mol } (NH_4)_2HPO_4 = (100.0 - X) \times \frac{1 \text{ mol } (NH_4)_2HPO_4}{132.06 \text{ g}}$$

$$\text{mass N} = \left(\left((X) \times \frac{1 \text{ mol } NH_4NO_3}{80.04 \text{ g}}\right) + \left((100.0 - X) \times \frac{1 \text{ mol } (NH_4)_2HPO_4}{132.06 \text{ g}}\right)\right) \times$$

$$\left(\frac{2 \text{ mol N}}{1 \text{ mol ammonium cmpds}}\right) \times \left(\frac{14.0067 \text{ g N}}{1 \text{ mol N}}\right) = 30.43 \text{ g}$$

Solve for X.

$$\left(\frac{X}{80.04} + \frac{100.0 - X}{132.06}\right)(2)(14.0067) = 30.43$$

$$\left(\frac{X}{80.04} + \frac{100.0 - X}{132.06}\right) = 1.08627$$

$$\frac{(132.06)(X) + (100.0 - X)(80.04)}{(80.04)(132.06)} = 1.08627$$

$(132.06)(X) + (100.0 - X)(80.04) = (1.08627)(80.04)(132.06)$

$132.06X + 8004 - 80.04X = 11481.96$

$132.06X - 80.04X = 11481.96 - 8004$

$52.02X = 3477.96$

$$X = \frac{3477.96}{52.02} = 66.86 \text{ g } NH_4NO_3$$

$(100.0 - X) = (100.0 - 66.86) = 33.14 \text{ g } (NH_4)_2HPO_4$

$$\frac{\text{mass}_{NH_4NO_3}}{\text{mass}_{(NH_4)_2HPO_4}} = \frac{66.86 \text{ g}}{33.14 \text{ g}} = 2.018$$

The mass ratio of NH_4NO_3 to $(NH_4)_2HPO_4$ in the mixture is 2 to 1.

3.130　(a)　56.0 mL = 0.0560 L

$$\text{mol } X_2 = (0.0560 \text{ L } X_2)\left(\frac{1 \text{ mol}}{22.41 \text{ L}}\right) = 0.00250 \text{ mol } X_2$$

$\text{mass } X_2 = 1.12 \text{ g } MX_2 - 0.720 \text{ g } MX = 0.40 \text{ g } X_2$

$$\text{molar mass } X_2 = \frac{0.40 \text{ g}}{0.00250 \text{ mol}} = 160 \text{ g/mol}$$

atomic mass of X = 160/2 = 80 amu;　X is Br.

(b) $\text{mol MX} = 0.00250 \text{ mol X}_2 \times \dfrac{2 \text{ mol MX}}{1 \text{ mol X}_2} = 0.00500 \text{ mol MX}$

$\text{mass of X in MX} = 0.00500 \text{ mol MX} \times \dfrac{1 \text{ mol X}}{1 \text{ mol MX}} \times \dfrac{80 \text{ g X}}{1 \text{ mol X}} = 0.40 \text{ g X}$

$\text{mass of M in MX} = 0.720 \text{ g MX} - 0.40 \text{ g X} = 0.32 \text{ g M}$

$\text{molar mass M} = \dfrac{0.32 \text{ g}}{0.00500 \text{ mol}} = 64 \text{ g/mol}$

atomic mass of X = 64 amu;   M is Cu.

4 Reactions in Aqueous Solution

4.1 (a) precipitation (b) redox (c) acid-base neutralization

4.2 $FeBr_3$ contains 3 Br^- ions.
The molar concentration of Br^- ions = 3 x 0.225 M = 0.675 M

4.3 A_2Y is the strongest electrolyte because it is completely dissociated into ions.
A_2X is the weakest electrolyte because it is the least dissociated of the three substances.

4.4 (a) Ionic equation:
$2\ Ag^+(aq) + 2\ NO_3^-(aq) + 2\ Na^+(aq) + CrO_4^{2-}(aq) \rightarrow Ag_2CrO_4(s) + 2\ Na^+(aq) + 2\ NO_3^-(aq)$
Delete spectator ions from the ionic equation to get the net ionic equation.
Net ionic equation: $2\ Ag^+(aq)\ +\ CrO_4^{2-}(aq)\ \rightarrow\ Ag_2CrO_4(s)$
(b) Ionic equation:
$2\ H^+(aq) + SO_4^{2-}(aq) + MgCO_3(s) \rightarrow H_2O(l) + CO_2(g) + Mg^{2+}(aq) + SO_4^{2-}(aq)$
Delete spectator ions from the ionic equation to get the net ionic equation.
Net ionic equation: $2\ H^+(aq) + MgCO_3(s)\ \rightarrow\ H_2O(l) + CO_2(g) + Mg^{2+}(aq)$
(c) Ionic equation:
$Hg_2^{2+}(aq) + 2\ NO_3^-(aq) + 2\ NH_4^+(aq) + 2\ Cl^-(aq)\ \rightarrow\ Hg_2Cl_2(s) + 2\ NH_4^+(aq) + 2\ NO_3^-(aq)$
Delete spectator ions from the ionic equation to get the net ionic equation.
Net ionic equation: $Hg_2^{2+}(aq)\ +\ 2\ Cl^-(aq)\ \rightarrow\ Hg_2Cl_2(s)$

4.5 (a) $CdCO_3$, insoluble (b) MgO, insoluble (c) Na_2S, soluble
(d) $PbSO_4$, insoluble (e) $(NH_4)_3PO_4$, soluble (f) $HgCl_2$, soluble

4.6 (a) Ionic equation:
$Ni^{2+}(aq) + 2\ Cl^-(aq) + 2\ NH_4^+(aq) + S^{2-}(aq) \rightarrow NiS(s) + 2\ NH_4^+(aq) + 2\ Cl^-(aq)$
Delete spectator ions from the ionic equation to get the net ionic equation.
Net ionic equation: $Ni^{2+}(aq)\ +\ S^{2-}(aq)\ \rightarrow\ NiS(s)$
(b) Ionic equation:
$2\ Na^+(aq) + CrO_4^{2-}(aq) + Pb^{2+}(aq) + 2\ NO_3^-(aq) \rightarrow PbCrO_4(s) + 2\ Na^+(aq) + 2\ NO_3^-(aq)$
Delete spectator ions from the ionic equation to get the net ionic equation.
Net ionic equation: $Pb^{2+}(aq)\ +\ CrO_4^{2-}(aq)\ \rightarrow\ PbCrO_4(s)$
(c) Ionic equation:
$2\ Ag^+(aq) + 2\ ClO_4^-(aq) + Ca^{2+}(aq) + 2\ Br^-(aq) \rightarrow 2\ AgBr(s) + Ca^{2+}(aq) + 2\ ClO_4^-(aq)$
Delete spectator ions from the ionic equation, and reduce coefficients to get the net ionic equation.
Net ionic equation: $Ag^+(aq)\ +\ Br^-(aq)\ \rightarrow\ AgBr(s)$

(d) Ionic equation:

$Zn^{2+}(aq) + 2\ Cl^-(aq) + 2\ K^+(aq) + CO_3^{2-}(aq) \rightarrow ZnCO_3(s) + 2\ K^+(aq) + 2\ Cl^-(aq)$

Delete spectator ions from the ionic equation to get the net ionic equation.

Net ionic equation: $Zn^{2+}(aq) + CO_3^{2-}(aq) \rightarrow ZnCO_3(s)$

4.7 $3\ CaCl_2(aq) + 2\ Na_3PO_4(aq) \rightarrow Ca_3(PO_4)_2(s) + 6\ NaCl(aq)$

Ionic equation:

$3\ Ca^{2+}(aq) + 6\ Cl^-(aq) + 6\ Na^+(aq) + 2\ PO_4^{3-}(aq) \rightarrow Ca_3(PO_4)_2(s) + 6\ Na^+(aq) + 6\ Cl^-(aq)$

Delete spectator ions from the ionic equation to get the net ionic equation.

Net ionic equation: $3\ Ca^{2+}(aq) + 2\ PO_4^{3-}(aq) \rightarrow Ca_3(PO_4)_2(s)$

4.8   A precipitate results from the reaction. The precipitate contains cations and anions in a 3:2 ratio. The precipitate is either $Mg_3(PO_4)_2$ or $Zn_3(PO_4)_2$.

4.9   (a) Ionic equation:

$2\ Cs^+(aq) + 2\ OH^-(aq) + 2\ H^+(aq) + SO_4^{2-}(aq) \rightarrow 2\ Cs^+(aq) + SO_4^{2-}(aq) + 2\ H_2O(l)$

Delete spectator ions from the ionic equation, and reduce coefficients to get the net ionic equation.

Net ionic equation: $H^+(aq) + OH^-(aq) \rightarrow H_2O(l)$

(b) Ionic equation:

$Ca^{2+}(aq) + 2\ OH^-(aq) + 2\ CH_3CO_2H(aq) \rightarrow Ca^{2+}(aq) + 2\ CH_3CO_2^-(aq) + 2\ H_2O(l)$

Delete spectator ions from the ionic equation, and reduce coefficients to get the net ionic equation.

Net ionic equation: $CH_3CO_2H(aq) + OH^-(aq) \rightarrow CH_3CO_2^-(aq) + H_2O(l)$

4.10   HY is the strongest acid because it is completely dissociated.
HX is the weakest acid because it is the least dissociated.

4.11   (a) $SnCl_4$:   Cl –1, Sn +4   (b) CrO_3:   O –2, Cr +6
(c) $VOCl_3$:   O –2, Cl –1, V +5   (d) V_2O_3:   O –2, V +3
(e) HNO_3:   O –2, H +1, N +5   (f) $FeSO_4$:   O –2, S +6, Fe +2

4.12 $2\ Cu^{2+}(aq) + 4\ I^-(aq) \rightarrow 2\ CuI(s) + I_2(aq)$
oxidation numbers: Cu^{2+} +2; I^- –1; CuI: Cu +1, I –1; I_2: 0
oxidizing agent (oxidation number decreases), Cu^{2+}
reducing agent (oxidation number increases), I^-

4.13   (a) $SnO_2(s) + 2\ C(s) \rightarrow Sn(s) + 2\ CO(g)$
C is oxidized (its oxidation number increases from 0 to +2). C is the reducing agent.
The Sn in SnO_2 is reduced (its oxidation number decreases from +4 to 0). SnO_2 is the oxidizing agent.
(b) $Sn^{2+}(aq) + 2\ Fe^{3+}(aq) \rightarrow Sn^{4+}(aq) + 2\ Fe^{2+}(aq)$
Sn^{2+} is oxidized (its oxidation number increases from +2 to +4). Sn^{2+} is the reducing agent.
Fe^{3+} is reduced (its oxidation number decreases from +3 to +2). Fe^{3+} is the oxidizing agent.

(c) $4 NH_3(g) + 5 O_2(g) \rightarrow 4 NO(g) + 6 H_2O(l)$
The N in NH_3 is oxidized (its oxidation number increases from –3 to +2). NH_3 is the reducing agent.
Each O in O_2 is reduced (its oxidation number decreases from 0 to –2). O_2 is the oxidizing agent.

4.14   (a) Pt is below H in the activity series; therefore NO REACTION.
(b) Mg is below Ca in the activity series; therefore NO REACTION.

4.15   Because B will reduce A^+, B is above A in the activity series. Because B will not reduce C^+, C is above B in the activity series. Therefore C must be above A in the activity series and C will reduce A^+.

4.16 $Cr_2O_7{}^{2-}(aq) + I^-(aq) \rightarrow Cr^{3+}(aq) + IO_3{}^-(aq)$

$$Cr_2O_7{}^{2-}(aq) + I^-(aq) \rightarrow 2 Cr^{3+}(aq) + IO_3{}^-(aq)$$

$$8 H^+(aq) + Cr_2O_7{}^{2-}(aq) + I^-(aq) \rightarrow 2 Cr^{3+}(aq) + IO_3{}^-(aq) + 4 H_2O(l)$$

4.17 $MnO_4{}^-(aq) + Br^-(aq) \rightarrow MnO_2(s) + BrO_3{}^-(aq)$

$$2 MnO_4{}^-(aq) + Br^-(aq) \rightarrow 2 MnO_2(s) + BrO_3{}^-(aq)$$
$$2 H^+(aq) + 2 MnO_4{}^-(aq) + Br^-(aq) \rightarrow 2 MnO_2(s) + BrO_3{}^-(aq) + H_2O(l)$$

$$2 H^+(aq) + 2 OH^-(aq) + 2 MnO_4{}^-(aq) + Br^-(aq) \rightarrow$$
$$2 MnO_2(s) + BrO_3{}^-(aq) + H_2O(l) + 2 OH^-(aq)$$
$$H_2O(l) + 2 MnO_4{}^-(aq) + Br^-(aq) \rightarrow 2 MnO_2(s) + BrO_3{}^-(aq) + 2 OH^-(aq)$$

4.18   (a) $MnO_4{}^-(aq) \rightarrow MnO_2(s)$ (reduction)
          $IO_3{}^-(aq) \rightarrow IO_4{}^-(aq)$ (oxidation)

(b) $NO_3{}^-(aq) \rightarrow NO_2(g)$ (reduction)
          $SO_2(aq) \rightarrow SO_4{}^{2-}(aq)$ (oxidation)

4.19 $NO_3{}^-(aq) + Cu(s) \rightarrow NO(g) + Cu^{2+}(aq)$
          $[Cu(s) \rightarrow Cu^{2+}(aq) + 2 e^-] \times 3$ (oxidation half reaction)

$NO_3^-(aq) \rightarrow NO(g)$

$NO_3^-(aq) \rightarrow NO(g) + 2 H_2O(l)$

$4 H^+(aq) + NO_3^-(aq) \rightarrow NO(g) + 2 H_2O(l)$

$[3 e^- + 4 H^+(aq) + NO_3^-(aq) \rightarrow NO(g) + 2 H_2O(l)] \times 2$ (reduction half reaction)

Combine the two half reactions.

$2 NO_3^-(aq) + 8 H^+(aq) + 3 Cu(s) \rightarrow 3 Cu^{2+}(aq) + 2 NO(g) + 4 H_2O(l)$

4.20 $Fe(OH)_2(s) + O_2(g) \rightarrow Fe(OH)_3(s)$

$[Fe(OH)_2(s) + OH^-(aq) \rightarrow Fe(OH)_3(s) + e^-] \times 4$ (oxidation half reaction)

$O_2(g) \rightarrow 2 H_2O(l)$

$4 H^+(aq) + O_2(g) \rightarrow 2 H_2O(l)$

$4 e^- + 4 H^+(aq) + O_2(g) \rightarrow 2 H_2O(l)$

$4 e^- + 4 H^+(aq) + 4 OH^-(aq) + O_2(g) \rightarrow 2 H_2O(l) + 4 OH^-(aq)$

$4 e^- + 4 H_2O(l) + O_2(g) \rightarrow 2 H_2O(l) + 4 OH^-(aq)$

$4 e^- + 2 H_2O(l) + O_2(g) \rightarrow 4 OH^-(aq)$ (reduction half reaction)

Combine the two half reactions.

$4 Fe(OH)_2(s) + 4 OH^-(aq) + 2 H_2O(l) + O_2(g) \rightarrow 4 Fe(OH)_3(s) + 4 OH^-(aq)$

$4 Fe(OH)_2(s) + 2 H_2O(l) + O_2(g) \rightarrow 4 Fe(OH)_3(s)$

4.21 $31.50 \text{ mL} = 0.031\ 50 \text{ L};\ 10.00 \text{ mL} = 0.010\ 00 \text{ L}$

$$0.031\ 50 \text{ L} \times \frac{0.105 \text{ mol BrO}_3^-}{1 \text{ L}} \times \frac{6 \text{ mol Fe}^{2+}}{1 \text{ mol BrO}_3^-} = 1.98 \times 10^{-2} \text{ mol Fe}^{2+}$$

$$\text{molarity} = \frac{1.98 \times 10^{-2} \text{ mol Fe}^{2+}}{0.010\ 00 \text{ L}} = 1.98 \text{ M Fe}^{2+} \text{ solution}$$

4.22   The $Na_2S_2O_3$, or hypo, is used to solubilize the remaining unreduced AgBr on the film so that it is no longer sensitive to light. The reaction is

$AgBr(s) + 2 S_2O_3^{2-}(aq) \rightarrow Ag(S_2O_3)_2^{3-}(aq) + Br^-(aq)$

4.23   To convert this negative image into the final printed photograph, the entire photographic procedure is repeated a second time. Light is passed through the negative image onto special photographic paper that is coated with the same kind of gelatin–AgBr emulsion used on the original film. Developing the photographic paper with hydroquinone and fixing the image with sodium thiosulfate reverses the negative image, and a final, positive image is produced.

Understanding Key Concepts

4.24   (a) $2 Na^+(aq) + CO_3^{2-}(aq)$ does not form a precipitate. This is represented by box (1).
       (b) $Ba^{2+}(aq) + CrO_4^{2-}(aq) \rightarrow BaCrO_4(s)$. This is represented by box (2).
       (c) $2 Ag^+(aq) + SO_4^{2-}(aq) \rightarrow Ag_2SO_4(s)$. This is represented by box (3).

4.26   One OH^- will react with each available H^+ on the acid forming H_2O. The acid is identified by how many of the 12 OH^- react with three molecules of each acid.
(a) Three HF's react with three OH^-, leaving nine OH^- unreacted (box 2).
(b) Three H_2SO_3's react with six OH^-, leaving six OH^- unreacted (box 3).
(c) Three H_3PO_4's react with nine OH^-, leaving three OH^- unreacted (box 1).

4.28   (a) Ionic equation:
$K^+(aq) + Cl^-(aq) + Ag^+(aq) + NO_3^-(aq) \rightarrow AgCl(s) + K^+(aq) + NO_3^-(aq)$
(b) Ionic equation:
$HF(aq) + K^+(aq) + OH^-(aq) \rightarrow K^+(aq) + F^-(aq) + H_2O(l)$
(c) Ionic equation:
$Ba^{2+}(aq) + 2\,Cl^-(aq) + 2\,Na^+(aq) + SO_4^{2-}(aq) \rightarrow BaSO_4(s) + 2\,Na^+(aq) + 2\,Cl^-(aq)$

Reaction (c) would have the highest initial conductivity because of the 3 net ions for each $BaCl_2$ (a strong electrolyte).
Reaction (b) would have have the lowest (almost zero) initial conductivity because HF is a very weak acid/electrolyte.
Reaction (a) would have an intermediate initial conductivity between that for reactions (b) and (c).
Figure (1) is for reaction (a); figure (2) is for reaction (b); and figure (3) is for reaction (c).

Additional Problems
Aqueous Reactions and Net Ionic Equations

4.30   (a) precipitation   (b) redox   (c) acid-base neutralization

4.32   (a) Ionic equation:
$Hg^{2+}(aq) + 2\,NO_3^-(aq) + 2\,Na^+(aq) + 2\,I^-(aq) \rightarrow 2\,Na^+(aq) + 2\,NO_3^-(aq) + HgI_2(s)$
Delete spectator ions from the ionic equation to get the net ionic equation.
Net ionic equation: $Hg^{2+}(aq) + 2\,I^-(aq) \rightarrow HgI_2(s)$

(b) $2\,HgO(s) \overset{\text{Heat}}{\rightarrow} 2\,Hg(l) + O_2(g)$
(c) Ionic equation:
$H_3PO_4(aq) + 3\,K^+(aq) + 3\,OH^-(aq) \rightarrow 3\,K^+(aq) + PO_4^{3-}(aq) + 3\,H_2O(l)$
Delete spectator ions from the ionic equation to get the net ionic equation.
Net ionic equation: $H_3PO_4(aq) + 3\,OH^-(aq) \rightarrow PO_4^{3-}(aq) + 3\,H_2O(l)$

4.34 $Ba(OH)_2$ is soluble in aqueous solution, dissociates into $Ba^{2+}(aq)$ and 2 $OH^-(aq)$, and conducts electricity. In aqueous solution H_2SO_4 dissociates into $H^+(aq)$ and $HSO_4^-(aq)$. H_2SO_4 solutions conduct electricity. When equal molar solutions of $Ba(OH)_2$ and H_2SO_4 are mixed, the insoluble $BaSO_4$ is formed along with two H_2O. In water $BaSO_4$ does not produce any appreciable amount of ions and the mixture does not conduct electricity.

4.36   (a) HBr, strong electrolyte   (b) HF, weak electrolyte
(c) $NaClO_4$, strong electrolyte   (d) $(NH_4)_2CO_3$, strong electrolyte
(e) NH_3, weak electrolyte   (f) C_2H_5OH, nonelectrolyte

4.38   (a) K_2CO_3 contains 3 ions (2 K^+ and 1 CO_3^{2-}).
        The molar concentration of ions = 3 × 0.750 M = 2.25 M.
        (b) $AlCl_3$ contains 4 ions (1 Al^{3+} and 3 Cl^-).
        The molar concentration of ions = 4 × 0.355 M = 1.42 M.

Precipitation Reactions and Solubility Rules

4.40   (a) Ag_2O, insoluble   (b) $Ba(NO_3)_2$, soluble
        (c) $SnCO_3$, insoluble   (d) Fe_2O_3, insoluble

4.42   (a) No precipitate will form.
        (b) $FeCl_2(aq)$ + 2 $KOH(aq)$ → $Fe(OH)_2(s)$ + 2 $KCl(aq)$
        (c) No precipitate will form.
        (d) No precipitate will form.

4.44   (a) $Pb(NO_3)_2(aq)$ + $Na_2SO_4(aq)$ → $PbSO_4(s)$ + 2 $NaNO_3(aq)$
        (b) 3 $MgCl_2(aq)$ + 2 $K_3PO_4(aq)$ → $Mg_3(PO_4)_2(s)$ + 6 $KCl(aq)$
        (c) $ZnSO_4(aq)$ + $Na_2CrO_4(aq)$ → $ZnCrO_4(s)$ + $Na_2SO_4(aq)$

4.46   Add $HCl(aq)$; it will selectively precipitate $AgCl(s)$.

4.48 Ag^+ is eliminated because it would have precipitated as $AgCl(s)$; Ba^{2+} is eliminated
        because it would have precipitated as $BaSO_4(s)$. The solution might contain Cs^+ and/or
        NH_4^+. Neither of these will precipitate with OH^-, SO_4^{2-}, or Cl^-.

Acids, Bases, and Neutralization Reactions

4.50   Add the solution to an active metal, such as magnesium. Bubbles of H_2 gas indicate the
        presence of an acid.

4.52   (a) 2 $H^+(aq)$ + 2 $ClO_4^-(aq)$ + $Ca^{2+}(aq)$ + 2 $OH^-(aq)$ → $Ca^{2+}(aq)$ + 2 $ClO_4^-(aq)$ + 2 $H_2O(l)$
        (b) $CH_3CO_2H(aq)$ + $Na^+(aq)$ + $OH^-(aq)$ → $CH_3CO_2^-(aq)$ + $Na^+(aq)$ + $H_2O(l)$

4.54   (a) $LiOH(aq)$ + $HI(aq)$ → $LiI(aq)$ + $H_2O(l)$
        Ionic equation: $Li^+(aq)$ + $OH^-(aq)$ + $H^+(aq)$ + $I^-(aq)$ → $Li^+(aq)$ + $I^-(aq)$ + $H_2O(l)$
        Delete spectator ions from the ionic equation to get the net ionic equation.
        Net ionic equation: $H^+(aq)$ + $OH^-(aq)$ → $H_2O(l)$

        (b) 2 $HBr(aq)$ + $Ca(OH)_2(aq)$ → $CaBr_2(aq)$ + 2 $H_2O(l)$
        Ionic equation:
        2 $H^+(aq)$ + 2 $Br^-(aq)$ + $Ca^{2+}(aq)$ + 2 $OH^-(aq)$ → $Ca^{2+}(aq)$ + 2 $Br^-(aq)$ + 2 $H_2O(l)$
        Delete spectator ions from the ionic equation to get the net ionic equation.
        Net ionic equation: $H^+(aq)$ + $OH^-(aq)$ → $H_2O(l)$

Redox Reactions and Oxidation Numbers

4.56   The best reducing agents are at the bottom left of the periodic table. The best oxidizing agents are at the top right of the periodic table (excluding the noble gases).

4.58   (a) An oxidizing agent gains electrons.
(b) A reducing agent loses electrons.
(c) A substance undergoing oxidation loses electrons.
(d) A substance undergoing reduction gains electrons.

4.60   (a) NO_2 O –2, N +4   (b) SO_3 O –2, S +6
(c) $COCl_2$ O –2, Cl –1, C +4   (d) CH_2Cl_2 Cl –1, H +1, C 0
(e) $KClO_3$ O –2, K +1, Cl +5   (f) HNO_3 O –2, H +1, N +5

4.62   (a) ClO_3^- O –2, Cl +5   (b) SO_3^{2-} O –2, S +4
(c) $C_2O_4^{2-}$ O –2, C +3   (d) NO_2^- O –2, N +3
(e) BrO^- O –2, Br +1   (f) AsO_4^{3-} O –2, As +5

4.64   (a) $Ca(s) + Sn^{2+}(aq) \rightarrow Ca^{2+}(aq) + Sn(s)$
$Ca(s)$ is oxidized (oxidation number increases from 0 to +2).
$Sn^{2+}(aq)$ is reduced (oxidation number decreases from +2 to 0).
(b) $ICl(s) + H_2O(l) \rightarrow HCl(aq) + HOI(aq)$
No oxidation numbers change. The reaction is not a redox reaction.

4.66   (a) Zn is below Na^+; therefore no reaction.
(b) Pt is below H^+; therefore no reaction.
(c) Au is below Ag^+; therefore no reaction.
(d) Ag is above Au^{3+}; the reaction is $Au^{3+}(aq) + 3\ Ag(s) \rightarrow 3\ Ag^+(aq) + Au(s)$.

4.68   (a) "Any element higher in the activity series will react with the ion of any element lower in the activity series."
$A + B^+ \rightarrow A^+ + B$; therefore A is higher than B.
$C^+ + D \rightarrow$ no reaction; therefore C is higher than D.
$B + D^+ \rightarrow B^+ + D$; therefore B is higher than D.
$B + C^+ \rightarrow B^+ + C$; therefore B is higher than C.
The net result is A > B > C > D.
(b) (1) C is below A^+; therefore no reaction.
    (2) D is below A^+; therefore no reaction.

Balancing Redox Reactions

4.70   (a) N oxidation number decreases from +5 to +2; reduction.
(b) Zn oxidation number increases from 0 to +2; oxidation.
(c) Ti oxidation number increases from +3 to +4; oxidation.
(d) Sn oxidation number decreases from +4 to +2; reduction.

4.72 (a) $NO_3^-(aq) \rightarrow NO(g)$
$NO_3^-(aq) \rightarrow NO(g) + 2\,H_2O(l)$
$4\,H^+(aq) + NO_3^-(aq) \rightarrow NO(g) + 2\,H_2O(l)$
$3\,e^- + 4\,H^+(aq) + NO_3^-(aq) \rightarrow NO(g) + 2\,H_2O(l)$

(b) $Zn(s) \rightarrow Zn^{2+}(aq) + 2\,e^-$

(c) $Ti^{3+}(aq) \rightarrow TiO_2(s)$
$Ti^{3+}(aq) + 2\,H_2O(l) \rightarrow TiO_2(s)$
$Ti^{3+}(aq) + 2\,H_2O(l) \rightarrow TiO_2(s) + 4\,H^+(aq)$
$Ti^{3+}(aq) + 2\,H_2O(l) \rightarrow TiO_2(s) + 4\,H^+(aq) + e^-$

(d) $Sn^{4+}(aq) + 2\,e^- \rightarrow Sn^{2+}(aq)$

4.74 (a) $Te(s) + NO_3^-(aq) \rightarrow TeO_2(s) + NO(g)$
oxidation: $Te(s) \rightarrow TeO_2(s)$
reduction: $NO_3^-(aq) \rightarrow NO(g)$

(b) $H_2O_2(aq) + Fe^{2+}(aq) \rightarrow Fe^{3+}(aq) + H_2O(l)$
oxidation: $Fe^{2+}(aq) \rightarrow Fe^{3+}(aq)$
reduction: $H_2O_2(aq) \rightarrow H_2O(l)$

4.76 (a) $Cr_2O_7^{2-}(aq) \rightarrow Cr^{3+}(aq)$
$Cr_2O_7^{2-}(aq) \rightarrow 2\,Cr^{3+}(aq)$
$Cr_2O_7^{2-}(aq) \rightarrow 2\,Cr^{3+}(aq) + 7\,H_2O(l)$
$14\,H^+(aq) + Cr_2O_7^{2-}(aq) \rightarrow 2\,Cr^{3+}(aq) + 7\,H_2O(l)$
$14\,H^+(aq) + Cr_2O_7^{2-}(aq) + 6\,e^- \rightarrow 2\,Cr^{3+}(aq) + 7\,H_2O(l)$

(b) $CrO_4^{2-}(aq) \rightarrow Cr(OH)_4^-(aq)$
$4\,H^+(aq) + CrO_4^{2-}(aq) \rightarrow Cr(OH)_4^-(aq)$
$4\,H^+(aq) + 4\,OH^-(aq) + CrO_4^{2-}(aq) \rightarrow Cr(OH)_4^-(aq) + 4\,OH^-(aq)$
$4\,H_2O(l) + CrO_4^{2-}(aq) \rightarrow Cr(OH)_4^-(aq) + 4\,OH^-(aq)$
$4\,H_2O(l) + CrO_4^{2-}(aq) + 3\,e^- \rightarrow Cr(OH)_4^-(aq) + 4\,OH^-(aq)$

(c) $Bi^{3+}(aq) \rightarrow BiO_3^-(aq)$
$Bi^{3+}(aq) + 3\,H_2O(l) \rightarrow BiO_3^-(aq)$
$Bi^{3+}(aq) + 3\,H_2O(l) \rightarrow BiO_3^-(aq) + 6\,H^+(aq)$
$Bi^{3+}(aq) + 3\,H_2O(l) + 6\,OH^-(aq) \rightarrow BiO_3^-(aq) + 6\,H^+(aq) + 6\,OH^-(aq)$
$Bi^{3+}(aq) + 3\,H_2O(l) + 6\,OH^-(aq) \rightarrow BiO_3^-(aq) + 6\,H_2O(l)$
$Bi^{3+}(aq) + 6\,OH^-(aq) \rightarrow BiO_3^-(aq) + 3\,H_2O(l)$
$Bi^{3+}(aq) + 6\,OH^-(aq) \rightarrow BiO_3^-(aq) + 3\,H_2O(l) + 2\,e^-$

(d) $ClO^-(aq) \rightarrow Cl^-(aq)$
$ClO^-(aq) \rightarrow Cl^-(aq) + H_2O(l)$
$2\,H^+(aq) + ClO^-(aq) \rightarrow Cl^-(aq) + H_2O(l)$
$2\,H^+(aq) + 2\,OH^-(aq) + ClO^-(aq) \rightarrow Cl^-(aq) + H_2O(l) + 2\,OH^-(aq)$

$$2 H_2O(l) + ClO^-(aq) \rightarrow Cl^-(aq) + H_2O(l) + 2 OH^-(aq)$$
$$H_2O(l) + ClO^-(aq) \rightarrow Cl^-(aq) + 2 OH^-(aq)$$
$$H_2O(l) + ClO^-(aq) + 2 e^- \rightarrow Cl^-(aq) + 2 OH^-(aq)$$

4.78   (a) $MnO_4^-(aq) \rightarrow MnO_2(s)$

$$MnO_4^-(aq) \rightarrow MnO_2(s) + 2 H_2O(l)$$
$$4 H^+(aq) + MnO_4^-(aq) \rightarrow MnO_2(s) + 2 H_2O(l)$$
$$[4 H^+(aq) + MnO_4^-(aq) + 3 e^- \rightarrow MnO_2(s) + 2 H_2O(l)] \times 2 \quad \text{(reduction half reaction)}$$

$$IO_3^-(aq) \rightarrow IO_4^-(aq)$$
$$H_2O(l) + IO_3^-(aq) \rightarrow IO_4^-(aq)$$
$$H_2O(l) + IO_3^-(aq) \rightarrow IO_4^-(aq) + 2 H^+(aq)$$
$$[H_2O(l) + IO_3^-(aq) \rightarrow IO_4^-(aq) + 2 H^+(aq) + 2 e^-] \times 3 \quad \text{(oxidation half reaction)}$$

Combine the two half reactions.
$$8 H^+(aq) + 3 H_2O(l) + 2 MnO_4^-(aq) + 3 IO_3^-(aq) \rightarrow$$
$$6 H^+(aq) + 4 H_2O(l) + 2 MnO_2(s) + 3 IO_4^-(aq)$$
$$2 H^+(aq) + 2 MnO_4^-(aq) + 3 IO_3^-(aq) \rightarrow 2 MnO_2(s) + 3 IO_4^-(aq) + H_2O(l)$$
$$2 H^+(aq) + 2 OH^-(aq) + 2 MnO_4^-(aq) + 3 IO_3^-(aq) \rightarrow$$
$$2 MnO_2(s) + 3 IO_4^-(aq) + H_2O(l) + 2 OH^-(aq)$$
$$2 H_2O(l) + 2 MnO_4^-(aq) + 3 IO_3^-(aq) \rightarrow$$
$$2 MnO_2(s) + 3 IO_4^-(aq) + H_2O(l) + 2 OH^-(aq)$$
$$H_2O(l) + 2 MnO_4^-(aq) + 3 IO_3^-(aq) \rightarrow 2 MnO_2(s) + 3 IO_4^-(aq) + 2 OH^-(aq)$$

   (b) $Cu(OH)_2(s) \rightarrow Cu(s)$

$$Cu(OH)_2(s) \rightarrow Cu(s) + 2 H_2O(l)$$
$$2 H^+(aq) + Cu(OH)_2(s) \rightarrow Cu(s) + 2 H_2O(l)$$
$$[2 H^+(aq) + Cu(OH)_2(s) + 2 e^- \rightarrow Cu(s) + 2 H_2O(l)] \times 2 \quad \text{(reduction half reaction)}$$

$$N_2H_4(aq) \rightarrow N_2(g)$$
$$N_2H_4(aq) \rightarrow N_2(g) + 4 H^+(aq)$$
$$N_2H_4(aq) \rightarrow N_2(g) + 4 H^+(aq) + 4 e^- \quad \text{(oxidation half reaction)}$$

Combine the two half reactions.
$$4 H^+(aq) + 2 Cu(OH)_2(s) + N_2H_4(aq) \rightarrow 2 Cu(s) + 4 H_2O(l) + N_2(g) + 4 H^+(aq)$$
$$2 Cu(OH)_2(s) + N_2H_4(aq) \rightarrow 2 Cu(s) + 4 H_2O(l) + N_2(g)$$

   (c) $Fe(OH)_2(s) \rightarrow Fe(OH)_3(s)$

$$Fe(OH)_2(s) + H_2O(l) \rightarrow Fe(OH)_3(s)$$
$$Fe(OH)_2(s) + H_2O(l) \rightarrow Fe(OH)_3(s) + H^+(aq)$$
$$[Fe(OH)_2(s) + H_2O(l) \rightarrow Fe(OH)_3(s) + H^+(aq) + e^-] \times 3 \quad \text{(oxidation half reaction)}$$

$$CrO_4^{2-}(aq) \rightarrow Cr(OH)_4^-(aq)$$
$$4 H^+(aq) + CrO_4^{2-}(aq) \rightarrow Cr(OH)_4^-(aq)$$
$$4 H^+(aq) + CrO_4^{2-}(aq) + 3 e^- \rightarrow Cr(OH)_4^-(aq) \quad \text{(reduction half reaction)}$$

Combine the two half reactions.

$3\ Fe(OH)_2(s) + 3\ H_2O(l) + 4\ H^+(aq) + CrO_4^{2-}(aq)\ \rightarrow$
$\qquad\qquad\qquad 3\ Fe(OH)_3(s) + 3\ H^+(aq) + Cr(OH)_4^-(aq)$
$3\ Fe(OH)_2(s) + 3\ H_2O(l) + H^+(aq) + CrO_4^{2-}(aq)\ \rightarrow\ 3\ Fe(OH)_3(s) + Cr(OH)_4^-(aq)$
$3\ Fe(OH)_2(s) + 3\ H_2O(l) + H^+(aq) + OH^-(aq) + CrO_4^{2-}(aq)\ \rightarrow$
$\qquad\qquad\qquad 3\ Fe(OH)_3(s) + Cr(OH)_4^-(aq) + OH^-(aq)$
$3\ Fe(OH)_2(s) + 4\ H_2O(l) + CrO_4^{2-}(aq)\ \rightarrow 3\ Fe(OH)_3(s) + Cr(OH)_4^-(aq) + OH^-(aq)$

(d) $ClO_4^-(aq)\ \rightarrow\ ClO_2^-(aq)$
$ClO_4^-(aq)\ \rightarrow\ ClO_2^-(aq) + 2\ H_2O(l)$
$4\ H^+(aq) + ClO_4^-(aq)\ \rightarrow\ ClO_2^-(aq) + 2\ H_2O(l)$
$4\ H^+(aq) + ClO_4^-(aq) + 4\ e^-\ \rightarrow\ ClO_2^-(aq) + 2\ H_2O(l)$ (reduction half reaction)

$H_2O_2(aq)\ \rightarrow\ O_2(g)$
$H_2O_2(aq)\ \rightarrow\ O_2(g) + 2\ H^+(aq)$
$[H_2O_2(aq)\ \rightarrow\ O_2(g) + 2\ H^+(aq) + 2\ e^-] \times 2$ (oxidation half reaction)

Combine the two half reactions.
$4\ H^+(aq) + ClO_4^-(aq) + 2\ H_2O_2(aq)\ \rightarrow\ ClO_2^-(aq) + 2\ H_2O(l) + 2\ O_2(g) + 4\ H^+(aq)$
$ClO_4^-(aq) + 2\ H_2O_2(aq)\ \rightarrow\ ClO_2^-(aq) + 2\ H_2O(l) + 2\ O_2(g)$

4.80 (a) $Zn(s)\ \rightarrow\ Zn^{2+}(aq)$
$Zn(s)\ \rightarrow\ Zn^{2+}(aq) + 2\ e^-$ (oxidation half reaction)

$VO^{2+}(aq)\ \rightarrow\ V^{3+}(aq)$
$VO^{2+}(aq)\ \rightarrow\ V^{3+}(aq) + H_2O(l)$
$2\ H^+(aq) + VO^{2+}(aq)\ \rightarrow\ V^{3+}(aq) + H_2O(l)$
$[2\ H^+(aq) + VO^{2+}(aq) + e^-\ \rightarrow\ V^{3+}(aq) + H_2O(l)] \times 2$ (reduction half reaction)

Combine the two half reactions.
$Zn(s) + 2\ VO^{2+}(aq) + 4\ H^+(aq)\ \rightarrow\ Zn^{2+}(aq) + 2\ V^{3+}(aq) + 2\ H_2O(l)$

(b) $Ag(s)\ \rightarrow\ Ag^+(aq)$
$Ag(s)\ \rightarrow\ Ag^+(aq) + e^-$ (oxidation half reaction)

$NO_3^-(aq)\ \rightarrow\ NO_2(g)$
$NO_3^-(aq)\ \rightarrow\ NO_2(g) + H_2O(l)$
$2\ H^+(aq) + NO_3^-(aq)\ \rightarrow\ NO_2(g) + H_2O(l)$
$2\ H^+(aq) + NO_3^-(aq) + e^-\ \rightarrow\ NO_2(g) + H_2O(l)$ (reduction half reaction)

Combine the two half reactions.
$2\ H^+(aq) + Ag(s) + NO_3^-(aq)\ \rightarrow\ Ag^+(aq) + NO_2(g) + H_2O(l)$

(c) $Mg(s)\ \rightarrow\ Mg^{2+}(aq)$
$[Mg(s)\ \rightarrow\ Mg^{2+}(aq) + 2\ e^-] \times 3$ (oxidation half reaction)

$$VO_4^{3-}(aq) \rightarrow V^{2+}(aq)$$
$$VO_4^{3-}(aq) \rightarrow V^{2+}(aq) + 4\ H_2O(l)$$
$$8\ H^+(aq) + VO_4^{3-}(aq) \rightarrow V^{2+}(aq) + 4\ H_2O(l)$$
$$[8\ H^+(aq) + VO_4^{3-}(aq) + 3\ e^- \rightarrow V^{2+}(aq) + 4\ H_2O(l)] \times 2 \quad \text{(reduction half reaction)}$$

Combine the two half reactions.
$$3\ Mg(s) + 16\ H^+(aq) + 2\ VO_4^{3-}(aq) \rightarrow 3\ Mg^{2+}(aq) + 2\ V^{2+}(aq) + 8\ H_2O(l)$$

(d) $$I^-(aq) \rightarrow I_3^-(aq)$$
$$3\ I^-(aq) \rightarrow I_3^-(aq)$$
$$[3\ I^-(aq) \rightarrow I_3^-(aq) + 2\ e^-] \times 8 \qquad \text{(oxidation half reaction)}$$

$$IO_3^-(aq) \rightarrow I_3^-(aq)$$
$$3\ IO_3^-(aq) \rightarrow I_3^-(aq)$$
$$3\ IO_3^-(aq) \rightarrow I_3^-(aq) + 9\ H_2O(l)$$
$$18\ H^+(aq) + 3\ IO_3^-(aq) \rightarrow I_3^-(aq) + 9\ H_2O(l)$$
$$18\ H^+(aq) + 3\ IO_3^-(aq) + 16\ e^- \rightarrow I_3^-(aq) + 9\ H_2O(l) \quad \text{(reduction half reaction)}$$

Combine the two half reactions.
$$18\ H^+(aq) + 3\ IO_3^-(aq) + 24\ I^-(aq) \rightarrow 9\ I_3^-(aq) + 9\ H_2O(l)$$
Divide each coefficient by 3.
$$6\ H^+(aq) + IO_3^-(aq) + 8\ I^-(aq) \rightarrow 3\ I_3^-(aq) + 3\ H_2O(l)$$

Redox Titrations

4.82 $I_2(aq) + 2\ S_2O_3^{2-}(aq) \rightarrow S_4O_6^{2-}(aq) + 2\ I^-(aq);$ $35.20\ \text{mL} = 0.032\ 50\ L$

$$0.035\ 20\ L \times \frac{0.150\ \text{mol}\ S_2O_3^{2-}}{L} \times \frac{1\ \text{mol}\ I_2}{2\ \text{mol}\ S_2O_3^{2-}} \times \frac{253.8\ g\ I_2}{1\ \text{mol}\ I_2} = 0.670\ g\ I_2$$

4.84 $3\ H_3AsO_3(aq) + BrO_3^-(aq) \rightarrow Br^-(aq) + 3\ H_3AsO_4(aq)$
$22.35\ \text{mL} = 0.022\ 35\ L$ and $50.00\ \text{mL} = 0.050\ 00\ L$

$$0.022\ 35\ L \times \frac{0.100\ \text{mol}\ BrO_3^-}{L} \times \frac{3\ \text{mol}\ H_3AsO_3}{1\ \text{mol}\ BrO_3^-} = 6.70 \times 10^{-3}\ \text{mol}\ H_3AsO_3$$

$$\text{molarity} = \frac{6.70 \times 10^{-3}\ \text{mol}}{0.050\ 00\ L} = 0.134\ M\ As(III)$$

4.86 $2\ Fe^{3+}(aq) + Sn^{2+}(aq) \rightarrow 2\ Fe^{2+}(aq) + Sn^{4+}(aq);$ $13.28\ \text{mL} = 0.013\ 28\ L$

$$0.013\ 28\ L \times \frac{0.1015\ \text{mol}\ Sn^{2+}}{L} \times \frac{2\ \text{mol}\ Fe^{3+}}{1\ \text{mol}\ Sn^{2+}} \times \frac{55.847\ g\ Fe^{3+}}{1\ \text{mol}\ Fe^{3+}} = 0.1506\ g\ Fe^{3+}$$

$$\text{mass \% Fe} = \frac{0.1506\ g}{0.1875\ g} \times 100\% = 80.32\%$$

Chapter 4 – Reactions in Aqueous Solutions

4.88 $C_2H_5OH(aq) + 2\ Cr_2O_7^{2-}(aq) + 16\ H^+(aq) \rightarrow 2\ CO_2(g) + 4\ Cr^{3+}(aq) + 11\ H_2O(l)$
C_2H_5OH, 46.07 amu; 8.76 mL = 0.008 76 L

$$0.008\ 76\ L\ x\ \frac{0.049\ 88\ mol\ Cr_2O_7^{2-}}{L}\ x\ \frac{1\ mol\ C_2H_5OH}{2\ mol\ Cr_2O_7^{2-}}\ x\ \frac{46.07\ g\ C_2H_5OH}{1\ mol\ C_2H_5OH}$$

$= 0.010\ 07\ g\ C_2H_5OH$

mass % $C_2H_5OH = \dfrac{0.010\ 07\ g}{10.002\ g}\ x\ 100\% = 0.101\%$

General Problems

4.90 (a) $[Fe(CN)_6]^{3-}(aq) \rightarrow Fe(CN)_6]^{4-}(aq)$
$([Fe(CN)_6]^{3-}(aq) + e^- \rightarrow [Fe(CN)_6]^{4-}(aq))\ x\ 4$ (reduction half reaction)

$N_2H_4(aq) \rightarrow N_2(g)$
$N_2H_4(aq) \rightarrow N_2(g) + 4\ H^+(aq)$
$N_2H_4(aq) \rightarrow N_2(g) + 4\ H^+(aq) + 4\ e^-$
$N_2H_4(aq) + 4\ OH^-(aq) \rightarrow N_2(g) + 4\ H^+(aq) + 4\ OH^-(aq) + 4\ e^-$
$N_2H_4(aq) + 4\ OH^-(aq) \rightarrow N_2(g) + 4\ H_2O(l) + 4\ e^-$ (oxidation half reaction)

Combine the two half reactions.
$4\ [Fe(CN)_6]^{3-}(aq) + N_2H_4(aq) + 4\ OH^-(aq) \rightarrow$
$\qquad\qquad\qquad 4\ [Fe(CN)_6]^{4-}(aq) + N_2(g) + 4\ H_2O(l)$

(b) $Cl_2(g) \rightarrow Cl^-(aq)$
$Cl_2(g) \rightarrow 2\ Cl^-(aq)$
$Cl_2(g) + 2\ e^- \rightarrow 2\ Cl^-(aq)$ (reduction half reaction)

$SeO_3^{2-}(aq) \rightarrow SeO_4^{2-}(aq)$
$SeO_3^{2-}(aq) + H_2O(l) \rightarrow SeO_4^{2-}(aq)$
$SeO_3^{2-}(aq) + H_2O(l) \rightarrow SeO_4^{2-}(aq) + 2\ H^+(aq)$
$SeO_3^{2-}(aq) + H_2O(l) \rightarrow SeO_4^{2-}(aq) + 2\ H^+(aq) + 2\ e^-$
$SeO_3^{2-}(aq) + H_2O(l) + 2\ OH^-(aq) \rightarrow SeO_4^{2-}(aq) + 2\ H^+(aq) + 2\ OH^-(aq) + 2\ e^-$
$SeO_3^{2-}(aq) + H_2O(l) + 2\ OH^-(aq) \rightarrow SeO_4^{2-}(aq) + 2\ H_2O(l) + 2\ e^-$
$SeO_3^{2-}(aq) + 2\ OH^-(aq) \rightarrow SeO_4^{2-}(aq) + H_2O(l) + 2\ e^-$ (oxidation half reaction)

Combine the two half reactions.
$SeO_3^{2-}(aq) + Cl_2(g) + 2\ OH^-(aq) \rightarrow SeO_4^{2-}(aq) + 2\ Cl^-(aq) + H_2O(l)$

(c) $CoCl_2(aq) \rightarrow Co(OH)_3(s) + Cl^-(aq)$
$CoCl_2(aq) \rightarrow Co(OH)_3(s) + 2\ Cl^-(aq)$
$CoCl_2(aq) + 3\ H_2O(l) \rightarrow Co(OH)_3(s) + 2\ Cl^-(aq)$
$CoCl_2(aq) + 3\ H_2O(l) \rightarrow Co(OH)_3(s) + 2\ Cl^-(aq) + 3\ H^+(aq)$
$[CoCl_2(aq) + 3\ H_2O(l) \rightarrow Co(OH)_3(s) + 2\ Cl^-(aq) + 3\ H^+(aq) + e^-]\ x\ 2$
$\qquad\qquad\qquad\qquad\qquad$ (oxidation half reaction)

$HO_2^-(aq) \rightarrow H_2O(l)$

$HO_2^-(aq) \rightarrow 2\,H_2O(l)$

$3\,H^+(aq) + HO_2^-(aq) \rightarrow 2\,H_2O(l)$

$3\,H^+(aq) + HO_2^-(aq) + 2\,e^- \rightarrow 2\,H_2O(l)$ (reduction half reaction)

Combine the two half reactions.

$2\,CoCl_2(aq) + 6\,H_2O(l) + 3\,H^+(aq) + HO_2^-(aq) \rightarrow$
$ 2\,Co(OH)_3(s) + 4\,Cl^-(aq) + 6\,H^+(aq) + 2\,H_2O(l)$

$2\,CoCl_2(aq) + 4\,H_2O(l) + HO_2^-(aq) \rightarrow 2\,Co(OH)_3(s) + 4\,Cl^-(aq) + 3\,H^+(aq)$

$2\,CoCl_2(aq) + 4\,H_2O(l) + HO_2^-(aq) + 3\,OH^-(aq) \rightarrow$
$ 2\,Co(OH)_3(s) + 4\,Cl^-(aq) + 3\,H^+(aq) + 3\,OH^-(aq)$

$2\,CoCl_2(aq) + 4\,H_2O(l) + HO_2^-(aq) + 3\,OH^-(aq) \rightarrow$
$ 2\,Co(OH)_3(s) + 4\,Cl^-(aq) + 3\,H_2O(l)$

$2\,CoCl_2(aq) + H_2O(l) + HO_2^-(aq) + 3\,OH^-(aq) \rightarrow 2\,Co(OH)_3(s) + 4\,Cl^-(aq)$

4.92   (a) C_2H_6 H +1,   C −3

        (b) $Na_2B_4O_7$ O −2,   Na +1,   B +3

        (c) Mg_2SiO_4 O −2,   Mg +2,   Si +4

4.94   (a) "Any element higher in the activity series will react with the ion of any element lower in the activity series."

        $C + B^+ \rightarrow C^+ + B$;   therefore C is higher than B.

        $A^+ + D \rightarrow$ no reaction;   therefore A is higher than D.

        $C^+ + A \rightarrow$ no reaction;   therefore C is higher than A.

        $D + B^+ \rightarrow D^+ + B$;   therefore D is higher than B.

        The net result is $C > A > D > B$.

        (b) (1) The reaction, $A^+ + C \rightarrow A + C^+$, will occur because C is above A in the activity series.

           (2) The reaction, $A^+ + B \rightarrow A + B^+$, will not occur because B is below A in the activity series.

4.96 $MgF_2(s) \rightleftarrows Mg^{2+}(aq) + 2\,F^-(aq)$

                   x   2x

        $[Mg^{2+}] = x = 2.6 \times 10^{-4}$ M and $[F^-] = 2x = 2(2.6 \times 10^{-4}$ M$) = 5.2 \times 10^{-4}$ M in a saturated solution.

        $K_{sp} = [Mg^{2+}][F^-]^2 = (2.6 \times 10^{-4}$ M$)(5.2 \times 10^{-4}$ M$)^2 = 7.0 \times 10^{-11}$

4.98   (a) Add HCl to precipitate Hg_2Cl_2. $Hg_2^{2+}(aq) + 2Cl^-(aq) \rightarrow Hg_2Cl_2(s)$

        (b) Add H_2SO_4 to precipitate $PbSO_4$. $Pb^{2+}(aq) + SO_4^{2-}(aq) \rightarrow PbSO_4(s)$

        (c) Add Na_2CO_3 to precipitate $CaCO_3$. $Ca^{2+}(aq) + CO_3^{2-}(aq) \rightarrow CaCO_3(s)$

        (d) Add Na_2SO_4 to precipitate $BaSO_4$. $Ba^{2+}(aq) + SO_4^{2-}(aq) \rightarrow BaSO_4(s)$

4.100 All four reactions are redox reactions.

(a) $Mn(OH)_2(s) \rightarrow Mn(OH)_3(s)$

$Mn(OH)_2(s) + OH^-(aq) \rightarrow Mn(OH)_3(s)$

$[Mn(OH)_2(s) + OH^-(aq) \rightarrow Mn(OH)_3(s) + e^-] \times 2$ (oxidation half reaction)

$H_2O_2(aq) \rightarrow 2 H_2O(l)$

$2 H^+(aq) + H_2O_2(aq) \rightarrow 2 H_2O(l)$

$2 e^- + 2 H^+(aq) + H_2O_2(aq) \rightarrow 2 H_2O(l)$

$2 e^- + 2 OH^-(aq) + 2 H^+(aq) + H_2O_2(aq) \rightarrow 2 H_2O(l) + 2 OH^-(aq)$

$2 e^- + 2 H_2O(l) + H_2O_2(aq) \rightarrow 2 H_2O(l) + 2 OH^-(aq)$

$2 e^- + H_2O_2(aq) \rightarrow 2 OH^-(aq)$ (reduction half reaction)

Combine the two half reactions.

$2 Mn(OH)_2(s) + 2 OH^-(aq) + H_2O_2(aq) \rightarrow 2 Mn(OH)_3(s) + 2 OH^-(aq)$

$2 Mn(OH)_2(s) + H_2O_2(aq) \rightarrow 2 Mn(OH)_3(s)$

(b) $[MnO_4^{2-}(aq) \rightarrow MnO_4^-(aq) + e^-] \times 2$ (oxidation half reaction)

$MnO_4^{2-}(aq) \rightarrow MnO_2(s)$

$MnO_4^{2-}(aq) \rightarrow MnO_2(s) + 2 H_2O(l)$

$4 H^+(aq) + MnO_4^{2-}(aq) \rightarrow MnO_2(s) + 2 H_2O(l)$

$2 e^- + 4 H^+(aq) + MnO_4^{2-}(aq) \rightarrow MnO_2(s) + 2 H_2O(l)$ (reduction half reaction)

Combine the two half reactions.

$4 H^+(aq) + 3 MnO_4^{2-}(aq) \rightarrow MnO_2(s) + 2 MnO_4^-(aq) + 2 H_2O(l)$

(c) $I^-(aq) \rightarrow I_3^-(aq)$

$3 I^-(aq) \rightarrow I_3^-(aq)$

$[3 I^-(aq) \rightarrow I_3^-(aq) + 2 e^-] \times 8$ (oxidation half reaction)

$IO_3^-(aq) \rightarrow I_3^-(aq)$

$3 IO_3^-(aq) \rightarrow I_3^-(aq)$

$3 IO_3^-(aq) \rightarrow I_3^-(aq) + 9 H_2O(l)$

$18 H^+(aq) + 3 IO_3^-(aq) \rightarrow I_3^-(aq) + 9 H_2O(l)$

$16 e^- + 18 H^+(aq) + 3 IO_3^-(aq) \rightarrow I_3^-(aq) + 9 H_2O(l)$ (reduction half reaction)

Combine the two half reactions.

$24 I^-(aq) + 3 IO_3^-(aq) + 18 H^+(aq) \rightarrow 9 I_3^-(aq) + 9 H_2O(l)$

Divide all coefficients by 3.

$8 I^-(aq) + IO_3^-(aq) + 6 H^+(aq) \rightarrow 3 I_3^-(aq) + 3 H_2O(l)$

(d) $P(s) \rightarrow HPO_3^{2-}(aq)$

$3 H_2O(l) + P(s) \rightarrow HPO_3^{2-}(aq)$

$3 H_2O(l) + P(s) \rightarrow HPO_3^{2-}(aq) + 5 H^+(aq)$

$[3 H_2O(l) + P(s) \rightarrow HPO_3^{2-}(aq) + 5 H^+(aq) + 3 e^-] \times 2$

(oxidation half reaction)

$PO_4^{3-}(aq) \rightarrow HPO_3^{2-}(aq)$

$PO_4^{3-}(aq) \rightarrow HPO_3^{2-}(aq) + H_2O(l)$

$3 H^+(aq) + PO_4^{3-}(aq) \rightarrow HPO_3^{2-}(aq) + H_2O(l)$

$[2 e^- + 3 H^+(aq) + PO_4^{3-}(aq) \rightarrow HPO_3^{2-}(aq) + H_2O(l)] \times 3$

(reduction half reaction)

Combine the two half reactions and add OH^-.

$6 H_2O(l) + 2 P(s) + 9 H^+(aq) + 3 PO_4^{3-}(aq) \rightarrow$
$\qquad\qquad 5 HPO_3^{2-}(aq) + 10 H^+(aq) + 3 H_2O(l)$

$3 H_2O(l) + 2 P(s) + 3 PO_4^{3-}(aq) \rightarrow 5 HPO_3^{2-}(aq) + H^+(aq)$

$3 H_2O(l) + 2 P(s) + 3 PO_4^{3-}(aq) + OH^-(aq) \rightarrow$
$\qquad\qquad 5 HPO_3^{2-}(aq) + H^+(aq) + OH^-(aq)$

$3 H_2O(l) + 2 P(s) + 3 PO_4^{3-}(aq) + OH^-(aq) \rightarrow 5 HPO_3^{2-}(aq) + H_2O(l)$

$2 H_2O(l) + 2 P(s) + 3 PO_4^{3-}(aq) + OH^-(aq) \rightarrow 5 HPO_3^{2-}(aq)$

4.102 (a) $S_4O_6^{2-}(aq) \rightarrow H_2S(aq)$

$S_4O_6^{2-}(aq) \rightarrow 4 H_2S(aq)$

$S_4O_6^{2-}(aq) \rightarrow 4 H_2S(aq) + 6 H_2O(l)$

$20 H^+(aq) + S_4O_6^{2-}(aq) \rightarrow 4 H_2S(aq) + 6 H_2O(l)$

$18 e^- + 20 H^+(aq) + S_4O_6^{2-}(aq) \rightarrow 4 H_2S(aq) + 6 H_2O(l)$ (reduction half reaction)

$Al(s) \rightarrow Al^{3+}(aq)$

$[Al(s) \rightarrow Al^{3+}(aq) + 3 e^-] \times 6$ (oxidation half reaction)

Combine the two half reactions.

$20 H^+(aq) + S_4O_6^{2-}(aq) + 6 Al(s) \rightarrow 4 H_2S(aq) + 6 Al^{3+}(aq) + 6 H_2O(l)$

(b) $S_2O_3^{2-}(aq) \rightarrow S_4O_6^{2-}(aq)$

$2 S_2O_3^{2-}(aq) \rightarrow S_4O_6^{2-}(aq)$

$[2 S_2O_3^{2-}(aq) \rightarrow S_4O_6^{2-}(aq) + 2 e^-] \times 3$ (oxidation half reaction)

$Cr_2O_7^{2-}(aq) \rightarrow Cr^{3+}(aq)$

$Cr_2O_7^{2-}(aq) \rightarrow 2 Cr^{3+}(aq)$

$Cr_2O_7^{2-}(aq) \rightarrow 2 Cr^{3+}(aq) + 7 H_2O(l)$

$14 H^+(aq) + Cr_2O_7^{2-}(aq) \rightarrow 2 Cr^{3+}(aq) + 7 H_2O(l)$

$6 e^- + 14 H^+(aq) + Cr_2O_7^{2-}(aq) \rightarrow 2 Cr^{3+}(aq) + 7 H_2O(l)$ (reduction half reaction)

Combine the two half reactions.

$14 H^+(aq) + 6 S_2O_3^{2-}(aq) + Cr_2O_7^{2-}(aq) \rightarrow 3 S_4O_6^{2-}(aq) + 2 Cr^{3+}(aq) + 7 H_2O(l)$

(c) $ClO_3^-(aq) \rightarrow Cl^-(aq)$

$ClO_3^-(aq) \rightarrow Cl^-(aq) + 3 H_2O(l)$

$6 H^+(aq) + ClO_3^-(aq) \rightarrow Cl^-(aq) + 3 H_2O(l)$

$[6 e^- + 6 H^+(aq) + ClO_3^-(aq) \rightarrow Cl^-(aq) + 3 H_2O(l)] \times 14$ (reduction half reaction)

$As_2S_3(s) \rightarrow H_2AsO_4^-(aq) + HSO_4^-(aq)$

$As_2S_3(s) \rightarrow 2 H_2AsO_4^-(aq) + 3 HSO_4^-(aq)$

$20 H_2O(l) + As_2S_3(s) \rightarrow 2 H_2AsO_4^-(aq) + 3 HSO_4^-(aq)$

$20 H_2O(l) + As_2S_3(s) \rightarrow 2 H_2AsO_4^-(aq) + 3 HSO_4^-(aq) + 33 H^+(aq)$

$[20 H_2O(l) + As_2S_3(s) \rightarrow 2 H_2AsO_4^-(aq) + 3 HSO_4^-(aq) + 33 H^+(aq) + 28 e^-] \times 3$

(oxidation half reaction)

Combine the two half reactions.

$84 H^+(aq) + 60 H_2O(l) + 14 ClO_3^-(aq) + 3 As_2S_3(s) \rightarrow$
$\qquad 14 Cl^-(aq) + 6 H_2AsO_4^-(aq) + 9 HSO_4^-(aq) + 42 H_2O(l) + 99 H^+(aq)$

$18 H_2O(l) + 14 ClO_3^-(aq) + 3 As_2S_3(s) \rightarrow$
$\qquad\qquad 14 Cl^-(aq) + 6 H_2AsO_4^-(aq) + 9 HSO_4^-(aq) + 15 H^+(aq)$

(d) $IO_3^-(aq) \rightarrow I^-(aq)$

$IO_3^-(aq) \rightarrow I^-(aq) + 3 H_2O(l)$

$6 H^+(aq) + IO_3^-(aq) \rightarrow I^-(aq) + 3 H_2O(l)$

$[6 e^- + 6 H^+(aq) + IO_3^-(aq) \rightarrow I^-(aq) + 3 H_2O(l)] \times 7$ (reduction half reaction)

$Re(s) \rightarrow ReO_4^-(aq)$

$4 H_2O(l) + Re(s) \rightarrow ReO_4^-(aq)$

$4 H_2O(l) + Re(s) \rightarrow ReO_4^-(aq) + 8 H^+(aq)$

$[4 H_2O(l) + Re(s) \rightarrow ReO_4^-(aq) + 8 H^+(aq) + 7 e^-] \times 6$ (oxidation half reaction)

Combine the two half reactions.

$42 H^+(aq) + 24 H_2O(l) + 7 IO_3^-(aq) + 6 Re(s) \rightarrow$
$\qquad\qquad\qquad 7 I^-(aq) + 6 ReO_4^-(aq) + 21 H_2O(l) + 48 H^+(aq)$

$3 H_2O(l) + 7 IO_3^-(aq) + 6 Re(s) \rightarrow 7 I^-(aq) + 6 ReO_4^-(aq) + 6 H^+(aq)$

(e) $HSO_4^-(aq) + Pb_3O_4(s) \rightarrow PbSO_4(s)$

$3 HSO_4^-(aq) + Pb_3O_4(s) \rightarrow 3 PbSO_4(s)$

$3 HSO_4^-(aq) + Pb_3O_4(s) \rightarrow 3 PbSO_4(s) + 4 H_2O(l)$

$5 H^+(aq) + 3 HSO_4^-(aq) + Pb_3O_4(s) \rightarrow 3 PbSO_4(s) + 4 H_2O(l)$

$[2 e^- + 5 H^+(aq) + 3 HSO_4^-(aq) + Pb_3O_4(s) \rightarrow 3 PbSO_4(s) + 4 H_2O(l)] \times 10$

(reduction half reaction)

$As_4(s) \rightarrow H_2AsO_4^-(aq)$

$As_4(s) \rightarrow 4 H_2AsO_4^-(aq)$

$16 H_2O(l) + As_4(s) \rightarrow 4 H_2AsO_4^-(aq)$

$16 H_2O(l) + As_4(s) \rightarrow 4 H_2AsO_4^-(aq) + 24 H^+(aq)$

$16 H_2O(l) + As_4(s) \rightarrow 4 H_2AsO_4^-(aq) + 24 H^+(aq) + 20 e^-$ (oxidation half reaction)

Combine the two half reactions.

$26 H^+(aq) + 30 HSO_4^-(aq) + As_4(s) + 10 Pb_3O_4(s) \rightarrow$
$\qquad\qquad\qquad 4 H_2AsO_4^-(aq) + 30 PbSO_4(s) + 24 H_2O(l)$

(f) $HNO_2(aq) \rightarrow NO_3^-(aq)$

$H_2O(l) + HNO_2(aq) \rightarrow NO_3^-(aq)$

$H_2O(l) + HNO_2(aq) \rightarrow NO_3^-(aq) + 3 H^+(aq)$

$H_2O(l) + HNO_2(aq) \rightarrow NO_3^-(aq) + 3 H^+(aq) + 2 e^-$ (oxidation half reaction)

$HNO_2(aq) \rightarrow NO(g)$

$HNO_2(aq) \rightarrow NO(g) + H_2O(l)$

$H^+(aq) + HNO_2(aq) \rightarrow NO(g) + H_2O(l)$

$[1\,e^- + H^+(aq) + HNO_2(aq) \rightarrow NO(g) + H_2O(l)] \times 2$　　(reduction half reaction)

Combine the two half reactions.

$3\,HNO_2(aq) \rightarrow NO_3^-(aq) + 2\,NO(g) + H_2O(l) + H^+(aq)$

4.104　CuO, 79.55 amu;　Cu_2O, 143.09 amu

Let X equal the mass of CuO and Y the mass of Cu_2O in the 10.50 g mixture.　Therefore, X + Y = 10.50 g.

$$\text{mol Cu} = 8.66 \text{ g} \times \frac{1 \text{ mol Cu}}{63.546 \text{ g Cu}} = 0.1363 \text{ mol Cu}$$

mol CuO + 2 × mol Cu_2O = 0.1363 mol Cu

$$X \times \frac{1 \text{ mol CuO}}{79.55 \text{ g CuO}} + 2 \times \left(Y \times \frac{1 \text{ mol Cu}_2\text{O}}{143.09 \text{ g Cu}_2\text{O}} \right) = 0.1363 \text{ mol Cu}$$

Rearrange to get X = 10.50 g – Y and then substitute it into the equation above to solve for Y.

$$(10.50 \text{ g} - Y) \times \frac{1 \text{ mol CuO}}{79.55 \text{ g CuO}} + 2 \times \left(Y \times \frac{1 \text{ mol Cu}_2\text{O}}{143.09 \text{ g Cu}_2\text{O}} \right) = 0.1363 \text{ mol Cu}$$

$$\frac{10.50 \text{ mol}}{79.55} - \frac{Y \text{ mol}}{79.55 \text{ g}} + \frac{2\,Y \text{ mol}}{143.09 \text{ g}} = 0.1363 \text{ mol}$$

$$-\frac{Y \text{ mol}}{79.55 \text{ g}} + \frac{2\,Y \text{ mol}}{143.09 \text{ g}} = 0.1363 \text{ mol} - \frac{10.50 \text{ mol}}{79.55} = 0.0043 \text{ mol}$$

$$\frac{(-Y \text{ mol})(143.09 \text{ g}) + (2\,Y \text{ mol})(79.55 \text{ g})}{(79.55 \text{ g})(143.09 \text{ g})} = 0.0043 \text{ mol}$$

$$\frac{16.01\,Y \text{ mol}}{11383 \text{ g}} = 0.0043 \text{ mol}; \quad \frac{16.01\,Y}{11383 \text{ g}} = 0.0043$$

Y = (0.0043)(11383 g)/16.01 = 3.06 g Cu_2O

X = 10.50 g – Y = 10.50 g – 3.06 g = 7.44 g CuO

Multi-Concept Problems

4.106　NaOH, 40.00 amu;　$Ba(OH)_2$, 171.34 amu

Let X equal the mass of NaOH and Y the mass of $Ba(OH)_2$ in the 10.0 g mixture. Therefore, X + Y = 10.0 g.

$$\text{mol HCl} = 108.9 \text{ mL} \times \frac{1 \text{ L}}{1000 \text{ mL}} \times \frac{1.50 \text{ mol HCl}}{1 \text{ L}} = 0.163 \text{ mol HCl}$$

mol NaOH + 2 × mol $Ba(OH)_2$ = 0.163 mol HCl

$$X \times \frac{1 \text{ mol NaOH}}{40.00 \text{ g NaOH}} + 2 \times \left(Y \times \frac{1 \text{ mol Ba(OH)}_2}{171.34 \text{ g Ba(OH)}_2} \right) = 0.163 \text{ mol HCl}$$

Rearrange to get X = 10.0 g – Y and then substitute it into the equation above to solve for Y.

$$(10.0\ g - Y) \times \frac{1\ mol\ NaOH}{40.00\ g\ NaOH} + 2 \times \left(Y \times \frac{1\ mol\ Ba(OH)_2}{171.34\ g\ Ba(OH)_2}\right) = 0.163\ mol\ HCl$$

$$\frac{10.00\ mol}{40.00} - \frac{Y\ mol}{40.00\ g} + \frac{2\ Y\ mol}{171.34\ g} = 0.163\ mol$$

$$- \frac{Y\ mol}{40.00\ g} + \frac{2\ Y\ mol}{171.34\ g} = 0.163\ mol - \frac{10.00\ mol}{40.00} = -0.087\ mol$$

$$\frac{(-Y\ mol)(171.34\ g) + (2\ Y\ mol)(40.00\ g)}{(40.00\ g)(171.34\ g)} = -0.087\ mol$$

$$\frac{-91.34\ Y\ mol}{6853.6\ g} = -0.087\ mol; \quad \frac{91.34\ Y}{6853.6\ g} = 0.087$$

Y = (0.087)(6853.6 g)/91.34 = 6.5 g Ba(OH)$_2$

X = 10.0 g – Y = 10.0 g – 6.5 g = 3.5 g NaOH

4.108 KNO$_3$, 101.10 amu; BaCl$_2$, 208.24 amu; NaCl, 58.44 amu; BaSO$_4$, 233.40 amu; AgCl, 143.32 amu

(a) The two precipitates are BaSO$_4$(s) and AgCl(s).

(b) H$_2$SO$_4$ only reacts with BaCl$_2$.

H$_2$SO$_4$(aq) + BaCl$_2$(aq) → BaSO$_4$(s) + 2 HCl(aq)

Calculate the number of moles of BaCl$_2$ in 100.0 g of the mixture.

$$mol\ BaCl_2 = 67.3\ g\ BaSO_4 \times \frac{1\ mol\ BaSO_4}{233.40\ g\ BaSO_4} \times \frac{1\ mol\ BaCl_2}{1\ mol\ BaSO_4} = 0.288\ mol\ BaCl_2$$

Calculate mass and moles of BaCl$_2$ in 250.0 g sample.

$$mass\ BaCl_2 = 0.288\ mol\ BaCl_2 \times \frac{208.24\ g\ BaCl_2}{1\ mol\ BaCl_2} \times \frac{250.0\ g}{100.0\ g} = 150.\ g\ BaCl_2$$

$$mol\ BaCl_2 = 150.\ g\ BaCl_2 \times \frac{1\ mol\ BaCl_2}{208.24\ g\ BaCl_2} = 0.720\ mol\ BaCl_2$$

AgNO$_3$ reacts with both NaCl and BaCl$_2$ in the remaining 150.0 g of the mixture.

3 AgNO$_3$(aq) + NaCl(aq) + BaCl$_2$(aq) → 3 AgCl(s) + NaNO$_3$(aq) + Ba(NO$_3$)$_2$(aq)

Calculate the moles of AgCl that would have been produced from the 250.0 g mixture.

$$mol\ AgCl = 197.6\ g\ AgCl \times \frac{1\ mol\ AgCl}{143.32\ g\ AgCl} \times \frac{250.0\ g}{150.0\ g} = 2.30\ mol\ AgCl$$

mol AgCl = 2 x (mol BaCl$_2$) + mol NaCl

Calculate the moles and mass of NaCl in the 250.0 g mixture.

2.30 mol AgCl = 2 x 0.720 mol BaCl$_2$ + mol NaCl

mol NaCl = 2.30 mol – 2(0.720 mol) = 0.86 mol NaCl

$$mass\ NaCl = 0.86\ mol\ NaCl \times \frac{58.44\ g\ NaCl}{1\ mol\ NaCl} = 50.\ g\ NaCl$$

Calculate the mass of KNO$_3$ in the 250.0 g mixture.

total mass = mass BaCl$_2$ + mass NaCl + mass KNO$_3$

250.0 g = 150. g BaCl$_2$ + 50. g NaCl + mass KNO$_3$

mass KNO$_3$ = 250.0 g – 150. g BaCl$_2$ – 50. g NaCl = 50. g KNO$_3$

4.110 (a) $Cr^{2+}(aq) + Cr_2O_7^{2-}(aq) \rightarrow Cr^{3+}(aq)$

$[Cr^{2+}(aq) \rightarrow Cr^{3+}(aq) + e^-] \times 6$ (oxidation half reaction)

$Cr_2O_7^{2-}(aq) \rightarrow Cr^{3+}(aq)$

$Cr_2O_7^{2-}(aq) \rightarrow 2\,Cr^{3+}(aq)$

$Cr_2O_7^{2-}(aq) \rightarrow 2\,Cr^{3+}(aq) + 7\,H_2O(l)$

$14\,H^+(aq) + Cr_2O_7^{2-}(aq) \rightarrow 2\,Cr^{3+}(aq) + 7\,H_2O(l)$

$6\,e^- + 14\,H^+(aq) + Cr_2O_7^{2-}(aq) \rightarrow 2\,Cr^{3+}(aq) + 7\,H_2O(l)$ (reduction half reaction)

Combine the two half reactions.

$14\,H^+(aq) + Cr_2O_7^{2-}(aq) + 6\,Cr^{2+}(aq) \rightarrow 8\,Cr^{3+}(aq) + 7\,H_2O(l)$

(b) total volume = 100.0 ml + 20.0 mL = 120.0 mL = 0.1200 L

Initial moles:

$0.120\ \dfrac{mol\ Cr(NO_3)_2}{1\ L} \times 0.1000\ L = 0.0120\ mol\ Cr(NO_3)_2$

$0.500\ \dfrac{mol\ HNO_3}{1\ L} \times 0.1000\ L = 0.0500\ mol\ HNO_3$

$0.250\ \dfrac{mol\ K_2Cr_2O_7}{1\ L} \times 0.0200\ L = 0.005\,00\ mol\ K_2Cr_2O_7$

Check for the limiting reactant. 0.0120 mol of Cr^{2+} requires (0.0120)/6 = 0.00200 mol $Cr_2O_7^{2-}$ and (14/6)(0.0120) = 0.0280 mol H^+. Both are in excess of the required amounts, so Cr^{2+} is the limiting reactant.

	$14\,H^+(aq)$	$+ Cr_2O_7^{2-}(aq)$	$+ 6\,Cr^{2+}(aq) \rightarrow$	$8\,Cr^{3+}(aq) + 7\,H_2O(l)$
Initial moles	0.0500	0.00500	0.0120	0
Change	−14x	−x	−6x	+8x

Because Cr^{2+} is the limiting reactant, 6x = 0.0120 and x = 0.00200

Final moles	0.0220	0.00300	0	0.00160

$mol\ K^+ = 0.00500\ mol\ K_2Cr_2O_7 \times \dfrac{2\ mol\ K^+}{1\ mol\ K_2Cr_2O_7} = 0.0100\ mol\ K^+$

$mol\ NO_3^- = 0.0120\ mol\ Cr(NO_3)_2 \times \dfrac{2\ mol\ NO_3^-}{1\ mol\ Cr(NO_3)_2}$

$+ 0.0500\ mol\ HNO_3 \times \dfrac{1\ mol\ NO_3^-}{1\ mol\ HNO_3} = 0.0740\ mol\ NO_3^-$

$mol\ H^+ = 0.0220\ mol;$ $mol\ Cr_2O_7^{2-} = 0.00300\ mol;$ $mol\ Cr^{3+} = 0.01600\ mol$

Check for charge neutrality.

Total moles of +charge = 0.0100 + 0.0220 + 3 x (0.01600) = 0.0800 mol +charge

Total moles of −charge = 0.0740 + 2 x (0.00300) = 0.0800 mol −charge

The charges balance and there is electrical neutrality in the solution after the reaction.

$$K^+ \text{ molarity} = \frac{0.0100 \text{ mol } K^+}{0.1200 \text{ L}} = 0.0833 \text{ M}$$

$$NO_3^- \text{ molarity} = \frac{0.0740 \text{ mol } NO_3^-}{0.1200 \text{ L}} = 0.617 \text{ M}$$

$$H^+ \text{ molarity} = \frac{0.0220 \text{ mol } H^+}{0.1200 \text{ L}} = 0.183 \text{ M}$$

$$Cr_2O_7^{2-} \text{ molarity} = \frac{0.00300 \text{ mol } Cr_2O_7^{2-}}{0.1200 \text{ L}} = 0.0250 \text{ M}$$

$$Cr^{3+} \text{ molarity} = \frac{0.0160 \text{ mol } Cr^{3+}}{0.1200 \text{ L}} = 0.133 \text{ M}$$

4.112 (a) (1) $Cu(s) \rightarrow Cu^{2+}(aq)$
    $[Cu(s) \rightarrow Cu^{2+}(aq) + 2 e^-] \times 3$ (oxidation half reaction)

$NO_3^-(aq) \rightarrow NO(g)$
$NO_3^-(aq) \rightarrow NO(g) + 2 H_2O(l)$
$4 H^+(aq) + NO_3^-(aq) \rightarrow NO(g) + 2 H_2O(l)$
$[3 e^- + 4 H^+(aq) + NO_3^-(aq) \rightarrow NO(g) + 2 H_2O(l)] \times 2$
                                              (reduction half reaction

Combine the two half reactions.
$3 Cu(s) + 8 H^+(aq) + 2 NO_3^-(aq) \rightarrow 3 Cu^{2+}(aq) + 2 NO(g) + 4 H_2O(l)$

(2) $Cu^{2+}(aq) + SCN^-(aq) \rightarrow CuSCN(s)$
    $[e^- + Cu^{2+}(aq) + SCN^-(aq) \rightarrow CuSCN(s)] \times 2$(reduction half reaction)

$HSO_3^-(aq) \rightarrow HSO_4^-(aq)$
$H_2O(l) + HSO_3^-(aq) \rightarrow HSO_4^-(aq)$
$H_2O(l) + HSO_3^-(aq) \rightarrow HSO_4^-(aq) + 2 H^+(aq)$
$H_2O(l) + HSO_3^-(aq) \rightarrow HSO_4^-(aq) + 2 H^+(aq) + 2 e^-$
                                           (oxidation half reaction)

Combine the two half reactions.
$2 Cu^{2+}(aq) + 2 SCN^-(aq) + H_2O(l) + HSO_3^-(aq) \rightarrow$
                        $2 CuSCN(s) + HSO_4^-(aq) + 2 H^+(aq)$

(3) $Cu^+(aq) \rightarrow Cu^{2+}(aq)$
    $[Cu^+(aq) \rightarrow Cu^{2+}(aq) + e^-] \times 10$ (oxidation half reaction)

$IO_3^-(aq) \rightarrow I_2(aq)$
$2 IO_3^-(aq) \rightarrow I_2(aq)$
$2 IO_3^-(aq) \rightarrow I_2(aq) + 6 H_2O(l)$
$12 H^+(aq) + 2 IO_3^-(aq) \rightarrow I_2(aq) + 6 H_2O(l)$
$10 e^- + 12 H^+(aq) + 2 IO_3^-(aq) \rightarrow I_2(aq) + 6 H_2O(l)$
                                         (reduction half reaction)

Combine the two half reactions.
$$10\ Cu^+(aq)\ +\ 12\ H^+(aq)\ +\ 2\ IO_3^-(aq)\ \rightarrow\ 10\ Cu^{2+}(aq)\ +\ I_2(aq)\ +\ 6\ H_2O(l)$$

(4) $I_2(aq)\ \rightarrow\ I^-(aq)$
$I_2(aq)\ \rightarrow\ 2\ I^-(aq)$
$2\ e^-\ +\ I_2(aq)\ \rightarrow\ 2\ I^-(aq)$ (reduction half reaction)

$S_2O_3^{2-}(aq)\ \rightarrow\ S_4O_6^{2-}(aq)$
$2\ S_2O_3^{2-}(aq)\ \rightarrow\ S_4O_6^{2-}(aq)$
$2\ S_2O_3^{2-}(aq)\ \rightarrow\ S_4O_6^{2-}(aq)\ +\ 2\ e^-$ (oxidation half reaction)

Combine the two half reactions.
$$I_2(aq)\ +\ 2\ S_2O_3^{2-}(aq)\ \rightarrow\ 2\ I^-(aq)\ +\ S_4O_6^{2-}(aq)$$

(5) $2\ ZnNH_4PO_4\ \rightarrow\ Zn_2P_2O_7\ +\ H_2O\ +\ 2\ NH_3$

(b) 10.82 mL = 0.01082 L
mol $S_2O_3^{2-}$ = (0.1220 mol/L)(0.01082 L) = 0.00132 mol $S_2O_3^{2-}$
$$mol\ I_2 = 0.00132\ mol\ S_2O_3^{2-}\ x\ \frac{1\ mol\ I_2}{2\ mol\ S_2O_3^{2-}} = 6.60\ x\ 10^{-4}\ mol\ I_2$$
$$mol\ Cu^+ = 6.60\ x\ 10^{-4}\ mol\ I_2\ x\ \frac{10\ mol\ Cu^+}{1\ mol\ I_2} = 6.60\ x\ 10^{-3}\ mol\ Cu^+\ (Cu)$$
g Cu = (6.60 x 10^{-3} mol)(63.546 g/mol) = 0.419 g Cu
$$mass\ \%\ Cu\ in\ brass\ =\ \frac{0.419\ g\ Cu}{0.544\ g\ brass}\ x\ 100\% = 77.1\%\ Cu$$

(c) $Zn_2P_2O_7$, 304.72 amu
$$mass\ \%\ Zn\ in\ Zn_2P_2O_7 = \frac{2\ x\ 65.39\ g}{304.72\ g}\ x\ 100\% = 42.92\%$$
mass of Zn in $Zn_2P_2O_7$ = (0.4292)(0.246 g) = 0.106 g Zn
$$mass\ \%\ Zn\ in\ brass\ =\ \frac{0.106\ g\ Zn}{0.544\ g\ brass}\ x\ 100\% = 19.5\%\ Zn$$

Periodicity and Atomic Structure

5.1 Gamma ray $\nu = \dfrac{c}{\lambda} = \dfrac{3.00 \times 10^8 \text{ m/s}}{3.56 \times 10^{-11} \text{ m}} = 8.43 \times 10^{18} \text{ s}^{-1} = 8.43 \times 10^{18} \text{ Hz}$

Radar wave $\nu = \dfrac{c}{\lambda} = \dfrac{3.00 \times 10^8 \text{ m/s}}{10.3 \times 10^{-2} \text{ m}} = 2.91 \times 10^9 \text{ s}^{-1} = 2.91 \times 10^9 \text{ Hz}$

5.2 $\nu = 102.5 \text{ MHz} = 102.5 \times 10^6 \text{ Hz} = 102.5 \times 10^6 \text{ s}^{-1}$

$\lambda = \dfrac{c}{\nu} = \dfrac{3.00 \times 10^8 \text{ m/s}}{102.5 \times 10^6 \text{ s}^{-1}} = 2.93 \text{ m}$

$\nu = 9.55 \times 10^{17} \text{ Hz} = 9.55 \times 10^{17} \text{ s}^{-1}$

$\lambda = \dfrac{c}{\nu} = \dfrac{3.00 \times 10^8 \text{ m/s}}{9.55 \times 10^{17} \text{ s}^{-1}} = 3.14 \times 10^{-10} \text{ m}$

5.3 The wave with the shorter wavelength (b) has the higher frequency. The wave with the larger amplitude (b) represents the more intense beam of light. The wave with the shorter wavelength (b) represents blue light. The wave with the longer wavelength (a) represents red light.

5.4 Balmer series: $m = 2$; $R = 1.097 \times 10^{-2} \text{ nm}^{-1}$

$\dfrac{1}{\lambda} = R\left[\dfrac{1}{m^2} - \dfrac{1}{n^2}\right]$; $\dfrac{1}{\lambda} = R\left[\dfrac{1}{2^2} - \dfrac{1}{7^2}\right]$; $\dfrac{1}{\lambda} = 2.519 \times 10^{-3} \text{ nm}^{-1}$; $\lambda = 397.0 \text{ nm}$

5.5 Paschen series: $m = 3$; $R = 1.097 \times 10^{-2} \text{ nm}^{-1}$

$\dfrac{1}{\lambda} = R\left[\dfrac{1}{m^2} - \dfrac{1}{n^2}\right]$; $\dfrac{1}{\lambda} = R\left[\dfrac{1}{3^2} - \dfrac{1}{4^2}\right]$; $\dfrac{1}{\lambda} = 5.333 \times 10^{-4} \text{ nm}^{-1}$; $\lambda = 1875 \text{ nm}$

5.6 Paschen series: $m = 3$; $R = 1.097 \times 10^{-2} \text{ nm}^{-1}$

$\dfrac{1}{\lambda} = R\left[\dfrac{1}{m^2} - \dfrac{1}{n^2}\right]$; $\dfrac{1}{\lambda} = R\left[\dfrac{1}{3^2} - \dfrac{1}{\infty^2}\right]$; $\dfrac{1}{\lambda} = 1.219 \times 10^{-3} \text{ nm}^{-1}$; $\lambda = 820.4 \text{ nm}$

5.7 $\lambda = 91.2 \text{ nm} = 91.2 \times 10^{-9} \text{ m}$

$\nu = \dfrac{c}{\lambda} = \dfrac{3.00 \times 10^8 \text{ m/s}}{91.2 \times 10^{-9} \text{ m}} = 3.29 \times 10^{15} \text{ s}^{-1}$

$E = h\nu = (6.626 \times 10^{-34} \text{ J·s})(3.29 \times 10^{15} \text{ s}^{-1}) = 2.18 \times 10^{-18} \text{ J/photon}$

$E = (2.18 \times 10^{-18} \text{ J/photon})(6.022 \times 10^{23} \text{ photons/mol}) = 1.31 \times 10^6 \text{ J/mol} = 1310 \text{ kJ/mol}$

Chapter 5 – Periodicity and Atomic Structure

5.8   IR, $\lambda = 1.55 \times 10^{-6}$ m

$$E = h\frac{c}{\lambda} = (6.626 \times 10^{-34} \text{ J·s}) \left(\frac{3.00 \times 10^8 \text{ m/s}}{1.55 \times 10^{-6} \text{ m}} \right) (6.022 \times 10^{23} / \text{mol})$$

$E = 7.72 \times 10^4$ J/mol = 77.2 kJ/mol

UV, $\lambda = 250$ nm $= 250 \times 10^{-9}$ m

$$E = h\frac{c}{\lambda} = (6.626 \times 10^{-34} \text{ J·s}) \left(\frac{3.00 \times 10^8 \text{ m/s}}{250 \times 10^{-9} \text{ m}} \right) (6.022 \times 10^{23} / \text{mol})$$

$E = 4.79 \times 10^5$ J/mol = 479 kJ/mol

X ray, $\lambda = 5.49$ nm $= 5.49 \times 10^{-9}$ m

$$E = h\frac{c}{\lambda} = (6.626 \times 10^{-34} \text{ J·s}) \left(\frac{3.00 \times 10^8 \text{ m/s}}{5.49 \times 10^{-9} \text{ m}} \right) (6.022 \times 10^{23} / \text{mol})$$

$E = 2.18 \times 10^7$ J/mol = 2.18×10^4 kJ/mol

5.9 $\lambda = \dfrac{h}{mv} = \dfrac{6.626 \times 10^{-34} \text{ kg m}^2 \text{ s}^{-1}}{(1150 \text{ kg})(24.6 \text{ m/s})} = 2.34 \times 10^{-38}$ m

5.10 $(\Delta x)(\Delta mv) \geq \dfrac{h}{4\pi};$ uncertainty in velocity = (45 m/s)(0.02) = 0.9 m/s

$$\Delta x \geq \frac{h}{4\pi(\Delta mv)} = \frac{6.626 \times 10^{-34} \text{ kg m}^2 \text{ s}^{-1}}{4\pi(0.120 \text{ kg})(0.9 \text{ m/s})} = 5 \times 10^{-34} \text{ m}$$

5.11

n	l	m_l	Orbital	No. of Orbitals
5	0	0	5s	1
	1	−1, 0, +1	5p	3
	2	−2, −1, 0, +1, +2	5d	5
	3	−3, −2, −1, 0, +1, +2, +3	5f	7
	4	−4, −3, −2, −1, 0, +1, +2, +3, +4	5g	9

There are 25 possible orbitals in the fifth shell.

5.12   (a) 2p   (b) 4f   (c) 3d

5.13   (a) 3s orbital: n = 3, l = 0, m_l = 0
     (b) 2p orbital: n = 2, l = 1, m_l = −1, 0, +1
     (c) 4d orbital: n = 4, l = 2, m_l = −2, −1, 0, +1, +2

5.14   The g orbitals have four nodal planes.

5.15   The figure represents a d orbital, n = 4 and l = 2.

5.16 $m = 1, n = \infty$; $R = 1.097 \times 10^{-2}$ nm^{-1}

$$\frac{1}{\lambda} = R\left[\frac{1}{m^2} - \frac{1}{n^2}\right]; \quad \frac{1}{\lambda} = R\left[\frac{1}{1^2} - \frac{1}{\infty^2}\right]; \quad \frac{1}{\lambda} = R\left[\frac{1}{1}\right] = 1.097 \times 10^{-2} \text{ nm}^{-1}; \quad \lambda = 91.2 \text{ nm}$$

$$E = (6.626 \times 10^{-34} \text{ J} \cdot \text{s})\left(\frac{3.00 \times 10^8 \text{ m/s}}{91.2 \times 10^{-9} \text{ m}}\right)(6.022 \times 10^{23}/\text{mol})$$

$$E = 1.31 \times 10^6 \text{ J/mol} = 1.31 \times 10^3 \text{ kJ/mol}$$

5.17 (a) Ti, $1s^2 2s^2 2p^6 3s^2 3p^6 4s^2 3d^2$ or $[Ar]\, 4s^2 3d^2$
[Ar] $\underline{\uparrow\downarrow}$ $\underline{\uparrow}\ \underline{\uparrow}\ \underline{}\ \underline{}\ \underline{}$
$\quad$ 4s $\qquad\quad$ 3d

(b) Zn, $1s^2 2s^2 2p^6 3s^2 3p^6 4s^2 3d^{10}$ or $[Ar]\, 4s^2 3d^{10}$
[Ar] $\underline{\uparrow\downarrow}$ $\underline{\uparrow\downarrow}\ \underline{\uparrow\downarrow}\ \underline{\uparrow\downarrow}\ \underline{\uparrow\downarrow}\ \underline{\uparrow\downarrow}$
$\quad$ 4s $\qquad\quad$ 3d

(c) Sn, $1s^2 2s^2 2p^6 3s^2 3p^6 4s^2 3d^{10} 4p^6 5s^2 4d^{10} 5p^2$ or $[Kr]\, 5s^2 4d^{10} 5p^2$
[Kr] $\underline{\uparrow\downarrow}$ $\underline{\uparrow\downarrow}\ \underline{\uparrow\downarrow}\ \underline{\uparrow\downarrow}\ \underline{\uparrow\downarrow}\ \underline{\uparrow\downarrow}$ $\underline{\uparrow}\ \underline{\uparrow}\ \underline{}$
$\quad$ 5s $\qquad\quad$ 4d $\qquad\qquad$ 5p

(d) Pb, $[Xe]\, 6s^2 4f^{14} 5d^{10} 6p^2$

5.18 For Na$^+$, $1s^2 2s^2 2p^6$; for Cl$^-$, $1s^2 2s^2 2p^6 3s^2 3p^6$

5.19 The ground-state electron configuration contains 28 electrons. The atom is Ni.

5.20 Cr, Cu, Nb, Mo, Ru, Rh, Pd, Ag, La, Ce, Gd, Pt, Au, Ac, Th, Pa, U, Np, Cm

5.21 (a) Ba; atoms get larger as you go down a group.
(b) W; atoms get smaller as you go across a period.
(c) Sn; atoms get larger as you go down a group.
(d) Ce; atoms get smaller as you go across a period.

5.22 The aurora borealis begins on the surface of the sun with a massive solar flare. These flares eject a solar "gas" of energetic protons and electrons that reach earth after about 2 days and are then attracted toward the north and south magnetic poles. The energetic electrons are deflected by the earth's magnetic field into a series of sheetlike beams. The electrons then collide with O_2 and N_2 molecules in the upper atmosphere, exciting them, ionizing them, and breaking them apart into O and N atoms. The energetically excited atoms, ions, and molecules generated by collisions with electrons emit energy of characteristic wavelengths when they decay to their ground states. The $O_2{}^+$ ions emit a red light around 630 nm; $N_2{}^+$ ions emit violet and blue light at 391.4 nm and 470.0 nm; and O atoms emit a greenish-yellow light at 557.7 nm and a deep red light at 630.0 nm.

Understanding Key Concepts

5.24

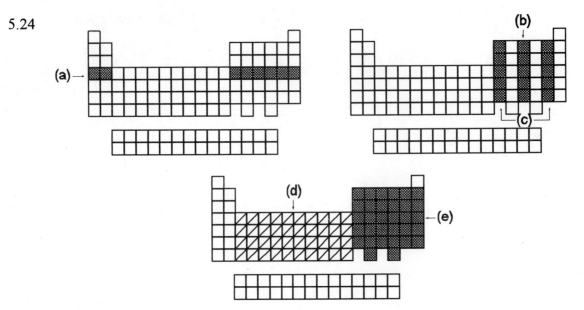

5.26 [Ar] $4s^2 3d^{10} 4p^1$ is Ga.

5.28 Ca and Br are in the same period, with Br to the far right of Ca. Ca is larger than Br. Sr is directly below Ca in the same group, and is larger than Ca. The result is
Sr (215 pm) > Ca (197 pm) > Br (114 pm)

Additional Problems
Electromagnetic Radiation

5.30 Violet has the higher frequency and energy. Red has the higher wavelength.

5.32 $\lambda = \dfrac{c}{\nu} = \dfrac{3.00 \times 10^8 \text{ m/s}}{5.5 \times 10^{15} \text{ s}^{-1}} = 5.5 \times 10^{-8} \text{ m}$

5.34 (a) $\nu = 99.5 \text{ MHz} = 99.5 \times 10^6 \text{ s}^{-1}$
$E = h\nu = (6.626 \times 10^{-34} \text{ J·s})(99.5 \times 10^6 \text{ s}^{-1})(6.022 \times 10^{23} \text{ /mol})$
$E = 3.97 \times 10^{-2} \text{ J/mol} = 3.97 \times 10^{-5} \text{ kJ/mol}$
$\nu = 1150 \text{ kHz} = 1150 \times 10^3 \text{ s}^{-1}$
$E = h\nu = (6.626 \times 10^{-34} \text{ J·s})(1150 \times 10^3 \text{ s}^{-1})(6.022 \times 10^{23} \text{ /mol})$
$E = 4.589 \times 10^{-4} \text{ J/mol} = 4.589 \times 10^{-7} \text{ kJ/mol}$
The FM radio wave (99.5 MHz) has the higher energy.

(b) $\lambda = 3.44 \times 10^{-9} \text{ m}$

$E = h\dfrac{c}{\lambda} = (6.626 \times 10^{-34} \text{ J·s})\left(\dfrac{3.00 \times 10^8 \text{ m/s}}{3.44 \times 10^{-9} \text{ m}} \right)(6.022 \times 10^{23} \text{ /mol})$

$E = 3.48 \times 10^7 \text{ J/mol} = 3.48 \times 10^4 \text{ kJ/mol}$

$$\lambda = 6.71 \times 10^{-2} \text{ m}$$

$$E = h\frac{c}{\lambda} = (6.626 \times 10^{-34} \text{ J·s}) \left(\frac{3.00 \times 10^8 \text{ m/s}}{6.71 \times 10^{-2} \text{ m}} \right) (6.022 \times 10^{23}/\text{mol})$$

$$E = 1.78 \text{ J/mol} = 1.78 \times 10^{-3} \text{ kJ/mol}$$

The X ray ($\lambda = 3.44 \times 10^{-9}$ m) has the higher energy.

5.36 (a) $E = 90.5 \text{ kJ/mol} \times \dfrac{1000 \text{ J}}{1 \text{ kJ}} \times \dfrac{1 \text{ mol}}{6.02 \times 10^{23}} = 1.50 \times 10^{-19} \text{ J}$

$$\nu = \frac{E}{h} = \frac{1.50 \times 10^{-19} \text{ J}}{6.626 \times 10^{-34} \text{ J·s}} = 2.27 \times 10^{14} \text{ s}^{-1}$$

$$\lambda = \frac{c}{\nu} = \frac{3.00 \times 10^8 \text{ m/s}}{2.27 \times 10^{14} \text{ s}^{-1}} = 1.32 \times 10^{-6} \text{ m} = 1320 \times 10^{-9} \text{ m} = 1320 \text{ nm, near IR}$$

(b) $E = 8.05 \times 10^{-4} \text{ kJ/mol} \times \dfrac{1000 \text{ J}}{1 \text{ kJ}} \times \dfrac{1 \text{ mol}}{6.02 \times 10^{23}} = 1.34 \times 10^{-24} \text{ J}$

$$\nu = \frac{E}{h} = \frac{1.34 \times 10^{-24} \text{ J}}{6.626 \times 10^{-34} \text{ J·s}} = 2.02 \times 10^9 \text{ s}^{-1}$$

$$\lambda = \frac{c}{\nu} = \frac{3.00 \times 10^8 \text{ m/s}}{2.02 \times 10^9 \text{ s}^{-1}} = 0.149 \text{ m, radio wave}$$

(c) $E = 1.83 \times 10^3 \text{ kJ/mol} \times \dfrac{1000 \text{ J}}{1 \text{ kJ}} \times \dfrac{1 \text{ mol}}{6.02 \times 10^{23}} = 3.04 \times 10^{-18} \text{ J}$

$$\nu = \frac{E}{h} = \frac{3.04 \times 10^{-18} \text{ J}}{6.626 \times 10^{-34} \text{ J·s}} = 4.59 \times 10^{15} \text{ s}^{-1}$$

$$\lambda = \frac{c}{\nu} = \frac{3.00 \times 10^8 \text{ m/s}}{4.59 \times 10^{15} \text{ s}^{-1}} = 6.54 \times 10^{-8} \text{ m} = 65.4 \times 10^{-9} \text{ m} = 65.4 \text{ nm, UV}$$

5.38 $\lambda = \dfrac{h}{mv} = \dfrac{6.626 \times 10^{-34} \text{ kg m}^2 \text{ s}^{-1}}{(9.11 \times 10^{-31} \text{ kg})(0.99 \times 3.00 \times 10^8 \text{ m/s})} = 2.45 \times 10^{-12} \text{ m, } \gamma \text{ ray}$

5.40 156 km/h = 156×10^3 m/3600 s = 43.3 m/s; 145 g = 0.145 kg

$$\lambda = \frac{h}{mv} = \frac{6.626 \times 10^{-34} \text{ kg m}^2 \text{ s}^{-1}}{(0.145 \text{ kg})(43.3 \text{ m/s})} = 1.06 \times 10^{-34} \text{ m}$$

5.42 145 g = 0.145 kg; 0.500 nm = 0.500×10^{-9} m

$$v = \frac{h}{m\lambda} = \frac{6.626 \times 10^{-34} \text{ kg m}^2 \text{ s}^{-1}}{(0.145 \text{ kg})(0.500 \times 10^{-9} \text{ m})} = 9.14 \times 10^{-24} \text{ m/s}$$

Atomic Spectra

5.44   For n = 3; $\lambda = 656.3$ nm = 656.3 x 10^{-9} m

$$E = h\frac{c}{\lambda} = (6.626 \times 10^{-34} \text{ J·s})\left(\frac{2.998 \times 10^8 \text{ m/s}}{656.3 \times 10^{-9} \text{ m}}\right)\left(\frac{1 \text{ kJ}}{1000 \text{ J}}\right)(6.022 \times 10^{23}/\text{mol})$$

E = 182.3 kJ/mol

For n = 4; $\lambda = 486.1$ nm = 486.1 x 10^{-9} m

$$E = h\frac{c}{\lambda} = (6.626 \times 10^{-34} \text{ J·s})\left(\frac{2.998 \times 10^8 \text{ m/s}}{486.1 \times 10^{-9} \text{ m}}\right)\left(\frac{1 \text{ kJ}}{1000 \text{ J}}\right)(6.022 \times 10^{23}/\text{mol})$$

E = 246.1 kJ/mol

For n = 5; $\lambda = 434.0$ nm = 434.0 x 10^{-9} m

$$E = h\frac{c}{\lambda} = (6.626 \times 10^{-34} \text{ J·s})\left(\frac{2.998 \times 10^8 \text{ m/s}}{434.0 \times 10^{-9} \text{ m}}\right)\left(\frac{1 \text{ kJ}}{1000 \text{ J}}\right)(6.022 \times 10^{23}/\text{mol})$$

E = 275.6 kJ/mol

5.46   From problem 5.45, for n = ∞, $\lambda = 364.6$ nm = 364.6 x 10^{-9} m

$$E = h\frac{c}{\lambda} = (6.626 \times 10^{-34} \text{ J·s})\left(\frac{2.998 \times 10^8 \text{ m/s}}{364.6 \times 10^{-9} \text{ m}}\right)\left(\frac{1 \text{ kJ}}{1000 \text{ J}}\right)(6.022 \times 10^{23}/\text{mol})$$

E = 328.1 kJ/mol

5.48 $\lambda = 330$ nm = 330 x 10^{-9} m

$$E = h\frac{c}{\lambda} = (6.626 \times 10^{-34} \text{ J·s})\left(\frac{3.00 \times 10^8 \text{ m/s}}{330 \times 10^{-9} \text{ m}}\right)\left(\frac{1 \text{ kJ}}{1000 \text{ J}}\right)(6.022 \times 10^{23}/\text{mol})$$

E = 363 kJ/mol

Orbitals and Quantum Numbers

5.50   n is the principal quantum number. The size and energy level of an orbital depends on n.
        l is the angular-momentum quantum number. l defines the three-dimensional shape of an orbital.
        m_l is the magnetic quantum number. m_l defines the spatial orientation of an orbital.
        m_s is the spin quantum number. m_s indicates the spin of the electron and can have either of two values, +½ or –½.

5.52   The probability of finding the electron drops off rapidly as distance from the nucleus increases, although it never drops to zero, even at large distances. As a result, there is no definite boundary or size for an orbital. However, we usually imagine the boundary surface of an orbital enclosing the volume where an electron spends 95% of its time.

5.54   Part of the electron-nucleus attraction is canceled by the electron-electron repulsion, an effect we describe by saying that the electrons are shielded from the nucleus by the other electrons. The net nuclear charge actually felt by an electron is called the effective nuclear charge, Z_{eff}, and is often substantially lower than the actual nuclear charge, Z_{actual}.

$Z_{eff} = Z_{actual} -$ electron shielding

5.56 (a) 4s $n = 4$; $l = 0$; $m_l = 0$; $m_s = \pm\frac{1}{2}$
(b) 3p $n = 3$; $l = 1$; $m_l = -1, 0, +1$; $m_s = \pm\frac{1}{2}$
(c) 5f $n = 5$; $l = 3$; $m_l = -3, -2, -1, 0, +1, +2, +3$; $m_s = \pm\frac{1}{2}$
(d) 5d $n = 5$; $l = 2$; $m_l = -2, -1, 0, +1, +2$; $m_s = \pm\frac{1}{2}$

5.58 (a) is not allowed because for $l = 0$, $m_l = 0$ only.
(b) is allowed.
(c) is not allowed because for $n = 4$, $l = 0, 1, 2,$ or 3 only.

5.60 For $n = 5$, the maximum number of electrons will occur when the 5g orbital is filled:
[Rn] $7s^2 5f^{14} 6d^{10} 7p^6 8s^2 5g^{18} = 138$ electrons

5.62 0.68 g $= 0.68 \times 10^{-3}$ kg

$$(\Delta x)(\Delta mv) \geq \frac{h}{4\pi}; \quad \Delta x \geq \frac{h}{4\pi(\Delta mv)} = \frac{6.626 \times 10^{-34} \text{ kg m}^2 \text{ s}^{-1}}{4\pi(0.68 \times 10^{-3} \text{ kg})(0.1 \text{ m/s})} = 8 \times 10^{-31} \text{ m}$$

Electron Configurations

5.64 The number of elements in successive periods of the periodic table increases by the progression 2, 8, 18, 32 because the principal quantum number n increases by 1 from one period to the next. As the principal quantum number increases, the number of orbitals in a shell increases. The progression of elements parallels the number of electrons in a particular shell.

5.66 (a) 5d   (b) 4s   (c) 6s

5.68 (a) 3d after 4s   (b) 4p after 3d   (c) 6d after 5f   (d) 6s after 5p

5.70 (a) Ti, Z = 22 $1s^2 2s^2 2p^6 3s^2 3p^6 4s^2 3d^2$
(b) Ru, Z = 44 $1s^2 2s^2 2p^6 3s^2 3p^6 4s^2 3d^{10} 4p^6 5s^2 4d^6$
(c) Sn, Z = 50 $1s^2 2s^2 2p^6 3s^2 3p^6 4s^2 3d^{10} 4p^6 5s^2 4d^{10} 5p^2$
(d) Sr, Z = 38 $1s^2 2s^2 2p^6 3s^2 3p^6 4s^2 3d^{10} 4p^6 5s^2$
(e) Se, Z = 34 $1s^2 2s^2 2p^6 3s^2 3p^6 4s^2 3d^{10} 4p^4$

5.72 (a) Rb, Z = 37   [Kr]   ↑
                                  5s

(b) W, Z = 74   [Xe]   ↑↓   ↑↓ ↑↓ ↑↓ ↑↓ ↑↓ ↑↓ ↑↓   ↑ ↑ ↑ ↑ _
                            6s   4f   5d

(c) Ge, Z = 32   [Ar] $\underset{4s}{\uparrow\downarrow}$ $\underset{3d}{\uparrow\downarrow\ \uparrow\downarrow\ \uparrow\downarrow\ \uparrow\downarrow\ \uparrow\downarrow}$ $\underset{4p}{\uparrow\ \ \uparrow\ \ \_}$

(d) Zr, Z = 40   [Kr] $\underset{5s}{\uparrow\downarrow}$ $\underset{4d}{\uparrow\ \ \uparrow\ \ \_\ \ \_\ \ \_}$

5.74   4s > 4d > 4f

5.76   Z = 116   [Rn] $7s^2 5f^{14} 6d^{10} 7p^4$

5.78   (a) O $1s^2 2s^2 2p^4$ $\underset{2p}{\uparrow\downarrow\ \uparrow\ \ \uparrow}$ 2 unpaired e⁻

(b) Si $1s^2 2s^2 2p^6 3s^2 3p^2$ $\underset{3p}{\uparrow\ \ \uparrow\ \ \_}$ 2 unpaired e⁻

(c) K   [Ar] $4s^1$ 1 unpaired e⁻

(d) As   [Ar] $4s^2 3d^{10} 4p^3$ $\underset{4p}{\uparrow\ \ \uparrow\ \ \uparrow}$ 3 unpaired e⁻

5.80   Order of orbital filling:
1s→2s→2p→3s→3p→4s→3d→4p→5s→4d→5p→6s→4f→5d→6p→7s→5f→6d→7p→8s→5g
Z = 121

Atomic Radii and Periodic Properties

5.82   Atomic radii increase down a group because the electron shells are farther away from the nucleus.

5.84   F < O < S

5.86   Mg has a higher ionization energy than Na because Mg has a higher Z_{eff} and a smaller size.

General Problems

5.88   Balmer series: m = 2; R = 1.097 x 10⁻² nm⁻¹

$$\frac{1}{\lambda} = R\left[\frac{1}{m^2} - \frac{1}{n^2}\right]; \quad \frac{1}{\lambda} = R\left[\frac{1}{2^2} - \frac{1}{6^2}\right] = 2.438 \times 10^{-3}\ nm^{-1}$$

λ = 410.2 nm = 410.2 x 10⁻⁹ m

$$E = h\frac{c}{\lambda} = (6.626 \times 10^{-34}\ J\cdot s)\left(\frac{2.998\times 10^8\ m/s}{410.2\times 10^{-9}\ m}\right)\left(\frac{1\ kJ}{1000\ J}\right)(6.022\times 10^{23}/mol)$$

E = 291.6 kJ/mol

5.90   Pfund series: m = 5, n = ∞; R = 1.097 x 10⁻² nm⁻¹

$$\frac{1}{\lambda} = R\left[\frac{1}{5^2} - \frac{1}{\infty^2}\right] = R\left[\frac{1}{25}\right] = 4.388 \times 10^{-4}\ nm^{-1}; \ \lambda = 2279\ nm$$

5.92 (a) $E = h\nu = (6.626 \times 10^{-34}\text{ J·s})(3.79 \times 10^{11}\text{ s}^{-1})\left(\dfrac{1\text{ kJ}}{1000\text{ J}}\right)(6.022 \times 10^{23}/\text{mol})$

$E = 0.151\text{ kJ/mol}$

(b) $E = h\nu = (6.626 \times 10^{-34}\text{ J·s})(5.45 \times 10^{4}\text{ s}^{-1})\left(\dfrac{1\text{ kJ}}{1000\text{ J}}\right)(6.022 \times 10^{23}/\text{mol})$

$E = 2.17 \times 10^{-8}\text{ kJ/mol}$

(c) $E = h\nu = (6.626 \times 10^{-34}\text{ J·s})\left(\dfrac{3.00 \times 10^{8}\text{ m/s}}{4.11 \times 10^{-5}\text{ m}}\right)\left(\dfrac{1\text{ kJ}}{1000\text{ J}}\right)(6.022 \times 10^{23}/\text{mol})$

$E = 2.91\text{ kJ/mol}$

5.94 (a) Ra   [Rn] $7s^2$ [Rn] ↑↓
7s

(b) Sc   [Ar] $4s^2 3d^1$ [Ar] ↑↓ | ↑ _ _ _ _
4s   3d

(c) Lr   [Rn] $7s^2 5f^{14} 6d^1$ [Rn] ↑↓ | ↑↓ ↑↓ ↑↓ ↑↓ ↑↓ ↑↓ ↑↓ | ↑ _ _ _ _
7s   5f   6d

(d) B   [He] $2s^2 2p^1$ [He] ↑↓ | ↑ _ _
2s   2p

(e) Te   [Kr] $5s^2 4d^{10} 5p^4$ [Kr] ↑↓ | ↑↓ ↑↓ ↑↓ ↑↓ ↑↓ | ↑↓ ↑ ↑
5s   4d   5p

5.96 $206.5\text{ kJ} = 206.5 \times 10^3\text{ J};$ $E = \dfrac{206.5 \times 10^3\text{ J}}{1\text{ mol}} \times \dfrac{1\text{ mol}}{6.022 \times 10^{23}} = 3.429 \times 10^{-19}\text{ J}$

$E = h\dfrac{c}{\lambda}, \quad \lambda = \dfrac{hc}{E} = \dfrac{(6.626 \times 10^{-34}\text{ J·s})(3.00 \times 10^{8}\text{ m/s})}{3.429 \times 10^{-19}\text{ J}} = 5.797 \times 10^{-7}\text{ m} = 580.\text{ nm}$

5.98 (a) Sr, Z = 38   [Kr] ↑↓
5s

(b) Cd, Z = 48   [Kr] ↑↓ | ↑↓ ↑↓ ↑↓ ↑↓ ↑↓
5s   4d

(c) Z = 22, Ti   [Ar] ↑↓ | ↑ ↑ _ _ _
4s   3d

(d) Z = 34, Se   [Ar] ↑↓ | ↑↓ ↑↓ ↑↓ ↑↓ ↑↓ | ↑↓ ↑ ↑
4s   3d   4p

5.100 For K, $Z_{eff} = \sqrt{\dfrac{(418.8\text{ kJ/mol})(4^2)}{1312\text{ kJ/mol}}} = 2.26$

For Kr, $Z_{eff} = \sqrt{\dfrac{(1350.7\text{ kJ/mol})(4^2)}{1312\text{ kJ/mol}}} = 4.06$

5.102 $q = (350 \text{ g})(4.184 \text{ J/g·°C})(95°C - 20°C) = 109,830 \text{ J}$

$\lambda = 15.0 \text{ cm} = 15.0 \times 10^{-2} \text{ m}$

$E = (6.626 \times 10^{-34} \text{ J·s})\left(\dfrac{3.00 \times 10^8 \text{ m/s}}{15.0 \times 10^{-2} \text{ m}}\right) = 1.33 \times 10^{-24} \text{ J/photon}$

$\text{number of photons} = \dfrac{109,830 \text{ J}}{1.33 \times 10^{-24} \text{ J/photon}} = 8.3 \times 10^{28} \text{ photons}$

5.104 $48.2 \text{ nm} = 48.2 \times 10^{-9} \text{ m}$

$E(\text{photon}) = 6.626 \times 10^{-34} \text{ J·s} \times \dfrac{3.00 \times 10^8 \text{ m/s}}{48.2 \times 10^{-9} \text{ m}} \times \dfrac{1 \text{ kJ}}{1000 \text{ J}} \times \dfrac{6.022 \times 10^{23}}{\text{mol}} = 2.48 \times 10^3 \text{ kJ/mol}$

$E_K = E(\text{electron}) = \tfrac{1}{2}(9.109 \times 10^{-31} \text{ kg})(2.371 \times 10^6 \text{ m/s})^2 \left(\dfrac{1 \text{ kJ}}{1000 \text{ J}}\right)\left(\dfrac{6.022 \times 10^{23}}{\text{mol}}\right)$

$E_K = 1.54 \times 10^3 \text{ kJ/mol}$

$E(\text{photon}) = E_i + E_K; \quad E_i = E(\text{photon}) - E_K = (2.48 \times 10^3) - (1.54 \times 10^3) = 940 \text{ kJ/mol}$

5.106 Substitute the equation for the orbit radius, r, into the equation for the energy level, E, to
get $E = \dfrac{-Ze^2}{2\left(\dfrac{n^2 a_o}{Z}\right)} = \dfrac{-Z^2 e^2}{2a_o n^2}$

Let E_1 be the energy of an electron in a lower orbit and E_2 the energy of an electron in a
higher orbit. The difference between the two energy levels is

$\Delta E = E_2 - E_1 = \dfrac{-Z^2 e^2}{2a_o n_2^2} - \dfrac{-Z^2 e^2}{2a_o n_1^2} = \dfrac{-Z^2 e^2}{2a_o n_2^2} + \dfrac{Z^2 e^2}{2a_o n_1^2} = \dfrac{Z^2 e^2}{2a_o n_1^2} - \dfrac{Z^2 e^2}{2a_o n_2^2}$

$\Delta E = \dfrac{Z^2 e^2}{2a_o}\left[\dfrac{1}{n_1^2} - \dfrac{1}{n_2^2}\right]$

Because Z, e, and a_o are constants, this equation shows that ΔE is proportional to
$\left[\dfrac{1}{n_1^2} - \dfrac{1}{n_2^2}\right]$ where n_1 and n_2 are integers with $n_2 > n_1$. This is similar to the Balmer-
Rydberg equation where $1/\lambda$ or ν for the emission spectra of atoms is proportional to
$\left[\dfrac{1}{m^2} - \dfrac{1}{n^2}\right]$ where m and n are integers with $n > m$.

5.108 (a) 3d, $n = 3$, $l = 2$

(b) 2p, $n = 2$, $l = 1$, $m_l = -1, 0, +1$

3p, $n = 3$, $l = 1$, $m_l = -1, 0, +1$

3d, $n = 3$, $l = 2$, $m_l = -2, -1, 0, +1, +2$

(c) N, $1s^2 2s^2 2p^3$ so the 3s, 3p, and 3d orbitals are empty.
(d) C, $1s^2 2s^2 2p^2$ so the 1s and 2s orbitals are filled.
(e) Be, $1s^2 2s^2$ so the 2s orbital contains the outermost electrons.
(f) 2p and 3p (↑ ↑ __) and 3d (↑ ↑ __ __ __).

5.110 (a) $E = h\nu$

$$\nu = \frac{E}{h} = \frac{7.21 \times 10^{-19}\,J}{6.626 \times 10^{-34}\,J\cdot s} = 1.09 \times 10^{15}\,s^{-1}$$

(b) $E(photon) = E_i + E_K$; from (a), $E_i = 7.21 \times 10^{-19}\,J$

$$E(photon) = h\frac{c}{\lambda} = (6.626 \times 10^{-34}\,J\cdot s)\left(\frac{3.00 \times 10^8\,m/s}{2.50 \times 10^{-7}\,m}\right) = 7.95 \times 10^{-19}\,J$$

$E_K = E(photon) - E_i = (7.95 \times 10^{-19}\,J) - (7.21 \times 10^{-19}\,J) = 7.4 \times 10^{-20}\,J$
Calculate the electron velocity from the kinetic energy, E_K.
$E_K = 7.4 \times 10^{-20}\,J = 7.4 \times 10^{-20}\,kg\cdot m^2/s^2 = \frac{1}{2}mv^2 = \frac{1}{2}(9.109 \times 10^{-31}\,kg)v^2$

$$v = \sqrt{\frac{2 \times (7.4 \times 10^{-20}\,kg\cdot m^2/s^2)}{9.109 \times 10^{-31}\,kg}} = 4.0 \times 10^5\,m/s$$

$$\text{deBroglie wavelength} = \frac{h}{mv} = \frac{6.626 \times 10^{-34}\,kg\cdot m^2/s}{(9.109 \times 10^{-31}\,kg)(4.0 \times 10^5\,m/s)} = 1.8 \times 10^{-9}\,m = 1.8\,nm$$

Multi-Concept Problem

5.112 (a) 5f subshell: $n = 5,\ l = 3,\ m_l = -3, -2, -1, 0, +1, +2, +3$
       3d subshell: $n = 3,\ l = 2,\ m_l = -2, -1, 0, +1, +2$
(b) In the H atom the subshells in a particular energy level are all degenerate, i.e., all have the same energy. Therefore, you only need to consider the principal quantum number, n, to calculate the wavelength emitted for an electron that drops from the 5f to the 3d subshell.
$m = 3,\ n = 5;\ R = 1.097 \times 10^{-2}\,nm^{-1}$

$$\frac{1}{\lambda} = R\left[\frac{1}{m^2} - \frac{1}{n^2}\right];\quad \frac{1}{\lambda} = R\left[\frac{1}{3^2} - \frac{1}{5^2}\right];\quad \frac{1}{\lambda} = 7.801 \times 10^{-4}\,nm^{-1};\quad \lambda = 1282\,nm$$

(c) $m = 3,\ n = \infty;\ R = 1.097 \times 10^{-2}\,nm^{-1}$

$$\frac{1}{\lambda} = R\left[\frac{1}{m^2} - \frac{1}{n^2}\right];\quad \frac{1}{\lambda} = R\left[\frac{1}{3^2} - \frac{1}{\infty^2}\right];\quad \frac{1}{\lambda} = R\left[\frac{1}{3^2}\right] = 1.219 \times 10^{-3}\,nm^{-1};\quad \lambda = 820.3\,nm$$

$$E = (6.626 \times 10^{-34}\,J\cdot s)\left(\frac{3.00 \times 10^8\,m/s}{820.3 \times 10^{-9}\,m}\right)(6.022 \times 10^{23}/mol)$$

$E = 1.46 \times 10^5\,J/mol = 146\,kJ/mol$

5.114 (a) Cl_2, 70.91 amu

$$M + Cl_2 \rightarrow MCl_2$$

$$\text{mol } Cl_2 = 0.8092 \text{ g } Cl_2 \times \frac{1 \text{ mol } Cl_2}{70.91 \text{ g } Cl_2} = 0.01141 \text{ mol } Cl_2$$

$$\text{mol } M = 0.01141 \text{ mol } Cl_2 \times \frac{1 \text{ mol } M}{1 \text{ mol } Cl_2} = 0.01141 \text{ mol } M$$

$$\text{molar mass of } M = \frac{1.000 \text{ g}}{0.01141 \text{ mol}} = 87.64 \text{ g/mol}$$

atomic mass of M = 87.64 amu;   M = Sr

(b) $q = \dfrac{9.46 \text{ kJ}}{0.01141 \text{ mol}} = 829 \text{ kJ/mol}$

6 Ionic Bonds and Some Main-Group Chemistry

6.1 (a) Ra^{2+} [Rn]   (b) La^{3+} [Xe]   (c) Ti^{4+} [Ar]   (d) N^{3-} [Ne]
Each ion has the ground-state electron configuration of the noble gas closest to it in the periodic table.

6.2 The neutral atom contains 30 e^- and is Zn. The ion is Zn^{2+}.

6.3 (a) O^{2-}; decrease in effective nuclear charge and an increase in electron-electron repulsions lead to the larger anion.
(b) S; atoms get larger as you go down a group.
(c) Fe; in Fe^{3+} electrons are removed from a larger valence shell and there is an increase in effective nuclear charge leading to the smaller cation.
(d) H^-; decrease in effective nuclear charge and an increase in electron-electron repulsions lead to the larger anion.

6.4 K^+ is smaller than neutral K because the ion has one less electron. K^+ and Cl^- are isoelectronic, but K^+ is smaller than Cl^- because of its higher effective nuclear charge. K is larger than Cl^- because K has one additional electron and that electron begins the next shell (period). K^+, r = 133 pm; Cl^-, r = 184 pm; K, r = 227 pm

6.5 (a) Br   (b) S   (c) Se   (d) Ne

6.6 (a) Be $1s^2 2s^2$ N $1s^2 2s^2 2p^3$
Be would have the larger third ionization energy because this electron would come from the 1s orbital.
(b) Ga   [Ar] $4s^2 3d^{10} 4p^1$ Ge   [Ar] $4s^2 3d^{10} 4p^2$
Ga would have the larger fourth ionization energy because this electron would come from the 3d orbitals.

6.7 (b) Cl has the highest E_{i1} and smallest E_{i4}.

6.8 Ca (red) would have the largest third ionization energy of the three because the electron being removed is from a filled valence shell. For Al (green) and Kr (blue), the electron being removed is from a partially filled valence shell. The third ionization energy for Kr would be larger than that for Al because the electron being removed from Kr is coming out of a set of filled 4p orbitals while the electron being removed from Al is coming out of a half-filled 3s orbital. In addition, Z_{eff} is larger for Kr than for Al. The ease of losing its third electron is Al < Kr < Ca.

75

6.9   Cr   [Ar] $4s^1 3d^5$ Mn   [Ar] $4s^2 3d^5$ Fe   [Ar] $4s^2 3d^6$
       Cr can accept an electron into a 4s orbital. The 4s orbital is lower in energy than a 3d
       orbital. Both Mn and Fe accept the added electron into a 3d orbital that contains an
       electron, but Mn has a lower value of Z_{eff}. Therefore, Mn has a less negative E_{ea} than
       either Cr or Fe.

6.10   The least favorable E_{ea} is for Kr (red) because it is a noble gas with filled set of 4p orbitals.
       The most favorable E_{ea} is for Ge (blue) because the 4p orbitals would become half filled.
       In addition, Z_{eff} is larger for Ge than it is for K (green).

6.11   (a) KCl has the higher lattice energy because of the smaller K^+.
       (b) CaF_2 has the higher lattice energy because of the smaller Ca^{2+}.
       (c) CaO has the higher lattice energy because of the higher charge on both the cation and
       anion.

6.12 $K(s) \rightarrow K(g)$ +89.2 kJ/mol
       $K(g) \rightarrow K^+(g) + e^-$ +418.8 kJ/mol
       $\frac{1}{2} [F_2(g) \rightarrow 2 F(g)]$ +79   kJ/mol
       $F(g) + e^- \rightarrow F^-(g)$ −328   kJ/mol
       $K^+(g) + F^-(g) \rightarrow KF(s)$ −821   kJ/mol
                                   Sum = −562   kJ/mol for $K(s) + \frac{1}{2} F_2(g) \rightarrow KF(s)$

6.13   The anions are larger than the cations. Cl^- is larger than O^{2-} because it is below it in the
       periodic table. Therefore, (a) is NaCl and (b) is MgO. Because of the higher ion charge
       and shorter cation – anion distance, MgO has the larger lattice energy.

6.14   (a) Li_2O,   O −2   (b) K_2O_2,   O −1   (c) CsO_2,   O −½

6.15   (a) $2 Cs(s) + 2 H_2O(l) \rightarrow 2 Cs^+(aq) + 2 OH^-(aq) + H_2(g)$
       (b) $Na(s) + N_2(g) \rightarrow$ N. R.
       (c) $Rb(s) + O_2(g) \rightarrow RbO_2(s)$
       (d) $2 K(s) + 2 NH_3(g) \rightarrow 2 KNH_2(s) + H_2(g)$
       (e) $2 Rb(s) + H_2(g) \rightarrow 2 RbH(s)$

6.16   (a) $Be(s) + Br_2(l) \rightarrow BeBr_2(s)$
       (b) $Sr(s) + 2 H_2O(l) \rightarrow Sr(OH)_2(aq) + H_2(g)$
       (c) $2 Mg(s) + O_2(g) \rightarrow 2 MgO(s)$

6.17 $BeCl_2(s) + 2 K(s) \rightarrow Be(s) + 2 KCl(s)$

6.18 $Mg(s) + S(s) \rightarrow MgS(s)$;   In MgS, the oxidation number of S is −2.

6.19 $2 Al(s) + 6 H^+(aq) \rightarrow 2 Al^{3+}(aq) + 3 H_2(g)$
       H^+ gains electrons and is the oxidizing agent. Al loses electrons and is the reducing agent.

6.20 $2 Al(s) + 3 S(s) \rightarrow Al_2S_3(s)$

6.21   (a) $Br_2(l) + Cl_2(g) \rightarrow 2\,BrCl(g)$
       (b) $2\,Al(s) + 3\,F_2(g) \rightarrow 2\,AlF_3(s)$
       (c) $H_2(g) + I_2(s) \rightarrow 2\,HI(g)$

6.22 $Br_2(l) + 2\,NaI(s) \rightarrow 2\,NaBr(s) + I_2(s)$
       Br_2 gains electrons and is the oxidizing agent. I^- (from NaI) loses electrons and is the reducing agent.

6.23   (a) XeF_2 F –1, Xe +2
       (b) XeF_4 F –1, Xe +4
       (c) $XeOF_4$ F –1, O –2, Xe +6

6.24   (a) Rb would lose one electron and adopt the Kr noble-gas configuration.
       (b) Ba would lose two electrons and adopt the Xe noble-gas configuration.
       (c) Ga would lose three electrons and adopt an Ar-like noble-gas configuration (note that Ga^{3+} has ten 3d electrons in addition to the two 3s and six 3p electrons).
       (d) F would gain one electron and adopt the Ne noble-gas configuration.

6.25   Group 6A elements will gain 2 electrons.

6.26   Only about 10% of current world salt production comes from evaporation of seawater. Most salt is obtained by mining the vast deposits of halite, or rock salt, formed by evaporation of ancient inland seas. These salt beds can be up to hundreds of meters thick and may occur anywhere from a few meters to thousands of meters below the earth's surface.

Understanding Key Concepts

6.28   (a) shows an extended array, which represents an ionic compound.
       (b) shows discrete units, which represent a covalent compound.

6.30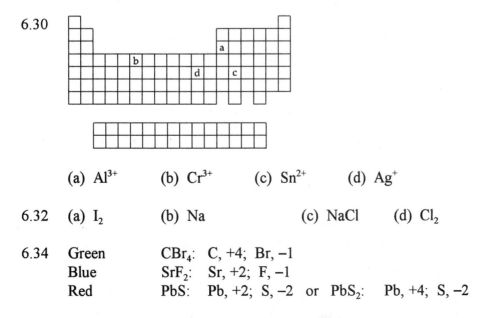

       (a) Al^{3+} (b) Cr^{3+} (c) Sn^{2+} (d) Ag^+

6.32   (a) I_2 (b) Na   (c) NaCl   (d) Cl_2

6.34   Green CBr_4: C, +4; Br, –1
       Blue SrF_2: Sr, +2; F, –1
       Red   PbS: Pb, +2; S, –2 or PbS_2:   Pb, +4; S, –2

Additional Problems
Ionization Energy and Electron Affinity

6.36   (a) La^{3+},   [Xe]   (b) Ag^+,   [Kr] $4d^{10}$ (c) Sn^{2+},   [Kr] $5s^2 4d^{10}$

6.38 Cr^{2+} [Ar] $3d^4$ $\underline{\uparrow}\ \underline{\uparrow}\ \underline{\uparrow}\ \underline{\uparrow}\ \underline{\ }$
                                          3d

        Fe^{2+} [Ar] $3d^6$ $\underline{\uparrow\downarrow}\ \underline{\uparrow}\ \underline{\uparrow}\ \underline{\uparrow}\ \underline{\uparrow}$
                                          3d

6.40   Ionization energies have a positive sign because energy is required to remove an electron from an atom of any element.

6.42   The largest E_{i1} are found in Group 8A because of the largest values of Z_{eff}.
       The smallest E_{i1} are found in Group 1A because of the smallest values of Z_{eff}.

6.44   (a) K   [Ar] $4s^1$ Ca   [Ar] $4s^2$
       Ca has the smaller second ionization energy because it is easier to remove the second 4s valence electron in Ca than it is to remove the second electron in K from the filled 3p orbitals.
       (b) Ca   [Ar] $4s^2$ Ga   [Ar] $4s^2 3d^{10} 4p^1$
       Ca has the larger third ionization energy because it is more difficult to remove the third electron in Ca from the filled 3p orbitals than it is to remove the third electron (second 4s valence electron) from Ga.

6.46   (a) $1s^2 2s^2 2p^6 3s^2 3p^3$ is P   (b) $1s^2 2s^2 2p^6 3s^2 3p^6$ is Ar   (c) $1s^2 2s^2 2p^6 3s^2 3p^6 4s^2$ is Ca
       Ar has the highest E_{i2}. Ar has a higher Z_{eff} than P. The 4s electrons in Ca are easier to remove than any 3p electrons.
       Ar has the lowest E_{i7}. It is difficult to remove 3p electrons from Ca, and it is difficult to remove 2p electrons from P.

6.48   Using Figure 6.3 as a reference:

	Lowest E_{i1}	Highest E_{i1}
(a)	K	Li
(b)	B	Cl
(c)	Ca	Cl

6.50   The relationship between the electron affinity of a univalent cation and the ionization energy of the neutral atom is that they have the same magnitude but opposite sign.

6.52 Na^+ has a more negative electron affinity than either Na or Cl because of its positive charge.

6.54   Energy is usually released when an electron is added to a neutral atom but absorbed when an electron is removed from a neutral atom because of the positive Z_{eff}.

6.56 (a) F; nonmetals have more negative electron affinities than metals.

(b) Na; Ne (noble gas) has a positive electron affinity.

(c) Br; nonmetals have more negative electron affinities than metals.

Lattice Energy and Ionic Bonds

6.58 $MgCl_2$ > $LiCl$ > KCl > KBr

6.60 $Li \rightarrow Li^+ + e^-$ +520 kJ/mol

     $Br + e^- \rightarrow Br^-$ $\underline{-325 \text{ kJ/mol}}$

                         +195 kJ/mol

6.62

$Li(s) \rightarrow Li(g)$	+159.4 kJ/mol
$Li(s) \rightarrow Li(g) + e^-$	+520 kJ/mol
$\frac{1}{2}[Br_2(l) \rightarrow Br_2(g)]$	+15.4 kJ/mol
$\frac{1}{2}[Br_2(g) \rightarrow 2\,Br(g)]$	+112 kJ/mol
$Br(g) + e^- \rightarrow Br^-(g)$	−325 kJ/mol
$Li^+(g) + Br^-(g) \rightarrow LiBr(s)$	$\underline{-807 \text{ kJ/mol}}$

Sum = −325 kJ/mol for $Li(s) + \frac{1}{2}Br_2(l) \rightarrow LiBr(s)$

6.64

$Na(s) \rightarrow Na(g)$	+107.3 kJ/mol
$Na(g) \rightarrow Na^+(g) + e^-$	+495.8 kJ/mol
$\frac{1}{2}[H_2(g) \rightarrow 2\,H(g)]$	$\frac{1}{2}$(+435.9) kJ/mol
$H(g) + e^- \rightarrow H^-(g)$	−72.8 kJ/mol
$Na^+(g) + H^-(g) \rightarrow NaH(s)$	$\underline{-U}$

Sum = −60 kJ/mol for $Na(s) + \frac{1}{2}H_2 \rightarrow NaH(s)$

$-U = -60 - 107.3 - 495.8 - \dfrac{435.9}{2} + 72.8 = -808$ kJ/mol;   U = 808 kJ/mol

6.66

$Cs(s) \rightarrow Cs(g)$	+76.1 kJ/mol
$Cs(g) \rightarrow Cs^+(g) + e^-$	+375.7 kJ/mol
$\frac{1}{2}[F_2(g) \rightarrow 2\,F(g)]$	+79 kJ/mol
$F(g) + e^- \rightarrow F^-(g)$	−328 kJ/mol
$Cs^+(g) + F^-(g) \rightarrow CsF(s)$	$\underline{-740 \text{ kJ/mol}}$

Sum = −537 kJ/mol for $Cs(s) + \frac{1}{2}F_2(g) \rightarrow CsF(s)$

6.68

$Ca(s) \rightarrow Ca(g)$	+178.2 kJ/mol
$Ca(g) \rightarrow Ca^+(g) + e^-$	+589.8 kJ/mol
$\frac{1}{2}[Cl_2(g) \rightarrow 2\,Cl(g)]$	+121.5 kJ/mol
$Cl(g) + e^- \rightarrow Cl^-(g)$	−348.6 kJ/mol
$Ca^+(g) + Cl^-(g) \rightarrow CaCl(s)$	$\underline{-717 \text{ kJ/mol}}$

Sum = −176 kJ/mol for $Ca(s) + \frac{1}{2}Cl_2(g) \rightarrow CaCl(s)$

6.70

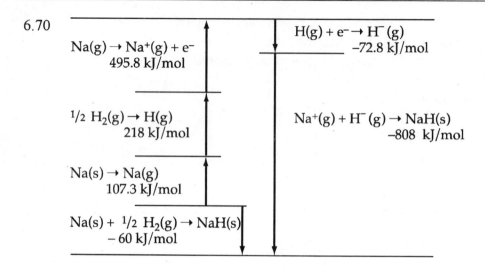

$Na(g) \rightarrow Na^+(g) + e^-$
495.8 kJ/mol

$H(g) + e^- \rightarrow H^-(g)$
–72.8 kJ/mol

$\frac{1}{2} H_2(g) \rightarrow H(g)$
218 kJ/mol

$Na^+(g) + H^-(g) \rightarrow NaH(s)$
–808 kJ/mol

$Na(s) \rightarrow Na(g)$
107.3 kJ/mol

$Na(s) + \frac{1}{2} H_2(g) \rightarrow NaH(s)$
– 60 kJ/mol

Main-Group Chemistry

6.72 Solids: I_2;   Liquids: Br_2;   Gases: F_2, Cl_2, He, Ne, Ar, Kr, Xe

6.74 (a) At is in Group 7A. The trend going down the group is gas $\rightarrow$ liquid $\rightarrow$ solid. At, being at the bottom of the group, should be a solid.
(b) At would likely be dark, like I_2, maybe with a metallic sheen.
(c) At is likely to react with·Na just like the other halogens, yielding NaAt.

6.76 (a) $2\,NaCl \xrightarrow[\substack{580°C}]{\substack{electrolysis \\ in\ CaCl_2}} 2\,Na(l) + Cl_2(g)$

(b) $2\,Al_2O_3 \xrightarrow[\substack{980°C}]{\substack{electrolysis \\ in\ Na_3AlF_6}} 4\,Al(l) + 3\,O_2(g)$

(c) Ar is obtained from the distillation of liquid air.
(d) $2\,Br^-(aq) + Cl_2(g) \rightarrow Br_2(l) + 2\,Cl^-(aq)$

6.78 Main-group elements tend to undergo reactions that leave them with eight valence electrons. That is, main-group elements react so that they attain a noble-gas electron configuration with filled s and p sublevels in their valence electron shell.
The octet rule works for valence-shell electrons because taking electrons away from a filled octet is difficult because they are tightly held by a high Z_{eff}; adding more electrons to a filled octet is difficult because, with s and p sublevels full, there is no low-energy orbital available.

6.80 (a) $2\,K(s) + H_2(g) \rightarrow 2\,KH(s)$
(b) $2\,K(s) + 2\,H_2O(l) \rightarrow 2\,K^+(aq) + 2\,OH^-(aq) + H_2(g)$
(c) $2\,K(s) + 2\,NH_3(g) \rightarrow 2\,KNH_2(s) + H_2(g)$

(d) $2 K(s) + Br_2(l) \rightarrow 2 KBr(s)$
(e) $K(s) + N_2(g) \rightarrow$ N. R.
(f) $K(s) + O_2(g) \rightarrow KO_2(s)$

6.82 (a) $Cl_2(g) + H_2(g) \rightarrow 2 HCl(g)$
(b) $Cl_2(g) + Ar(g) \rightarrow$ N. R.
(c) $Cl_2(g) + Br_2(l) \rightarrow 2 BrCl(g)$
(d) $Cl_2(g) + N_2(g) \rightarrow$ N. R.

6.84 $AlCl_3 + 3 Na \rightarrow Al + 3 NaCl$
Al^{3+} (from $AlCl_3$) gains electrons and is reduced. Na loses electrons and is oxidized.

6.86 $CaIO_3$, 215.0 amu; 1.00 kg = 1000 g
$$\% I = \frac{126.9 \text{ g}}{215.0 \text{ g}} \times 100\% = 59.02\%; \quad (0.5902)(1000 \text{ g}) = 590 \text{ g } I_2$$

6.88 $Ca(s) + H_2(g) \rightarrow CaH_2(s)$; $\quad H_2$, 2.016 amu; CaH_2, 42.09 amu
$$5.65 \text{ g Ca} \times \frac{1 \text{ mol Ca}}{40.08 \text{ g Ca}} = 0.141 \text{ mol Ca}$$

$$3.15 \text{ L } H_2 \times 0.0893 \frac{\text{g } H_2}{1 \text{ L}} \times \frac{1 \text{ mol } H_2}{2.016 \text{ g } H_2} = 0.140 \text{ mol } H_2$$

Because the reaction stoichiometry between Ca and H_2 is one to one, H_2 is the limiting reactant.
$$0.140 \text{ mol } H_2 \times \frac{1 \text{ mol } CaH_2}{1 \text{ mol } H_2} \times \frac{42.09 \text{ g } CaH_2}{1 \text{ mol } CaH_2} \times 0.943 = 5.56 \text{ g } CaH_2$$

6.90 (a) $Mg(s) + 2 H^+(aq) \rightarrow Mg^{2+}(aq) + H_2(g)$
H^+ gains electrons and is the oxidizing agent. Mg loses electrons and is the reducing agent.
(b) $Kr(g) + F_2(g) \rightarrow KrF_2(s)$
F_2 gains electrons and is the oxidizing agent. Kr loses electrons and is the reducing agent.
(c) $I_2(s) + 3 Cl_2(g) \rightarrow 2 ICl_3(l)$
Cl_2 gains electrons and is the oxidizing agent. I_2 loses electrons and is the reducing agent.

General Problems

6.92 Cu^{2+} has fewer electrons and a larger effective nuclear charge; therefore it has the smaller ionic radius.

6.94
$Mg(s) \rightarrow Mg(g)$	+147.7 kJ/mol
$Mg(g) \rightarrow Mg^+(g) + e^-$	+737.7 kJ/mol
$\frac{1}{2} [F_2(g) \rightarrow 2 F(g)]$	+79 kJ/mol
$F(g) + e^- \rightarrow F^-(g)$	−328 kJ/mol
$Mg^+(g) + F^-(g) \rightarrow MgF(s)$	<u>−930</u> kJ/mol

Sum = −294 kJ/mol for $Mg(s) + \frac{1}{2} F_2(g) \rightarrow MgF(s)$

$Mg(s) \rightarrow Mg(g)$ +147.7   kJ/mol

$Mg(g) \rightarrow Mg^+(g) + e^-$ +737.7   kJ/mol

$Mg^+(g) \rightarrow Mg^{2+}(g) + e^-$ +1450.7   kJ/mol

$F_2(g) \rightarrow 2 F(g)$ +158   kJ/mol

$2[F(g) + e^- \rightarrow F^-(g)]$ 2(−328)   kJ/mol

$Mg^{2+}(g) + 2 F^-(g) \rightarrow MgF_2(s)$ −2952   kJ/mol

                Sum = −1114   kJ/mol for $Mg(s) + F_2(g) \rightarrow MgF_2(s)$

6.96   (a) Na is used in table salt (NaCl), glass, rubber, and pharmaceutical agents.

         (b) Mg is used as a structural material when alloyed with Al.

         (c) F_2 is used in the manufacture of Teflon, $(C_2F_4)_n$, and in toothpaste as SnF_2.

6.98   (a) $2 Li(s) + H_2(g) \rightarrow 2 LiH(s)$

         (b) $2 Li(s) + 2 H_2O(l) \rightarrow 2 Li^+(aq) + 2 OH^-(aq) + H_2(g)$

         (c) $2 Li(s) + 2 NH_3(g) \rightarrow 2 LiNH_2(s) + H_2(g)$

         (d) $2 Li(s) + Br_2(l) \rightarrow 2 LiBr(s)$

         (e) $6 Li(s) + N_2(g) \rightarrow 2 Li_3N(s)$

         (f) $4 Li(s) + O_2(g) \rightarrow 2 Li_2O(s)$

6.100   When moving diagonally down and right on the periodic table, the increase in atomic radius caused by going to a larger shell is offset by a decrease caused by a higher Z_{eff}. Thus, there is little net change.

6.102

$Cl(g) \rightarrow Cl^+(g) + e^-$
1251 kJ/mol

$Na(g) + e^- \rightarrow Na^-(g)$
−52.9 kJ/mol

$Na^-(g) + Cl^+(g) \rightarrow ClNa(s)$
−787 kJ/mol

$\frac{1}{2} Cl_2(g) \rightarrow Cl(g)$
122 kJ/mol

$Na(s) + \frac{1}{2} Cl_2(g) \rightarrow ClNa(s)$
640 kJ/mol

$Na(s) \rightarrow Na(g)$
107.3 kJ/mol

6.104 $Mg(s) \rightarrow Mg(g)$ +147.7 kJ/mol

         $Mg(g) \rightarrow Mg^+(g) + e^-$ +738   kJ/mol

         $Mg^+(g) \rightarrow Mg^{2+}(g) + e^-$ +1451   kJ/mol

         $\frac{1}{2}[O_2(g) \rightarrow 2 O(g)]$ +249.2 kJ/mol

         $O(g) + e^- \rightarrow O^-(g)$ −141.0 kJ/mol

         $O^-(g) + e^- \rightarrow O^{2-}(g)$ E_{ea2}

         $Mg^{2+}(g) + O^{2-}(g) \rightarrow MgO(s)$ −3791   kJ/mol

         $Mg(s) + \frac{1}{2}O_2(g) \rightarrow MgO(s)$ −601.7 kJ/mol

         $147.7 + 738 + 1451 + 249.2 − 141.0 + E_{ea2} − 3791 = −601.7$

$E_{ea2} = -147.7 - 738 - 1451 - 249.2 + 141.0 + 3791 - 601.7 = +744$ kJ/mol

Because E_{ea2} is positive, O^{2-} is not stable in the gas phase. It is stable in MgO because of the large lattice energy that results from the +2 and –2 charge of the ions and their small size.

6.106 (a) The more negative the E_{ea}, the greater the tendency of the atom to accept an electron, and the more stable the anion that results. Be, N, O, and F are all second row elements. F has the most negative E_{ea} of the group because the anion that forms, F^-, has a complete octet of electrons and its nucleus has the highest effective nuclear charge.
(b) Se^{2-} and Rb^+ are below O^{2-} and F^- in the periodic table and are the larger of the four. Se^{2-} and Rb^+ are isoelectronic, but Rb^+ has the higher effective nuclear charge so it is smaller. Therefore Se^{2-} is the largest of the four ions.

6.108

$Cr(s) \rightarrow Cr(g)$	+397 kJ/mol
$Cr(g) \rightarrow Cr^+(g)$	+652 kJ/mol
$Cr^+(g) \rightarrow Cr^{2+}(g)$	+1588 kJ/mol
$Cr^{2+}(g) \rightarrow Cr^{3+}(g)$	+2882 kJ/mol
$\frac{1}{2}(I_2(s) \rightarrow I_2(g))$	+62/2 kJ/mol
$\frac{1}{2}(I_2(g) \rightarrow 2\,I(g))$	+151/2 kJ/mol
$I(g) + e^- \rightarrow I^-(g)$	−295 kJ/mol
$Cl_2(g) \rightarrow 2\,Cl(g)$	+243 kJ/mol
$2(Cl(g) + e^- \rightarrow Cl^-(g))$	2(−349) kJ/mol
$Cr^{3+}(g) + 2\,Cl^-(g) + I^-(g) \rightarrow CrCl_2I(s)$	−U
$Cr(s) + Cl_2(g) + \frac{1}{2}\,I_2(g) \rightarrow CrCl_2I(s)$	− 420 kJ/mol

$-U = -420 - 397 - 652 - 1588 - 2882 - 62/2 - 151/2 + 295 - 243 + 2(349) = -5295.5$ kJ/mol
$U = 5295$ kJ/mol

Multi-Concept Problems

6.110 (a)

Fe	[Ar] $4s^2 3d^6$
Fe^{2+}	[Ar] $3d^6$
Fe^{3+}	[Ar] $3d^5$

(b) A 3d electron is removed on going from Fe^{2+} to Fe^{3+}. For the 3d electron, n = 3 and $l = 2$.

(c) E(J/photon) = 2952 kJ/mol $\times \dfrac{1\ \text{mol photons}}{6.022 \times 10^{23}\ \text{photons}} \times \dfrac{1000\ \text{J}}{1\ \text{kJ}} = 4.90 \times 10^{-18}$ J/photon

$E = \dfrac{hc}{\lambda}$

$\lambda = \dfrac{hc}{E} = \dfrac{(6.626 \times 10^{-34}\ \text{J}\cdot\text{s})(3.00 \times 10^8\ \text{m/s})}{4.90 \times 10^{-18}\ \text{J}} = 4.06 \times 10^{-8}\ \text{m} = 40.6 \times 10^{-9}\ \text{m} = 40.6$ nm

(d) Ru is directly below Fe in the periodic table and the two metals have similar electron configurations. The electron removed from Ru to go from Ru^{2+} to Ru^{3+} is a 4d electron. The electron with the higher principal quantum number, n = 4, is farther from the nucleus, less tightly held, and requires less energy to remove.

6.112 AgCl, 143.32 amu

(a) mass Cl in AgCl = $1.126 \text{ g AgCl} \times \dfrac{35.453 \text{ g Cl}}{143.32 \text{ g AgCl}} = 0.279 \text{ g Cl}$

%Cl in alkaline earth chloride = $\dfrac{0.279 \text{ g Cl}}{0.436 \text{ g}} \times 100\% = 64.0\% \text{ Cl}$

(b) Because M is an alkaline earth metal, M is a 2+ cation.
For MCl_2, mass of M = 0.436 g – 0.279 g = 0.157 g M

mol M = $0.279 \text{ g Cl} \times \dfrac{1 \text{ mol Cl}}{35.453 \text{ g Cl}} \times \dfrac{1 \text{ mol M}}{2 \text{ mol Cl}} = 0.003\,93 \text{ mol M}$

molar mass for M = $\dfrac{0.157 \text{ g}}{0.003\,93 \text{ mol}} = 39.9 \text{ g/mol}; \quad M = Ca$

(c) $Ca(s) + Cl_2(g) \rightarrow CaCl_2(s)$
$CaCl_2(aq) + 2\,AgNO_3(aq) \rightarrow 2\,AgCl(s) + Ca(NO_3)_2(aq)$

(d) $1.005 \text{ g Ca} \times \dfrac{1 \text{ mol Ca}}{40.078 \text{ g Ca}} = 0.0251 \text{ mol Ca}$

$1.91 \times 10^{22} \text{ Cl}_2 \text{ molecules} \times \dfrac{1 \text{ mol Cl}_2}{6.022 \times 10^{23} \text{ Cl}_2 \text{ molecules}} = 0.0317 \text{ mol Cl}_2$

Because the stoichiometry between Ca and Cl_2 is one to one, the Cl_2 is in excess.

Mass Cl_2 unreacted = $(0.0317 - 0.0251) \text{ mol Cl}_2 \times \dfrac{70.91 \text{ g Cl}_2}{1 \text{ mol Cl}_2} = 0.47 \text{ g Cl}_2$ unreacted

6.114 (a)

$Sr(s) \rightarrow Sr(g)$	+164.44 kJ/mol
$Sr(g) \rightarrow Sr^+(g) + e^-$	+549.5 kJ/mol
$Sr^+(g) \rightarrow Sr^{2+}(g) + e^-$	+1064.2 kJ/mol
$Cl_2(g) \rightarrow 2\,Cl(g)$	+243 kJ/mol
$2[Cl(g) + e^- \rightarrow Cl^-(g)]$	2(–348.6) kJ/mol
$Sr^{2+}(g) + 2\,Cl^-(g) \rightarrow SrCl_2(s)$	<u>–2156 kJ/mol</u>

Sum = –832   kJ/mol for $Sr(s) + Cl_2(g) \rightarrow SrCl_2(s)$

(b) Sr, 87.62 amu; Cl_2, 70.91 amu; $SrCl_2$, 158.53 amu
$Sr(s) + Cl_2(g) \rightarrow SrCl_2(s)$

$20.0 \text{ g Sr} \times \dfrac{1 \text{ mol Sr}}{87.62 \text{ g Sr}} = 0.228 \text{ mol Sr}$ and $25.0 \text{ g Cl}_2 \times \dfrac{1 \text{ mol Cl}_2}{70.91 \text{ g Cl}_2} = 0.353 \text{ mol Cl}_2$

Because there is a 1:1 stoichiometry between the reactants, the one with the smaller mole amount is the limiting reactant. Sr is the limiting reactant.

$0.228 \text{ mol Sr} \times \dfrac{1 \text{ mol SrCl}_2}{1 \text{ mol Sr}} \times \dfrac{158.53 \text{ g SrCl}_2}{1 \text{ mol SrCl}_2} = 36.1 \text{ g SrCl}_2$

(c) $0.228 \text{ mol SrCl}_2 \times \dfrac{-832 \text{ kJ}}{1 \text{ mol SrCl}_2} = -190 \text{ kJ}$

190 kJ is released during the reaction of 20.0 g of Sr with 25.0 g Cl_2.

7 Covalent Bonds and Molecular Structure

7.1   (a) $SiCl_4$ chlorine   EN = 3.0
                 silicon $\underline{EN = 1.8}$
                          $\Delta EN = 1.2$ The Si–Cl bond is polar covalent.

       (b) CsBr   bromine   EN = 2.8
                 cesium $\underline{EN = 0.7}$
                          $\Delta EN = 2.1$ The Cs^+Br^- bond is ionic.

       (c) $FeBr_3$ bromine   EN = 2.8
                 iron $\underline{EN = 1.8}$
                          $\Delta EN = 1.0$ The Fe–Br bond is polar covalent.

       (d) CH_4 carbon   EN = 2.5
                 hydrogen $\underline{EN = 2.1}$
                          $\Delta EN = 0.4$ The C–H bond is polar covalent.

7.2   (a) CCl_4 chlorine   EN = 3.0
                 carbon $\underline{EN = 2.5}$
                          $\Delta EN = 0.5$

       (b) $BaCl_2$ chlorine   EN = 3.0
                 barium $\underline{EN = 0.9}$
                          $\Delta EN = 2.1$

       (c) $TiCl_3$ chlorine   EN = 3.0
                 titanium $\underline{EN = 1.5}$
                          $\Delta EN = 1.5$

       (d) Cl_2O oxygen   EN = 3.5
                 chlorine $\underline{EN = 3.0}$
                          $\Delta EN = 0.5$

       Increasing ionic character: $CCl_4 \sim ClO_2 < TiCl_3 < BaCl_2$.

7.3   H is positively polarized (blue). O is negatively polarized (red). This is consistent with the electronegativity values for O (3.5) and H (2.1). The more negatively polarized atom should be the one with the larger electronegativity.

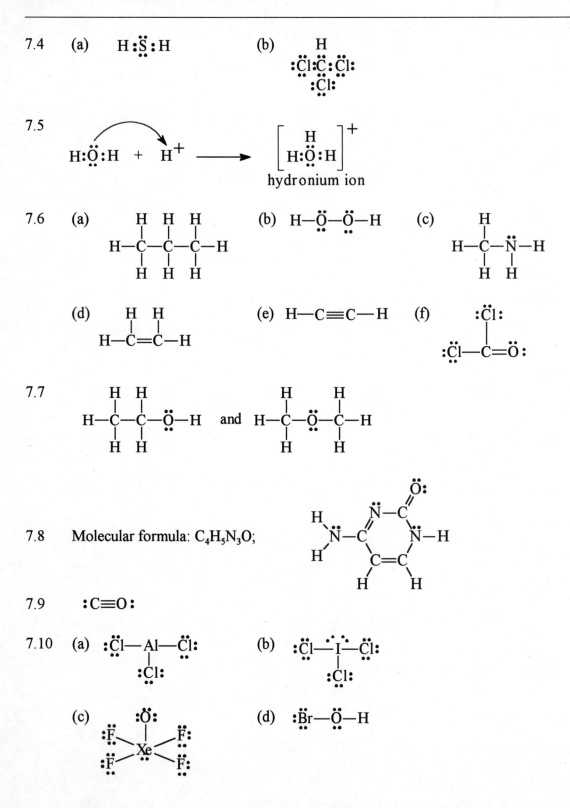

7.4 (a) H:S̈:H

(b) structure with H on top, :C̈l:C̈:C̈l: and :C̈l: below

7.5 reaction: H:Ö:H + H⁺ ⟶ [H:Ö:H with H on top]⁺ hydronium ion

7.6 (a) propane structure (b) H—Ö—Ö—H (c) H—C—N̈—H structure

(d) H—C=C—H structure (e) H—C≡C—H (f) :C̈l: and :C̈l—C=Ö:

7.7 two structures with "and" between them

7.8 Molecular formula: C₄H₅N₃O;

7.9 :C≡O:

7.10 (a) :C̈l—Al—C̈l: with :C̈l: below (b) :C̈l—I—C̈l: with :C̈l: below

(c) Xe structure with O and F atoms (d) :B̈r—Ö—H

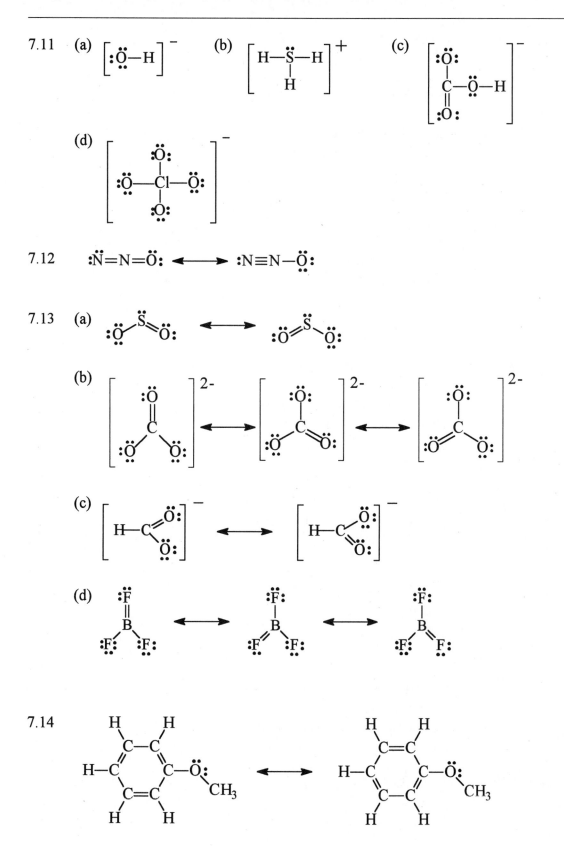

7.15   For nitrogen:

Isolated nitrogen valence electrons	5
Bound nitrogen bonding electrons	8
Bound nitrogen nonbonding electrons	0

Formal charge = $5 - \frac{1}{2}(8) - 0 = +1$

For singly bound oxygen:

Isolated oxygen valence electrons	6
Bound oxygen bonding electrons	2
Bound oxygen nonbonding electrons	6

Formal charge = $6 - \frac{1}{2}(2) - 6 = -1$

For doubly bound oxygen:

Isolated oxygen valence electrons	6
Bound oxygen bonding electrons	4
Bound oxygen nonbonding electrons	4

Formal charge = $6 - \frac{1}{2}(4) - 4 = 0$

7.16   (a) $\left[:\ddot{N}=C=\ddot{O}: \right]^{-}$

For nitrogen:

Isolated nitrogen valence electrons	5
Bound nitrogen bonding electrons	4
Bound nitrogen nonbonding electrons	4

Formal charge = $5 - \frac{1}{2}(4) - 4 = -1$

For carbon:

Isolated carbon valence electrons	4
Bound carbon bonding electrons	8
Bound carbon nonbonding electrons	0

Formal charge = $4 - \frac{1}{2}(8) - 0 = 0$

For oxygen:

Isolated oxygen valence electrons	6
Bound oxygen bonding electrons	4
Bound oxygen nonbonding electrons	4

Formal charge = $6 - \frac{1}{2}(4) - 4 = 0$

(b) $:\ddot{O}-\ddot{O}=\ddot{O}:$

For left oxygen:

Isolated oxygen valence electrons	6
Bound oxygen bonding electrons	2
Bound oxygen nonbonding electrons	6

Formal charge = $6 - \frac{1}{2}(2) - 6 = -1$

For central oxygen:

Isolated oxygen valence electrons	6
Bound oxygen bonding electrons	6
Bound oxygen nonbonding electrons	2

Formal charge = $6 - \frac{1}{2}(6) - 2 = +1$

	For right oxygen:	Isolated oxygen valence electrons	6
		Bound oxygen bonding electrons	4
		Bound oxygen nonbonding electrons	4

Formal charge = 6 – ½(4) – 4 = 0

7.17

	Number of Bonded Atoms	Number of Lone Pairs	Shape
(a) O_3	2	1	bent
(b) H_3O^+	3	1	trigonal pyramidal
(c) XeF_2	2	3	linear
(d) PF_6^-	6	0	octahedral
(e) $XeOF_4$	5	1	square pyramidal
(f) AlH_4^-	4	0	tetrahedral
(g) BF_4^-	4	0	tetrahedral
(h) $SiCl_4$	4	0	tetrahedral
(i) ICl_4^-	4	2	square planar
(j) $AlCl_3$	3	0	trigonal planar

7.18

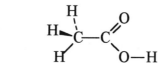

7.19 (a) tetrahedral   (b) seesaw

7.20

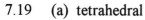

Each C is sp³ hybridized. The C–C bond is formed by the overlap of one singly occupied sp³ hybrid orbital from each C. The C–H bonds are formed by the overlap of one singly occupied sp³ orbital on C with a singly occupied H 1s orbital.

7.21

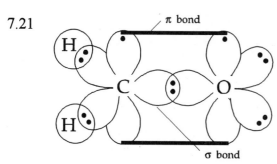

The carbon in formaldehyde is sp² hybridized.

7.22   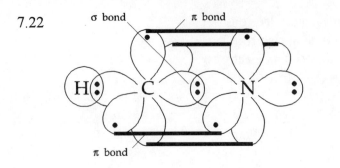   In HCN the carbon is sp hybridized.

7.23   The central I in I_3^- has two single bonds and three lone pairs of electrons. The hybridization of the central I is sp^3d. A sketch of the ion showing the orbitals involved in bonding is shown below.

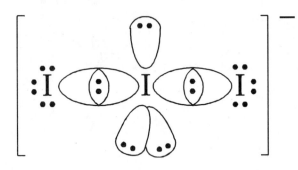

7.24
	Single Bonds	Lone Pairs	Hybridization of S
SF_2	2	2	sp^3
SF_4	4	1	sp^3d
SF_6	6	0	sp^3d^2

7.25   (a) sp   (b) sp^3d

7.26   For He_2^+

σ^*_{1s} ↑

σ_{1s} ↑↓

He_2^+ Bond order $= \dfrac{\left(\begin{array}{c}\text{number of}\\\text{bonding electrons}\end{array}\right) - \left(\begin{array}{c}\text{number of}\\\text{antibonding electrons}\end{array}\right)}{2} = \dfrac{2-1}{2} = 1/2$

He_2^+ should be stable with a bond order of 1/2.

7.27   For B_2

σ^*_{2p} ___

π^*_{2p} ___   ___

σ_{2p} ___

π_{2p} ↑   ↑

σ^*_{2s} ↑↓

σ_{2s} ↑↓

$$B_2 \text{ Bond order } = \frac{\left(\begin{array}{c}\text{number of}\\ \text{bonding electrons}\end{array}\right) - \left(\begin{array}{c}\text{number of}\\ \text{antibonding electrons}\end{array}\right)}{2} = \frac{4-2}{2} = 1$$

B_2 is paramagnetic because it has two unpaired electrons in the π_{2p} molecular orbitals.

For C_2

σ^*_{2p} —

π^*_{2p} —   —

σ_{2p} —

π_{2p} ↑↓   ↑↓

σ^*_{2s} ↑↓

σ_{2s} ↑↓

C_2 Bond order $= \dfrac{6-2}{2} = 2;$ C_2 is diamagnetic because all electrons are paired.

7.28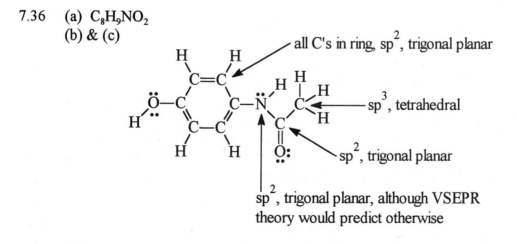

7.29   Handed biomolecules have specific shapes that only match complementary-shaped receptor sites in living systems. The mirror-image forms of the molecules can't fit into the receptor sites and thus don't elicit the same biological response.

7.30   The mirror image of molecule (a) has the same shape as (a) and is identical to it in all respects so there is no handedness associated with it. The mirror image of molecule (b) is different than (b) so there is a handedness to this molecule.

Understanding Key Concepts

7.32   (a) square pyramidal   (b) trigonal pyramidal
        (c) square planar   (d) trigonal planar

7.34   Molecular model (c) does not have a tetrahedral central atom. It is square planar.

7.36   (a) $C_8H_9NO_2$
        (b) & (c)

all C's in ring, sp^2, trigonal planar

sp^3, tetrahedral

sp^2, trigonal planar

sp^2, trigonal planar, although VSEPR theory would predict otherwise

Additional Problems
Electronegativity and Polar Covalent Bonds

7.38   Electronegativity increases from left to right across a period and decreases down a group.

7.40   K < Li < Mg < Pb < C < Br

7.42   (a) HF   fluorine   EN = 4.0
                     hydrogen $\underline{EN = 2.1}$
                                  $\Delta EN = 1.9$ HF is polar covalent.

        (b) HI   iodine   EN = 2.5
                     hydrogen $\underline{EN = 2.1}$
                                  $\Delta EN = 0.4$ HI is polar covalent.

        (c) $PdCl_2$ chlorine   EN = 3.0
                     palladium $\underline{EN = 2.2}$
                                  $\Delta EN = 0.8$ $PdCl_2$ is polar covalent.

        (d) BBr_3 bromine   EN = 2.8
                     boron $\underline{EN = 2.0}$
                                  $\Delta EN = 0.8$ BBr_3 is polar covalent.

        (e) NaOH $Na^+ - OH^-$ is ionic
            OH^- oxygen   EN = 3.5
                     hydrogen $\underline{EN = 2.1}$
                                  $\Delta EN = 1.4$ OH^- is polar covalent.

        (f) CH_3Li lithium   EN = 1.0
                     carbon $\underline{EN = 2.5}$
                                  $\Delta EN = 1.5$ CH_3Li is polar covalent.

7.44   (a) $\overset{\delta-}{C} - \overset{\delta+}{H}$ $\overset{\delta+}{C} - \overset{\delta-}{Cl}$ (b) $\overset{\delta-}{Si} - \overset{\delta+}{Li}$ $\overset{\delta+}{Si} - \overset{\delta-}{Cl}$

        (c) N – Cl $\overset{\delta-}{N} - \overset{\delta+}{Mg}$

Electron-Dot Structures and Resonance

7.46   The octet rule states that main-group elements tend to react so that they attain a noble gas electron configuration with filled s and p sublevels (8 electrons) in their valence electron shells. The transition metals are characterized by partially filled d orbitals that can be used to expand their valence shell beyond the normal octet of electrons.

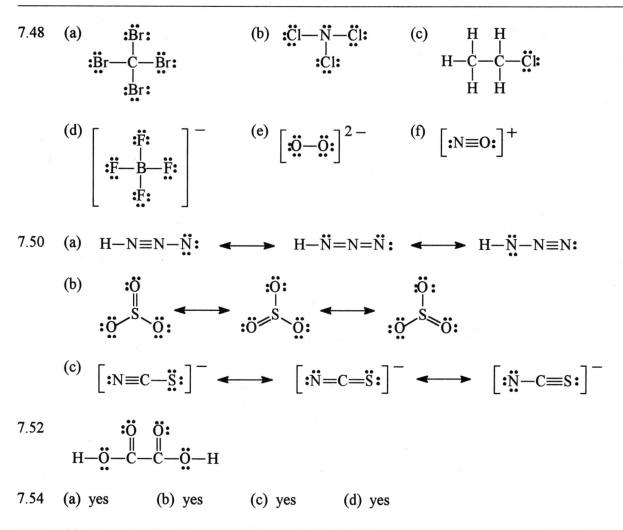

7.54 (a) yes   (b) yes   (c) yes   (d) yes

7.56 (a) The anion has 32 valence electrons. Each Cl has seven valence electrons (28 total). The minus one charge on the anion accounts for one valence electron. This leaves three valence electrons for X. X is Al.

(b) The cation has eight valence electrons. Each H has one valence electron (4 total). X is left with four valence electrons. Since this is a cation, one valence electron was removed from X. X has five valence electrons. X is P.

7.58 (a)   (b)

Formal Charges

7.60 $:C\equiv O:$

       For carbon:

Isolated carbon valence electrons	4
Bound carbon bonding electrons	6
Bound carbon nonbonding electrons	2

Formal charge = $4 - \frac{1}{2}(6) - 2 = -1$

For oxygen:

Isolated oxygen valence electrons	6
Bound oxygen bonding electrons	6
Bound oxygen nonbonding electrons	2

Formal charge = $6 - \frac{1}{2}(6) - 2 = +1$

7.62

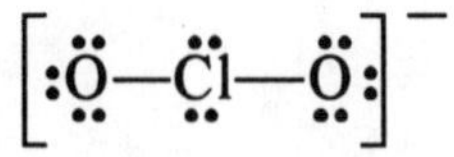

$$\left[:\ddot{O}\!-\!\ddot{Cl}\!-\!\ddot{O}: \right]^{-}$$

For both oxygens:

Isolated oxygen valence electrons	6
Bound oxygen bonding electrons	2
Bound oxygen nonbonding electrons	6

Formal charge = $6 - \frac{1}{2}(2) - 6 = -1$

For chlorine:

Isolated chlorine valence electrons	7
Bound chlorine bonding electrons	4
Bound chlorine nonbonding electrons	4

Formal charge = $7 - \frac{1}{2}(4) - 4 = +1$

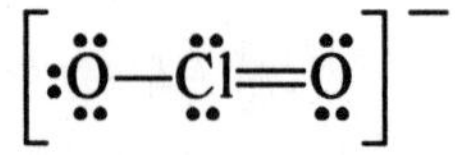

$$\left[:\ddot{O}\!-\!\ddot{Cl}\!=\!\ddot{O} \right]^{-}$$

For left oxygen:

Isolated oxygen valence electrons	6
Bound oxygen bonding electrons	2
Bound oxygen nonbonding electrons	6

Formal charge = $6 - \frac{1}{2}(2) - 6 = -1$

For right oxygen:

Isolated oxygen valence electrons	6
Bound oxygen bonding electrons	4
Bound oxygen nonbonding electrons	4

Formal charge = $6 - \frac{1}{2}(4) - 4 = 0$

For chlorine:

Isolated chlorine valence electrons	7
Bound chlorine bonding electrons	6
Bound chlorine nonbonding electrons	4

Formal charge = $7 - \frac{1}{2}(6) - 4 = 0$

7.64 (a)

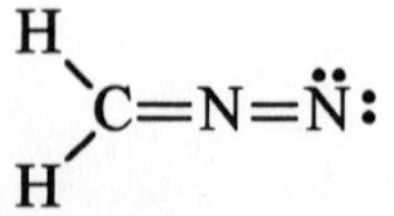

$$\begin{array}{c} H \\ \diagdown \\ \diagup \\ H \end{array} C\!=\!N\!=\!\ddot{N}:$$

For hydrogen:

Isolated hydrogen valence electrons	1
Bound hydrogen bonding electrons	2
Bound hydrogen nonbonding electrons	0

Formal charge = $1 - \frac{1}{2}(2) - 0 = 0$

For nitrogen: (central)	Isolated nitrogen valence electrons	5
	Bound nitrogen bonding electrons	8
	Bound nitrogen nonbonding electrons	0
	Formal charge = $5 - \frac{1}{2}(8) - 0 = +1$	

For nitrogen: (terminal)	Isolated nitrogen valence electrons	5
	Bound nitrogen bonding electrons	4
	Bound nitrogen nonbonding electrons	4
	Formal charge = $5 - \frac{1}{2}(4) - 4 = -1$	

For carbon:	Isolated carbon valence electrons	4
	Bound carbon bonding electrons	8
	Bound carbon nonbonding electrons	0
	Formal charge = $4 - \frac{1}{2}(8) - 0 = 0$	

(b)

$$\begin{array}{c} H \\ \diagdown \\ \diagup C - \ddot{N} = \ddot{N} : \\ H \end{array}$$

For hydrogen:	Isolated hydrogen valence electrons	1
	Bound hydrogen bonding electrons	2
	Bound hydrogen nonbonding electrons	0
	Formal charge = $1 - \frac{1}{2}(2) - 0 = 0$	

For nitrogen: (central)	Isolated nitrogen valence electrons	5
	Bound nitrogen bonding electrons	6
	Bound nitrogen nonbonding electrons	2
	Formal charge = $5 - \frac{1}{2}(6) - 2 = 0$	

For nitrogen: (terminal)	Isolated nitrogen valence electrons	5
	Bound nitrogen bonding electrons	4
	Bound nitrogen nonbonding electrons	4
	Formal charge = $5 - \frac{1}{2}(4) - 4 = -1$	

For carbon:	Isolated carbon valence electrons	4
	Bound carbon bonding electrons	6
	Bound carbon nonbonding electrons	0
	Formal charge = $4 - \frac{1}{2}(6) - 0 = +1$	

Structure (a) is more important because of the octet of electrons around carbon.

The VSEPR Model

7.66   From data in Table 7.4:
       (a) trigonal planar   (b) trigonal bipyramidal   (c) linear   (d) octahedral

7.68   From data in Table 7.4:
  (a) tetrahedral, 4   (b) octahedral, 6   (c) bent, 3 or 4
  (d) linear, 2 or 5   (e) square pyramidal, 6 (f) trigonal pyramidal, 4

7.70

	Number of Bonded Atoms	Number of Lone Pairs	Shape
(a) H_2Se	2	2	bent
(b) $TiCl_4$	4	0	tetrahedral
(c) O_3	2	1	bent
(d) GaH_3	3	0	trigonal planar

7.72

	Number of Bonded Atoms	Number of Lone Pairs	Shape
(a) SbF_5	5	0	trigonal bipyramidal
(b) IF_4^+	4	1	see saw
(c) SeO_3^{2-}	3	1	trigonal pyramidal
(d) CrO_4^{2-}	4	0	tetrahedral

7.74

	Number of Bonded Atoms	Number of Lone Pairs	Shape
(a) PO_4^{3-}	4	0	tetrahedral
(b) MnO_4^-	4	0	tetrahedral
(c) SO_4^{2-}	4	0	tetrahedral
(d) SO_3^{2-}	3	1	trigonal pyramidal
(e) ClO_4^-	4	0	tetrahedral
(f) SCN^-	2	0	linear

   (C is the central atom)

7.76   (a) In SF_2 the sulfur is bound to two fluorines and contains two lone pairs of electrons. SF_2 is bent and the F–S–F bond angle is approximately 109°.
(b) In N_2H_2 each nitrogen is bound to the other nitrogen and one hydrogen. Each nitrogen has one lone pair of electrons. The H–N–N bond angle is approximately 120°.
(c) In KrF_4 the krypton is bound to four fluorines and contains two lone pairs of electrons. KrF_4 is square planar, and the F–Kr–F bond angle is 90°.
(d) In NOCl the nitrogen is bound to one oxygen and one chlorine and contains one lone pair of electrons. NOCl is bent, and the Cl–N–O bond angle is approximately 120°.

7.78

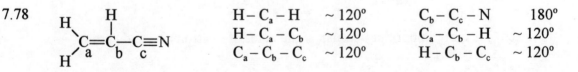

$H–C_a–H$	~ 120°	$C_b–C_c–N$	180°
$H–C_a–C_b$	~ 120°	$C_a–C_b–H$	~ 120°
$C_a–C_b–C_c$	~ 120°	$H–C_b–C_c$	~ 120°

7.80   All six carbons in cyclohexane are bonded to two other carbons and two hydrogens (i.e. four charge clouds). The geometry about each carbon is tetrahedral with a C–C–C bond angle of approximately 109°. Because the geometry about each carbon is tetrahedral, the cyclohexane ring cannot be flat.

Hybrid Orbitals and Molecular Orbital Theory

7.82   In a π bond, the shared electrons occupy a region above and below a line connecting the two nuclei. A σ bond has its shared electrons located along the axis between the two nuclei.

7.84   See Table 7.5.
  (a) sp   (b) sp^3d (c) sp^3d^2 (d) sp^3

7.86   See Table 7.5.
  (a) sp^3 (b) sp^3d^2 (c) sp^2 or sp^3 (d) sp or sp^3d (e) sp^3d^2

7.88   (a) sp^2 (b) sp^3 (c) sp^3d^2 (d) sp^2

7.90   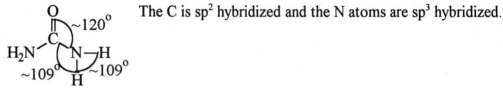   The C is sp^2 hybridized and the N atoms are sp^3 hybridized.

7.92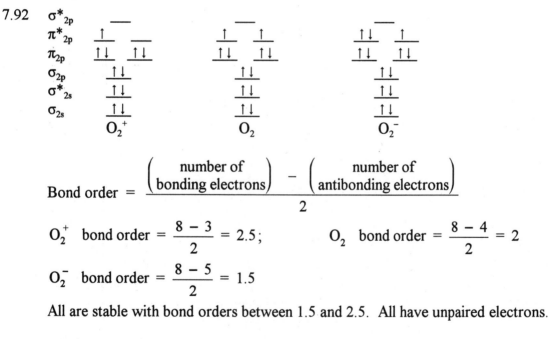

$$\text{Bond order} = \frac{\left(\begin{array}{c}\text{number of}\\\text{bonding electrons}\end{array}\right) - \left(\begin{array}{c}\text{number of}\\\text{antibonding electrons}\end{array}\right)}{2}$$

O_2^+ bond order $= \dfrac{8-3}{2} = 2.5$; O_2 bond order $= \dfrac{8-4}{2} = 2$

O_2^- bond order $= \dfrac{8-5}{2} = 1.5$

All are stable with bond orders between 1.5 and 2.5. All have unpaired electrons.

7.94   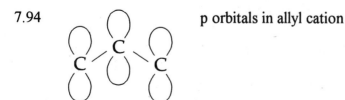   p orbitals in allyl cation

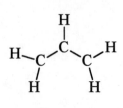

allyl cation showing only the σ bonds (each C is sp² hybridized)

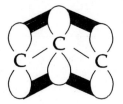

delocalized MO model for π bonding in the allyl cation

General Problems

7.96

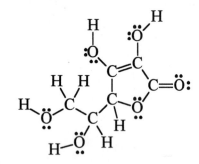

7.98   Every carbon is sp² hybridized. There are 18 σ bonds and 5 π bonds.

7.100

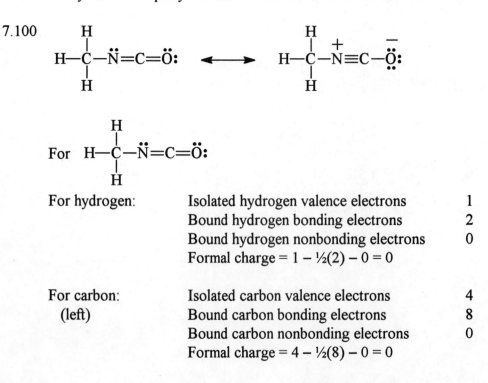

For hydrogen:	Isolated hydrogen valence electrons	1
	Bound hydrogen bonding electrons	2
	Bound hydrogen nonbonding electrons	0
	Formal charge = $1 - \frac{1}{2}(2) - 0 = 0$

For carbon: (left)	Isolated carbon valence electrons	4
	Bound carbon bonding electrons	8
	Bound carbon nonbonding electrons	0
	Formal charge = $4 - \frac{1}{2}(8) - 0 = 0$

For nitrogen: Isolated nitrogen valence electrons   5
                Bound nitrogen bonding electrons   6
                Bound nitrogen nonbonding electrons   2
                Formal charge = $5 - \frac{1}{2}(6) - 2 = 0$

For carbon: Isolated carbon valence electrons   4
  (right)   Bound carbon bonding electrons   8
                Bound carbon nonbonding electrons   0
                Formal charge = $4 - \frac{1}{2}(8) - 0 = 0$

For oxygen: Isolated oxygen valence electrons   6
                Bound oxygen bonding electrons   4
                Bound oxygen nonbonding electrons   4
                Formal charge = $6 - \frac{1}{2}(4) - 4 = 0$

For
$$\text{H}-\overset{\overset{\displaystyle \text{H}}{|}}{\underset{\underset{\displaystyle \text{H}}{|}}{\text{C}}}-\overset{+}{\text{N}}\equiv\text{C}-\overset{\displaystyle \overline{}}{\ddot{\underset{..}{\text{O}}}}\!:$$

For hydrogen: Isolated hydrogen valence electrons   1
                Bound hydrogen bonding electrons   2
                Bound hydrogen nonbonding electrons   0
                Formal charge = $1 - \frac{1}{2}(2) - 0 = 0$

For carbon: Isolated carbon valence electrons   4
  (left)   Bound carbon bonding electrons   8
                Bound carbon nonbonding electrons   0
                Formal charge = $4 - \frac{1}{2}(8) - 0 = 0$

For nitrogen: Isolated nitrogen valence electrons   5
                Bound nitrogen bonding electrons   8
                Bound nitrogen nonbonding electrons   0
                Formal charge = $5 - \frac{1}{2}(8) - 0 = +1$

For carbon: Isolated carbon valence electrons   4
  (right)   Bound carbon bonding electrons   8
                Bound carbon nonbonding electrons   0
                Formal charge = $4 - \frac{1}{2}(8) - 0 = 0$

For oxygen: Isolated oxygen valence electrons   6
                Bound oxygen bonding electrons   2
                Bound oxygen nonbonding electrons   6
                Formal charge = $6 - \frac{1}{2}(2) - 6 = -1$

7.102 (a)

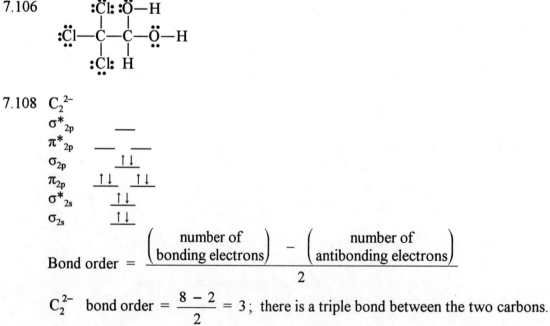

For boron:	Isolated boron valence electrons	3
	Bound boron bonding electrons	8
	Bound boron nonbonding electrons	0
	Formal charge $= 3 - \frac{1}{2}(8) - 0 = -1$	

For oxygen:	Isolated oxygen valence electrons	6
	Bound oxygen bonding electrons	6
	Bound oxygen nonbonding electrons	2
	Formal charge $= 6 - \frac{1}{2}(6) - 2 = +1$	

(b) In BF_3 the B has three bonding pairs of electrons and no lone pairs. The B is sp^2 hybridized and BF_3 is trigonal planar.

$H_3C\!-\!\overset{..}{\underset{..}{O}}\!-\!CH_3$ is bent about the oxygen because of two bonding pairs and two lone pairs of electrons. The O is sp^3 hybridized.

In the product, B is sp^3 hybridized (with four bonding pairs of electrons), and the geometry about it is tetrahedral. The O is also sp^3 hybridized (with three bonding pairs and one lone pair of electrons), and the geometry about it is trigonal pyramidal.

7.104 The triply bonded carbon atoms are sp hybridized. The theoretical bond angle for $C-C\equiv C$ is 180°. Benzyne is so reactive because the $C-C\equiv C$ bond angle is closer to 120° and is very strained.

7.106

$$\begin{array}{ccc}
 & :\overset{..}{\underset{..}{Cl}}: & :\overset{..}{O}\!-\!H \\
 & | & | \\
:\overset{..}{\underset{..}{Cl}}\!-\! & C\!-\!C & \!-\!\overset{..}{\underset{..}{O}}\!-\!H \\
 & | & | \\
 & :\overset{..}{\underset{..}{Cl}}: & H
\end{array}$$

7.108 C_2^{2-}

σ^*_{2p}	—
π^*_{2p}	— —
σ_{2p}	↑↓
π_{2p}	↑↓ ↑↓
σ^*_{2s}	↑↓
σ_{2s}	↑↓

$$\text{Bond order} = \frac{\left(\begin{array}{c}\text{number of} \\ \text{bonding electrons}\end{array}\right) - \left(\begin{array}{c}\text{number of} \\ \text{antibonding electrons}\end{array}\right)}{2}$$

C_2^{2-} bond order $= \dfrac{8-2}{2} = 3$; there is a triple bond between the two carbons.

7.110 $CH_4(g) + Cl_2(g) \rightarrow CH_3Cl(g) + HCl(g)$
Energy change = D (Reactant bonds) – D (Product bonds)
Energy change = $[4\ D_{C-H} + D_{Cl-Cl}] - [3\ D_{C-H} + D_{C-Cl} + D_{H-Cl}]$
Energy change = $[(4\ mol)(410\ kJ/mol) + (1\ mol)(243\ kJ/mol]$
  $- [(3\ mol)(410\ kJ/mol) + (1\ mol)(330\ kJ/mol) + (1\ mol)(432\ kJ/mol)] = -109\ kJ$

7.112 (a)   (b)   (c)

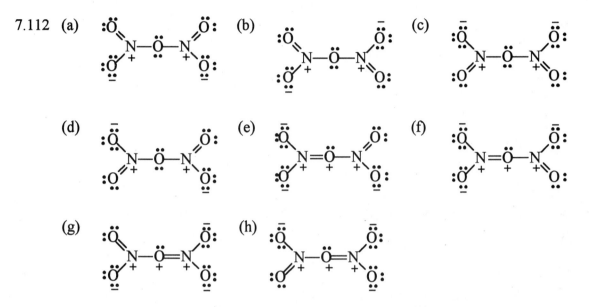

(d)   (e)   (f)

(g)   (h)

Structures (a) - (d) make more important contributions to the resonance hybrid because of only −1 and 0 formal charges on the oxygens.

7.114

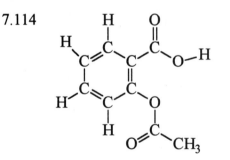

21 σ bonds
5 π bonds
Each C with a double bond is sp² hybridized.
The –CH₃ carbon is sp³ hybridized.

7.116 (a)

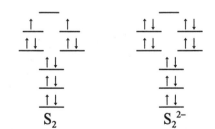

(b) S₂ would be paramagnetic with two unpaired electrons in the π^*_{3p} MOs.

(c)   Bond order = $\dfrac{\left(\begin{array}{c}\text{number of} \\ \text{bonding electrons}\end{array}\right) - \left(\begin{array}{c}\text{number of} \\ \text{antibonding electrons}\end{array}\right)}{2}$

S_2 bond order = $\dfrac{8-4}{2} = 2$

(d) S_2^{2-} bond order = $\dfrac{8-6}{2} = 1$

The two added electrons go into the antibonding π^*_{3p} MOs, the bond order drops from 2 to 1, and the bond length in S_2^{2-} should be longer than the bond length in S_2.

7.118 (a)

$$\overset{\displaystyle :\ddot{F}:}{\underset{\displaystyle :\ddot{F}:}{\;\;|\;\;}} \;\; :\ddot{F}-\overset{|}{\underset{|}{S}}-\ddot{\underset{..}{S}}-\ddot{F}:$$

The left S has 5 electron clouds (4 bonding, 1 lone pair). The S is sp³d hybridized and the geometry about this S is see-saw. The right S has 4 electron clouds (2 bonding, 2 lone pairs). The S is sp³ hybridized and the geometry about this S is bent.

(b)

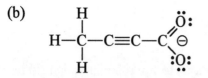

The left C has 4 electron clouds (4 bonding, 0 lone pairs). This C is sp³ hybridized and its geometry is tetrahedral. The right C has 3 electron clouds (3 bonding, 0 lone pairs). This C is sp² hybridized and its geometry is trigonal planar. The central two C's have 2 electron clouds (2 bonding, 0 lone pairs). These C's are sp hybridized and the geometry about both is linear.

Multi-Concept Problems

7.120 (a) $\left[:\ddot{O}-H\right]^-$ $:\dot{\ddot{O}}-H$

(b) The oxygen in OH has a half-filled 2p orbital that can accept the additional electron. For a 2p orbital, n = 2 and l = 1.
(c) The electron affinity for OH is slightly more negative than for an O atom because when OH gains an additional electron, it achieves an octet configuration.

7.122 (a)

$$\left[:O=\overset{\displaystyle :\ddot{O}\;\;\|}{\underset{\displaystyle :\ddot{O}:}{Cr}}-\ddot{\underset{..}{O}}-\overset{\displaystyle :\ddot{O}\;\;\|}{\underset{\displaystyle :\ddot{O}:}{Cr}}=O: \right]^{2-}$$

(b) Each Cr atom has 6 pairs of electrons around it. The likely geometry about each Cr atom is tetrahedral because each Cr has 4 charge clouds.

7.124 (a)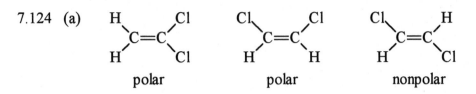

     polar   polar   nonpolar

(b) All three molecules are planar. The first two structures are polar because they both have an unsymmetrical distribution of atoms about the center of the molecule (the middle of the double bond), and bond polarities do not cancel. Structure 3 is nonpolar because the H's and Cl's, respectively, are symmetrically distributed about the center of the molecule, both being opposite each other. In this arrangement, bond polarities cancel.

(c) 200 nm = 200 x 10^{-9} m

$$E = \frac{hc}{\lambda} = \frac{(6.626 \times 10^{-34} \text{ J·s})(3.00 \times 10^8 \text{ m/s})}{200 \times 10^{-9} \text{ m}} (6.022 \times 10^{23} / \text{mol})$$

E = 5.99 x 10^5 J/mol = 599 kJ/mol

(d)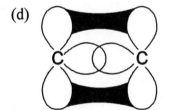

The π bond must be broken before rotation can occur.

Thermochemistry: Chemical Energy

8.1 Convert lb to kg.
$$2300 \text{ lb} \times \frac{453.59 \text{ g}}{1 \text{ lb}} \times \frac{1 \text{ kg}}{1000 \text{ g}} = 1043 \text{ kg}$$

Convert mi/h to m/s.
$$55 \frac{\text{mi}}{\text{h}} \times \frac{1 \text{ km}}{0.62137 \text{ mi}} \times \frac{1000 \text{ m}}{1 \text{km}} \times \frac{1 \text{ h}}{3600 \text{ s}} = 24.6 \text{ m/s}$$

$1 \text{ kg} \cdot \text{m}^2/\text{s}^2 = 1 \text{ J}$; $E = \frac{1}{2}mv^2 = \frac{1}{2}(1043 \text{ kg})(24.6 \text{ m/s})^2 = 3.2 \times 10^5 \text{ J}$

$E = 3.2 \times 10^5 \text{ J} \times \dfrac{1 \text{ kJ}}{1000 \text{ J}} = 3.2 \times 10^2 \text{ kJ}$

8.2 (a) and (b) are state functions; (c) is not.

8.3 $\Delta V = (4.3 \text{ L} - 8.6 \text{ L}) = -4.3 \text{ L}$
$w = -P\Delta V = -(44 \text{ atm})(-4.3 \text{ L}) = +189.2 \text{ L} \cdot \text{atm}$

$w = (189.2 \text{ L} \cdot \text{atm})(101 \dfrac{\text{J}}{\text{L} \cdot \text{atm}}) = +1.9 \times 10^4 \text{ J}$

The positive sign for the work indicates that the surroundings does work on the system. Energy flows into the system.

8.4 $w = -P\Delta V = -(2.5 \text{ atm})(3 \text{ L} - 2 \text{ L}) = -2.5 \text{ L} \cdot \text{atm}$

$w = (-2.5 \text{ L} \cdot \text{atm})\left(101 \dfrac{\text{J}}{\text{L} \cdot \text{atm}}\right) = -252.5 \text{ J} = -250 \text{ J} = -0.25 \text{ kJ}$

The negative sign indicates that the expanding system loses work energy and does work on the surroundings.

8.5 (a) $w = -P\Delta V$ is positive and $P\Delta V$ is negative for this reaction because the system volume is decreased at constant pressure.
(b) $P\Delta V$ is small compared to ΔE.
$\Delta H = \Delta E + P\Delta V$; ΔH is negative. Its value is slightly more negative than ΔE.

8.6 $\Delta H° = -484 \dfrac{\text{kJ}}{2 \text{ mol H}_2}$

$P\Delta V = (1.00 \text{ atm})(-5.6 \text{ L}) = -5.6 \text{ L} \cdot \text{atm}$

$P\Delta V = (-5.6 \text{ L} \cdot \text{atm})(101 \dfrac{\text{J}}{\text{L} \cdot \text{atm}}) = -565.6 \text{ J} = -570 \text{ J} = -0.57 \text{ kJ}$

$w = -P\Delta V = 570 \text{ J} = 0.57 \text{ kJ}$

$\Delta H = -121 \dfrac{\text{kJ}}{0.50 \text{ mol H}_2}$

$\Delta E = \Delta H - P\Delta V = -121 \text{ kJ} - (-0.57 \text{ kJ}) = -120.43 \text{ kJ} = -120 \text{ kJ}$

8.7 $\Delta V = 448$ L and assume P = 1.00 atm

$w = -P\Delta V = -(1.00 \text{ atm})(448 \text{ L}) = -448$ L· atm

$w = -(448 \text{ L· atm})(101 \dfrac{J}{\text{L· atm}}) = -4.52 \times 10^4$ J

$w = -4.52 \times 10^4$ J $\times \dfrac{1 \text{ kJ}}{1000 \text{ J}} = -45.2$ kJ

8.8   (a) C_3H_8, 44.10 amu; $\Delta H° = -2219$ kJ/mol C_3H_8

15.5 g x $\dfrac{1 \text{ mol } C_3H_8}{44.10 \text{ g } C_3H_8}$ x $\dfrac{2219 \text{ kJ}}{1 \text{ mol } C_3H_8} = -780.$ kJ

780. kJ of heat is evolved.

(b) $Ba(OH)_2 \cdot 8 H_2O$, 315.5 amu; $\Delta H° = +80.3$ kJ/mol $Ba(OH)_2 \cdot 8 H_2O$

4.88 g x $\dfrac{1 \text{ mol } Ba(OH)_2 \cdot 8 H_2O}{315.5 \text{ g } Ba(OH)_2 \cdot 8 H_2O}$ x $\dfrac{80.3 \text{ kJ}}{1 \text{ mol } Ba(OH)_2 \cdot 8 H_2O} = +1.24$ kJ

1.24 kJ of heat is absorbed.

8.9 CH_3NO_2, 61.04 amu

$q = 100.0 \text{ g } CH_3NO_2$ x $\dfrac{1 \text{ mol } CH_3NO_2}{61.04 \text{ g } CH_3NO_2}$ x $\dfrac{2441.6 \text{ kJ}}{4 \text{ mol } CH_3NO_2} = 1.000 \times 10^3$ kJ

8.10 $q = (\text{specific heat}) \times m \times \Delta T = (4.18 \dfrac{J}{g \cdot °C})(350 \text{ g})(3°C - 25°C) = -3.2 \times 10^4$ J

$q = -3.2 \times 10^4 \times \dfrac{1 \text{ kJ}}{1000 \text{ J}} = -32$ kJ

8.11 $q = (\text{specific heat}) \times m \times \Delta T$;   specific heat $= \dfrac{q}{m \times \Delta T} = \dfrac{96 \text{ J}}{(75 \text{ g})(10°C)} = 0.13$ J/(g · °C)

8.12   25.0 mL = 0.0250 L and 50.0 mL = 0.0500 L

mol $H_2SO_4 = (1.00 \text{ mol/L})(0.0250 \text{ L}) = 0.0250$ mol H_2SO_4

mol NaOH $= (1.00 \text{ mol/L})(0.0500 \text{ L}) = 0.0500$ mol NaOH

NaOH and H_2SO_4 are present in a 2:1 mol ratio. This matches the stoichiometric ratio in the balanced equation.

$q = (\text{specific heat}) \times m \times \Delta T$

$m = (25.0 \text{ mL} + 50.0 \text{ mL})(1.00 \dfrac{g}{mL}) = 75.0$ g

$q = (4.18 \dfrac{J}{g \cdot °C})(75.0 \text{ g})(33.9°C - 25.0°C) = 2790$ J

mol $H_2SO_4 = 0.0250 \text{ L} \times 1.00 \dfrac{mol}{L} \; H_2SO_4 = 0.0250$ mol H_2SO_4

Heat evolved per mole of H_2SO_4 = $\dfrac{2.79 \times 10^3 \text{ J}}{0.0250 \text{ mol } H_2SO_4}$ = 1.1×10^5 J/mol H_2SO_4

Since the reaction evolves heat, the sign for ΔH is negative.

$\Delta H = -1.1 \times 10^5$ J $\times \dfrac{1 \text{ kJ}}{1000 \text{ J}} = -1.1 \times 10^2$ kJ

8.13 $CH_4(g) + Cl_2(g) \rightarrow CH_3Cl(g) + HCl(g)$ $\Delta H^\circ_1 = -98.3$ kJ
            $\underline{CH_3Cl(g) + Cl_2(g) \rightarrow CH_2Cl_2(g) + HCl(g)}$ $\Delta H^\circ_2 = -104$ kJ
    Sum $CH_4(g) + 2\,Cl_2(g) \rightarrow CH_2Cl_2(g) + 2\,HCl(g)$
            $\Delta H^\circ = \Delta H^\circ_1 + \Delta H^\circ_2 = -202$ kJ

8.14   (a) A + 2 B $\rightarrow$ D; $\Delta H^\circ = -100$ kJ + (-50 kJ) = -150 kJ
       (b) The red arrow corresponds to step 1:   A + B $\rightarrow$ C
            The green arrow corresponds to step 2:   C + B $\rightarrow$ D
            The blue arrow corresponds to the overall reaction.
       (c) The top energy level represents A + 2 B.
            The middle energy level represents C + B.
            The bottom energy level represents D.

8.15   Reactants $CH_4 + 2\,Cl_2$

                                         $\Delta H^\circ = -98.3$ kJ

                                         $CH_3Cl + HCl + Cl_2$

          $\Delta H^\circ = -202$ kJ

                                         $\Delta H^\circ = -104$ kJ

    Products $CH_2Cl_2 + 2\,HCl$

8.16 $4\,NH_3(g) + 5\,O_2(g) \rightarrow 4\,NO(g) + 6\,H_2O(g)$
       $\Delta H^\circ_{rxn} = [4\,\Delta H^\circ_f\,(NO) + 6\,\Delta H^\circ_f\,(H_2O)] - [4\,\Delta H^\circ_f\,(NH_3)]$
       $\Delta H^\circ_{rxn} = [(4 \text{ mol})(90.2 \text{ kJ/mol}) + (6 \text{ mol})(-241.8 \text{ kJ/mol})] - [(4 \text{ mol})(-46.1 \text{ kJ/mol})]$
       $\Delta H^\circ_{rxn} = -905.6$ kJ

8.17 $6\,CO_2(g) + 6\,H_2O(l) \rightarrow C_6H_{12}O_6(s) + 6\,O_2(g)$
       $\Delta H^\circ_{rxn} = \Delta H^\circ_f(C_6H_{12}O_6) - [6\,\Delta H^\circ_f(CO_2) + 6\,\Delta H^\circ_f(H_2O(l))]$
       $\Delta H^\circ_{rxn} = [(1 \text{ mol})(-1260 \text{ kJ/mol})] - [(6 \text{ mol})(-393.5 \text{ kJ/mol}) + (6 \text{ mol})(-285.8 \text{ kJ/mol})]$
       $\Delta H^\circ_{rxn} = +2815.8$ kJ = +2816 kJ

8.18 $H_2C=CH_2(g) + H_2O(g) \rightarrow C_2H_5OH(g)$

$\Delta H°_{rxn} = D$ (Reactant bonds) $- D$ (Product bonds)

$\Delta H°_{rxn} = (D_{C=C} + 4 D_{C-H} + 2 D_{O-H}) - (D_{C-C} + D_{C-O} + 5 D_{C-H} + D_{O-H})$

$\Delta H°_{rxn} = [(1 \text{ mol})(611 \text{ kJ/mol}) + (4 \text{ mol})(410 \text{ kJ/mol}) + (2 \text{ mol})(460 \text{ kJ/mol})]$

$- [(1 \text{ mol})(350 \text{ kJ/mol}) + (1 \text{ mol})(350 \text{ kJ/mol}) + (5 \text{ mol})(410 \text{ kJ/mol}) + (1 \text{ mol})(460 \text{ kJ/mol})]$

$= -39 \text{ kJ}$

8.19 $2 NH_3(g) + Cl_2(g) \rightarrow N_2H_4(g) + 2 HCl(g)$

$\Delta H°_{rxn} = D$ (Reactant bonds) $- D$ (Product bonds)

$\Delta H°_{rxn} = (6 D_{N-H} + D_{Cl-Cl}) - (D_{N-N} + 4 D_{N-H} + 2 D_{H-Cl})$

$\Delta H°_{rxn} = [(6 \text{ mol})(390 \text{ kJ/mol}) + (1 \text{ mol})(243 \text{ kJ/mol})]$

$- [(1 \text{ mol})(240 \text{ kJ/mol}) + (4 \text{ mol})(390 \text{ kJ/mol}) + (2 \text{ mol})(432 \text{ kJ/mol})] = -81 \text{ kJ}$

8.20 $C_4H_{10}(l) + \dfrac{13}{2} O_2(g) \rightarrow 4 CO_2(g) + 5 H_2O(g)$

$\Delta H°_{rxn} = [4 \Delta H°_f (CO_2) + 5 \Delta H°_f (H_2O)] - \Delta H°_f (C_4H_{10})$

$\Delta H°_{rxn} = [(4 \text{ mol})(-393.5 \text{ kJ/mol}) + (5 \text{ mol})(-241.8 \text{ kJ/mol})] - [(1 \text{ mol})(-147.5 \text{ kJ/mol})]$

$\Delta H°_{rxn} = -2635.5 \text{ kJ}$

$\Delta H°_c = -2635.5 \text{ kJ/mol}$

C_4H_{10}, 58.12 amu; $\Delta H°_c = \left(-2635.5 \dfrac{\text{kJ}}{\text{mol}}\right)\left(\dfrac{1 \text{ mol}}{58.12 \text{ g}}\right) = -45.35 \text{ kJ/g}$

$\Delta H°_c = \left(-45.35 \dfrac{\text{kJ}}{\text{g}}\right)\left(0.579 \dfrac{\text{g}}{\text{mL}}\right) = -26.3 \text{ kJ/mL}$

8.21 $\Delta S°$ is negative because the reaction decreases the number of moles of gaseous molecules.

8.22 The reaction proceeds from a solid and a gas (reactants) to all gas (product). This is more disordered and $\Delta S°$ is positive.

8.23 (a) Because $\Delta G°$ is negative, the reaction is spontaneous.
(b) Because $\Delta G°$ is positive, the reaction is nonspontaneous.

8.24 $\Delta G° = \Delta H° - T\Delta S° = (-92.2 \text{ kJ}) - (298 \text{ K})(-0.199 \text{ kJ/K}) = -32.9 \text{ kJ}$
Because $\Delta G°$ is negative, the reaction is spontaneous.
Set $\Delta G° = 0$ and solve for T.
$\Delta G° = 0 = \Delta H° - T\Delta S°$
$T = \dfrac{\Delta H°}{\Delta S°} = \dfrac{-92.2 \text{ kJ}}{-0.199 \text{ kJ/K}} = 463 \text{ K} = 190°C$

8.25 (a) $2 A_2 + B_2 \rightarrow 2 A_2B$
(b) Because the reaction is exothermic, ΔH is negative. There are more reactant molecules than product molecules. The randomness of the system decreases on going from reactant to product, therefore ΔS is negative.

(c) Because $\Delta G = \Delta H - T\Delta S$, a reaction with both ΔH and ΔS negative is favored at low temperatures where the negative ΔH term is larger than the positive $-T\Delta S$, and ΔG is negative.

Understanding Key Concepts

8.26. (a) $w = -P\Delta V$, $\Delta V > 0$; therefore $w < 0$ and the system is doing work on the surroundings.
(b) Since the temperature has increased there has been an enthalpy change. The system evolved heat, the reaction is exothermic, and $\Delta H < 0$.

8.28

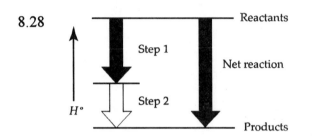

8.30 $\Delta H = \Delta E + P\Delta V$

$\Delta H - \Delta E = P\Delta V$

$$\Delta V = \frac{\Delta H - \Delta E}{P} = \frac{[-35.0\ \text{kJ} - (-34.8\ \text{kJ})]}{1\ \text{atm}} \times \frac{1\ \text{L} \cdot \text{atm}}{101 \times 10^{-3}\ \text{kJ}} = -2\ \text{L}$$

$\Delta V = -2\ \text{L} = V_{final} - V_{initial} = V_{final} - 5\ \text{L}; \qquad V_{final} = -2\ \text{L} - (-5\ \text{L}) = 3\ \text{L}$
The volume decreases from 5 L to 3 L.

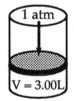

8.32 The change is the spontaneous conversion of a liquid to a gas. ΔG is negative because the change is spontaneous. The conversion of a liquid to a gas is endothermic, therefore ΔH is positive. ΔS is positive because the gas is more disordered than the liquid.

Additional Problems
Heat, Work, and Energy

8.34 Heat is the energy transferred from one object to another as the result of a temperature difference between them. Temperature is a measure of the kinetic energy of molecular motion.
Energy is the capacity to do work or supply heat. Work is defined as the distance moved times the force that opposes the motion ($w = d \times F$).
Kinetic energy is the energy of motion. Potential energy is stored energy.

8.36   Car: $E_K = \frac{1}{2}(1400 \text{ kg})\left(\dfrac{115 \times 10^3 \text{ m}}{3600 \text{ s}}\right)^2 = 7.1 \times 10^5 \text{ J}$

Truck: $E_K = \frac{1}{2}(12{,}000 \text{ kg})\left(\dfrac{38 \times 10^3 \text{ m}}{3600 \text{ s}}\right)^2 = 6.7 \times 10^5 \text{ J}$

The car has more kinetic energy.

8.38 $w = -P\Delta V = -(3.6 \text{ atm})(3.4 \text{ L} - 3.2 \text{ L}) = -0.72 \text{ L} \cdot \text{atm}$

$w = (-0.72 \text{ L} \cdot \text{atm})\left(\dfrac{101 \text{ J}}{1 \text{ L} \cdot \text{atm}}\right) = -72.7 \text{ J} = -70 \text{ J};$ The energy change is negative.

Energy and Enthalpy

8.40 $\Delta E = q_v$ is the heat change associated with a reaction at constant volume. Since $\Delta V = 0$, no PV work is done.
$\Delta H = q_p$ is the heat change associated with a reaction at constant pressure. Since $\Delta V \neq 0$, PV work can also be done.

8.42 $\Delta H = \Delta E + P\Delta V$; ΔH and ΔE are nearly equal when there are no gases involved in a chemical reaction, or, if gases are involved, $\Delta V = 0$ (that is, there are the same number of reactant and product gas molecules).

8.44 $P\Delta V = -7.6 \text{ J}$ (from Problem 8.39)
$\Delta H = \Delta E + P\Delta V$
$\Delta E = \Delta H - P\Delta V = -0.31 \text{ kJ} - (-7.6 \times 10^{-3} \text{ kJ}) = -0.30 \text{ kJ}$

8.46 $\Delta H = -1255.5 \text{ kJ/mol C}_2\text{H}_2$; C_2H_2, 26.04 amu
$w = -P\Delta V = -(1.00 \text{ atm})(-2.80 \text{ L}) = 2.80 \text{ L} \cdot \text{atm}$

$w = (2.80 \text{ L} \cdot \text{atm})\left(\dfrac{101 \text{ J}}{1 \text{ L} \cdot \text{atm}}\right) = 283 \text{ J} = 0.283 \text{ kJ}$

$6.50 \text{ g} \times \dfrac{1 \text{ mol C}_2\text{H}_2}{26.04 \text{ g C}_2\text{H}_2} = 0.250 \text{ mol C}_2\text{H}_2$

$q = (-1255.5 \text{ kJ/mol})(0.250 \text{ mol}) = -314 \text{ kJ}$
$\Delta E = \Delta H - P\Delta V = -314 \text{ kJ} - (-0.283 \text{ kJ}) = -314 \text{ kJ}$

8.48 $\text{C}_4\text{H}_{10}\text{O}$, 74.12 amu; mass of $\text{C}_4\text{H}_{10}\text{O} = \left(0.7138 \dfrac{\text{g}}{\text{mL}}\right)(100 \text{ mL}) = 71.38 \text{ g}$

$\text{mol C}_4\text{H}_{10}\text{O} = 71.38 \text{ g} \times \dfrac{1 \text{ mol}}{74.12 \text{ g}} = 0.9626 \text{ mol}$

$q = n \times \Delta H_{vap} = 0.9626 \text{ mol} \times 26.5 \text{ kJ/mol} = 25.5 \text{ kJ}$

8.50   Al, 26.98 amu

$\text{mol Al} = 5.00 \text{ g} \times \dfrac{1 \text{ mol}}{26.98 \text{ g}} = 0.1853 \text{ mol}$

$q = n \times \Delta H° = 0.1853 \text{ mol Al} \times \dfrac{-1408.4 \text{ kJ}}{2 \text{ mol Al}} = -131 \text{ kJ}; \quad 131 \text{ kJ is released.}$

8.52 Fe_2O_3, 159.7 amu

$\text{mol } Fe_2O_3 = 2.50 \text{ g} \times \dfrac{1 \text{ mol}}{159.7 \text{ g}} = 0.015 \ 65 \text{ mol}$

$q = n \times \Delta H° = 0.015 \ 65 \text{ mol } Fe_2O_3 \times \dfrac{-24.8 \text{ kJ}}{1 \text{ mol } Fe_2O_3} = -0.388 \text{ kJ}; \quad 0.388 \text{ kJ is evolved.}$

Because ΔH is negative, the reaction is exothermic.

Calorimetry and Heat Capacity

8.54   Heat capacity is the amount of heat required to raise the temperature of a substance a given amount. Specific heat is the amount of heat necessary to raise the temperature of exactly 1 g of a substance by exactly 1°C.

8.56   Na, 22.99 amu

$\text{specific heat} = 28.2 \ \dfrac{J}{\text{mol} \cdot °C} \times \dfrac{1 \text{ mol}}{22.99 \text{ g}} = 1.23 \text{ J/(g} \cdot °\text{C)}$

8.58   Mass of solution = 50.0 g + 1.045 g = 51.0 g

$q = \text{(specific heat)} \times m \times \Delta T$

$q = \left(4.18 \ \dfrac{J}{\text{g} \cdot °C} \right) (51.0 \text{ g})(32.3 °C - 25.0 °C) = 1.56 \times 10^3 \text{ J} = 1.56 \text{ kJ}$

CaO, 56.08 amu; $\text{mol CaO} = 1.045 \text{ g} \times \dfrac{1 \text{ mol}}{56.08 \text{ g}} = 0.018 \ 63 \text{ mol}$

$\text{Heat evolved per mole of CaO} = \dfrac{1.56 \text{ kJ}}{0.018 \ 63 \text{ mol}} = 83.7 \text{ kJ/mol CaO}$

Because the reaction evolves heat, the sign for ΔH is negative. ΔH = −83.7 kJ

8.60   NaOH, 40.00 amu;   HCl, 36.46 amu

$8.00 \text{ g NaOH} \times \dfrac{1 \text{ mol NaOH}}{40.00 \text{ g NaOH}} = 0.200 \text{ mol NaOH}$

$8.00 \text{ g HCl} \times \dfrac{1 \text{ mol HCl}}{36.46 \text{ g HCl}} = 0.219 \text{ mol HCl}$

Because the reaction stoichiometry between NaOH and HCl is one to one, the NaOH is the limiting reactant.

$q_P = -q_{soln} = -\text{(specific heat)} \times m \times \Delta T = - \left(4.18 \ \dfrac{J}{\text{g} \cdot °C} \right) (316 \text{ g})(33.5 °C - 25.0 °C) = -11.2 \text{ kJ}$

$$\Delta H = \frac{q_p}{n} = (-11.2 \text{ kJ})/(0.200 \text{ mol}) = -56 \text{ kJ/mol}$$

When 10.00 g of HCl in 248.0 g of water is added the same temperature increase is observed because the mass of NaOH is the same and it is still the limiting reactant. The mass of the solution is also the same.

Hess's Law and Heats of Formation

8.62   The standard state of an element is its most stable form at 1 atm and the specified temperature, usually 25°C.

8.64   Hess's Law -- the overall enthalpy change for a reaction is equal to the sum of the enthalpy changes for the individual steps in the reaction.
Hess's Law works because of the law of conservation of energy.

8.66

	$S(s) + O_2(g) \rightarrow SO_2(g)$	$\Delta H^\circ_1 = -296.8 \text{ kJ}$
	$\underline{SO_2 + \frac{1}{2} O_2(g) \rightarrow SO_3(g)}$	$\Delta H^\circ_2 = -98.9 \text{ kJ}$
Sum	$S(s) + 3/2\, O_2(g) \rightarrow SO_3(g)$	$\Delta H^\circ_3 = \Delta H^\circ_1 + \Delta H^\circ_2$

$$\Delta H^\circ_f = \Delta H^\circ_3 = -296.8 \text{ kJ} + (-98.9 \text{ kJ}) = -395.7 \text{ kJ/mol}$$

8.68

	$SO_3(g) + H_2O(l) \rightarrow H_2SO_4(aq)$	$\Delta H^\circ_1 = -227.8 \text{ kJ}$
	$H_2(g) + \frac{1}{2} O_2(g) \rightarrow H_2O(l)$	$\Delta H^\circ_2 = \Delta H^\circ_f = -285.8 \text{ kJ}$
	$\underline{S(s) + 3/2\, O_2(g) \rightarrow SO_3(g)}$	$\Delta H^\circ_3 = \Delta H^\circ_f = -395.7$
Sum	$S(s) + H_2(g) + 2\, O_2(g) \rightarrow H_2SO_4(aq)$	$\Delta H^\circ_f (H_2SO_4) = ?$

$$\Delta H^\circ_f (H_2SO_4) = \Delta H^\circ_1 + \Delta H^\circ_2 + \Delta H^\circ_3 = -909.3 \text{ kJ}$$

8.70 $C_8H_8(l) + 10\, O_2(g) \rightarrow 8\, CO_2(g) + 4\, H_2O(l)$
$\Delta H^\circ_{rxn} = \Delta H^\circ_c = -4395.2 \text{ kJ}$
$\Delta H^\circ_{rxn} = [8\, \Delta H^\circ_f(CO_2) + 4\, \Delta H^\circ_f(H_2O)] - \Delta H^\circ_f(C_8H_8)$
$-4395.2 \text{ kJ} = [(8 \text{ mol})(-393.5 \text{ kJ/mol}) + (4 \text{ mol})(-285.8 \text{ kJ/mol})] - [(1 \text{ mol})(\Delta H^\circ_f(C_8H_8))]$
Solve for $\Delta H^\circ_f(C_8H_8)$
$-4395.2 \text{ kJ} = -4291.2 \text{ kJ} - (1 \text{ mol})(\Delta H^\circ_f(C_8H_8))$
$-104.0 \text{ kJ} = -(1 \text{ mol})(\Delta H^\circ_f(C_8H_8))$
$$\Delta H^\circ_f(C_8H_8) = \frac{-104.0 \text{ kJ}}{-1 \text{ mol}} = +104.0 \text{ kJ/mol}$$

8.72 $\Delta H^\circ_{rxn} = \Delta H^\circ_f(\text{MTBE}) - [\Delta H^\circ_f(2\text{-Methylpropene}) + \Delta H^\circ_f(CH_3OH)]$
$-57.8 \text{ kJ} = -313.6 \text{ kJ} - [(1 \text{ mol})(\Delta H^\circ_f(2\text{-Methylpropene})) + (-238.7 \text{ kJ})]$
Solve for $\Delta H^\circ_f(2\text{-Methylpropene})$.
$-17.1 \text{ kJ} = (1 \text{ mol})(\Delta H^\circ_f(2\text{-Methylpropene}))$
$\Delta H^\circ_f(2\text{-Methylpropene}) = -17.1 \text{ kJ/mol}$

Bond Dissociation Energies

8.74 $H_2C=CH_2(g) + H_2(g) \rightarrow CH_3CH_3(g)$

$\Delta H°_{rxn} = D$ (Reactant bonds) $- D$ (Product bonds)

$\Delta H°_{rxn} = (D_{C=C} + 4 D_{C-H} + D_{H-H}) - (6 D_{C-H} + D_{C-C})$

$\Delta H°_{rxn} = [(1\ mol)(611\ kJ/mol) + (4\ mol)(410\ kJ/mol) + (1\ mol)(436\ kJ/mol)]$
$- [(6\ mol)(410\ kJ/mol) + (1\ mol)(350\ kJ/mol)] = -123\ kJ$

8.76 $C_4H_{10} + 13/2\ O_2 \rightarrow 4\ CO_2 + 5\ H_2O$

$\Delta H°_{rxn} = D$ (Reactant bonds) $- D$ (Product bonds)

$\Delta H°_{rxn} = (3 D_{C-C} + 10 D_{C-H} + 13/2 D_{O=O}) - (8 D_{C=O} + 10 D_{O-H})$

$\Delta H°_{rxn} = [(3\ mol)(350\ kJ/mol) + (10\ mol)(410\ kJ/mol) + (13/2\ mol)(498\ kJ/mol)]$
$- [(8\ mol)(804\ kJ/mol) + (10\ mol)(460\ kJ/mol)] = -2645\ kJ$

Free Energy and Entropy

8.78 Entropy is a measure of molecular disorder.

8.80 A reaction can be spontaneous yet endothermic if ΔS is positive (more disorder) and the $T\Delta S$ term is larger than ΔH.

8.82 (a) positive (more disorder) (b) negative (more order)

8.84 (a) zero (equilibrium) (b) zero (equilibrium)
(c) negative (spontaneous)

8.86 ΔS is positive. The reaction increases the total number of molecules.

8.88 $\Delta G = \Delta H - T\Delta S$
(a) $\Delta G = -48\ kJ - (400\ K)(135 \times 10^{-3}\ kJ/K) = -102\ kJ$
$\Delta G < 0$, spontaneous; $\Delta H < 0$, exothermic.
(b) $\Delta G = -48\ kJ - (400\ K)(-135 \times 10^{-3}\ kJ/K) = +6\ kJ$
$\Delta G > 0$, nonspontaneous; $\Delta H < 0$, exothermic.
(c) $\Delta G = +48\ kJ - (400\ K)(135 \times 10^{-3}\ kJ/K) = -6\ kJ$
$\Delta G < 0$, spontaneous; $\Delta H > 0$, endothermic.
(d) $\Delta G = +48\ kJ - (400\ K)(-135 \times 10^{-3}\ kJ/K) = +102\ kJ$
$\Delta G > 0$, nonspontaneous; $\Delta H > 0$, endothermic.

8.90 $\Delta G = \Delta H - T\Delta S$; Set $\Delta G = 0$ and solve for T (the crossover temperature).

$T = \dfrac{\Delta H}{\Delta S} = \dfrac{-33\ kJ}{-0.058\ kJ/K} = 570\ K$

8.92 (a) $\Delta H < 0$ and $\Delta S > 0$; reaction is spontaneous at all temperatures.
(b) $\Delta H < 0$ and $\Delta S < 0$; reaction has a crossover temperature.

(c) $\Delta H > 0$ and $\Delta S > 0$; reaction has a crossover temperature.

(d) $\Delta H > 0$ and $\Delta S < 0$; reaction is nonspontaneous at all temperatures.

8.94 $T = -114.1°C = 273.15 + (-114.1) = 159.0 \text{ K}$

$\Delta G_{fus} = \Delta H_{fus} - T\Delta S_{fus}$; $\Delta G = 0$ at the melting point temperature.

Set $\Delta G = 0$ and solve for ΔS_{fus}.

$\Delta G = 0 = \Delta H_{fus} - T\Delta S_{fus}$

$$\Delta S_{fus} = \frac{\Delta H_{fus}}{T} = \frac{5.02 \text{ kJ/mol}}{159.0 \text{ K}} = 0.0316 \text{ kJ/(K·mol)} = 31.6 \text{ J/(K·mol)}$$

General Problems

8.96 $\text{Mg(s)} + 2\,\text{HCl(aq)} \rightarrow \text{MgCl}_2\text{(aq)} + \text{H}_2\text{(g)}$

$\text{mol Mg} = 1.50 \text{ g} \times \dfrac{1 \text{ mol}}{24.3 \text{ g}} = 0.0617 \text{ mol Mg}$

$\text{mol HCl} = 0.200 \text{ L} \times 6.00\, \dfrac{\text{mol}}{\text{L}} = 1.20 \text{ mol HCl}$

There is an excess of HCl. Mg is the limiting reactant.

$q = \left(4.18\,\dfrac{\text{J}}{\text{g·°C}}\right)(200 \text{ g})(42.9\,°C - 25.0\,°C) + \left(776\,\dfrac{\text{J}}{°C}\right)(42.9\,°C - 25.0\,°C) = 2.89 \times 10^4 \text{ J}$

$q = 2.89 \times 10^4 \text{ J} \times \dfrac{1 \text{ kJ}}{1000 \text{ J}} = 28.9 \text{ kJ}$

$\text{Heat evolved per mole of Mg} = \dfrac{28.9 \text{ kJ}}{0.0617 \text{ mol}} = 468 \text{ kJ/mol}$

Because the reaction evolves heat, the sign for ΔH is negative. $\Delta H = -468 \text{ kJ}$

8.98

$2\,\text{NO(g)} + \text{O}_2\text{(g)} \rightarrow 2\,\text{NO}_2\text{ (g)}$	$\Delta H°_1 = 2(-57.0 \text{ kJ})$
$2\,\text{NO}_2\text{ (g)} \rightarrow \text{N}_2\text{O}_4\text{(g)}$	$\Delta H°_2 = -57.2 \text{ kJ}$

Sum $\quad 2\,\text{NO(g)} + \text{O}_2\text{(g)} \rightarrow \text{N}_2\text{O}_4\text{(g)}$

$\Delta H° = \Delta H°_1 + \Delta H°_2 = -171.2 \text{ kJ}$

8.100 $\Delta G_{fus} = \Delta H_{fus} - T\Delta S_{fus}$; at the melting point $\Delta G = 0$. Set $\Delta G = 0$ and solve for T (the melting point).

$\Delta G = 0 = \Delta H_{fus} - T\Delta S_{fus}$

$T = \dfrac{\Delta H_{fus}}{\Delta S_{fus}} = \dfrac{9.95 \text{ kJ}}{0.0357 \text{ kJ/K}} = 279 \text{ K}$

8.102 $\Delta H°_{rxn} = D \text{ (Reactant bonds)} - D \text{ (Product bonds)}$

(a) $2\,\text{CH}_4\text{(g)} \rightarrow \text{C}_2\text{H}_6\text{(g)} + \text{H}_2\text{(g)}$

$\Delta H°_{rxn} = (8\,D_{C-H}) - (D_{C-C} + 6\,D_{C-H} + D_{H-H})$

$\Delta H°_{rxn} = [(8 \text{ mol})(410 \text{ kJ/mol})] - [(1 \text{ mol})(350 \text{ kJ/mol}) + (6 \text{ mol})(410 \text{ kJ/mol})$

$+ (1 \text{ mol})(436 \text{ kJ/mol})] = +34 \text{ kJ}$

(b) $C_2H_6(g) + F_2(g) \rightarrow C_2H_5F(g) + HF(g)$

$\Delta H°_{rxn} = (6\,D_{C-H} + D_{C-C} + D_{F-F}) - (5\,D_{C-H} + D_{C-C} + D_{C-F} + D_{H-F})$

$\Delta H°_{rxn} = [(6\text{ mol})(410\text{ kJ/mol}) + (1\text{ mol})(350\text{ kJ/mol}) + (1\text{ mol})(159\text{ kJ/mol})]$
$\qquad - [(5\text{ mol})(410\text{ kJ/mol}) + (1\text{ mol})(350\text{ kJ/mol}) + (1\text{ mol})(450\text{ kJ/mol})$
$\qquad + (1\text{ mol})(570\text{ kJ/mol})] = -451\text{ kJ}$

(c) $N_2(g) + 3\,H_2(g) \rightarrow 2\,NH_3(g)$

The bond dissociation energy for N_2 is 945 kJ/mol.

$\Delta H°_{rxn} = (D_{N_2} + 3\,D_{H-H}) - (6\,D_{N-H})$

$\Delta H°_{rxn} = [(1\text{ mol})(945\text{ kJ/mol}) + (3\text{ mol})(436\text{ kJ/mol})] - [(6\text{ mol})(390\text{ kJ/mol})] = -87\text{ kJ}$

8.104 (a) $2\,C_8H_{18}(l) + 25\,O_2(g) \rightarrow 16\,CO_2(g) + 18\,H_2O(g)$

(b) $C_8H_{18}(l) + 25/2\,O_2(g) \rightarrow 8\,CO_2(g) + 9\,H_2O(g)$

$\Delta H°_{rxn} = \Delta H°_c = -5456.6\text{ kJ}$

$\Delta H°_{rxn} = [8\,\Delta H°_f(CO_2) + 9\,\Delta H°_f(H_2O)] - \Delta H°_f(C_8H_{18})$

$-5456.6\text{ kJ} = [(8\text{ mol})(-393.5\text{ kJ/mol}) + (9\text{ mol})(-241.8\text{ kJ/mol})] - [(1\text{ mol})(\Delta H°_f(C_8H_{18}))]$

Solve for $\Delta H°_f(C_8H_{18})$.

$-5456.6\text{ kJ} = -5324\text{ kJ} - [(1\text{ mol})(\Delta H°_f(C_8H_{18}))]$

$-132.4\text{ kJ} = -(1\text{ mol})(\Delta H°_f(C_8H_{18}))$

$\Delta H°_f(C_8H_{18}) = +132.4\text{ kJ/mol}$

8.106 (a) $\Delta S_{total} = \Delta S_{system} + \Delta S_{surr}$ and $\Delta S_{surr} = -\Delta H/T$

$\Delta S_{total} = \Delta S_{system} + (-\Delta H/T) = \Delta S_{system} - \Delta H/T$

$\Delta S_{system} = \Delta S_{total} + \Delta H/T$

$\Delta G = \Delta H - T\Delta S$ (substitute ΔS_{system} for ΔS in this equation)

$\Delta G = \Delta H - T(\Delta S_{total} + \Delta H/T) = -T\Delta S_{total}$

$\Delta G = -T\Delta S_{total}$ For a spontaneous reaction, if $\Delta S_{total} > 0$ then $\Delta G < 0$.

(b) $\Delta G° = \Delta H° - T\Delta S°$

$\Delta H° = \Delta G° + T\Delta S°$

$\Delta S_{surr} = -\dfrac{\Delta H°}{T} = -\dfrac{[\Delta G° + T\Delta S°]}{T} = -\dfrac{[2879 \times 10^3 \text{ J/mol} + (298\text{ K})(-210\text{ J/(K}\cdot\text{mol}))]}{298\text{ K}}$

$\Delta S_{surr} = -9451\text{ J/(K}\cdot\text{mol})$

8.108

$2\,CH_4(g) + 4\,O_2(g) \rightarrow 2\,CO_2(g) + 4\,H_2O(l)$	$\Delta H°_1 = 2(-890.3\text{ kJ})$
$C_2H_6(g) \rightarrow C_2H_4(g) + H_2(g)$	$\Delta H°_2 = +137.0\text{ kJ}$
$2\,CO_2(g) + 3\,H_2O(l) \rightarrow C_2H_6(g) + 7/2\,O_2(g)$	$\Delta H°_3 = \dfrac{3119.4\text{ kJ}}{2}$
$H_2O(l) \rightarrow H_2(g) + 1/2\,O_2(g)$	$\Delta H°_4 = +285.8\text{ kJ}$
Sum $2\,CH_4(g) \rightarrow C_2H_4(g) + 2\,H_2(g)$	$\Delta H° = +201.9\text{ kJ}$

8.110 $q_{ice\,tea} = -q_{ice}$

$q_{ice\,tea} = (4.18\,\dfrac{J}{g\cdot°C})(400.0\text{ g})(10.0°C - 80.0°C) = -1.17 \times 10^5\text{ J}$

H_2O, 18.02 amu

$$q_{ice} = 1.17 \times 10^5 \text{ J} = (6.01 \text{ kJ/mol})\left(\frac{1000 \text{ J}}{1 \text{ kJ}}\right)\left(m_{ice} \times \frac{1 \text{ mol H}_2\text{O}}{18.02 \text{ g H}_2\text{O}}\right)$$

$$+ \left(4.18 \frac{\text{J}}{(\text{g} \cdot {}^\circ\text{C})}\right)(m_{ice})(10.0{}^\circ\text{C} - 0.0{}^\circ\text{C})$$

Solve for the mass of ice, m_{ice}.
$$1.17 \times 10^5 \text{ J} = (3.34 \times 10^2 \text{ J/g})(m_{ice}) + (41.8 \text{ J/g})(m_{ice}) = (3.76 \times 10^2 \text{ J/g})(m_{ice})$$

$$m_{ice} = \frac{1.17 \times 10^5 \text{ J}}{3.76 \times 10^2 \text{ J/g}} = 311 \text{ g of ice}$$

8.112 $CsOH(aq) + HCl(aq) \rightarrow CsCl(aq) + H_2O(l)$

$$\text{mol CsOH} = 0.100 \text{ L} \times \frac{0.200 \text{ mol CsOH}}{1.00 \text{ L}} = 0.0200 \text{ mol CsOH}$$

$$\text{mol HCl} = 0.050 \text{ L} \times \frac{0.400 \text{ mol HCl}}{1.00 \text{ L}} = 0.0200 \text{ mol HCl}$$

The reactants were mixed in equal mole amounts.
Total volume = 150 mL and has a mass of 150 g

$$q_{solution} = \left(4.2 \frac{\text{J}}{(\text{g} \cdot {}^\circ\text{C})}\right)(150 \text{ g})(24.28{}^\circ\text{C} - 22.50{}^\circ\text{C}) = 1121 \text{ J}$$

$$q_{reaction} = -q_{solution} = -1121 \text{ J}$$

$$\Delta H = \frac{q_{reaction}}{\text{mol CsOH}} = \frac{-1121 \text{ J}}{0.0200 \text{ mol CsOH}} \times \frac{1 \text{ kJ}}{1000 \text{ J}} = -56 \text{ kJ/mol CsOH}$$

8.114

	ΔH
$4 \text{ CO}(g) + 2 \text{ O}_2(g) \rightarrow 4 \text{ CO}_2(g)$	$2(-566.0 \text{ kJ})$
$2 \text{ NO}_2(g) \rightarrow 2 \text{ NO}(g) + \text{O}_2(g)$	$+114.0 \text{ kJ}$
$2 \text{ NO}(g) \rightarrow \text{O}_2(g) + \text{N}_2(g)$	$2(-90.2 \text{ kJ})$
$4 \text{ CO}(g) + 2 \text{ NO}_2(g) \rightarrow 4 \text{ CO}_2(g) + \text{N}_2(g)$	-1198.4 kJ

Multi-Concept Problems

8.116 (a)

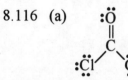

(b) $C(g) + \frac{1}{2} O_2(g) + Cl_2(g) \rightarrow COCl_2(g)$
$\Delta H^\circ_f = \Delta H^\circ_f(C(g)) + (\frac{1}{2} D_{O=O} + D_{Cl-Cl}) - (D_{C=O} + 2 D_{C-Cl})$
$\Delta H^\circ_f = (716.7 \text{ kJ}) + [(\frac{1}{2} \text{ mol})(498 \text{ kJ/mol}) + (1 \text{ mol})(243 \text{ kJ/mol})]$
$\qquad\qquad\qquad - [(1 \text{ mol})(732 \text{ kJ/mol}) + (2 \text{ mol})(330 \text{ kJ/mol})]$
$\Delta H^\circ_f = -183 \text{ kJ per mol COCl}_2$ (calculated from bond energies)

From Appendix B, $\Delta H°_f(COCl_2) = -219.1$ kJ/mol
The calculation of $\Delta H°_f$ from bond energies is only an estimate because the bond energies are average values derived from many different compounds.

8.118 (a) $2\,K(s) + 2\,H_2O(l) \rightarrow 2\,KOH(aq) + H_2(g)$

(b) $\Delta H°_{rxn} = [2\,\Delta H°_f(KOH)] - [2\,\Delta H°_f(H_2O)]$
$\Delta H°_{rxn} = [(2\ mol)(-482.4\ kJ/mol)] - [(2\ mol)(-285.8\ kJ/mol)] = -393.2$ kJ

(c) The reaction produces 393.2 kJ/ 2 mol K = 196.6 kJ/ mol K.
Assume that the mass of the water does not change and that the specific heat = 4.18 J/(g·°C) for the solution that is produced.

$$q = 7.55\ g\ K \times \frac{1\ mol\ K}{39.10\ g\ K} \times \frac{196.6\ kJ}{1\ mol\ K} \times \frac{1000\ J}{1\ kJ} = 3.80 \times 10^4\ J$$

$q = (specific\ heat) \times m \times \Delta T$

$$\Delta T = \frac{q}{(specific\ heat) \times m} = \frac{3.80 \times 10^4\ J}{[4.18\ J/(g \cdot °C)](400.0\ g)} = 22.7°C$$

$\Delta T = T_{final} - T_{initial}$
$T_{final} = \Delta T + T_{initial} = 22.7°C + 25.0°C = 47.7°C$

(d) $7.55\ g\ K \times \dfrac{1\ mol\ K}{39.10\ g\ K} \times \dfrac{2\ mol\ KOH}{2\ mol\ K} = 0.193\ mol\ KOH$

Assume that the mass of the solution does not change during the reaction and that the solution has a density of 1.00 g/mL.

$$solution\ volume = 400.0\ g \times \frac{1.00\ mL}{1\ g} \times \frac{1\ L}{1000\ mL} = 0.400\ L$$

$$molarity = \frac{0.193\ mol\ KOH}{0.400\ L} = 0.483\ M$$

$2\,KOH(aq) + H_2SO_4(aq) \rightarrow K_2SO_4(aq) + 2\,H_2O(l)$

$$0.193\ mol\ KOH \times \frac{1\ mol\ H_2SO_4}{2\ mol\ KOH} \times \frac{1000\ mL}{0.554\ mol\ H_2SO_4} = 174\ mL\ of\ 0.554\ M\ H_2SO_4$$

8.120 Assume 100.0 g of Y.

$$mol\ F = 61.7\ g\ F \times \frac{1\ mol\ F}{19.00\ g\ F} = 3.25\ mol\ F$$

$$mol\ Cl = 38.3\ g\ Cl \times \frac{1\ mol\ Cl}{35.45\ g\ Cl} = 1.08\ mol\ Cl$$

$Cl_{1.08}F_{3.25}$, divide each subscript by the smaller of the two, 1.08.
$Cl_{1.08/1.08}F_{3.25/1.08}$
ClF_3

(a) Y is ClF_3 and X is ClF

(b) 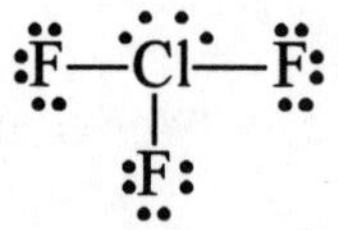 :F̈—C̈l—F̈:
       |
      :F̈:

There are five electron clouds around the Cl (3 bonding and 2 lone pairs). The geometry is T-shaped.

(c) ΔH

$$Cl_2O(g) + 3\ OF_2(g) \rightarrow 2\ O_2(g) + 2\ ClF_3(g) \qquad -533.4\ kJ$$
$$2\ ClF(g) + O_2(g) \rightarrow Cl_2O(g) + OF_2(g) \qquad +205.6\ kJ$$
$$\underline{O_2(g) + 2\ F_2(g) \rightarrow 2\ OF_2(g)} \qquad\qquad \underline{2(24.7\ kJ)}$$
$$2\ ClF(g) + 2\ F_2(g) \rightarrow 2\ ClF_3(g) \qquad -278.4\ kJ$$

Divide reaction coefficients and ΔH by 2.

$$ClF(g) + F_2(g) \rightarrow ClF_3(g) \qquad \Delta H = -278.4\ kJ/2 = -139.2\ kJ/mol\ ClF_3$$

(d) ClF, 54.45 amu

$$q = 25.0\ g\ ClF \times \frac{1\ mol\ ClF}{54.45\ g\ ClF} \times \frac{-139.2\ kJ}{1\ mol\ ClF} \times 0.875 = -55.9\ kJ$$

55.9 kJ is released in this reaction.

9

Gases: Their Properties and Behavior

9.1   1.00 atm = 14.7 psi

$$1.00 \text{ mm Hg} \times \frac{1 \text{ atm}}{760 \text{ mm Hg}} \times \frac{14.7 \text{ psi}}{1 \text{ atm}} = 1.93 \times 10^{-2} \text{ psi}$$

9.2   1.00 atmosphere pressure can support a column of Hg 0.760 m high. Because the density of H_2O is 1.00 g/mL and that of Hg is 13.6 g/mL, 1.00 atmosphere pressure can support a column of H_2O 13.6 times higher than that of Hg. The column of H_2O supported by 1.00 atmosphere will be (0.760 m)(13.6) = 10.3 m.

9.3   The pressure in the flask is less than 0.975 atm because the liquid level is higher on the side connected to the flask. The 24.7 cm of Hg is the difference between the two pressures.

$$\text{Pressure difference} = 24.7 \text{ cm Hg} \times \frac{1.00 \text{ atm}}{76.0 \text{ cm Hg}} = 0.325 \text{ atm}$$

Pressure in flask = 0.975 atm – 0.325 atm = 0.650 atm

9.4   The pressure in the flask is greater than 750 mm Hg because the liquid level is lower on the side connected to the flask.

$$\text{Pressure difference} = 25 \text{ cm Hg} \times \frac{10 \text{ mm Hg}}{1 \text{ cm Hg}} = 250 \text{ mm Hg}$$

Pressure in flask = 750 mm Hg + 250 mm Hg = 1000 mm Hg

9.5   (a) Assume an initial volume of 1.00 L.
First consider the volume change resulting from a change in the number of moles with the pressure and temperature constant.

$$\frac{V_i}{n_i} = \frac{V_f}{n_f}; \quad V_f = \frac{V_i n_f}{n_i} = \frac{(1.00 \text{ L})(0.225 \text{ mol})}{0.3 \text{ mol}} = 0.75 \text{ L}$$

Now consider the volume change from 0.75 L as a result of a change in temperature with the number of moles and the pressure constant.

$$\frac{V_i}{T_i} = \frac{V_f}{T_f}; \quad V_f = \frac{V_i T_f}{T_i} = \frac{(0.75 \text{ L})(400 \text{ K})}{300 \text{ K}} = 1.0 \text{ L}$$

There is no net change in the volume as a result of the decrease in the number of moles of gas and a temperature increase.

(b) Assume an initial volume of 1.00 L.

First consider the volume change resulting from a change in the number of moles with the pressure and temperature constant.

$$\frac{V_i}{n_i} = \frac{V_f}{n_f}; \quad V_f = \frac{V_i\,n_f}{n_i} = \frac{(1.00\ L)(0.225\ mol)}{0.3\ mol} = 0.75\ L$$

Now consider the volume change from 0.75 L as a result of a change in temperature with the number of moles and the pressure constant.

$$\frac{V_i}{T_i} = \frac{V_f}{T_f}; \quad V_f = \frac{V_i\,T_f}{T_i} = \frac{(0.75\ L)(200\ K)}{300\ K} = 0.5\ L$$

The volume would be cut in half as a result of the decrease in the number of moles of gas and a temperature decrease.

9.6 $n = \dfrac{PV}{RT} = \dfrac{(1.000\ atm)(1.000 \times 10^5\ L)}{\left(0.082\ 06\ \dfrac{L \cdot atm}{K \cdot mol}\right)(273.15\ K)} = 4.461 \times 10^3\ mol\ CH_4$

CH_4, 16.04 amu; $mass\ CH_4 = (4.461 \times 10^3\ mol)\left(\dfrac{16.04\ g}{1\ mol}\right) = 7.155 \times 10^4\ g\ CH_4$

9.7 C_3H_8, 44.10 amu;   V = 350 mL = 0.350 L;   T = 20°C = 293 K

$$n = 3.2\ g \times \frac{1\ mol\ C_3H_8}{44.10\ g\ C_3H_8} = 0.073\ mol\ C_3H_8$$

$$P = \frac{nRT}{V} = \frac{(0.073\ mol)\left(0.082\ 06\ \dfrac{L \cdot atm}{K \cdot mol}\right)(293\ K)}{0.350\ L} = 5.0\ atm$$

9.8 $P = 1.51 \times 10^4\ kPa \times \dfrac{1\ atm}{101.325\ kPa} = 149\ atm;$ T = 25.0°C = 298 K

$$n = \frac{PV}{RT} = \frac{(149\ atm)(43.8\ L)}{\left(0.082\ 06\ \dfrac{L \cdot atm}{K \cdot mol}\right)(298\ K)} = 267\ mol\ He$$

9.9   The volume and number of moles of gas remain constant.

$$\frac{nR}{V} = \frac{P_i}{T_i} = \frac{P_f}{T_f}; \quad T_f = \frac{P_f\,T_i}{P_i} = \frac{(2.37\ atm)(273\ K)}{2.15\ atm} = 301\ K = 28°C$$

9.10   (a) The temperature has increased by about 10% (from 300 K to 325 K) while the amount and the pressure are unchanged. Thus, the volume should increase by about 10%.

(b) The temperature has increased by a factor of 1.5 (from 300 K to 450 K) and the pressure has increased by a factor of 3 (from 0.9 atm to 2.7 atm) while the amount is unchanged. Thus, the volume should decrease by half (1.5/3 = 0.5).

(c) Both the amount and the pressure have increased by a factor of 3 (from 0.075 mol to 0.22 mol and from 0.9 atm to 2.7 atm) while the temperature is unchanged. Thus, the volume is unchanged.

9.11 $CaCO_3(s) + 2\,HCl(aq) \rightarrow CaCl_2(aq) + CO_2(g) + H_2O(l)$
       $CaCO_3$, 100.1 amu; CO_2, 44.01 amu

$$\text{mole } CO_2 = 33.7 \text{ g } CaCO_3 \times \frac{1 \text{ mol } CaCO_3}{100.1 \text{ g } CaCO_3} \times \frac{1 \text{ mol } CO_2}{1 \text{ mol } CaCO_3} = 0.337 \text{ mol } CO_2$$

$$\text{mass } CO_2 = 0.337 \text{ mol } CO_2 \times \frac{44.01 \text{ g } CO_2}{1 \text{ mol } CO_2} = 14.8 \text{ g } CO_2$$

$$V = \frac{nRT}{P} = \frac{(0.337 \text{ mol})\left(0.082\,06\,\dfrac{L \cdot atm}{K \cdot mol}\right)(273 \text{ K})}{1.00 \text{ atm}} = 7.55 \text{ L}$$

9.12 $C_3H_8(g) + 5\,O_2(g) \rightarrow 3\,CO_2(g) + 4\,H_2O(l)$

$$n_{propane} = \frac{PV}{RT} = \frac{(4.5 \text{ atm})(15.0 \text{ L})}{\left(0.082\,06\,\dfrac{L \cdot atm}{K \cdot mol}\right)(298 \text{ K})} = 2.76 \text{ mol } C_3H_8$$

$$2.76 \text{ mol } C_3H_8 \times \frac{3 \text{ mol } CO_2}{1 \text{ mol } C_3H_8} = 8.28 \text{ mol } CO_2$$

$$V = \frac{nRT}{P} = \frac{(8.28 \text{ mol})\left(0.082\,06\,\dfrac{L \cdot atm}{K \cdot mol}\right)(273 \text{ K})}{1.00 \text{ atm}} = 186 \text{ L} = 190 \text{ L}$$

9.13 $n = \dfrac{PV}{RT} = \dfrac{(1.00 \text{ atm})(1.00 \text{ L})}{\left(0.082\ 06\ \dfrac{\text{L} \cdot \text{atm}}{\text{K} \cdot \text{mol}}\right)(273 \text{ K})} = 0.0446 \text{ mol}$

molar mass $= \dfrac{1.52 \text{ g}}{0.0446 \text{ mol}} = 34.1 \text{ g/mol};$ molecular mass $= 34.1$ amu

$Na_2S(aq) + 2\ HCl(aq) \rightarrow H_2S(g) + 2\ NaCl(aq)$
The foul-smelling gas is H_2S, hydrogen sulfide.

9.14 $12.45 \text{ g } H_2 \ \times \ \dfrac{1 \text{ mol } H_2}{2.016 \text{ g } H_2} = 6.176 \text{ mol } H_2$

$60.67 \text{ g } N_2 \ \times \ \dfrac{1 \text{ mol } N_2}{28.01 \text{ g } N_2} = 2.166 \text{ mol } N_2$

$2.38 \text{ g } NH_3 \ \times \ \dfrac{1 \text{ mol } NH_3}{17.03 \text{ g } NH_3} = 0.140 \text{ mol } NH_3$

$n_{total} = n_{H_2} + n_{N_2} + n_{NH_3} = 6.176 \text{ mol} + 2.166 \text{ mol} + 0.140 \text{ mol} = 8.482 \text{ mol}$

$X_{H_2} = \dfrac{6.176 \text{ mol}}{8.482 \text{ mol}} = 0.7281;$ $X_{N_2} = \dfrac{2.166 \text{ mol}}{8.482 \text{ mol}} = 0.2554;$ $X_{NH_3} = \dfrac{0.140 \text{ mol}}{8.482 \text{ mol}} = 0.0165$

9.15 $n_{total} = 8.482$ mol (from Problem 9.14). $T = 90°C = 363$ K

$P_{total} = \dfrac{n_{total}RT}{V} = \dfrac{(8.482 \text{ mol})\left(0.082\ 06\ \dfrac{\text{L} \cdot \text{atm}}{\text{K} \cdot \text{mol}}\right)(363 \text{ K})}{10.00 \text{ L}} = 25.27 \text{ atm}$

$P_{H_2} = X_{H_2} \cdot P_{total} = (0.7281)(25.27 \text{ atm}) = 18.4 \text{ atm}$

$P_{N_2} = X_{N_2} \cdot P_{total} = (0.2554)(25.27 \text{ atm}) = 6.45 \text{ atm}$

$P_{NH_3} = X_{NH_3} \cdot P_{total} = (0.0165)(25.27 \text{ atm}) = 0.417 \text{ atm}$

9.16 $P_{H_2O} = X_{H_2O} \cdot P_{Total} = (0.0287)(0.977 \text{ atm}) = 0.0280 \text{ atm}$

9.17   The number of moles of each gas is proportional to the number of each of the different gas molecules in the container.

$n_{total} = n_{red} + n_{yellow} + n_{green} = 6 + 2 + 4 = 12$

$X_{red} = \dfrac{n_{red}}{n_{total}} = \dfrac{6}{12} = 0.500;$ $X_{yellow} = \dfrac{n_{yellow}}{n_{total}} = \dfrac{2}{12} = 0.167;$ $X_{green} = \dfrac{n_{green}}{n_{total}} = \dfrac{4}{12} = 0.333$

$P_{red} = X_{red} \cdot P_{total} = (0.500)(600 \text{ mm Hg}) = 300 \text{ mm Hg}$
$P_{yellow} = X_{yellow} \cdot P_{total} = (0.167)(600 \text{ mm Hg}) = 100 \text{ mm Hg}$
$P_{green} = X_{green} \cdot P_{total} = (0.333)(600 \text{ mm Hg}) = 200 \text{ mm Hg}$

9.18 $\quad u = \sqrt{\dfrac{3RT}{M}}$, M = molar mass, R = 8.314 J/(K · mol), 1 J = 1 kg · m²/s²

at 37°C = 310 K, $\quad u = \sqrt{\dfrac{3 \times 8.314 \text{ kg m}^2/(\text{s}^2 \text{ K mol}) \times 310 \text{ K}}{28.01 \times 10^{-3} \text{ kg/mol}}} = 525$ m/s

at –25°C = 248 K, $\quad u = \sqrt{\dfrac{3 \times 8.314 \text{ kg m}^2/(\text{s}^2 \text{ K mol}) \times 248 \text{ K}}{28.01 \times 10^{-3} \text{ kg/mol}}} = 470$ m/s

9.19 $\quad u = \sqrt{\dfrac{3RT}{M}}$, M = molar mass, R = 8.314 J/(K · mol), 1 J = 1 kg · m²/s²

O_2, 32.00 amu, 32.00 x 10⁻³ kg/mol

$u = 580 \text{ mi/h} \times \dfrac{1.6093 \text{ km}}{1 \text{ mi}} \times \dfrac{1000 \text{ m}}{1 \text{ km}} \times \dfrac{1 \text{ hr}}{60 \text{ min}} \times \dfrac{1 \text{ min}}{60 \text{ s}} = 259$ m/s

$u = \sqrt{\dfrac{3RT}{M}}; \quad u^2 = \dfrac{3RT}{M}$

$T = \dfrac{u^2 M}{3R} = \dfrac{(259 \text{ m/s})^2 (32.00 \times 10^{-3} \text{ kg/mol})}{(3)(8.314 \text{ kg} \cdot \text{m}^2/\text{s}^2 \cdot \text{K} \cdot \text{mol})} = 86.1$ K

T = 86.1 – 273.15 = –187.0°C

9.20 (a) $\quad \dfrac{\text{rate } O_2}{\text{rate Kr}} = \sqrt{\dfrac{M_{Kr}}{M_{O_2}}} = \sqrt{\dfrac{83.8}{32.0}}; \quad \dfrac{\text{rate } O_2}{\text{rate Kr}} = 1.62$

O_2 diffuses 1.62 times faster than Kr.

(b) $\quad \dfrac{\text{rate } C_2H_2}{\text{rate } N_2} = \sqrt{\dfrac{M_{N_2}}{M_{C_2H_2}}} = \sqrt{\dfrac{28.0}{26.0}}; \quad \dfrac{\text{rate } C_2H_2}{\text{rate } N_2} = 1.04$

C_2H_2 diffuses 1.04 times faster than N_2.

9.21 $\quad \dfrac{\text{rate } ^{20}\text{Ne}}{\text{rate} ^{22}\text{Ne}} = \sqrt{\dfrac{M\ ^{22}\text{Ne}}{M\ ^{20}\text{Ne}}} = \sqrt{\dfrac{22}{20}} = 1.05; \quad \dfrac{\text{rate } ^{21}\text{Ne}}{\text{rate } ^{22}\text{Ne}} = \sqrt{\dfrac{M\ ^{22}\text{Ne}}{M\ ^{21}\text{Ne}}} = \sqrt{\dfrac{22}{21}} = 1.02$

Thus, the relative rates of diffusion are $^{20}\text{Ne}(1.05) > ^{21}\text{Ne}(1.02) > ^{22}\text{Ne}(1.00)$.

9.22 $\quad P = \dfrac{nRT}{V} = \dfrac{(0.500 \text{ mol})\left(0.082\ 06 \dfrac{\text{L} \cdot \text{atm}}{\text{K} \cdot \text{mol}}\right)(300 \text{ K})}{(0.600 \text{ L})} = 20.5$ atm

$P = \dfrac{nRT}{V - nb} - \dfrac{an^2}{V^2}$

$$P = \frac{(0.500 \text{ mol})\left(0.082\ 06\ \dfrac{L \cdot atm}{K \cdot mol}\right)(300\ K)}{[(0.600\ L) - (0.500\ mol)(0.0387\ L/mol)]} - \frac{\left(1.35\ \dfrac{L^{2} \cdot atm}{mol^{2}}\right)(0.500\ mol)^{2}}{(0.600\ L)^{2}} = 20.3\ atm$$

9.23 The amount of ozone is assumed to be constant.

Therefore $nR = \dfrac{P_i V_i}{T_i} = \dfrac{P_f V_f}{T_f}$

Because $V \propto h$, then $\dfrac{P_i h_i}{T_i} = \dfrac{P_f h_f}{T_f}$ where h is the thickness of the O_3 layer.

$$h_f = \frac{P_i}{P_f} \times \frac{T_f}{T_i} \times h_i = \left(\frac{1.6 \times 10^{-9}\ atm}{1\ atm}\right)\left(\frac{273\ K}{230\ K}\right)(20 \times 10^3\ m) = 3.8 \times 10^{-5}\ m$$

(Actually, $V = 4\pi r^2 h$, where r = the radius of the earth. When you go out ~30 km to get to the ozone layer, the change in r^2 is less than 1%. Therefore you can neglect the change in r^2 and assume that V is proportional to h.)

9.24 For ether, the MAC = $\dfrac{15\ mm\ Hg}{760\ mm\ Hg} \times 100\% = 2.0\%$

9.25 (a) Let X = partial pressure of chloroform.

MAC = $\dfrac{X}{760\ mm\ Hg} \times 100\% = 0.77\%$

Solve for X. $X = 760\ mm\ Hg \times \dfrac{0.77\%}{100\%} = 5.9\ mm\ Hg$

(b) $CHCl_3$, 119.4 amu

$$PV = nRT; \quad n = \frac{PV}{RT} = \frac{\left(5.9\ mm\ Hg \times \dfrac{1.00\ atm}{760\ mm\ Hg}\right)(10.0\ L)}{\left(0.082\ 06\ \dfrac{L \cdot atm}{K \cdot mol}\right)(273\ K)} = 0.00347\ mol\ CHCl_3$$

mass $CHCl_3$ = 0.00347 mol $CHCl_3 \times \dfrac{119.4\ g\ CHCl_3}{1\ mol\ CHCl_3} = 0.41\ g\ CHCl_3$

Understanding Key Concepts

9.26 (a)

The volume of a gas is proportional to the kelvin temperature at constant pressure. As the temperature increases from 300 K to 450 K, the volume will increase by a factor of 1.5.

(b) The volume of a gas is inversely proportional to pressure at constant temperature. As the pressure increases from 1 atm to 2 atm, the volume will decrease by a factor of 2.

(c) 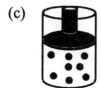 $PV = nRT$; The amount of gas (n) is constant.

Therefore $nR = \dfrac{P_i V_i}{T_i} = \dfrac{P_f V_f}{T_f}$.

Assume $V_i = 1$ L and solve for V_f.

$$\dfrac{P_i V_i T_f}{T_i P_f} = \dfrac{(3 \text{ atm})(1 \text{ L})(200 \text{ K})}{(300 \text{ K})(2 \text{ atm})} = V_f = 1 \text{ L}$$

There is no change in volume.

9.28   The two gases should mix randomly and homogeneously and this is best represented by drawing (c).

9.30   The gas pressure in the bulb in mm Hg is equal to the difference in the height of the Hg in the two arms of the manometer.

9.32   (a) Because there are more yellow gas molecules than there are blue, the yellow gas molecules have the higher average speed.
(b) Each rate is proportional to the number of effused gas molecules of each type.
$M_{yellow} = 25$ amu

$$\dfrac{rate_{blue}}{rate_{yellow}} = \sqrt{\dfrac{M_{yellow}}{M_{blue}}}; \quad \dfrac{5}{6} = \sqrt{\dfrac{25 \text{ amu}}{M_{blue}}}; \quad \left(\dfrac{5}{6}\right)^2 = \dfrac{25 \text{ amu}}{M_{blue}}$$

$$M_{blue} = \dfrac{25 \text{ amu}}{\left(\dfrac{5}{6}\right)^2} = 36 \text{ amu}$$

Additional Problems
Gases and Gas Pressure

9.34   Temperature is a measure of the average kinetic energy of gas particles.

9.36 $P = 480 \text{ mm Hg} \times \dfrac{1.00 \text{ atm}}{760 \text{ mm Hg}} = 0.632 \text{ atm}$

$P = 480 \text{ mm Hg} \times \dfrac{101,325 \text{ Pa}}{760 \text{ mm Hg}} = 6.40 \times 10^4 \text{ Pa}$

9.38 $P_{flask} > 754.3$ mm Hg; $P_{flask} = 754.3$ mm Hg $+ 176$ mm Hg $= 930$ mm Hg

9.40 P_{flask} > 752.3 mm Hg (see Figure 9.4)

If the pressure in the flask can support a column of ethyl alcohol (d = 0.7893 g/mL) 55.1 cm high, then it can only support a column of Hg that is much shorter because of the higher density of Hg.

$$55.1 \text{ cm } \times \frac{0.7893 \text{ g/mL}}{13.546 \text{ g/mL}} = 3.21 \text{ cm Hg} = 32.1 \text{ mm Hg}$$

P_{flask} = 752.3 mm Hg + 32.1 mm Hg = 784.4 mm Hg

$$P_{flask} = 784.4 \text{ mm Hg} \times \frac{101{,}325 \text{ Pa}}{760 \text{ mm Hg}} = 1.046 \times 10^5 \text{ Pa}$$

9.42

	% Volume
N_2	78.08
O_2	20.95
Ar	0.93
CO_2	0.037

The % volume for a particular gas is proportional to the number of molecules of that gas in a mixture of gases.

Average molecular mass of air

= (0.7808)(mol. mass N_2) + (0.2095)(mol. mass O_2)

$\qquad$ + (0.0093)(at. mass Ar) + (0.000 37)(mol. mass CO_2)

= (0.7808)(28.01 amu) + (0.2095)(32.00 amu)

$\qquad$ + (0.0093)(39.95 amu) + (0.000 37)(44.01 amu) = 28.96 amu

The Gas Laws

9.44 (a) $\dfrac{nR}{V} = \dfrac{P_i}{T_i} = \dfrac{P_f}{T_f}$; $\qquad$ $\dfrac{P_i T_f}{T_i} = P_f$

$\qquad$ Let P_i = 1 atm, T_i = 100 K, T_f = 300 K

$$P_f = \frac{P_i T_f}{T_i} = \frac{(1 \text{ atm})(300 \text{ K})}{(100 \text{ K})} = 3 \text{ atm}$$

$\qquad$ The pressure would triple.

$\quad$ (b) $\dfrac{RT}{V} = \dfrac{P_i}{n_i} = \dfrac{P_f}{n_f}$; $\qquad$ $\dfrac{P_i n_f}{n_i} = P_f$

$\qquad$ Let P_i = 1 atm, n_i = 3 mol, n_f = 1 mol

$$P_f = \frac{P_i n_f}{n_i} = \frac{(1 \text{ atm})(1 \text{ mol})}{(3 \text{ mol})} = \frac{1}{3} \text{ atm}$$

$\qquad$ The pressure would be $\dfrac{1}{3}$ the initial pressure.

(c) $nRT = P_iV_i = P_fV_f$; $\qquad \dfrac{P_iV_i}{V_f} = P_f$

Let $P_i = 1$ atm, $V_i = 1$ L, $V_f = 1 - 0.45$ L $= 0.55$ L

$$P_f = \frac{P_iV_i}{V_f} = \frac{(1\text{ atm})(1\text{ L})}{(0.55\text{ L})} = 1.8 \text{ atm}$$

The pressure would increase by 1.8 times.

(d) $nR = \dfrac{P_iV_i}{T_i} = \dfrac{P_fV_f}{T_f}$; $\qquad \dfrac{P_iV_iT_f}{T_iV_f} = P_f$

Let $P_i = 1$ atm, $V_i = 1$ L, $T_i = 200$ K, $V_f = 3$ L, $T_i = 100$ K

$$P_f = \frac{P_iV_iT_f}{T_iV_f} = \frac{(1\text{ atm})(1\text{ L})(100\text{ K})}{(200\text{ K})(3\text{ L})} = 0.17 \text{ atm}$$

The pressure would be 0.17 times the initial pressure.

9.46   They all contain the same number of gas molecules.

9.48 n and T are constant; therefore $nRT = P_iV_i = P_fV_f$

$$V_f = \frac{P_iV_i}{P_f} = \frac{(150\text{ atm})(49.0\text{ L})}{(1.02\text{ atm})} = 7210 \text{ L}$$

n and P are constant; therefore $\dfrac{nR}{P} = \dfrac{V_i}{T_i} = \dfrac{V_f}{T_f}$

$$V_f = \frac{V_iT_f}{T_i} = \frac{(49.0\text{ L})(308\text{ K})}{(293\text{ K})} = 51.5 \text{ L}$$

9.50 $15.0\text{ g CO}_2 \times \dfrac{1\text{ mol CO}_2}{44.0\text{ g CO}_2} = 0.341 \text{ mol CO}_2$

$$P = \frac{nRT}{V} = \frac{(0.341\text{ mol})\left(0.082\ 06\ \dfrac{\text{L}\cdot\text{atm}}{\text{K}\cdot\text{mol}}\right)(300\text{ K})}{(0.30\text{ L})} = 27.98 \text{ atm}$$

27.98 atm $\times \dfrac{760\text{ mm Hg}}{1\text{ atm}} = 2.1 \times 10^4$ mm Hg

9.52 $\dfrac{1\text{ H atom}}{\text{cm}^3} \times \dfrac{1\text{ mol H}}{6.02 \times 10^{23}\text{ atoms}} \times \dfrac{1000\text{ cm}^3}{1\text{ L}} = 1.7 \times 10^{-21}$ mol H/L

$$P = \frac{nRT}{V} = \frac{(1.7 \times 10^{-21}\text{ mol})\left(0.082\ 06\ \dfrac{\text{L}\cdot\text{atm}}{\text{K}\cdot\text{mol}}\right)(100\text{ K})}{(1\text{ L})} = 1.4 \times 10^{-20} \text{ atm}$$

$$P = 1.4 \times 10^{-20} \text{ atm} \times \frac{760 \text{ mm Hg}}{1.0 \text{ atm}} = 1 \times 10^{-17} \text{ mm Hg}$$

9.54 $\quad n = \dfrac{PV}{RT} = \dfrac{\left(17{,}180 \text{ kPa} \times \dfrac{1000 \text{ Pa}}{1 \text{ kPa}} \times \dfrac{1 \text{ atm}}{101{,}325 \text{ Pa}}\right)(43.8 \text{ L})}{\left(0.082\ 06 \dfrac{\text{L} \cdot \text{atm}}{\text{K} \cdot \text{mol}}\right)(293\text{K})} = 308.9 \text{ mol}$

$$\text{mass Ar} = 308.9 \text{ mol} \times \frac{39.948 \text{ g}}{1 \text{ mol}} = 12340 \text{ g} = 1.23 \times 10^4 \text{ g}$$

Gas Stoichiometry

9.56 $\quad$ For steam, $T = 123.0°C = 396 \text{ K}$

$$n = \frac{PV}{RT} = \frac{(0.93 \text{ atm})(15.0 \text{ L})}{\left(0.082\ 06 \dfrac{\text{L} \cdot \text{atm}}{\text{K} \cdot \text{mol}}\right)(396 \text{ K})} = 0.43 \text{ mol steam}$$

For ice, H_2O, 18.02 amu; $\quad n = 10.5 \text{ g} \times \dfrac{1 \text{ mol}}{18.02 \text{ g}} = 0.583 \text{ mol ice}$

Because the number of moles of ice is larger than the number of moles of steam, the ice contains more H_2O molecules.

9.58 $\quad$ The containers are identical. Both containers contain the same number of gas molecules. Weigh the containers. Because the molecular mass for O_2 is greater than the molecular mass for H_2, the heavier container contains O_2.

9.60 $\quad$ room volume $= 4.0 \text{ m} \times 5.0 \text{ m} \times 2.5 \text{ m} = 50 \text{ m}^3$

$$\text{room volume} = 50 \text{ m}^3 \times \frac{1 \text{ L}}{10^{-3} \text{ m}^3} = 5.0 \times 10^4 \text{ L}$$

$$n_{\text{total}} = \frac{PV}{RT} = \frac{(1.0 \text{ atm})(5.0 \times 10^4 \text{ L})}{\left(0.082\ 06 \dfrac{\text{L} \cdot \text{atm}}{\text{K} \cdot \text{mol}}\right)(273 \text{ K})} = 2.23 \times 10^3 \text{ mol}$$

$$n_{O_2} = (0.2095)n_{\text{total}} = (0.2095)(2.23 \times 10^3 \text{ mol}) = 467 \text{ mol } O_2$$

$$\text{mass } O_2 = 467 \text{ mol} \times \frac{32.0 \text{ g}}{1 \text{ mol}} = 1.5 \times 10^4 \text{ g } O_2$$

9.62 $\quad$ (a) CH_4, 16.04 amu; $\qquad d = \dfrac{16.04 \text{ g}}{22.4 \text{ L}} = 0.716 \text{ g/L}$

$\qquad$ (b) CO_2, 44.01 amu; $\qquad d = \dfrac{44.01 \text{ g}}{22.4 \text{ L}} = 1.96 \text{ g/L}$

(c) O_2, 32.00 amu; $d = \dfrac{32.00\ g}{22.4\ L} = 1.43\ g/L$

(d) UF_6, 352.0 amu; $d = \dfrac{352.0\ g}{22.4\ L} = 15.7\ g/L$

9.64 $n = \dfrac{PV}{RT} = \dfrac{\left(356\ mm\ Hg\ \times\ \dfrac{1.00\ atm}{760\ mm\ Hg}\right)(1.500\ L)}{\left(0.082\ 06\ \dfrac{L\cdot atm}{K\cdot mol}\right)(295.5\ K)} = 0.0290\ mol$

molar mass $= \dfrac{0.9847\ g}{0.0290\ mol} = 34.0\ g/mol;$ molecular mass $= 34.0$ amu

9.66 $2\ HgO(s)\ \rightarrow\ 2\ Hg(l)\ +\ O_2(g);$ HgO, 216.59 amu

$10.57\ g\ HgO \times \dfrac{1\ mol\ HgO}{216.59\ g\ HgO} \times \dfrac{1\ mol\ O_2}{2\ mol\ HgO} = 0.024\ 40\ mol\ O_2$

$V = \dfrac{nRT}{P} = \dfrac{(0.024\ 40\ mol)\left(0.082\ 06\ \dfrac{L\cdot atm}{K\cdot mol}\right)(273.15\ K)}{1.000\ atm} = 0.5469\ L$

9.68 $Zn(s)\ +\ 2\ HCl(aq)\ \rightarrow\ ZnCl_2(aq)\ +\ H_2(g)$

(a) $25.5\ g\ Zn \times \dfrac{1\ mol\ Zn}{65.39\ g\ Zn} \times \dfrac{1\ mol\ H_2}{1\ mol\ Zn} = 0.390\ mol\ H_2$

$V = \dfrac{nRT}{P} = \dfrac{(0.390\ mol)\left(0.082\ 06\ \dfrac{L\cdot atm}{K\cdot mol}\right)(288\ K)}{\left(742\ mm\ Hg\ \times\ \dfrac{1.00\ atm}{760\ mm\ Hg}\right)} = 9.44\ L$

(b) $n = \dfrac{PV}{RT} = \dfrac{\left(350\ mm\ Hg\ \times\ \dfrac{1.00\ atm}{760\ mm\ Hg}\right)(5.00\ L)}{\left(0.082\ 06\ \dfrac{L\cdot atm}{K\cdot mol}\right)(303.15\ K)} = 0.092\ 56\ mol\ H_2$

$0.092\ 56\ mol\ H_2 \times \dfrac{1\ mol\ Zn}{1\ mol\ H_2} \times \dfrac{65.39\ g\ Zn}{1\ mol\ Zn} = 6.05\ g\ Zn$

9.70 (a) $V_{24h} = (4.50\ L/min)(60\ min/h)(24\ h/day) = 6480\ L$

$V_{CO_2} = (0.034)V_{24h} = (0.034)(6480\ L) = 220\ L$

$$n = \frac{PV}{RT} = \frac{\left(735 \text{ mm Hg} \times \dfrac{1.00 \text{ atm}}{760 \text{ mm Hg}}\right)(220 \text{ L})}{\left(0.082\ 06 \dfrac{\text{L} \cdot \text{atm}}{\text{K} \cdot \text{mol}}\right)(298 \text{ K})} = 8.70 \text{ mol } CO_2$$

$$8.70 \text{ mol } CO_2 \times \frac{44.01 \text{ g } CO_2}{1 \text{ mol } CO_2} = 383 \text{ g} = 380 \text{ g } CO_2$$

(b) $2 Na_2O_2(s) + 2 CO_2(g) \rightarrow 2 Na_2CO_3(s) + O_2(g)$; Na_2O_2, 77.98 amu

3.65 kg = 3650 g

$$3650 \text{ g } Na_2O_2 \times \frac{1 \text{ mol } Na_2O_2}{77.98 \text{ g } Na_2O_2} \times \frac{2 \text{ mol } CO_2}{2 \text{ mol } Na_2O_2} \times \frac{1 \text{ day}}{8.70 \text{ mol } CO_2} = 5.4 \text{ days}$$

Dalton's Law and Mole Fraction

9.72   Because of Avogadro's Law ($V \propto n$), the % volumes are also % moles.

	% mole
N_2	78.08
O_2	20.95
Ar	0.93
CO_2	0.037

In decimal form, % mole = mole fraction.

$P_{N_2} = X_{N_2} \cdot P_{total} = (0.7808)(1.000 \text{ atm}) = 0.7808 \text{ atm}$

$P_{O_2} = X_{O_2} \cdot P_{total} = (0.2095)(1.000 \text{ atm}) = 0.2095 \text{ atm}$

$P_{Ar} = X_{Ar} \cdot P_{total} = (0.0093)(1.000 \text{ atm}) = 0.0093 \text{ atm}$

$P_{CO_2} = X_{CO_2} \cdot P_{total} = (0.000\ 37)(1.000 \text{ atm}) = 0.000\ 37 \text{ atm}$

Pressures of the rest are negligible.

9.74   Assume a 100.0 g sample.   g CO_2 = 1.00 g and g O_2 = 99.0 g

$$\text{mol } CO_2 = 1.00 \text{ g } CO_2 \times \frac{1 \text{ mol } CO_2}{44.01 \text{ g } CO_2} = 0.0227 \text{ mol } CO_2$$

$$\text{mol } O_2 = 99.0 \text{ g } O_2 \times \frac{1 \text{ mol } O_2}{32.00 \text{ g } O_2} = 3.094 \text{ mol } O_2$$

n_{total} = 3.094 mol + 0.0227 mol = 3.117 mol

$X_{O_2} = \dfrac{3.094 \text{ mol}}{3.117 \text{ mol}} = 0.993$ $X_{CO_2} = \dfrac{0.0227 \text{ mol}}{3.117 \text{ mol}} = 0.007\ 28$

$P_{O_2} = X_{O_2} \cdot P_{total} = (0.993)(0.977 \text{ atm}) = 0.970 \text{ atm}$

$P_{CO_2} = X_{CO_2} \cdot P_{total} = (0.007\ 28)(0.977 \text{ atm}) = 0.007\ 11 \text{ atm}$

9.76   Assume a 100.0 g sample.

g HCl = (0.0500)(100.0 g) = 5.00 g; $5.00 \text{ g HCl} \times \dfrac{1 \text{ mol HCl}}{36.5 \text{ g HCl}} = 0.137 \text{ mol HCl}$

g H_2 = (0.0100)(100.0 g) = 1.00 g; $1.00 \text{ g } H_2 \times \dfrac{1 \text{ mol } H_2}{2.016 \text{ g } H_2} = 0.496 \text{ mol } H_2$

g Ne = (0.94)(100.0 g) = 94 g; $94 \text{ g Ne} \times \dfrac{1 \text{ mol Ne}}{20.18 \text{ g Ne}} = 4.66 \text{ mol Ne}$

n_{total} = 0.137 + 0.496 + 4.66 = 5.3 mol

$X_{HCl} = \dfrac{0.137 \text{ mol}}{5.3 \text{ mol}} = 0.026$ $X_{H_2} = \dfrac{0.496 \text{ mol}}{5.3 \text{ mol}} = 0.094$ $X_{Ne} = \dfrac{4.66 \text{ mol}}{5.3 \text{ mol}} = 0.88$

9.78 $P_{total} = P_{H_2} + P_{H_2O};$ $P_{H_2} = P_{total} - P_{H_2O} = 747 \text{ mm Hg} - 23.8 \text{ mm Hg} = 723 \text{ mm Hg}$

$$n = \dfrac{PV}{RT} = \dfrac{\left(723 \text{ mm Hg} \times \dfrac{1.00 \text{ atm}}{760 \text{ mm Hg}}\right)(3.557 \text{ L})}{\left(0.082\ 06 \dfrac{\text{L} \cdot \text{atm}}{\text{K} \cdot \text{mol}}\right)(298 \text{ K})} = 0.1384 \text{ mol } H_2$$

$0.1384 \text{ mol } H_2 \times \dfrac{1 \text{ mol Mg}}{1 \text{ mol } H_2} \times \dfrac{24.3 \text{ g Mg}}{1 \text{ mol Mg}} = 3.36 \text{ g Mg}$

Kinetic-Molecular Theory and Graham's Law

9.80   The kinetic-molecular theory is based on the following assumptions:
1. A gas consists of tiny particles, either atoms or molecules, moving about at random.
2. The volume of the particles themselves is negligible compared with the total volume of the gas; most of the volume of a gas is empty space.
3. The gas particles act independently; there are no attractive or repulsive forces between particles.
4. Collisions of the gas particles, either with other particles or with the walls of the container, are elastic; that is, the total kinetic energy of the gas particles is constant at constant T.
5. The average kinetic energy of the gas particles is proportional to the Kelvin temperature of the sample.

9.82   Heat is the energy transferred from one object to another as the result of a temperature difference between them.
Temperature is a measure of the kinetic energy of molecular motion.

9.84 $u = \sqrt{\dfrac{3 \text{ RT}}{M}} = \sqrt{\dfrac{3 \times 8.314 \text{ kg m}^2/(\text{s}^2 \text{ K mol}) \times 220 \text{ K}}{28.0 \times 10^{-3} \text{ kg/mol}}} = 443 \text{ m/s}$

9.86   For H_2, $u = \sqrt{\dfrac{3\,RT}{M}} = \sqrt{\dfrac{3 \times 8.314\,kg\,m^2/(s^2\,K\,mol) \times 150\,K}{2.02 \times 10^{-3}\,kg/mol}} = 1360\,m/s$

For He, $u = \sqrt{\dfrac{3 \times 8.314\,kg\,m^2/(s^2\,K\,mol) \times 648\,K}{4.00 \times 10^{-3}\,kg/mol}} = 2010\,m/s$

He at 375°C has the higher average speed.

9.88 $\dfrac{rate_{H_2}}{rate_X} = \sqrt{\dfrac{M_X}{M_{H_2}}}\,;\qquad \dfrac{2.92}{1} = \dfrac{\sqrt{M_X}}{\sqrt{2.02}}\,;\qquad 2.92\,\sqrt{2.02} = \sqrt{M_X}$

$M_X = (2.92\,\sqrt{2.02}\,)^2 = 17.2\,g/mol;$ molecular mass = 17.2 amu

9.90   HCl, 36.5 amu; F_2, 38.0 amu;   Ar, 39.9 amu

$\dfrac{rate\ HCl}{rate\ Ar} = \sqrt{\dfrac{M_{Ar}}{M_{HCl}}} = \sqrt{\dfrac{39.9}{36.5}} = 1.05 \qquad \dfrac{rate\ F_2}{rate\ Ar} = \sqrt{\dfrac{M_{Ar}}{M_{F_2}}} = \sqrt{\dfrac{39.9}{38.0}} = 1.02$

The relative rates of diffusion are HCl(1.05) > F_2(1.02) > Ar(1.00).

9.92 $u = 45\,m/s = \sqrt{\dfrac{3 \times 8.314\,kg\,m^2/(s^2\,K\,mol) \times T}{4.00 \times 10^{-3}\,kg/mol}}$

Square both sides of the equation and solve for T.

$2025\,m^2/s^2 = \dfrac{3 \times 8.314\,kg\,m^2/(s^2\,K\,mol) \times T}{4.00 \times 10^{-3}\,kg/mol}$

$T = 0.325\,K = -272.83°C$ (near absolute zero)

General Problems

9.94 $\dfrac{rate\ {}^{35}Cl_2}{rate\ {}^{37}Cl_2} = \sqrt{\dfrac{M\ {}^{37}Cl_2}{M\ {}^{35}Cl_2}} = \sqrt{\dfrac{74.0}{70.0}} = 1.03$

$\dfrac{rate\ {}^{35}Cl\,{}^{37}Cl}{rate\ {}^{37}Cl_2} = \sqrt{\dfrac{M\ {}^{37}Cl_2}{M\ {}^{35}Cl\,{}^{37}Cl}} = \sqrt{\dfrac{74.0}{72.0}} = 1.01$

The relative rates of diffusion are ${}^{35}Cl_2$(1.03) > ${}^{35}Cl\,{}^{37}Cl$(1.01) > ${}^{37}Cl_2$(1.00).

9.96 $V = \dfrac{nRT}{P} = \dfrac{(1.00\,mol)\left(0.082\ 06\ \dfrac{L \cdot atm}{K \cdot mol}\right)(1050\,K)}{(75\,atm)} = 1.1\,L$

9.98 $\quad n = \dfrac{PV}{RT} = \dfrac{(2.15\ \text{atm})(7.35\ \text{L})}{\left(0.082\ 06\ \dfrac{\text{L}\cdot\text{atm}}{\text{K}\cdot\text{mol}}\right)(293\ \text{K})} = 0.657\ \text{mol Ar}$

$0.657\ \text{mol Ar} \times \dfrac{39.948\ \text{g Ar}}{1\ \text{mol Ar}} = 26.2\ \text{g Ar}$

$m_{total} = 478.1\ \text{g} + 26.2\ \text{g} = 504.3\ \text{g}$

9.100 (a) Bulb A contains $CO_2(g)$ and $N_2(g)$; Bulb B contains $CO_2(g)$, $N_2(g)$, and $H_2O(s)$.

(b) Initial moles of gas $= n = \dfrac{PV}{RT} = \dfrac{\left(564\ \text{mm Hg}\ \times\ \dfrac{1.00\ \text{atm}}{760\ \text{mm Hg}}\right)(1.000\ \text{L})}{\left(0.082\ 06\ \dfrac{\text{L}\cdot\text{atm}}{\text{K}\cdot\text{mol}}\right)(298\ \text{K})}$

Initial moles of gas $= 0.030\ 35\ \text{mol}$

mol gas in Bulb A $= n = \dfrac{PV}{RT} = \dfrac{\left(219\ \text{mm Hg}\ \times\ \dfrac{1.00\ \text{atm}}{760\ \text{mm Hg}}\right)(1.000\ \text{L})}{\left(0.082\ 06\ \dfrac{\text{L}\cdot\text{atm}}{\text{K}\cdot\text{mol}}\right)(298\ \text{K})} = 0.011\ 78\ \text{mol}$

mol gas in Bulb B $= n = \dfrac{PV}{RT} = \dfrac{\left(219\ \text{mm Hg}\ \times\ \dfrac{1.00\ \text{atm}}{760\ \text{mm Hg}}\right)(1.000\ \text{L})}{\left(0.082\ 06\ \dfrac{\text{L}\cdot\text{atm}}{\text{K}\cdot\text{mol}}\right)(203\ \text{K})} = 0.017\ 29\ \text{mol}$

$n_{H_2O} = n_{initial} - n_A - n_B = 0.030\ 35 - 0.011\ 78 - 0.017\ 29 = 0.001\ 28\ \text{mol} = 0.0013\ \text{mol}\ H_2O$

(c) Bulb A contains $N_2(g)$.
Bulb B contains $N_2(g)$ and $H_2O(s)$.
Bulb C contains $N_2(g)$ and $CO_2(s)$.

(d) $n_A = \dfrac{PV}{RT} = \dfrac{\left(33.5\ \text{mm Hg}\ \times\ \dfrac{1.00\ \text{atm}}{760\ \text{mm Hg}}\right)(1.000\ \text{L})}{\left(0.082\ 06\ \dfrac{\text{L}\cdot\text{atm}}{\text{K}\cdot\text{mol}}\right)(298\ \text{K})} = 0.001\ 803\ \text{mol}$

$n_B = \dfrac{PV}{RT} = \dfrac{\left(33.5\ \text{mm Hg}\ \times\ \dfrac{1.00\ \text{atm}}{760\ \text{mm Hg}}\right)(1.000\ \text{L})}{\left(0.082\ 06\ \dfrac{\text{L}\cdot\text{atm}}{\text{K}\cdot\text{mol}}\right)(203\ \text{K})} = 0.002\ 646\ \text{mol}$

$n_C = \dfrac{PV}{RT} = \dfrac{\left(33.5\ \text{mm Hg}\ \times\ \dfrac{1.00\ \text{atm}}{760\ \text{mm Hg}}\right)(1.000\ \text{L})}{\left(0.082\ 06\ \dfrac{\text{L}\cdot\text{atm}}{\text{K}\cdot\text{mol}}\right)(83\ \text{K})} = 0.006\ 472\ \text{mol}$

$$n_{N_2} = n_A + n_B + n_C = 0.001\ 803 + 0.002\ 646 + 0.006\ 472 = 0.010\ 92 \text{ mol } N_2$$

(e) $n_{CO_2} = n_{initial} - n_{H_2O} - n_{N_2} = 0.030\ 35 - 0.0013 - 0.010\ 92 = 0.0181 \text{ mol } CO_2$

9.102 NH_3, 17.03 amu; $\quad$ mol $NH_3 = 45.0 \text{ g} \times \dfrac{1 \text{ mol}}{17.03 \text{ g}} = 2.64$ mol

$$P = \frac{nRT}{V} \quad \text{or} \quad P = \frac{nRT}{(V - nb)} - \frac{an^2}{V^2}$$

(a) At $T = 0°C = 273$ K

$$P = \frac{(2.64 \text{ mol})\left(0.082\ 06\ \dfrac{L \cdot atm}{K \cdot mol}\right)(273 \text{ K})}{(1.000 \text{ L})} = 59.1 \text{ atm}$$

$$P = \frac{(2.64 \text{ mol})\left(0.082\ 06\ \dfrac{L \cdot atm}{K \cdot mol}\right)(273 \text{ K})}{[(1.000 \text{ L}) - (2.64 \text{ mol})(0.0371 \text{ L/mol})]} - \frac{\left(4.17\ \dfrac{L^2 \cdot atm}{mol^2}\right)(2.64 \text{ mol})^2}{(1.000 \text{ L})^2}$$

$P = 65.6 \text{ atm} - 29.1 \text{ atm} = 36.5 \text{ atm}$

(b) At $T = 50°C = 323$ K

$$P = \frac{(2.64 \text{ mol})\left(0.082\ 06\ \dfrac{L \cdot atm}{K \cdot mol}\right)(323 \text{ K})}{(1.000 \text{ L})} = 70.0 \text{ atm}$$

$$P = \frac{(2.64 \text{ mol})\left(0.082\ 06\ \dfrac{L \cdot atm}{K \cdot mol}\right)(323 \text{ K})}{[(1.000 \text{ L}) - (2.64 \text{ mol})(0.0371 \text{ L/mol})]} - \frac{\left(4.17\ \dfrac{L^2 \cdot atm}{mol^2}\right)(2.64 \text{ mol})^2}{(1.000 \text{ L})^2}$$

$P = 77.6 \text{ atm} - 29.1 \text{ atm} = 48.5 \text{ atm}$

(c) At $T = 100°C = 373$ K

$$P = \frac{(2.64 \text{ mol})\left(0.082\ 06\ \dfrac{L \cdot atm}{K \cdot mol}\right)(373 \text{ K})}{(1.000 \text{ L})} = 80.8 \text{ atm}$$

$$P = \frac{(2.64 \text{ mol})\left(0.082\ 06\ \dfrac{L \cdot atm}{K \cdot mol}\right)(373 \text{ K})}{[(1.000 \text{ L}) - (2.64 \text{ mol})(0.0371 \text{ L/mol})]} - \frac{\left(4.17\ \dfrac{L^2 \cdot atm}{mol^2}\right)(2.64 \text{ mol})^2}{(1.000 \text{ L})^2}$$

$P = 89.6 \text{ atm} - 29.1 \text{ atm} = 60.5 \text{ atm}$

At the three temperatures, the van der Waals equation predicts a much lower pressure than does the ideal gas law. This is likely due to the fact that NH_3 can hydrogen bond leading to strong intermolecular forces.

9.104 CO_2, 44.01 amu

$$\text{mol } CO_2 = 500.0 \text{ g } CO_2 \times \frac{1 \text{ mol } CO_2}{44.01 \text{ g } CO_2} = 11.36 \text{ mol } CO_2$$

$PV = nRT$

$$P = \frac{nRT}{V} = \frac{(11.36 \text{ mol})\left(0.082\ 06\ \frac{L \cdot atm}{K \cdot mol}\right)(700 \text{ K})}{(0.800 \text{ L})} = 816 \text{ atm}$$

9.106   (a) average molecular mass for natural gas
$$= (0.915)(16.04 \text{ amu}) + (0.085)(30.07 \text{ amu}) = 17.2 \text{ amu}$$

$$\text{total moles of gas} = 15.50 \text{ g} \times \frac{1 \text{ mol gas}}{17.2 \text{ g gas}} = 0.901 \text{ mol gas}$$

(b) $$P = \frac{(0.901 \text{ mol})\left(0.082\ 06\ \frac{L \cdot atm}{K \cdot mol}\right)(293 \text{ K})}{(15.00 \text{ L})} = 1.44 \text{ atm}$$

(c) $P_{CH_4} = X_{CH_4} \cdot P_{total} = (1.44 \text{ atm})(0.915) = 1.32 \text{ atm}$

$P_{C_2H_6} = X_{C_2H_6} \cdot P_{total} = (1.44 \text{ atm})(0.085) = 0.12 \text{ atm}$

(d) $\Delta H_{combustion}(CH_4) = -802.3$ kJ/mol and $\Delta H_{combustion}(C_2H_6) = -1427.7$ kJ/mol
Heat liberated $= (0.915)(0.901 \text{ mol})(-802.3 \text{ kJ/mol})$
$$+ \ (0.085)(0.901)(-1427.7 \text{ kJ/mol}) = -771 \text{ kJ}$$

9.108   (a) T = 0°C = 273 K; $PV = nRT$

$$n_Q = \frac{PV}{RT} = \frac{(0.229 \text{ atm})(0.0500 \text{ L})}{\left(0.082\ 06\ \frac{L \cdot atm}{K \cdot mol}\right)(273 \text{ K})} = 5.11 \times 10^{-4} \text{ mol Q}$$

$$Q \text{ molar mass} = \frac{0.100 \text{ g Q}}{5.11 \times 10^{-4} \text{ mol Q}} = 196 \text{ g/mol}$$

Xe molar mass = 131.3 g/mol
O_n molar mass = 196 g/mol – 131.3 g/mol = 65 g/mol
So, n = 4 and XeO_4 is the likely formula for Q.
(b) $XeO_4(g) \rightarrow Xe(g) + 2\ O_2(g)$
After decomposition, $P_{Total} = P_{Xe} + P_{O_2}$.

Because of the stoichiometry of the decomposition reaction, the partial pressure of O_2 is twice the partial pressure of Xe.
Let x = P_{Xe} and 2x = P_{O_2}. $P_{Total} = P_{Xe} + P_{O_2} = x + 2x = 3x = 0.941$ atm

$$x = \frac{0.941 \text{ atm}}{3} = 0.314 \text{ atm}$$

$P_{Xe} = x = 0.314$ atm; $P_{O_2} = P_{Total} - P_{Xe} = 0.941$ atm $- 0.314$ atm $= 0.627$ atm

9.110 PCl_3, 137.3 amu; O_2, 32.00 amu; $POCl_3$, 153.3 amu

$$2\ PCl_3(g) + O_2(g) \rightarrow 2\ POCl_3(g)$$

$$\text{mol } PCl_3 = 25.0\ g \times \frac{1\ mol\ PCl_3}{137.3\ g\ PCl_3} = 0.182\ mol\ PCl_3$$

$$\text{mol } O_2 = 3.00\ g \times \frac{1\ mol\ O_2}{32.00\ g\ O_2} = 0.0937\ mol\ O_2$$

Check for limiting reactant.

$$\text{mol } O_2\ needed = 0.182\ mol\ PCl_3 \times \frac{1\ mol\ O_2}{2\ mol\ PCl_3} = 0.0910\ mol\ O_2\ needed$$

There is a slight excess of O_2. PCl_3 is the limiting reactant.

$$\text{mol } POCl_3 = 0.182\ mol\ PCl_3 \times \frac{2\ mol\ POCl_3}{2\ mol\ PCl_3} = 0.182\ mol\ POCl_3$$

mol O_2 left over = 0.0937 mol – 0.0910 mol = 0.0027 mol O_2 left over

T = 200.0°C = 200.0 + 273.15 = 473.1 K; PV = nRT

$$P = \frac{nRT}{V} = \frac{(0.182\ mol + 0.0027\ mol)\left(0.082\ 06\ \dfrac{L \cdot atm}{K \cdot mol}\right)(473.1\ K)}{(5.00\ L)} = 1.43\ atm$$

9.112 O_2, 32.00 amu; O_3, 48.00 amu

	$3\ O_2(g)$	$\rightarrow$	$2\ O_3(g)$
initial	32.00 atm		0
change	–3x		+2x
after rxn	32.00 – 3x		2x

$$P_{Total} = P_{O_2} + P_{O_3} = 30.64\ atm = 32.00\ atm - 3x + 2x = 32.00\ atm - x$$

x = 32.00 atm – 30.64 atm = 1.36 atm

$$P_{O_2} = 32.00 - 3x = 32.00 - 3(1.36\ atm) = 27.92\ atm$$

$$P_{O_3} = 2x = 2(1.36\ atm) = 2.72\ atm$$

T = 25°C = 25 + 273 = 298 K; PV = nRT

$$n_{O_2} = \frac{PV}{RT} = \frac{(27.92\ atm)(10.00\ L)}{\left(0.082\ 06\ \dfrac{L \cdot atm}{K \cdot mol}\right)(298\ K)} = 11.42\ mol\ O_2$$

$$n_{O_3} = \frac{PV}{RT} = \frac{(2.72\ atm)(10.00\ L)}{\left(0.082\ 06\ \dfrac{L \cdot atm}{K \cdot mol}\right)(298\ K)} = 1.11\ mol\ O_3$$

$$\text{mass } O_2 = 11.42\ mol\ O_2 \times \frac{32.00\ g\ O_2}{1\ mol\ O_2} = 365.4\ g\ O_2$$

$$\text{mass } O_3 = 1.11 \text{ mol } O_3 \times \frac{48.00 \text{ g } O_3}{1 \text{ mol } O_3} = 53.3 \text{ g } O_3$$

total mass $= 365.4 \text{ g} + 53.3 \text{ g} = 418.7 \text{ g}$

$$\text{mass \% } O_3 = \frac{\text{mass } O_3}{\text{total mass}} = \frac{53.3 \text{ g}}{418.7 \text{ g}} \times 100\% = 12.7\%$$

Multi-Concept Problems

9.114 CO_2, 44.01 amu

$CH_4(g) + 2 O_2(g) \rightarrow CO_2(g) + 2 H_2O(g)$ $\Delta H° = -802 \text{ kJ}$

(a) 1.00 atm of CH_4 only requires 2.00 atm O_2, therefore O_2 is in excess.

$T = 300°C = 300 + 273 = 573 \text{ K}$; $PV = nRT$

$$n_{CH_4} = \frac{PV}{RT} = \frac{(1.00 \text{ atm})(4.00 \text{ L})}{\left(0.082\ 06 \dfrac{\text{L} \cdot \text{atm}}{\text{K} \cdot \text{mol}}\right)(573 \text{ K})} = 0.0851 \text{ mol } CH_4$$

$$n_{O_2} = \frac{PV}{RT} = \frac{(4.00 \text{ atm})(4.00 \text{ L})}{\left(0.082\ 06 \dfrac{\text{L} \cdot \text{atm}}{\text{K} \cdot \text{mol}}\right)(573 \text{ K})} = 0.340 \text{ mol } O_2$$

$$\text{mass } CO_2 = 0.0851 \text{ mol } CH_4 \times \frac{1 \text{ mol } CO_2}{1 \text{ mol } CH_4} \times \frac{44.01 \text{ g } CO_2}{1 \text{ mol } CO_2} = 3.75 \text{ g } CO_2$$

(b) $q_{rxn} = 0.0851 \text{ mol } CH_4 \times \dfrac{-802 \text{ kJ}}{1 \text{ mol } CH_4} = -68.3 \text{ kJ}$

	$CH_4(g)$	+	$2 O_2(g)$	$\rightarrow$	$CO_2(g)$	+	$2 H_2O(g)$
initial (mol)	0.0851		0.340		0		0
change (mol)	−0.0851		−2(0.0851)		+0.0851		+2(0.0851)
after rxn (mol)	0		0.340 − 2(0.0851)		0.0851		0.170

total moles of gas = 0.340 mol − 2(0.0851) mol + 0.0851 mol + 0.170 mol = 0.425 mol gas

$$q_{rxn} = -68.3 \text{ kJ} \times \frac{1000 \text{ J}}{1 \text{ kJ}} = -68,300 \text{ J}$$

$q_{vessel} = -q_{rxn} = 68,300 \text{ J} = (0.425 \text{ mol})(21 \text{ J/(mol} \cdot °C))(t_f - 300°C) +$

$$(14.500 \text{ kg})\left(\frac{1000 \text{ g}}{1 \text{ kg}}\right)(0.449 \text{ J/(g} \cdot °C))(t_f - 300°C)$$

Solve for t_f.

68,300 J = (8.925 J/°C + 6510 J/°C)$(t_f - 300°C)$ = (6519 J/°C)$(t_f - 300°C)$

$$\frac{68,300 \text{ J}}{6519 \text{ J/°C}} = 10.5°C = (t_f - 300°C)$$

$300°C + 10.5°C = t_f$

$t_f = 310°C$

(c) $T = 310°C = 310 + 273 = 583\ K$

$$P_{CO_2} = \frac{nRT}{V} = \frac{(0.0851\ mol)\left(0.082\ 06\ \frac{L\cdot atm}{K\cdot mol}\right)(583\ K)}{(4.00\ L)} = 1.02\ atm$$

9.116 (a) $2\ C_8H_{18}(l) + 25\ O_2(g) \rightarrow 16\ CO_2(g) + 18\ H_2O(g)$

(b) $4.6 \times 10^{10}\ L\ C_8H_{18} \times \dfrac{1000\ mL}{1\ L} \times \dfrac{0.792\ g}{1\ mL} = 3.64 \times 10^{13}\ g\ C_8H_{18}$

$3.64 \times 10^{13}\ g\ C_8H_{18} \times \dfrac{1\ mol\ C_8H_{18}}{114.2\ g\ C_8H_{18}} \times \dfrac{16\ mol\ CO_2}{2\ mol\ C_8H_{18}} = 2.55 \times 10^{12}\ mol\ CO_2$

$2.55 \times 10^{12}\ mol\ CO_2 \times \dfrac{44.0\ g\ CO_2}{1\ mol\ CO_2} \times \dfrac{1\ kg}{1000\ g} = 1.1 \times 10^{11}\ kg\ CO_2$

(c) $V = \dfrac{nRT}{P} = \dfrac{(2.55 \times 10^{12}\ mol)\left(0.082\ 06\ \frac{L\cdot atm}{K\cdot mol}\right)(273\ K)}{(1.00\ atm)} = 5.7 \times 10^{13}\ L\ of\ CO_2$

(d) 12.5 moles of O_2 are needed for each mole of isooctane (from part a).

$12.5\ mol\ O_2 = (0.210)(n_{air});$ $\qquad n_{air} = \dfrac{12.5\ mol}{0.210} = 59.5\ mol\ air$

$$V = \frac{nRT}{P} = \frac{(59.5\ mol)\left(0.082\ 06\ \frac{L\cdot atm}{K\cdot mol}\right)(273\ K)}{(1.00\ atm)} = 1.33 \times 10^3\ L$$

9.118 $n = \dfrac{PV}{RT} = \dfrac{(1\ atm)(1323\ L)}{\left(0.082\ 06\ \frac{L\cdot atm}{K\cdot mol}\right)(2223\ K)} = 7.25\ mol\ of\ all\ gases$

(a) $0.004\ 00\ mol\ "nitro" \times \dfrac{7.25\ mol\ gases}{1\ mol\ "nitro"} = 0.0290\ mol\ hot\ gases$

(b) $n = \dfrac{PV}{RT} = \dfrac{\left(623\ mm\ Hg \times \frac{1.00\ atm}{760\ mm\ Hg}\right)(0.500\ L)}{\left(0.082\ 06\ \frac{L\cdot atm}{K\cdot mol}\right)(263\ K)} = 0.0190\ mol\ B + C + D$

$n_A = n_{total} - n_{(B+C+D)} = 0.0290 - 0.0190 = 0.0100\ mol\ A;\ A = H_2O$

(c) $n = \dfrac{PV}{RT} = \dfrac{\left(260\ mm\ Hg \times \frac{1.00\ atm}{760\ mm\ Hg}\right)(0.500\ L)}{\left(0.082\ 06\ \frac{L\cdot atm}{K\cdot mol}\right)(298\ K)} = 0.007\ 00\ mol\ C + D$

$n_B = n_{(B+C+D)} - n_{(C+D)} = 0.0190 - 0.007\ 00 = 0.0120\ mol\ B;\ B = CO_2$

(d) $n = \dfrac{PV}{RT} = \dfrac{\left(223 \text{ mm Hg} \times \dfrac{1.00 \text{ atm}}{760 \text{ mm Hg}}\right)(0.500 \text{ L})}{\left(0.082\ 06 \dfrac{\text{L} \cdot \text{atm}}{\text{K} \cdot \text{mol}}\right)(298 \text{ K})} = 0.006\ 00 \text{ mol D}$

$n_C = n_{(C+D)} - n_D = 0.007\ 00 - 0.006\ 00 = 0.001\ 00 \text{ mol C}; \quad C = O_2$

molar mass D $= \dfrac{0.168 \text{ g}}{0.006\ 00 \text{ mol}} = 28.0 \text{ g/mol}; \quad D = N_2$

(e) $0.004 \text{ C}_3\text{H}_5\text{N}_3\text{O}_9(l) \rightarrow 0.0100 \text{ H}_2\text{O}(g) + 0.012 \text{ CO}_2(g) + 0.001 \text{ O}_2(g) + 0.006 \text{ N}_2(g)$
Multiply each coefficient by 1000 to obtain integers.
$4 \text{ C}_3\text{H}_5\text{N}_3\text{O}_9(l) \rightarrow 10 \text{ H}_2\text{O}(g) + 12 \text{ CO}_2(g) + \text{O}_2(g) + 6 \text{ N}_2(g)$

Liquids, Solids, and Changes of State

10.1 $\mu = Q \times r = (1.60 \times 10^{-19} \text{ C})(92 \times 10^{-12} \text{ m})\left(\dfrac{1 \text{ D}}{3.336 \times 10^{-30} \text{ C} \cdot \text{m}}\right) = 4.41 \text{ D}$

% ionic character for HF $= \dfrac{1.82 \text{ D}}{4.41 \text{ D}} \times 100\% = 41\%$

HF has more ionic character than HCl. HCl has only 17% ionic character.

10.2 (a) SF_6 has polar covalent bonds but the molecule is symmetrical (octahedral). The individual bond polarities cancel, and the molecule has no dipole moment.
(b) $H_2C=CH_2$ can be assumed to have nonpolar C–H bonds. In addition, the molecule is symmetrical. The molecule has no dipole moment.
(c) The C–Cl bonds in $CHCl_3$ are polar covalent bonds, and the molecule is polar.

(d) The C–Cl bonds in CH_2Cl_2 are polar covalent bonds, and the molecule is polar.

10.3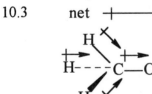

10.4 The N atom is electron rich (red) because of its high electronegativity. The H atoms are electron poor (blue) because they are less electronegative.

10.5 (a) Of the four substances, only HNO_3 has a net dipole moment.
(b) Only HNO_3 can hydrogen bond.
(c) Ar has fewer electrons than Cl_2 and CCl_4, and has the smallest dispersion forces.

10.6
H_2S	dipole-dipole, dispersion
CH_3OH	hydrogen bonding, dipole-dipole, dispersion
C_2H_6	dispersion
Ar	dispersion

$Ar < C_2H_6 < H_2S < CH_3OH$

10.7   (a) $CO_2(s) \rightarrow CO_2(g)$, ΔS is positive.
        (b) $H_2O(g) \rightarrow H_2O(l)$, ΔS is negative.
        (c) ΔS is positive (more disorder).

10.8 $\Delta G = \Delta H - T\Delta S$; at the boiling point (phase change), $\Delta G = 0$.

$$\Delta H = T\Delta S; \quad T = \frac{\Delta H_{vap}}{\Delta S_{vap}} = \frac{29.2 \text{ kJ/mol}}{87.5 \times 10^{-3} \text{ kJ/(K} \cdot \text{mol)}} = 334 \text{ K}$$

10.9   The boiling point is the temperature where the vapor pressure of a liquid equals the external pressure.
$P_1 = 760$ mm Hg; $P_2 = 260$ mm Hg; $T_1 = 80.1°C$
$\Delta H_{vap} = 30.8$ kJ/mol

$$\ln P_2 = \ln P_1 + \frac{\Delta H_{vap}}{R}\left(\frac{1}{T_1} - \frac{1}{T_2}\right)$$

$$(\ln P_2 - \ln P_1)\left(\frac{R}{\Delta H_{vap}}\right) = \frac{1}{T_1} - \frac{1}{T_2}$$

Solve for T_2 (the boiling point for benzene at 260 mm Hg).

$$\frac{1}{T_1} - (\ln P_2 - \ln P_1)\left(\frac{R}{\Delta H_{vap}}\right) = \frac{1}{T_2}$$

$$\frac{1}{353.2 \text{ K}} - [\ln(260) - \ln(760)]\left(\frac{8.3145 \dfrac{J}{K \cdot mol}}{30,800 \text{ J/mol}}\right) = \frac{1}{T_2}$$

$$\frac{1}{T_2} = 0.003\ 121 \text{ K}^{-1}; \ T_2 = 320 \text{ K} = 47°C \quad \text{(boiling point is lower at lower pressure)}$$

10.10 $$\Delta H_{vap} = \frac{(\ln P_2 - \ln P_1)(R)}{\left(\dfrac{1}{T_1} - \dfrac{1}{T_2}\right)}$$

$P_1 = 400$ mm Hg; $T_1 = 41.0\ °C = 314.2$ K
$P_2 = 760$ mm Hg; $T_2 = 331.9$ K

$$\Delta H_{vap} = \frac{[\ln(760) - \ln(400)]\left(8.3145 \dfrac{J}{K \cdot mol}\right)}{\left(\dfrac{1}{314.2 \text{ K}} - \dfrac{1}{331.9 \text{ K}}\right)} = 31,442 \text{ J/mol} = 31.4 \text{ kJ/mol}$$

10.11   (a) 1/8 atom at 8 corners and 1 atom at body center = 2 atoms
        (b) 1/8 atom at 8 corners and 1/2 atom at 6 faces = 4 atoms

10.12   For a simple cube, $d = 2r$; $r = \dfrac{d}{2} = \dfrac{334 \text{ pm}}{2} = 167 \text{ pm}$

10.13   For a simple cube, there is one atom per unit cell.

mass of one Po atom $= 209 \text{ g/mol} \times \dfrac{1 \text{ mol}}{6.022 \times 10^{23} \text{ atoms}} = 3.4706 \times 10^{-22} \text{ g/atom}$

unit cell edge $= d = 334 \text{ pm} = 334 \times 10^{-12} \text{ m} = 3.34 \times 10^{-8} \text{ cm}$

unit cell volume $= d^3 = (3.34 \times 10^{-8} \text{ cm})^3 = 3.7260 \times 10^{-23} \text{ cm}^3$

density $= \dfrac{\text{mass}}{\text{volume}} = \dfrac{3.4706 \times 10^{-22} \text{ g}}{3.7260 \times 10^{-23} \text{ cm}^3} = 9.31 \text{ g/cm}^3$

10.14   There are several possibilities. Here's one.

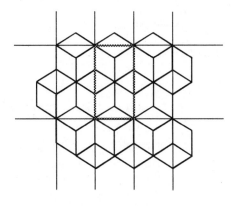

10.15   For CuCl:
1/8 Cl⁻ at 8 corners and 1/2 Cl⁻ at 6 faces = 4 Cl⁻ (4 minuses)
4 Cu⁺ inside (4 pluses)
For $BaCl_2$:
1/8 Ba^{2+} at 8 corners and 1/2 Ba^{2+} at 6 faces = 4 Ba^{2+} (8 pluses)
8 Cl⁻ inside (8 minuses)

10.16   (a) In the unit cell there is a rhenium atom at each corner of the cube. The number of rhenium atoms in the unit cell = 1/8 Re at 8 corners = 1 Re atom.
In the unit cell there is an oxygen atom in the center of each edge of the cube. The number of oxygen atoms in the unit cell = 1/4 O on 12 edges = 3 O atoms.
(b) ReO_3
(c) Each oxide has a –2 charge and there are three of them for a total charge of –6. The charge (oxidation state) of rhenium must be +6 to balance the negative charge of the oxides.
(d) Each oxygen atom is surrounded by two rhenium atoms. The geometry is linear.
(e) Each rhenium atom is surrounded by six oxygen atoms. The geometry is octahedral.

10.17   The minimum pressure at which liquid CO_2 can exist is its triple point pressure of 5.11 atm.

10.18   (a) $CO_2(s) \rightarrow CO_2(g)$
        (b) $CO_2(l) \rightarrow CO_2(g)$
        (c) $CO_2(g) \rightarrow CO_2(l) \rightarrow$ supercritical CO_2

10.19   (a)

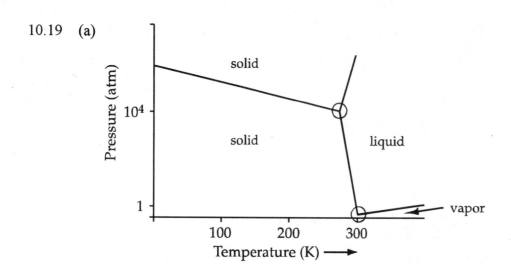

(b) Gallium has two triple points. The one below 1 atm is a solid, liquid, vapor triple point. The one at 10^4 atm is a solid(1), solid(2), liquid triple point.
(c) Increasing the pressure favors the liquid phase, giving the solid/liquid boundary a negative slope. At 1 atm pressure the liquid phase is more dense than the solid phase.

10.20   The molecules in a liquid crystal can move around, as in viscous liquids, but they have a restricted range of motion, as in solids.

10.21   Liquid crystal molecules have a rigid rodlike shape with a length four to eight times greater than their diameter.

Understanding Key Concepts

10.22   The electronegative O atoms are electron rich (red), while the rest of the molecule is electron poor (blue).

10.24   (a) cubic closest-packed
        (b) $1/8$ S^{2-} at 8 corners and $1/2$ S^{2-} at 6 faces = 4 S^{2-};   4 Zn^{2+} inside

10.26   (a) normal boiling point $\approx$ 300 K; normal melting point $\approx$ 180 K
        (b) (i) solid   (ii) gas   (iii) supercritical fluid

10.28   Here are two possibilities.

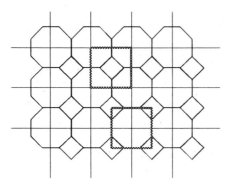

Additional Problems
Dipole Moments and Intermolecular Forces

10.30   If a molecule has polar covalent bonds, the molecular shape (and location of lone pairs of electrons) determines whether a molecule has a dipole moment or not. The molecular shape will determine whether the bond dipoles cancel or not.

10.32   (a) $CHCl_3$ has a permanent dipole moment. Dipole-dipole intermolecular forces are important. London dispersion forces are also present.
(b) O_2 has no dipole moment. London dispersion intermolecular forces are important.
(c) polyethylene, C_nH_{2n+2}. London dispersion intermolecular forces are important.
(d) CH_3OH has a permanent dipole moment. Dipole-dipole intermolecular forces and hydrogen bonding are important. London dispersion forces are also present.

10.34   For CH_3OH and CH_4, dispersion forces are small. CH_3OH can hydrogen bond; CH_4 cannot. This accounts for the large difference in boiling points.
For 1-decanol and decane, dispersion forces are comparable and relatively large along the C–H chain. 1-decanol can hydrogen bond; decane cannot. This accounts for the 55°C higher boiling point for 1-decanol.

10.36   (a)

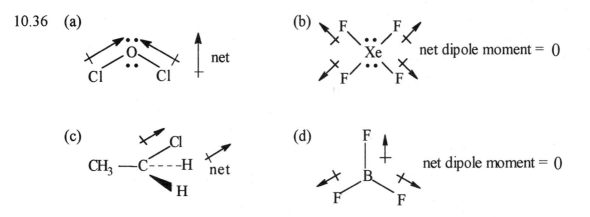

145

10.38

SO_2 is bent and the individual bond dipole moments add to give the molecule a net dipole moment.

CO_2 is linear and the individual bond dipole moments point in opposite directions to cancel each other out. CO_2 has no net dipole moment.

10.40

Vapor Pressure and Changes of State

10.42 ΔH_{vap} is usually larger than ΔH_{fusion} because ΔH_{vap} is the heat required to overcome all intermolecular forces.

10.44   (a) $Hg(l) \rightarrow Hg(g)$
(b) no change of state, Hg remains a liquid
(c) $Hg(g) \rightarrow Hg(l) \rightarrow Hg(s)$

10.46   As the pressure over the liquid H_2O is lowered, H_2O vapor is removed by the pump. As H_2O vapor is removed, more of the liquid H_2O is converted to H_2O vapor. This conversion is an endothermic process and the temperature decreases. The combination of both a decrease in pressure and temperature takes the system across the liquid/solid boundary in the phase diagram so the H_2O that remains turns to ice.

10.48 H_2O, 18.02 amu; $5.00 \text{ g } H_2O \times \dfrac{1 \text{ mol } H_2O}{18.02 \text{ g } H_2O} = 0.2775 \text{ mol } H_2O$

$q_1 = (0.2775 \text{ mol})[36.6 \times 10^{-3} \text{ kJ/(K} \cdot \text{mol})](273 \text{ K} - 263 \text{ K}) = 0.1016 \text{ kJ}$
$q_2 = (0.2775 \text{ mol})(6.01 \text{ kJ/mol}) = 1.668 \text{ kJ}$
$q_3 = (0.2775 \text{ mol})(75.3 \times 10^{-3} \text{ kJ/(K} \cdot \text{mol})](303 \text{ K} - 273 \text{ K}) = 0.6269 \text{ kJ}$
$q_{total} = q_1 + q_2 + q_3 = 2.40 \text{ kJ};$ 2.40 kJ of heat is required.

10.50 H_2O, 18.02 amu; $7.55 \text{ g } H_2O \times \dfrac{1 \text{ mol } H_2O}{18.02 \text{ g } H_2O} = 0.4190 \text{ mol } H_2O$

$q_1 = (0.4190 \text{ mol})[75.3 \times 10^{-3} \text{ kJ/(K} \cdot \text{mol})](273.15 \text{ K} - 306.65 \text{ K}) = -1.057 \text{ kJ}$
$q_2 = -(0.4190 \text{ mol})(6.01 \text{ kJ/mol}) = -2.518 \text{ kJ}$
$q_3 = (0.4190 \text{ mol})[36.6 \times 10^{-3} \text{ kJ/(K} \cdot \text{mol})](263.15 \text{ K} - 273.15 \text{ K}) = -0.1534 \text{ kJ}$
$q_{total} = q_1 + q_2 + q_3 = -3.73 \text{ kJ};$ 3.73 kJ of heat is released.

10.52

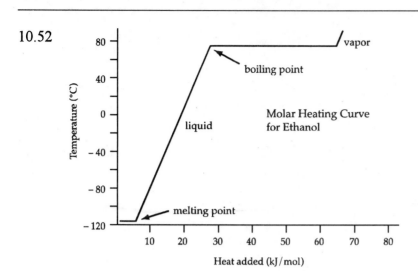

10.54 boiling point = 218°C = 491 K

$\Delta G = \Delta H_{vap} - T\Delta S_{vap};$ At the boiling point (phase change), $\Delta G = 0$

$\Delta H_{vap} = T\Delta S_{vap};$ $\Delta S_{vap} = \dfrac{\Delta H_{vap}}{T} = \dfrac{43.3 \text{ kJ/mol}}{491 \text{ K}} = 0.0882 \text{ kJ/(K} \cdot \text{mol)} = 88.2 \text{ J/(K} \cdot \text{mol)}$

10.56 $\Delta H_{vap} = \dfrac{(\ln P_2 - \ln P_1)(R)}{\left(\dfrac{1}{T_1} - \dfrac{1}{T_2}\right)}$

$T_1 = -5.1°C = 268.0 \text{ K};$ $P_1 = 100 \text{ mm Hg}$
$T_2 = 46.5°C = 319.6 \text{ K};$ $P_2 = 760 \text{ mm Hg}$

$\Delta H_{vap} \doteq \dfrac{[\ln(760) - \ln(100)][8.3145 \times 10^{-3} \text{ kJ/(K} \cdot \text{mol)}]}{\left(\dfrac{1}{268.0 \text{ K}} - \dfrac{1}{319.6 \text{ K}}\right)} = 28.0 \text{ kJ/mol}$

10.58 $\ln P_2 = \ln P_1 + \dfrac{\Delta H_{vap}}{R}\left(\dfrac{1}{T_1} - \dfrac{1}{T_2}\right)$

$\Delta H_{vap} = 28.0 \text{ kJ/mol}$
$P_1 = 100 \text{ mm Hg};$ $T_1 = -5.1°C = 268.0 \text{ K};$ $T_2 = 20.0°C = 293.2 \text{ K}$
Solve for P_2.

$\ln P_2 = \ln(100) + \dfrac{28.0 \text{ kJ/mol}}{[8.3145 \times 10^{-3} \text{ kJ/(K} \cdot \text{mol)}]}\left(\dfrac{1}{268.0 \text{ K}} - \dfrac{1}{293.2 \text{ K}}\right)$

$\ln P_2 = 5.6852$
$P_2 = e^{5.6852} = 294.5 \text{ mm Hg} = 294 \text{ mm Hg}$

10.60

T(K)	P_{vap}(mm Hg)	ln P_{vap}	1/T
263	80.1	4.383	0.003 802
273	133.6	4.8949	0.003 663
283	213.3	5.3627	0.003 534
293	329.6	5.7979	0.003 413
303	495.4	6.2054	0.003 300
313	724.4	6.5853	0.003 195

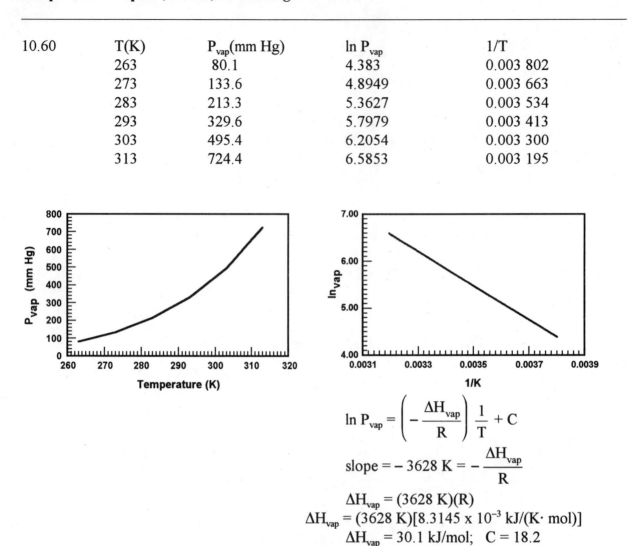

$$\ln P_{vap} = \left(-\frac{\Delta H_{vap}}{R} \right) \frac{1}{T} + C$$

$$\text{slope} = -3628\ K = -\frac{\Delta H_{vap}}{R}$$

$$\Delta H_{vap} = (3628\ K)(R)$$
$$\Delta H_{vap} = (3628\ K)[8.3145 \times 10^{-3}\ kJ/(K \cdot mol)]$$
$$\Delta H_{vap} = 30.1\ kJ/mol; \quad C = 18.2$$

10.62 $\Delta H_{vap} = 30.1\ kJ/mol$

10.64 $$\Delta H_{vap} = \frac{(\ln P_2 - \ln P_1)(R)}{\left(\dfrac{1}{T_1} - \dfrac{1}{T_2} \right)}$$

$P_1 = 80.1$ mm Hg; $\qquad$ $T_1 = 263$ K
$P_2 = 724.4$ mm Hg; $\qquad$ $T_2 = 313$ K

$$\Delta H_{vap} = \frac{[\ln(724.4) - \ln(80.1)][8.3145 \times 10^{-3}\ kJ/(K \cdot mol)]}{\left(\dfrac{1}{263\ K} - \dfrac{1}{313\ K} \right)} = 30.1\ kJ/mol$$

The calculated ΔH_{vap} and that obtained from the plot in Problem 10.62 are the same.

Structures of Solids

10.66   molecular solid, CO_2, I_2
       metallic solid, any metallic element
       covalent network solid, diamond
       ionic solid, NaCl

10.68   The unit cell is the smallest repeating unit in a crystal.

10.70   Cu is face-centered cubic. d = 362 pm; $r = \sqrt{\dfrac{d^2}{8}} = \sqrt{\dfrac{(362 \text{ pm})^2}{8}} = 128$ pm

    362 pm = 362 x 10^{-12} m = 3.62 x 10^{-8} cm
    unit cell volume = (3.62 x 10^{-8} cm)^3 = 4.74 x 10^{-23} cm^3

    mass of one Cu atom = 63.55 g/mol x $\dfrac{1 \text{ mol}}{6.022 \times 10^{23} \text{ atom}}$ = 1.055 x 10^{-22} g/atom

    Cu is face-centered cubic; there are therefore four Cu atoms in the unit cell.
    unit cell mass = (4 atoms)(1.055 x 10^{-22} g/atom) = 4.22 x 10^{-22} g

    density = $\dfrac{\text{mass}}{\text{volume}} = \dfrac{4.22 \times 10^{-22}\text{g}}{4.74 \times 10^{-23} \text{ cm}^3}$ = 8.90 g/cm^3

10.72   mass of one Al atom = 26.98 g/mol x $\dfrac{1 \text{ mol}}{6.022 \times 10^{23} \text{ atom}}$ = 4.480 x 10^{-23} g/atom

    Al is face-centered cubic; there are therefore four Al atoms in the unit cell.
    unit cell mass = (4 atoms)(4.480 x 10^{-23} g/atom) = 1.792 x 10^{-22} g

    density = $\dfrac{\text{mass}}{\text{volume}}$

    unit cell volume = $\dfrac{\text{unit cell mass}}{\text{density}} = \dfrac{1.792 \times 10^{-22} \text{ g}}{2.699 \text{ g/cm}^3}$ = 6.640 x 10^{-23} cm^3

    unit cell edge = d = $\sqrt[3]{6.640 \times 10^{-23} \text{ cm}^3}$ = 4.049 x 10^{-8} cm

    d = 4.049 x 10^{-8} cm x $\dfrac{1 \text{m}}{100 \text{ cm}}$ = 4.049 x 10^{-10} m = 404.9 x 10^{-12} m = 404.9 pm

10.74   unit cell body diagonal = 4r = 549 pm
    For W, r = $\dfrac{549 \text{ pm}}{4}$ = 137 pm

10.76   mass of one Ti atom = 47.88 g/mol x $\dfrac{1 \text{ mol}}{6.022 \times 10^{23} \text{ atoms}}$ = 7.951 x 10^{-23} g/atom

    r = 144.8 pm = 144.8 x 10^{-12} m
    r = 144.8 x 10^{-12} m x $\dfrac{100 \text{ cm}}{1 \text{ m}}$ = 1.448 x 10^{-8} cm

Calculate the volume and then the density for Ti assuming it is primitive cubic, body-centered cubic, and face-centered cubic. Compare the calculated density with the actual density to identify the unit cell.

For primitive cubic:

$$d = 2r; \text{ volume} = d^3 = [2(1.448 \times 10^{-8} \text{ cm})]^3 = 2.429 \times 10^{-23} \text{ cm}^3$$

$$\text{density} = \frac{\text{unit cell mass}}{\text{volume}} = \frac{7.951 \times 10^{-23} \text{ g}}{2.429 \times 10^{-23} \text{ cm}^3} = 3.273 \text{ g/cm}^3$$

For face-centered cubic:

$$d = 2\sqrt{2}r; \text{ volume} = d^3 = [2\sqrt{2}(1.448 \times 10^{-8} \text{ cm})]^3 = 6.870 \times 10^{-23} \text{ cm}^3$$

$$\text{density} = \frac{4(7.951 \times 10^{-23} \text{ g})}{6.870 \times 10^{-23} \text{ cm}^3} = 4.630 \text{ g/cm}^3$$

For body-centered cubic:

From Problems 10.73 and 10.74,

$$d = \frac{4r}{\sqrt{3}}; \text{ volume} = d^3 = \left[\frac{4(1.448 \times 10^{-8} \text{ cm})}{\sqrt{3}}\right]^3 = 3.739 \times 10^{-23} \text{ cm}^3$$

$$\text{density} = \frac{2(7.951 \times 10^{-23} \text{ g})}{3.739 \times 10^{-23} \text{ cm}^3} = 4.253 \text{ g/cm}^3$$

The calculated density for a face-centered cube (4.630 g/cm³) is closest to the actual density of 4.54 g/cm³. Ti crystallizes in the face-centered cubic unit cell.

10.78   Six Na^+ ions touch each H^- ion and six H^- ions touch each Na^+ ion.

10.80 Na^+ H^- Na^+
              ← 488 pm →   unit cell edge = d = 488 pm; Na–H bond = d/2 = 244 pm

Phase Diagrams

10.82   (a) gas   (b) liquid   (c) solid

10.84

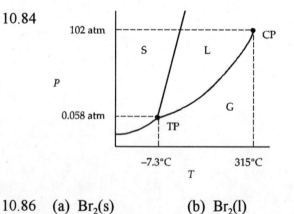

10.86   (a) $Br_2(s)$ (b) $Br_2(l)$

10.88   Solid O_2 does not melt when pressure is applied because the solid is denser than the liquid and the solid/liquid boundary in the phase diagram slopes to the right.

10.90

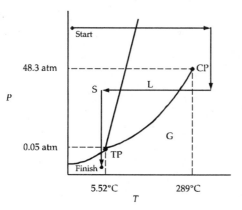

The starting phase is benzene as a solid, and the final phase is benzene as a gas.

10.92   solid → liquid → supercritical fluid → liquid → solid → gas

General Problems

10.94   Because chlorine is larger than fluorine, the charge separation is larger in CH_3Cl compared to CH_3F resulting in CH_3Cl having a slightly larger dipole moment.

10.96 $7.50 \text{ g} \times \dfrac{1 \text{ mol}}{200.6 \text{ g}} = 0.037\ 39 \text{ mol Hg}$

$q_1 = (0.037\ 39 \text{ mol})[28.2 \times 10^{-3} \text{ kJ/(K} \cdot \text{mol)}](234.2 \text{ K} - 223.2 \text{ K}) = 0.011\ 60 \text{ kJ}$
$q_2 = (0.037\ 39 \text{ mol})(2.33 \text{ kJ/mol}) = 0.087\ 12 \text{ kJ}$
$q_3 = (0.037\ 39 \text{ mol})[27.9 \times 10^{-3} \text{ kJ/(K} \cdot \text{mol)}](323.2 \text{ K} - 234.2 \text{ K}) = 0.092\ 84 \text{ kJ}$
$q_{total} = q_1 + q_2 + q_3 = 0.192 \text{ kJ};$ 0.192 kJ of heat is required.

10.98 $\ln P_2 = \ln P_1 + \dfrac{\Delta H_{vap}}{R}\left(\dfrac{1}{T_1} - \dfrac{1}{T_2}\right)$

$\Delta H_{vap} = 40.67 \text{ kJ/mol}$
At 1 atm, H_2O boils at 100°C; therefore set
$T_1 = 100°C = 373 \text{ K, and } P_1 = 1.00 \text{ atm.}$
Let $T_2 = 95°C = 368 \text{ K, and solve for } P_2.$ (P_2 is the atmospheric pressure in Denver.)

$\ln P_2 = \ln(1) + \dfrac{40.67 \text{ kJ/mol}}{[8.3145 \times 10^{-3} \text{ kJ/(K} \cdot \text{mol)}]}\left(\dfrac{1}{373 \text{ K}} - \dfrac{1}{368 \text{ K}}\right)$

$\ln P_2 = -0.1782$
$P_2 = e^{-0.1782} = 0.837 \text{ atm}$

10.100 $\Delta G = \Delta H - T\Delta S$; at the melting point (phase change), $\Delta G = 0$.

$\Delta H = T\Delta S; \quad T = \dfrac{\Delta H_{fus}}{\Delta S_{fus}} = \dfrac{9.037 \text{ kJ/mol}}{9.79 \times 10^{-3} \text{ kJ/(K} \cdot \text{mol)}} = 923 \text{ K} = 650°C$

10.102 $\Delta H_{vap} = \dfrac{(\ln P_2 - \ln P_1)(R)}{\left(\dfrac{1}{T_1} - \dfrac{1}{T_2}\right)}$

$P_1 = 40.0$ mm Hg; $\qquad$ $T_1 = -81.6°C = 191.6$ K

$P_2 = 400$ mm Hg; $\qquad$ $T_2 = -43.9°C = 229.2$ K

$\Delta H_{vap} = \dfrac{[\ln(400) - \ln(40.0)]\left(8.3145 \times 10^{-3}\dfrac{kJ}{K \cdot mol}\right)}{\left(\dfrac{1}{191.6\,K} - \dfrac{1}{229.2\,K}\right)} = 22.36$ kJ/mol

Using $\Delta H_{vap} = 22.36$ kJ/mol

$\ln P_2 = \ln P_1 + \dfrac{\Delta H_{vap}}{R}\left(\dfrac{1}{T_1} - \dfrac{1}{T_2}\right)$

$(\ln P_2 - \ln P_1)\left(\dfrac{R}{\Delta H_{vap}}\right) = \dfrac{1}{T_1} - \dfrac{1}{T_2}$

$\dfrac{1}{T_1} - (\ln P_2 - \ln P_1)\left(\dfrac{R}{\Delta H_{vap}}\right) = \dfrac{1}{T_2}$

$P_1 = 40.0$ mm Hg; $\quad$ $T_1 = 191.6$ K

$P_2 = 760$ mm Hg

Solve for T_2 (the normal boiling point).

$\dfrac{1}{191.6\,K} - [\ln(760) - \ln(40.0)]\left(\dfrac{8.3145 \times 10^{-3}\dfrac{kJ}{K \cdot mol}}{22.36\,kJ/mol}\right) = \dfrac{1}{T_2}$

$\dfrac{1}{T_2} = 0.004\ 124\ 33$; $T_2 = 242.46$ K $= -30.7°C$

10.104 $\Delta H_{vap} = \dfrac{(\ln P_2 - \ln P_1)(R)}{\left(\dfrac{1}{T_1} - \dfrac{1}{T_2}\right)}$

$P_1 = 100$ mm Hg; $\qquad$ $T_1 = -110.3°C = 162.85$ K

$P_2 = 760$ mm Hg; $\qquad$ $T_2 = -88.5°C = 184.65$ K

$\Delta H_{vap} = \dfrac{[\ln(760) - \ln(100)]\left(8.3145 \times 10^{-3}\dfrac{kJ}{K \cdot mol}\right)}{\left(\dfrac{1}{162.85\,K} - \dfrac{1}{184.65\,K}\right)} = 23.3$ kJ/mol

10.106

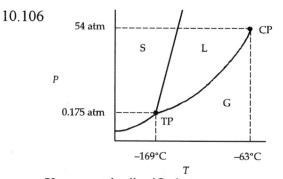

Kr cannot be liquified at room temperature because room temperature is above T_c (−63°C).

10.108 For a body-centered cube

$$4r = \sqrt{3} \text{ edge}; \qquad \text{edge} = \frac{4r}{\sqrt{3}}$$

$$\text{volume of sphere} = \frac{4}{3}\pi r^3$$

$$\text{volume of unit cell} = \left(\frac{4r}{\sqrt{3}}\right)^3 = \frac{64\, r^3}{3\sqrt{3}}$$

$$\text{volume of 2 spheres} = 2\left(\frac{4}{3}\pi r^3\right) = \frac{8}{3}\pi r^3$$

$$\text{\% volume occupied} = \frac{\left(\dfrac{8}{3}\pi r^3\right)}{\left(\dfrac{64\, r^3}{3\sqrt{3}}\right)} \text{ x } 100\% = 68\%$$

10.110 unit cell edge = d = 287 pm = 287 x 10^{-12} m = 2.87 x 10^{-8} cm
unit cell volume = d^3 = (2.87 x 10^{-8} cm)^3 = 2.364 x 10^{-23} cm^3
unit cell mass = (2.364 x 10^{-23} cm^3)(7.86 g/cm^3) = 1.858 x 10^{-22} g
Fe is body-centered cubic; therefore there are two Fe atoms per unit cell.

$$\text{mass of one Fe atom} = \frac{1.858 \text{ x } 10^{-22} \text{ g}}{2 \text{ Fe atoms}} = 9.290 \text{ x } 10^{-23} \text{ g/atom}$$

$$\text{Avogadro's number} = 55.85 \text{ g/mol x } \frac{1 \text{ atom}}{9.290 \text{ x } 10^{-23} \text{ g}} = 6.01 \text{ x } 10^{23} \text{ atoms/mol}$$

10.112 (a) unit cell edge = $2r_{Cl^-} + 2r_{Na^+}$ = 2(181 pm) + 2(97 pm) = 556 pm

(b) unit cell edge = d = 556 pm = 556 x 10^{-12} m = 5.56 x 10^{-8} cm
unit cell volume = (5.56 x 10^{-8} cm)^3 = 1.719 x 10^{-22} cm^3
The unit cell contains 4 Na$^+$ ions and 4 Cl$^-$ ions.

$$\text{mass of one Na}^+ \text{ ion} = 22.99 \text{ g/mol x } \frac{1 \text{ mol}}{6.022 \text{ x } 10^{23} \text{ ions}} = 3.818 \text{ x } 10^{-23} \text{ g/Na}^+$$

$$\text{mass of one Cl}^-\text{ ion} = 35.45 \text{ g/mol x } \frac{1 \text{ mol}}{6.022 \text{ x } 10^{23} \text{ ions}} = 5.887 \text{ x } 10^{-23} \text{ g/Cl}^-$$

$$\text{unit cell mass} = 4(3.818 \text{ x } 10^{-23} \text{ g}) + 4(5.887 \text{ x } 10^{-23} \text{ g}) = 3.882 \text{ x } 10^{-22} \text{ g}$$

$$\text{density} = \frac{\text{unit cell mass}}{\text{unit cell volume}} = \frac{3.882 \text{ x } 10^{-22} \text{ g}}{1.719 \text{ x } 10^{-22} \text{ cm}^3} = 2.26 \text{ g/cm}^3$$

10.114 Al_2O_3, ionic (greater lattice energy than NaCl because of higher ion charges); F_2, dispersion; H_2O, dipole-dipole, H–bonding; Br_2, dispersion (larger and more polarizable than F_2), ICl, dipole-dipole, NaCl, ionic

rank according to normal boiling points: $F_2 < Br_2 < ICl < H_2O < NaCl < Al_2O_3$

10.116 (a)

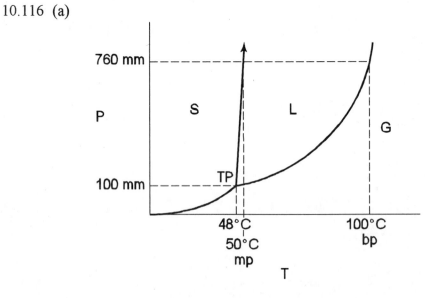

(b) (i) solid   (ii) gas   (iii) liquid   (iv) liquid   (v) solid

Multi-Concept Problems

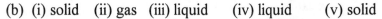

10.118 (a) Let the formula of magnetite be Fe_xO_y, then $Fe_xO_y + y \, CO \rightarrow x \, Fe + y \, CO_2$

$$n_{CO_2} = y = \frac{PV}{RT} = \frac{\left(751 \text{ mm Hg x } \dfrac{1.00 \text{ atm}}{760 \text{ mm Hg}}\right)(1.136 \text{ L})}{\left(0.082 \ 06 \ \dfrac{\text{L} \cdot \text{atm}}{\text{K} \cdot \text{mol}}\right)(298 \text{ K})} = 0.04590 \text{ mol } CO_2$$

$0.04590 \text{ mol } CO_2 = \text{mol of O in } Fe_xO_y$

$$\text{mass of O in } Fe_xO_y = 0.04590 \text{ mol O x } \frac{16.0 \text{ g O}}{1 \text{ mol O}} = 0.7345 \text{ g O}$$

$\text{mass of Fe in } Fe_xO_y = 2.660 \text{ g} - 0.7345 \text{ g} = 1.926 \text{ g Fe}$

(b) $\text{mol Fe in magnetite} = 1.926 \text{ g Fe x } \dfrac{1 \text{ mol Fe}}{55.85 \text{ g Fe}} = 0.0345 \text{ mol Fe}$

formula of magnetite: $Fe_{0.0345} O_{0.0459}$ (divide each subscript by the smaller)

154

$Fe_{0.0345 / 0.0345} O_{0.0459 / 0.0345}$

$FeO_{1.33}$ (multiply both subscripts by 3)

$Fe_{(1 \times 3)} O_{(1.33 \times 3)};$ Fe_3O_4

(c) unit cell edge = d = 839 pm = 839×10^{-12} m

$$d = 839 \times 10^{-12} \text{ m} \times \frac{100 \text{ cm}}{1 \text{ m}} = 8.39 \times 10^{-8} \text{ cm}$$

unit cell volume = $d^3 = (8.39 \times 10^{-8} \text{ cm})^3 = 5.91 \times 10^{-22} \text{ cm}^3$

unit cell mass = $(5.91 \times 10^{-22} \text{ cm}^3)(5.20 \text{ g/cm}^3) = 3.07 \times 10^{-21}$ g

$$\text{mass of Fe in unit cell} = \left(\frac{1.926 \text{ g Fe}}{2.660 \text{ g}} \right)(3.07 \times 10^{-21} \text{ g}) = 2.22 \times 10^{-21} \text{ g Fe}$$

$$\text{mass of O in unit cell} = \left(\frac{0.7345 \text{ g O}}{2.660 \text{ g}} \right)(3.07 \times 10^{-21} \text{ g}) = 8.47 \times 10^{-22} \text{ g O}$$

$$\text{Fe atoms in unit cell} = 2.22 \times 10^{-21} \text{ g} \times \frac{6.022 \times 10^{23} \text{ atoms/mol}}{55.847 \text{ g/mol}} = 24 \text{ Fe atoms}$$

$$\text{O atoms in unit cell} = 8.47 \times 10^{-22} \text{ g} \times \frac{6.022 \times 10^{23} \text{ atoms/mol}}{16.00 \text{ g/mol}} = 32 \text{ O atoms}$$

10.120 (a) M = alkali metal; 500.0 mL = 0.5000 L; 802°C = 1075 K

$$n_M = \frac{PV}{RT} = \frac{\left(12.5 \text{ mm Hg} \times \dfrac{1.00 \text{ atm}}{760 \text{ mm Hg}} \right)(0.5000 \text{ L})}{\left(0.082\,06 \dfrac{\text{L} \cdot \text{atm}}{\text{K} \cdot \text{mol}} \right)(1075 \text{ K})} = 9.32 \times 10^{-5} \text{ mol M}$$

1.62 mm = 1.62×10^{-3} m; crystal volume = $(1.62 \times 10^{-3} \text{ m})^3 = 4.25 \times 10^{-9} \text{ m}^3$

M atoms in crystal = $(9.32 \times 10^{-5} \text{ mol})(6.022 \times 10^{23} \text{ atoms/mol}) = 5.61 \times 10^{19}$ M atoms

Because M is body-centered cubic, only 68% (Table 10.10) of the total volume is occupied by M atoms.

$$\text{volume of M atom} = \frac{(0.68)(4.25 \times 10^{-9} \text{ m})}{5.61 \times 10^{19} \text{ M atoms}} = 5.15 \times 10^{-29} \text{ m}^3/\text{M atom}$$

$$\text{volume of a sphere} = \frac{4}{3}\pi r^3$$

$$r_M = \sqrt[3]{\frac{3(\text{volume})}{4\pi}} = \sqrt[3]{\frac{3(5.15 \times 10^{-29} \text{ m}^3)}{4\pi}} = 2.31 \times 10^{-10} \text{ m} = 231 \times 10^{-12} \text{ m} = 231 \text{ pm}$$

(b) The radius of 231 pm is closest to that of K.

(c) 1.62 mm = 0.162 cm

$$\text{density of solid} = \frac{(9.32 \times 10^{-5} \text{ mol})(39.1 \text{ g/mol})}{(0.162 \text{ cm})^3} = 0.857 \text{ g/cm}^3$$

$$\text{density of vapor} = \frac{(9.32 \times 10^{-5} \text{ mol})(39.1 \text{ g/mol})}{500.0 \text{ cm}^3} = 7.29 \times 10^{-6} \text{ g/cm}^3$$

Solutions and Their Properties

11.1 Toluene is nonpolar and is insoluble in water.
Br_2 is nonpolar but because of its size is polarizable and is soluble in water.
KBr is an ionic compound and is very soluble in water.
toluene $<$ Br_2 $<$ KBr   (solubility in H_2O)

11.2 (a) Na^+ has the larger (more negative) hydration energy because the Na^+ ion is smaller than the Cs^+ ion and water molecules can approach more closely and bind more tightly to the Na^+ ion.
(b) Ba^{2+} has the larger (more negative) hydration energy because of its higher charge.

11.3 NaCl, 58.44 amu; 1.00 mol NaCl = 58.44 g
1.00 L H_2O = 1000 mL = 1000 g (assuming a density of 1.00 g/mL)

$$\text{mass \% NaCl} = \frac{58.44 \text{ g}}{1000 \text{ g} + 58.44 \text{ g}} \times 100\% = 5.52 \text{ mass \%}$$

11.4 $$\text{ppm} = \frac{\text{mass of } CO_2}{\text{total mass of solution}} \times 10^6 \text{ ppm}$$

total mass of solution = density x volume = (1.3 g/L)(1.0 L) = 1.3 g

$$35 \text{ ppm} = \frac{\text{mass of } CO_2}{1.3 \text{ g}} \times 10^6 \text{ ppm}$$

$$\text{mass of } CO_2 = \frac{(35 \text{ ppm})(1.3 \text{ g})}{10^6 \text{ ppm}} = 4.6 \times 10^{-5} \text{ g } CO_2$$

11.5 Assume 1.00 L of sea water.
mass of 1.00 L = (1000 mL)(1.025 g/mL) = 1025 g

$$\frac{\text{mass NaCl}}{1025 \text{ g}} \times 100\% = 3.50 \text{ mass \%}; \qquad \text{mass NaCl} = \frac{1025 \text{ g} \times 3.50}{100} = 35.88 \text{ g}$$

There are 35.88 g NaCl per 1.00 L of solution.

$$M = \frac{\left(35.88 \text{ g NaCl} \times \dfrac{1 \text{ mol NaCl}}{58.44 \text{ g NaCl}}\right)}{1.00 \text{ L}} = 0.614 \text{ M}$$

11.6 $C_{27}H_{46}O$, 386.7 amu; $CHCl_3$, 119.4 amu; $40.0 \text{ g} \times \dfrac{1 \text{ kg}}{1000 \text{ g}} = 0.0400 \text{ kg}$

$$\text{molality} = \frac{\text{mol } C_{27}H_{46}O}{\text{kg } CHCl_3} = \frac{\left(0.385 \text{ g} \times \dfrac{1 \text{ mol}}{386.7 \text{ g}}\right)}{0.0400 \text{ kg}} = 0.0249 \text{ mol/kg} = 0.0249 \text{ } m$$

$$X_{C_{27}H_{46}O} = \frac{mol\ C_{27}H_{46}O}{mol\ C_{27}H_{46}O + mol\ CHCl_3}$$

$$X_{C_{27}H_{46}O} = \frac{\left(0.385\ g\ \times\ \dfrac{1\ mol}{386.7\ g}\right)}{\left[\left(0.385 \cdot g\ \times\ \dfrac{1\ mol}{386.7\ g}\right) + \left(40.0\ g\ \times\ \dfrac{1\ mol}{119.4\ g}\right)\right]} = 2.96\ \times\ 10^{-3}$$

11.7 CH_3CO_2Na, 82.03 amu

$$kg\ H_2O = (0.150\ mol\ CH_3CO_2Na)\left(\frac{1\ kg\ H_2O}{0.500\ mol\ CH_3CO_2Na}\right) = 0.300\ kg\ H_2O$$

$$mass\ CH_3CO_2Na = 0.150\ mol\ CH_3CO_2Na\ \times\ \frac{82.03\ g\ CH_3CO_2Na}{1\ mol\ CH_3CO_2Na} = 12.3\ g\ CH_3CO_2Na$$

mass of solution needed = 300 g + 12.3 g = 312 g

11.8   Assume you have a solution with 1.000 kg (1000 g) of H_2O. If this solution is 0.258 m, then it must also contain 0.258 mol glucose.

$$mass\ of\ glucose = 0.258\ mol\ \times\ \frac{180.2\ g}{1\ mol} = 46.5\ g\ glucose$$

mass of solution = 1000 g + 46.5 g = 1046.5 g
density = 1.0173 g/mL

$$volume\ of\ solution = 1046.5\ g\ \times\ \frac{1\ mL}{1.0173\ g} = 1028.7\ mL$$

$$volume = 1028.7\ mL\ \times\ \frac{1\ L}{1000\ mL} = 1.029\ L; \qquad molarity = \frac{0.258\ mol}{1.029\ L} = 0.251\ M$$

11.9   Assume 1.00 L of solution.
mass of 1.00 L = (1.0042 g/mL)(1000 mL) = 1004.2 g of solution

$$0.500\ mol\ CH_3CO_2H\ \times\ \frac{60.05\ g\ CH_3CO_2H}{1\ mol\ CH_3CO_2H} = 30.02\ g\ CH_3CO_2H$$

$$1004.2\ g - 30.02\ g = 974.2\ g = 0.9742\ kg\ of\ H_2O; \qquad molality = \frac{0.500\ mol}{0.9742\ kg} = 0.513\ m$$

11.10   Assume you have 100.0 g of seawater.
mass NaCl = (0.0350)(100.0 g) = 3.50 g NaCl
mass H_2O = 100.0 g – 3.50 g = 96.5 g H_2O

$$NaCl,\ 58.44\ amu; \qquad mol\ NaCl = 3.50\ g\ \times\ \frac{1\ mol}{58.44\ g} = 0.0599\ mol\ NaCl$$

$$mass\ H_2O = 96.5\ g\ \times\ \frac{1\ kg}{1000\ g} = 0.0965\ kg\ H_2O; \qquad molality = \frac{0.0599\ mol}{0.0965\ kg} = 0.621\ m$$

11.11 $M = k \cdot P$; $k = \dfrac{M}{P} = \dfrac{3.2 \times 10^{-2} \, M}{1.0 \, \text{atm}} = 3.2 \times 10^{-2} \, \text{mol/(L} \cdot \text{atm)}$

11.12   (a) $M = k \cdot P = [3.2 \times 10^{-2} \, \text{mol/(L} \cdot \text{atm)}](2.5 \, \text{atm}) = 0.080 \, M$
        (b) $M = k \cdot P = [3.2 \times 10^{-2} \, \text{mol/(L} \cdot \text{atm)}](4.0 \times 10^{-4} \, \text{atm}) = 1.3 \times 10^{-5} \, M$

11.13 $C_7H_6O_2$, 122.1 amu; C_2H_6O, 46.07 amu

$$X_{solv} = \frac{\text{mol } C_2H_6O}{\text{mol } C_2H_6O + \text{mol } C_7H_6O_2} = \frac{\left(100 \, g \times \dfrac{1 \, \text{mol}}{46.07 \, g}\right)}{\left(100 \, g \times \dfrac{1 \, \text{mol}}{46.07 \, g}\right) + \left(5.00 \, g \times \dfrac{1 \, \text{mol}}{122.1 \, g}\right)} = 0.981$$

$P_{soln} = P_{solv} \cdot X_{solv} = (100.5 \, \text{mm Hg})(0.981) = 98.6 \, \text{mm Hg}$

11.14 $P_{soln} = P_{solv} \cdot X_{solv}$; $X_{solv} = \dfrac{P_{soln}}{P_{solv}} = \dfrac{(55.3 - 1.30) \, \text{mm Hg}}{55.3 \, \text{mm Hg}} = 0.976$

NaBr dissociates into two ions in aqueous solution.

$$X_{solv} = \frac{\text{mol } H_2O}{\text{mol } H_2O + \text{mol } Na^+ + \text{mol } Br^-}$$

$$X_{solv} = 0.976 = \frac{\left(250 \, g \times \dfrac{1 \, \text{mol}}{18.02 \, g}\right)}{\left(250 \, g \times \dfrac{1 \, \text{mol}}{18.02 \, g}\right) + x \, \text{mol } Na^+ + x \, \text{mol } Br^-}$$

$0.976 = \dfrac{13.9 \, \text{mol}}{13.9 \, \text{mol} + 2x \, \text{mol}}$;   solve for x.

$0.976(13.9 \, \text{mol} + 2x \, \text{mol}) = 13.9 \, \text{mol}$
$13.566 \, \text{mol} + 1.952 \, x \, \text{mol} = 13.9 \, \text{mol}$
$1.952 \, x \, \text{mol} = 13.9 \, \text{mol} - 13.566 \, \text{mol}$
$x \, \text{mol} = \dfrac{13.9 \, \text{mol} - 13.566 \, \text{mol}}{1.952} = 0.171 \, \text{mol}$

$x = 0.171 \, \text{mol } Na^+ = 0.171 \, \text{mol } Br^- = 0.171 \, \text{mol NaBr}$

NaBr, 102.9 amu;   mass NaBr $= 0.171 \, \text{mol} \times \dfrac{102.9 \, g}{1 \, \text{mol}} = 17.6 \, g \, \text{NaBr}$

11.15   At any temperature, the vapor pressure of a solution is lower than the vapor pressure of the pure solvent. The upper curve represents the vapor pressure of the pure solvent. The lower curve represents the vapor pressure of the solution.

11.16 C_2H_5OH, 46.07 amu; H_2O, 18.02 amu

(a) $25.0 \, g \, C_2H_5OH \times \dfrac{1 \, \text{mol } C_2H_5OH}{46.07 \, g \, C_2H_5OH} = 0.5426 \, \text{mol } C_2H_5OH$

$$100.0 \text{ g H}_2\text{O} \times \frac{1 \text{ mol H}_2\text{O}}{18.02 \text{ g H}_2\text{O}} = 5.549 \text{ mol H}_2\text{O}$$

$$X_{\text{C}_2\text{H}_5\text{OH}} = \frac{0.5426 \text{ mol}}{0.5426 \text{ mol} + 5.549 \text{ mol}} = 0.08907$$

$$X_{\text{H}_2\text{O}} = \frac{5.549 \text{ mol}}{0.5426 \text{ mol} + 5.549 \text{ mol}} = 0.9109$$

$$P_{\text{soln}} = X_{\text{C}_2\text{H}_5\text{OH}} P^{\text{o}}_{\text{C}_2\text{H}_5\text{OH}} + X_{\text{H}_2\text{O}} P^{\text{o}}_{\text{H}_2\text{O}}$$

$$P_{\text{soln}} = (0.08907)(61.2 \text{ mm Hg}) + (0.9109)(23.8 \text{ mm Hg}) = 27.1 \text{ mm Hg}$$

(b) $\quad 100 \text{ g C}_2\text{H}_5\text{OH} \times \dfrac{1 \text{ mol C}_2\text{H}_5\text{OH}}{46.07 \text{ g C}_2\text{H}_5\text{OH}} = 2.171 \text{ mol C}_2\text{H}_6\text{O}$

$$25.0 \text{ g H}_2\text{O} \times \frac{1 \text{ mol H}_2\text{O}}{18.02 \text{ g H}_2\text{O}} = 1.387 \text{ mol H}_2\text{O}$$

$$X_{\text{C}_2\text{H}_5\text{OH}} = \frac{2.171 \text{ mol}}{2.171 \text{ mol} + 1.387 \text{ mol}} = 0.6102$$

$$X_{\text{H}_2\text{O}} = \frac{1.387 \text{ mol}}{2.171 \text{ mol} + 1.387 \text{ mol}} = 0.3898$$

$$P_{\text{soln}} = X_{\text{C}_2\text{H}_5\text{OH}} P^{\text{o}}_{\text{C}_2\text{H}_5\text{OH}} + X_{\text{H}_2\text{O}} P^{\text{o}}_{\text{H}_2\text{O}}$$

$$P_{\text{soln}} = (0.6102)(61.2 \text{ mm Hg}) + (0.3898)(23.8 \text{ mm Hg}) = 46.6 \text{ mm Hg}$$

11.17   (a) Because the vapor pressure of the solution (red curve) is higher than that of the first liquid (green curve), the vapor pressure of the second liquid must be higher than that of the solution (red curve). Because the second liquid has a higher vapor pressure than the first liquid, the second liquid has a lower boiling point.

(b)

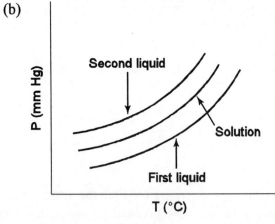

11.18 $\text{C}_9\text{H}_8\text{O}_4$, 180.2 amu; CHCl_3 is the solvent. For CHCl_3, $K_b = 3.63 \dfrac{^{\circ}\text{C} \cdot \text{kg}}{\text{mol}}$

$$75.00 \text{ g} \times \frac{1 \text{ kg}}{1000 \text{ g}} = 0.075 \, 00 \text{ kg}$$

$$\Delta T_b = K_b \cdot m = \left(3.63 \ \frac{°C \cdot kg}{mol} \right) \left(\frac{\left(1.50 \ g \times \frac{1 \ mol}{180.2 \ g} \right)}{0.075 \ 00 \ kg} \right) = 0.40°C$$

Solution boiling point = $61.7°C + \Delta T_b = 61.7°C + 0.40°C = 62.1°C$

11.19 $MgCl_2$, 95.21 amu

$$110 \ g \ \times \ \frac{1 \ kg}{1000 \ g} = 0.110 \ kg$$

$$\Delta T_f = K_f \cdot m \cdot i = \left(1.86 \ \frac{°C \cdot kg}{mol} \right) \left(\frac{\left(7.40 \ g \times \frac{1 \ mol}{95.21 \ g} \right)}{0.110 \ kg} \right) (2.7) = 3.55°C$$

Solution freezing point = $0.00°C - \Delta T_f = 0.00°C - 3.55°C = -3.55°C$

11.20 $\Delta T_f = K_f \cdot m \cdot i$;   For KBr, i = 2.

Solution freezing point = $-2.95°C = 0.00°C - \Delta T_f$; $\Delta T_f = 2.95°C$

$$m = \frac{\Delta T_f}{K_f \cdot i} = \frac{2.95°C}{\left(1.86 \ \frac{°C \cdot kg}{mol} \right) (2)} = 0.793 \ mol/kg = 0.793 \ m$$

11.21   HCl, 36.46 amu; $\Delta T_f = K_f \cdot m \cdot i$

$$190 \ g \ \times \ \frac{1 \ kg}{1000 \ g} = 0.190 \ kg$$

Solution freezing point = $-4.65°C = 0.00°C - \Delta T_f$; $\Delta T_f = 4.65°C$

$$i = \frac{\Delta T_f}{K_f \cdot m} = \frac{4.65°C}{\left(1.86 \ \frac{°C \cdot kg}{mol} \right) \left(\frac{9.12 \ g \times \frac{1 \ mol}{36.46 \ g}}{0.190 \ kg} \right)} = 1.9$$

11.22   The red curve represents the vapor pressure of pure chloroform.

(a) The normal boiling point for a liquid is the temperature where the vapor pressure of the liquid equals 1 atm (760 mm Hg). The approximate boiling point of pure chloroform is 62°C.

(b) The approximate boiling point of the solution is 69°C.

$\Delta T_b = 69°C - 62°C = 7°C$

$$\Delta T_b = K_b \cdot m; \qquad m = \frac{\Delta T_b}{K_b} = \frac{7°C}{3.63 \ \frac{°C \cdot kg}{mol}} = 2 \ mol/kg = 2 \ m$$

11.23 For $CaCl_2$ there are 3 ions (solute particles)/$CaCl_2$
$\Pi = MRT$; For $CaCl_2$, $\Pi = 3MRT$

$$\Pi = (3)(0.125 \text{ mol/L})\left(0.082\ 06\ \frac{L \cdot atm}{K \cdot mol}\right)(310 \text{ K}) = 9.54 \text{ atm}$$

11.24 $\Pi = MRT$; $M = \dfrac{\Pi}{RT} = \dfrac{(3.85 \text{ atm})}{\left(0.082\ 06\ \dfrac{L \cdot atm}{K \cdot mol}\right)(300 \text{ K})} = 0.156 \text{ M}$

11.25 $\Delta T_f = K_f \cdot m$; $m = \dfrac{\Delta T_f}{K_f} = \dfrac{2.10^{\circ}C}{37.7\ \dfrac{^{\circ}C \cdot kg}{mol}} = 0.0557 \text{ mol/kg} = 0.0557\ m$

$$35.00 \text{ g} \times \frac{1 \text{ kg}}{1000 \text{ g}} = 0.03500 \text{ kg}$$

$$\text{mol} = 0.0557\ \frac{mol}{kg} \times 0.03500 \text{ kg} = 0.001\ 95 \text{ mol naphthalene}$$

$$\text{molar mass of naphthalene} = \frac{0.250 \text{ g naphthalene}}{0.001\,95 \text{ mol naphthalene}} = 128 \text{ g/mol}$$

11.26 $\Pi = MRT$; $M = \dfrac{\Pi}{RT} = \dfrac{\left(149 \text{ mm Hg} \times \dfrac{1 \text{ atm}}{760 \text{ mm Hg}}\right)}{\left(0.08206\ \dfrac{L \cdot atm}{K \cdot mol}\right)(298 \text{ K})} = 8.02 \times 10^{-3} \text{ M}$

$300.0 \text{ mL} = 0.3000 \text{ L}$
$(8.02 \times 10^{-3} \text{ mol/L})(0.3000 \text{ L}) = 0.002\ 406 \text{ mol sucrose}$

$$\text{molar mass of sucrose} = \frac{0.822 \text{ g sucrose}}{0.002\,406 \text{ mol sucrose}} = 342 \text{ g/mol}$$

11.27 (a) and (c)

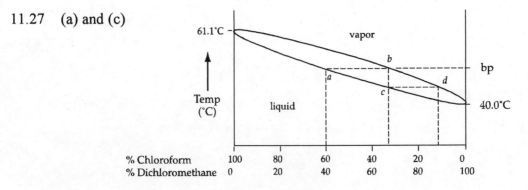

(b) The mixture will begin to boil at ~50°C.

(d) After two cycles of boiling and condensing, the approximate composition of the liquid would 90% dichloromethane and 10% chloroform.

11.28 Both solvent molecules and small solute particles can pass through a semipermeable dialysis membrane. Only large colloidal particles such as proteins can't pass through. Only solvent molecules can pass through a semipermeable membrane used for osmosis.

Understanding Key Concepts

11.30 (a) < (b) < (c)

11.32 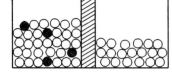   Assume that only the blue (open) spheres (solvent) can pass through the semipermeable membrane. There will be a net transfer of solvent from the right compartment (pure solvent) to the left compartment (solution) to achieve equilibrium.

11.34 The vapor pressure of the NaCl solution is lower than that of pure H_2O. More H_2O molecules will go into the vapor from the pure H_2O than from the NaCl solution. More H_2O vapor molecules will go into the NaCl solution than into pure H_2O. The result is represented by (b).

Additional Problems
Solutions and Energy Changes

11.36 The surface area of a solid plays an important role in determining how rapidly a solid dissolves. The larger the surface area, the more solid-solvent interactions, and the more rapidly the solid will dissolve. Powdered NaCl has a much larger surface area than a large block of NaCl, and it will dissolve more rapidly.

11.38 Substances tend to dissolve when the solute and solvent have the same type and magnitude of intermolecular forces; thus the rule of thumb "like dissolves like."

11.40 Energy is required to overcome intermolecular forces holding solute particles together in the crystal. For an ionic solid, this is the lattice energy. Substances with higher lattice energies tend to be less soluble than substances with lower lattice energies.

11.42 Ethyl alcohol and water are both polar with small dispersion forces. They both can hydrogen bond, and are miscible.
Pentyl alcohol is slightly polar and can hydrogen bond. It has, however, a relatively large dispersion force because of its size, which limits its water solubility.

11.44 $CaCl_2$, 110.98 amu
For a 1.00 m solution:
　　heat released = 81,300 J
　　mass of solution = 1000 g H_2O + 110.98 g $CaCl_2$ = 1110.98 g

$$\Delta T = \frac{q}{\text{(specific heat)(mass of solution)}} = \frac{81,300 \text{ J}}{[4.18 \text{ J/(K} \cdot \text{g)}](1110.98 \text{ g})} = 17.5 \text{ K} = 17.5°C$$

Final temperature = $25.0°C + 17.5°C = 42.5°C$

Units of Concentration

11.46 $\text{molarity} = \dfrac{\text{moles of solute}}{\text{liters of solution}}$; $\quad \text{molality} = \dfrac{\text{moles of solute}}{\text{kg of solvent}}$

11.48 (a) Dissolve 0.150 mol of glucose in water; dilute to 1.00 L.
(b) Dissolve 1.135 mol of KBr in 1.00 kg of H_2O.
(c) Mix together 0.15 mol of CH_3OH with 0.85 mol of H_2O.

11.50 $C_7H_6O_2$, 122.12 amu, 165 mL = 0.165 L
mol $C_7H_6O_2$ = (0.0268 mol/L)(0.165 L) = 0.004 42 mol

$\text{mass } C_7H_6O_2 = 0.004 \; 42 \text{ mol} \times \dfrac{122.12 \text{ g}}{1 \text{ mol}} = 0.540 \text{ g}$

Dissolve 4.42×10^{-3} mol (0.540 g) of $C_7H_6O_2$ in enough $CHCl_3$ to make 165 mL of solution.

11.52 (a) KCl, 74.6 amu
A 0.500 M KCl solution contains 37.3 g of KCl per 1.00 L of solution.
A 0.500 mass % KCl solution contains 5.00 g of KCl per 995 g of water.
The 0.500 M KCl solution is more concentrated (that is, it contains more solute per amount of solvent).
(b) Both solutions contain the same amount of solute. The 1.75 M solution contains less solvent than the 1.75 m solution. The 1.75 M solution is more concentrated.

11.54 (a) $C_6H_8O_7$, 192.12 amu

$0.655 \text{ mol } C_6H_8O_7 \times \dfrac{192.12 \text{ g } C_6H_8O_7}{1 \text{ mol } C_6H_8O_7} = 126 \text{ g } C_6H_8O_7$

$\text{mass \% } C_6H_8O_7 = \dfrac{126 \text{ g}}{126 \text{ g} + 1000 \text{ g}} \times 100\% = 11.2 \text{ mass \%}$

(b) 0.135 mg = 0.135×10^{-3} g
(5.00 mL H_2O)(1.00 g/mL) = 5.00 g H_2O

$\text{mass \% KBr} = \dfrac{0.135 \times 10^{-3} \text{ g}}{(0.135 \times 10^{-3} \text{ g}) + 5.00 \text{ g}} \times 100\% = 0.002 \; 70 \text{ mass \% KBr}$

(c) $\text{mass \% aspirin} = \dfrac{5.50 \text{ g}}{5.50 \text{ g} + 145 \text{ g}} \times 100\% = 3.65 \text{ mass \% aspirin}$

11.56 $P_{O_3} = P_{total} \cdot X_{O_3}$

$$X_{O_3} = \frac{P_{O_3}}{P_{total}} = \frac{1.6 \times 10^{-9}\ atm}{1.3 \times 10^{-2}\ atm} = 1.2 \times 10^{-7}$$

Assume one mole of air (29 g/mol)

mol $O_3 = n_{air} \cdot X_{O_3} = (1\ mol)(1.2 \times 10^{-7}) = 1.2 \times 10^{-7}$ mol O_3

O_3, 48.00 amu;   mass $O_3 = 1.2 \times 10^{-7}$ mol $\times \dfrac{48.0\ g}{1\ mol} = 5.8 \times 10^{-6}$ g O_3

ppm $O_3 = \dfrac{5.8 \times 10^{-6}\ g}{29\ g} \times 10^6 = 0.20$ ppm

11.58 (a) H_2SO_4, 98.08 amu;   molality $= \dfrac{\left(25.0\ g \times \dfrac{1\ mol}{98.08\ g}\right)}{1.30\ kg} = 0.196$ mol/kg $= 0.196\ m$

(b) $C_{10}H_{14}N_2$, 162.23 amu; CH_2Cl_2, 84.93 amu

2.25 g $C_{10}H_{14}N_2 \times \dfrac{1\ mol\ C_{10}H_{14}N_2}{162.23\ g\ C_{10}H_{14}N_2} = 0.0139$ mol $C_{10}H_{14}N_2$

80.0 g $CH_2Cl_2 \times \dfrac{1\ mol\ CH_2Cl_2}{84.93\ g\ CH_2Cl_2} = 0.942$ mol CH_2Cl_2

$X_{C_{10}H_{14}N_2} = \dfrac{0.0139\ mol}{0.942\ mol + 0.0139\ mol} = 0.0145$

$X_{CH_2Cl_2} = \dfrac{0.942\ mol}{0.942\ mol + 0.0139\ mol} = 0.985$

11.60 16.0 mass % $= \dfrac{16.0\ g\ H_2SO_4}{16.0\ g\ H_2SO_4 + 84.0\ g\ H_2O}$

H_2SO_4, 98.08 amu;   density = 1.1094 g/mL

volume of solution = 100.0 g $\times \dfrac{1\ mL}{1.1094\ g} = 90.14$ mL = 0.090 14 L

molarity $= \dfrac{\left(16.0\ g \times \dfrac{1\ mol}{98.08\ g}\right)}{0.090\ 14\ L} = 1.81$ M

11.62 molality $= \dfrac{\left(40.0\ g \times \dfrac{1\ mol}{62.07\ g}\right)}{0.0600\ kg} = 10.7$ mol/kg = 10.7 m

11.64 $C_{19}H_{21}NO_3$, 311.34 amu;   1.5 mg = 1.5 x 10^{-3} g

$$1.3 \times 10^{-3} \text{ mol/kg} = \frac{\left(1.5 \times 10^{-3} \text{ g} \times \dfrac{1 \text{ mol}}{311.34 \text{ g}}\right)}{\text{kg of solvent}}; \quad \text{solve for kg of solvent.}$$

$$\text{kg of solvent} = \frac{\left(1.5 \times 10^{-3} \text{ g} \times \dfrac{1 \text{ mol}}{311.34 \text{ g}}\right)}{1.3 \times 10^{-3} \text{ mol/kg}} = 0.0037 \text{ kg}$$

Because the solution is very dilute, kg of solvent $\approx$ kg of solution.

$$\text{g of solution} = (0.0037 \text{ kg})\left(\frac{1000 \text{ g}}{1 \text{ kg}}\right) = 3.7 \text{ g}$$

11.66 $C_6H_{12}O_6$, 180.16 amu; H_2O, 18.02 amu;   Assume 1.00 L of solution.

mass of solution = (1000 mL)(1.0624 g/mL) = 1062.4 g

$$\text{mass of solute} = 0.944 \text{ mol} \times \frac{180.16 \text{ g}}{1 \text{ mol}} = 170.1 \text{ g } C_6H_{12}O_6$$

mass of H_2O = 1062.4 g – 170.1 g = 892.3 g H_2O

$$\text{mol } C_6H_{12}O_6 = 0.944 \text{ mol}; \quad \text{mol } H_2O = 892.3 \text{ g} \times \frac{1 \text{ mol}}{18.02 \text{ g}} = 49.5 \text{ mol}$$

(a) $X_{C_6H_{12}O_6} = \dfrac{\text{mol } C_6H_{12}O_6}{\text{mol } C_6H_{12}O_6 + \text{mol } H_2O} = \dfrac{0.944 \text{ mol}}{0.944 \text{ mol} + 49.5 \text{ mol}} = 0.0187$

(b)   mass % $= \dfrac{\text{mass } C_6H_{12}O_6}{\text{total mass of solution}} \times 100\% = \dfrac{170.1 \text{ g}}{1062.4 \text{ g}} \times 100\% = 16.0\%$

(c)   molality $= \dfrac{\text{mol } C_6H_{12}O_6}{\text{kg } H_2O} = \dfrac{0.944 \text{ mol}}{0.8923 \text{ kg}} = 1.06 \text{ mol/kg} = 1.06 \ m$

Solubility and Henry's Law

11.68 $M = k \cdot P = (0.091 \ \dfrac{\text{mol}}{\text{L} \cdot \text{atm}})(0.75 \text{ atm}) = 0.068 \text{ M}$

11.70 $M = k \cdot P$

Calculate k: $k = \dfrac{M}{P} = \dfrac{2.21 \times 10^{-3} \text{ mol/L}}{1.00 \text{ atm}} = 2.21 \times 10^{-3} \ \dfrac{\text{mol}}{\text{L} \cdot \text{atm}}$

Convert 4 mg/L to mol/L:

4 mg = 4 x 10^{-3} g

$$O_2 \text{ molarity} = \frac{\left(4 \times 10^{-3} \text{ g} \times \dfrac{1 \text{ mol}}{32.00 \text{ g}}\right)}{1.00 \text{ L}} = 1.25 \times 10^{-4} \text{ M}$$

$$P_{O_2} = \frac{M}{k} = \frac{1.25 \times 10^{-4} \frac{mol}{L}}{2.21 \times 10^{-3} \frac{mol}{L \cdot atm}} = 0.06 \text{ atm}$$

11.72 $[Xe] = 10 \text{ mmol/L} = 10 \times 10^{-3} \text{ mol/L} = 0.010 \text{ M at STP}$

$$M = k \cdot P; \quad k = \frac{M}{P} = \frac{0.010 \text{ M}}{1.0 \text{ atm}} = 0.010 \text{ mol/(L} \cdot \text{atm)}$$

Colligative Properties

11.74   The difference in entropy between the solvent in a solution and a pure solvent is responsible for colligative properties.

11.76   NaCl is a nonvolatile solute. Methyl alcohol is a volatile solute. When NaCl is added to water, the vapor pressure of the solution is decreased, which means that the boiling point of the solution will increase. When methyl alcohol is added to water, the vapor pressure of the solution is increased which means that the boiling point of the solution will decrease.

11.78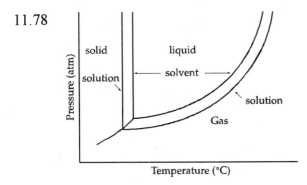

11.80   (a) CH_4N_2O, 60.06 amu; H_2O, 18.02 amu

$$10.0 \text{ g } CH_4N_2O \times \frac{1 \text{ mol } CH_4N_2O}{60.06 \text{ g } CH_4N_2O} = 0.167 \text{ mol } CH_4N_2O$$

$$150.0 \text{ g } H_2O \times \frac{1 \text{ mol } H_2O}{18.02 \text{ g } H_2O} = 8.32 \text{ mol } H_2O$$

$$X_{H_2O} = \frac{8.32 \text{ mol}}{8.32 \text{ mol} + 0.167 \text{ mol}} = 0.980$$

$$P_{soln} = P^{o}_{H_2O} \cdot X_{H_2O} = (71.93 \text{ mm Hg})(0.980) = 70.5 \text{ mm Hg}$$

(b) LiCl, 42.39 amu; $10.0 \text{ g LiCl} \times \frac{1 \text{ mol LiCl}}{42.39 \text{ g LiCl}} = 0.236 \text{ mol LiCl}$

LiCl dissociates into $Li^+(aq)$ and $Cl^-(aq)$ in H_2O.

mol Li$^+$ = mol Cl$^-$ = mol LiCl = 0.236 mol

$$150.0 \text{ g H}_2\text{O} \times \frac{1 \text{ mol H}_2\text{O}}{18.02 \text{ g H}_2\text{O}} = 8.32 \text{ mol H}_2\text{O}$$

$$X_{\text{H}_2\text{O}} = \frac{8.32 \text{ mol}}{8.32 \text{ mol} + 0.236 \text{ mol} + 0.236 \text{ mol}} = 0.946$$

$$P_{\text{soln}} = P^{\circ}_{\text{H}_2\text{O}} \cdot X_{\text{H}_2\text{O}} = (71.93 \text{ mm Hg})(0.946) = 68.0 \text{ mm Hg}$$

11.82   For H$_2$O, $K_b = 0.51 \frac{^{\circ}\text{C} \cdot \text{kg}}{\text{mol}}$;   150.0 g = 0.1500 kg

(a) $\Delta T_b = K_b \cdot m = \left(0.51 \frac{^{\circ}\text{C} \cdot \text{kg}}{\text{mol}}\right)\left(\frac{0.167 \text{ mol}}{0.1500 \text{ kg}}\right) = 0.57^{\circ}\text{C}$

Solution boiling point = $100.00^{\circ}\text{C} + \Delta T_b = 100.00^{\circ}\text{C} + 0.57^{\circ}\text{C} \doteq 100.57^{\circ}\text{C}$

(b) $\Delta T_b = K_b \cdot m = \left(0.51 \frac{^{\circ}\text{C} \cdot \text{kg}}{\text{mol}}\right)\left(\frac{2(0.236 \text{ mol})}{0.1500 \text{ kg}}\right) = 1.6^{\circ}\text{C}$

Solution boiling point = $100.00^{\circ}\text{C} + \Delta T_b = 100.00^{\circ}\text{C} + 1.6^{\circ}\text{C} = 101.6^{\circ}\text{C}$

11.84 $\Delta T_f = K_f \cdot m \cdot i$

Solution freezing point = $-4.3^{\circ}\text{C} = 0.00^{\circ}\text{C} - \Delta T_f$; $\Delta T_f = 4.3^{\circ}\text{C}$

$$i = \frac{\Delta T_f}{K_f \cdot m} = \frac{4.3^{\circ}\text{C}}{\left(1.86 \frac{^{\circ}\text{C} \cdot \text{kg}}{\text{mol}}\right)(1.0 \text{ mol/kg})} = 2.3$$

11.86   Acetone, C$_3$H$_6$O, 58.08 amu, $P^{\circ}_{\text{C}_3\text{H}_6\text{O}} = 285$ mm Hg

Ethyl acetate, C$_4$H$_8$O$_2$, 88.11 amu, $P^{\circ}_{\text{C}_4\text{H}_8\text{O}_2} = 118$ mm Hg

$$25.0 \text{ g C}_3\text{H}_6\text{O} \times \frac{1 \text{ mol C}_3\text{H}_6\text{O}}{58.08 \text{ g C}_3\text{H}_6\text{O}} = 0.430 \text{ mol C}_3\text{H}_6\text{O}$$

$$25.0 \text{ g C}_4\text{H}_8\text{O}_2 \times \frac{1 \text{ mol C}_4\text{H}_8\text{O}_2}{88.11 \text{ g C}_4\text{H}_8\text{O}_2} = 0.284 \text{ mol C}_4\text{H}_8\text{O}_2$$

$$X_{\text{C}_3\text{H}_6\text{O}} = \frac{0.430 \text{ mol}}{0.430 \text{ mol} + 0.284 \text{ mol}} = 0.602; \quad X_{\text{C}_4\text{H}_8\text{O}_2} = \frac{0.284 \text{ mol}}{0.430 \text{ mol} + 0.284 \text{ mol}} = 0.398$$

$$P_{\text{soln}} = P^{\circ}_{\text{C}_3\text{H}_6\text{O}} \cdot X_{\text{C}_3\text{H}_6\text{O}} + P^{\circ}_{\text{C}_4\text{H}_8\text{O}_2} \cdot X_{\text{C}_4\text{H}_8\text{O}_2}$$

$$P_{\text{soln}} = (285 \text{ mm Hg})(0.602) + (118 \text{ mm Hg})(0.398) = 219 \text{ mm Hg}$$

11.88   In the liquid, $X_{\text{acetone}} = 0.602$ and $X_{\text{ethyl acetate}} = 0.398$

In the vapor, $P_{\text{Total}} = 219$ mm Hg

$$P_{\text{acetone}} = P^{\circ}_{\text{acetone}} \cdot X_{\text{acetone}} = (285 \text{ mm Hg})(0.602) = 172 \text{ mm Hg}$$

$$P_{\text{ethyl acetate}} = P^{\circ}_{\text{ethyl acetate}} \cdot X_{\text{ethyl acetate}} = (118 \text{ mm Hg})(0.398) = 47 \text{ mm Hg}$$

$$X_{acetone} = \frac{P_{acetone}}{P_{total}} = \frac{172 \text{ mm Hg}}{219 \text{ mm Hg}} = 0.785; \quad X_{ethyl\ acetate} = \frac{P_{ethyl\ acetate}}{P_{total}} = \frac{47 \text{ mm Hg}}{219 \text{ mm Hg}} = 0.215$$

11.90 $C_9H_8O_4$, 180.16 amu; 215 g = 0.215 kg

$$\Delta T_b = K_b \cdot m = 0.47^\circ C; \qquad K_b = \frac{\Delta T_b}{m} = \frac{0.47^\circ C}{\left(\dfrac{5.00 \text{ g} \times \dfrac{1 \text{ mol}}{180.16 \text{ g}}}{0.215 \text{ kg}}\right)} = 3.6 \frac{^\circ C \cdot kg}{mol}$$

11.92 $\Delta T_b = K_b \cdot m = 1.76^\circ C; \qquad m = \dfrac{\Delta T_b}{K_b} = \dfrac{1.76^\circ C}{3.07 \dfrac{^\circ C \cdot kg}{mol}} = 0.573 \ m$

11.94 $\Pi = MRT$

(a) NaCl 58.44 amu; 350.0 mL = 0.3500 L

There are 2 moles of ions/mole of NaCl

$$\Pi = (2)\left(\frac{5.00 \text{ g} \times \dfrac{1 \text{mol}}{58.44 \text{ g}}}{0.3500 \text{ L}}\right)\left(0.082\ 06 \ \frac{L \cdot atm}{K \cdot mol}\right)(323 \text{ K}) = 13.0 \text{ atm}$$

(b) CH_3CO_2Na, 82.03 amu; 55.0 mL = 0.0550 L

There are 2 moles of ions/mole of CH_3CO_2Na

$$\Pi = (2)\left(\frac{6.33 \text{ g} \times \dfrac{1 \text{ mol}}{82.03 \text{ g}}}{0.0550 \text{ L}}\right)\left(0.082\ 06 \ \frac{L \cdot atm}{K \cdot mol}\right)(283\text{K}) = 65.2 \text{ atm}$$

11.96 $\Pi = MRT; \qquad M = \dfrac{\Pi}{RT} = \dfrac{4.85 \text{ atm}}{\left(0.082\ 06 \ \dfrac{L \cdot atm}{K \cdot mol}\right)(300 \text{ K})} = 0.197 \text{ M}$

Uses of Colligative Properties

11.98 Osmotic pressure is most often used for the determination of molecular mass because, of the four colligative properties, osmotic pressure gives the largest colligative property change per mole of solute.

11.100 $\Pi = 407.2 \text{ mm Hg} \times \dfrac{1 \text{ atm}}{760 \text{ mm Hg}} = 0.5358 \text{ atm}$

$$\Pi = MRT; \ M = \frac{\Pi}{RT} = \frac{0.5358 \text{ atm}}{\left(0.082\ 06 \ \dfrac{L \cdot atm}{K \cdot mol}\right)(298.15 \text{ K})} = 0.021\ 90 \text{ M}$$

$$200.0 \text{ mL} \times \frac{1 \text{ L}}{1000 \text{ mL}} = 0.2000 \text{ L}$$

mol cellobiose = (0.2000 L)(0.021 90 mol/L) = 4.380 x 10^{-3} mol

molar mass of cellobiose = $\dfrac{1.500 \text{ g cellobiose}}{4.380 \times 10^{-3} \text{ mol cellobiose}}$ = 342.5 g/mol

molecular mass = 342.5 amu

11.102 HCl is a strong electrolyte in H_2O and completely dissociates into two solute particles per each HCl.
HF is a weak electrolyte in H_2O. Only a few percent of the HF molecules dissociates into ions.

11.104 First, determine the empirical formula:
Assume 100.0 g of β-carotene.

10.51% H   10.51 g H x $\dfrac{1 \text{ mol H}}{1.008 \text{ g H}}$ = 10.43 mol H

89.49% C   89.49 g C x $\dfrac{1 \text{ mol C}}{12.01 \text{ g C}}$ = 7.45 mol C

$C_{7.45}H_{10.43}$;   Divide each subscript by the smaller, 7.45.
$C_{7.45/7.45}H_{10.43/7.45}$
$CH_{1.4}$
Multiply each subscript by 5 to obtain integers. Empirical formula is C_5H_7, 67.1 amu.

Second, calculate the molecular mass:

$\Delta T_f = K_f \cdot m$; $m = \dfrac{\Delta T_f}{K_f} = \dfrac{1.17°C}{37.7 \dfrac{°C \cdot kg}{mol}}$ = 0.0310 mol/kg = 0.0310 m

1.50 g x $\dfrac{1 kg}{1000 \text{ g}}$ = 1.50 x 10^{-3} kg

mol β-carotene = (1.50 x 10^{-3} kg)(0.0310 mol/kg) = 4.65 x 10^{-5} mol

molar mass of β-carotene = $\dfrac{0.0250 \text{ g β-carotene}}{4.65 \times 10^{-5} \text{ mol β-carotene}}$ = 538 g/mol

molecular mass = 538 amu

Finally, determine the molecular formula:
Divide the molecular mass by the empirical formula mass.
$\dfrac{538 \text{ amu}}{67.1 \text{ amu}}$ = 8;   molecular formula is $C_{(8 \times 5)}H_{(8 \times 7)}$, or $C_{40}H_{56}$

General Problems

11.106 K_f for snow (H_2O) is 1.86 $\dfrac{°C \cdot kg}{mol}$. Reasonable amounts of salt are capable of lowering

the freezing point (ΔT_f) of the snow below an air temperature of –2°C. Reasonable amounts of salt, however, are not capable of causing a ΔT_f of more than 30°C which would be required if it is to melt snow when the air temperature is –30°C.

11.108 $C_2H_6O_2$, 62.07 amu; $\Delta T_f = 22.0°C$

$$\Delta T_f = K_f \cdot m; \quad m = \frac{\Delta T_f}{K_f} = \frac{22.0°C}{1.86 \dfrac{°C \cdot kg}{mol}} = 11.8 \text{ mol/kg} = 11.8 \ m$$

mol $C_2H_6O_2$ = (3.55 kg)(11.8 mol/kg) = 41.9 mol $C_2H_6O_2$

mass $C_2H_6O_2$ = 41.9 mol $C_2H_6O_2$ x $\dfrac{62.07 \text{ g } C_2H_6O_2}{1 \text{ mol } C_2H_6O_2}$ = 2.60 x 10^3 g $C_2H_6O_2$

11.110 When solid $CaCl_2$ is added to liquid water, the temperature rises because ΔH_{soln} for $CaCl_2$ is exothermic.
When solid $CaCl_2$ is added to ice at 0°C, some of the ice will melt (an endothermic process) and the temperature will fall because the $CaCl_2$ lowers the freezing point of an ice/water mixture.

11.112 $C_{10}H_8$, 128.17 amu; $\Delta T_f = 0.35°C$

$$\Delta T_f = K_f \cdot m; \quad m = \frac{\Delta T_f}{K_f} = \frac{0.35°C}{5.12 \dfrac{°C \cdot kg}{mol}} = 0.0684 \text{ mol/kg} = 0.0684 \ m$$

150.0 g x $\dfrac{1 kg}{1000 g}$ = 0.1500 kg

mol $C_{10}H_8$ = (0.1500 kg)(0.0684 mol/kg) = 0.0103 mol $C_{10}H_8$

mass $C_{10}H_8$ = 0.0103 mol $C_{10}H_8$ x $\dfrac{128.17 \text{ g } C_{10}H_8}{1 \text{ mol } C_{10}H_8}$ = 1.3 g $C_{10}H_8$

11.114 NaCl, 58.44 amu; there are 2 ions/NaCl
A 3.5 mass % aqueous solution of NaCl contains 3.5 g NaCl and 96.5 g H_2O.

$$\text{molality} = \frac{\left(3.5 \text{ g} \ x \ \dfrac{1 \text{ mol}}{58.44 \text{ g}} \right)}{0.0965 \text{ kg}} = 0.62 \text{ mol/kg} = 0.62 \ m$$

$$\Delta T_f = K_f \cdot 2 \cdot m = \left(1.86 \ \dfrac{°C \cdot kg}{mol} \right)(2)(0.62 \text{ mol/kg}) = 2.3°C$$

Solution freezing point = 0.0°C – ΔT_f = 0.0°C – 2.3°C = –2.3°C

$$\Delta T_b = K_b \cdot 2 \cdot m = \left(0.51 \ \frac{°C \cdot kg}{mol}\right)(2)(0.62 \ mol/kg) = 0.63°C$$

Solution boiling point = $100.00°C + \Delta T_b = 100.00°C + 0.63°C = 100.63°C$

11.116 (a) 90 mass % isopropyl alcohol = $\dfrac{10.5 \ g}{10.5 \ g + \ mass \ of \ H_2O} \times 100\%$

Solve for the mass of H_2O.

mass of $H_2O = \left(10.5 \ g \times \dfrac{100}{90}\right) - 10.5 \ g = 1.2 \ g$

mass of solution = $10.5 \ g + 1.2 \ g = 11.7 \ g$

11.7 g of rubbing alcohol contains 10.5 g of isopropyl alcohol.

(b) C_3H_8O, 60.10 amu

mass $C_3H_8O = (0.90)(50.0 \ g) = 45 \ g$

$45 \ g \ C_3H_8O \times \dfrac{1 \ mol \ C_3H_8O}{60.10 \ g \ C_3H_8O} = 0.75 \ mol \ C_3H_8O$

11.118 First, determine the empirical formula.

3.47 mg = 3.47×10^{-3} g sample

10.10 mg = 10.10×10^{-3} g CO_2

2.76 mg = 2.76×10^{-3} g H_2O

mass C = $10.10 \times 10^{-3} \ g \ CO_2 \times \dfrac{12.01 \ g \ C}{44.01 \ g \ CO_2} = 2.76 \times 10^{-3} \ g \ C$

mass H = $2.76 \times 10^{-3} \ g \ H_2O \times \dfrac{2 \times 1.008 \ g \ H}{18.02 \ g \ H_2O} = 3.09 \times 10^{-4} \ g \ H$

mass O = $3.47 \times 10^{-3} \ g - 2.76 \times 10^{-3} \ g \ C - 3.09 \times 10^{-4} \ g \ H = 4.01 \times 10^{-4} \ g \ O$

$2.76 \times 10^{-3} \ g \ C \times \dfrac{1 \ mol \ C}{12.01 \ g \ C} = 2.30 \times 10^{-4} \ mol \ C$

$3.09 \times 10^{-4} \ g \ H \times \dfrac{1 \ mol \ H}{1.008 \ g \ H} = 3.07 \times 10^{-4} \ mol \ H$

$4.01 \times 10^{-4} \ g \ O \times \dfrac{1 \ mol \ O}{16.00 \ g \ O} = 2.51 \times 10^{-5} \ mol \ O = 0.251 \times 10^{-4} \ mol \ O$

To simplify the empirical formula, divide each mol quantity by 10^{-4}.

$C_{2.30}H_{3.07}O_{0.251}$; Divide all subscripts by the smallest, 0.251.

$C_{2.30/0.251}H_{3.07/0.251}O_{0.251/0.251}$

$C_{9.16}H_{12.23}O$, empirical formula is $C_9H_{12}O$ (136 amu)

Second, determine the molecular mass.

7.55 mg = 7.55×10^{-3} g estradiol; $0.500 \ g \times \dfrac{1 \ kg}{1000 \ g} = 5.00 \times 10^{-4} \ kg$ camphor

$$\Delta T_f = K_f \cdot m; \quad m = \frac{\Delta T_f}{K_f} = \frac{2.10°C}{37.7\ \frac{°C \cdot kg}{mol}} = 0.0557\ mol/kg = 0.0557\ m$$

$$m = \frac{mol\ estradiol}{kg\ solvent}$$

mol estradiol = m x (kg solvent) = (0.0557 mol/kg)(5.00 x 10^{-4} kg) = 2.79 x 10^{-5} mol

$$molar\ mass = \frac{7.55 \times 10^{-3}\ g\ estradiol}{2.79 \times 10^{-5}\ mol\ estradiol} = 271\ g/mol; \quad molecular\ mass = 271\ amu$$

Finally, determine the molecular formula:
Divide the molecular mass by the empirical formula mass.

$$\frac{271\ amu}{136\ amu} = 2; \qquad molecular\ formula\ is\ C_{(2\times 9)}H_{(2\times 12)}O_{(2\times 1)},\ or\ C_{18}H_{24}O_2$$

11.120   (a) H_2SO_4, 98.08 amu; $2.238\ mol\ H_2SO_4 \times \frac{98.08\ g\ H_2SO_4}{1\ mol\ H_2SO_4} = 219.50\ g\ H_2SO_4$

mass of 2.238 m solution = 219.50 g H_2SO_4 + 1000 g H_2O = 1219.50 g

volume of 2.238 m solution = $1219.50\ g \times \frac{1.0000\ mL}{1.1243\ g} = 1084.68\ mL = 1.0847\ L$

molarity of 2.238 m solution = $\frac{2.238\ mol}{1.0847\ L} = 2.063\ M$

The molarity of the H_2SO_4 solution is less than the molarity of the $BaCl_2$ solution. Because equal volumes of the two solutions are mixed, H_2SO_4 is the limiting reactant and the number of moles of H_2SO_4 determines the number of moles of $BaSO_4$ produced as the white precipitate.

$(0.05000\ L) \times (2.063\ mol\ H_2SO_4/L) \times \frac{1\ mol\ BaSO_4}{1\ mol\ H_2SO_4} \times \frac{233.39\ g\ BaSO_4}{1\ mol\ BaSO_4} = 24.07\ g\ BaSO_4$

(b) More precipitate will form because of the excess $BaCl_2$ in the solution.

11.122   Let x = X_{H_2O} and y = X_{CH_3OH} and assume n_{total} = 1.00 mol

(14.5 mm Hg)x + (82.5 mm Hg)y = 39.4 mm Hg
(26.8 mm Hg)x + (140.3 mm Hg)y = 68.2 mm Hg

$$x = \frac{68.2 - 140.3y}{26.8}$$

$$\frac{14.5(68.2 - 140.3y)}{26.8} + 82.5y = 39.4$$

$$\frac{(988.9 - 2034.35y)}{26.8} + 82.5y = 39.4$$

$$36.90 - 75.91y + 82.5y = 39.4; \quad 6.59 = 2.5; \quad y = \frac{2.5}{6.59} = 0.3794$$

$$x = \frac{[68.2 - 140.3(0.3794)]}{26.8} = 0.5586$$

$$X_{LiCl} = 1 - X_{H_2O} - X_{CH_3OH} = 1 - 0.5586 - 0.3794 = 0.0620$$

The mole fraction equals the number of moles of each component because $n_{total} = 1.00$ mol.

$$\text{mass LiCl} = 0.0620 \text{ mol LiCl} \times \frac{42.39 \text{ g LiCl}}{1 \text{ mol LiCl}} = 2.6 \text{ g LiCl}$$

$$\text{mass } H_2O = 0.5588 \text{ mol } H_2O \times \frac{18.02 \text{ g } H_2O}{1 \text{ mol } H_2O} = 10.1 \text{ g } H_2O$$

$$\text{mass } CH_3OH = 0.3794 \text{ mol } CH_3OH \times \frac{32.04 \text{ g } CH_3OH}{1 \text{ mol } CH_3OH} = 12.2 \text{ g } CH_3OH$$

total mass = 2.6 g + 10.1 g + 12.2 g = 24.9 g

$$\text{mass \% LiCl} = \frac{2.6 \text{ g}}{24.9 \text{ g}} \times 100\% = 10\%$$

$$\text{mass \% } H_2O = \frac{10.1 \text{ g}}{24.9 \text{ g}} \times 100\% = 41\%$$

$$\text{mass \% } CH_3OH = \frac{12.2 \text{ g}}{24.9 \text{ g}} \times 100\% = 49\%$$

11.124 Solution freezing point = $-1.03°C = 0.00°C - \Delta T_f$; $\Delta T_f = 1.03°C$

$$\Delta T_f = K_f \cdot m; \quad m = \frac{\Delta T_f}{K_f} = \frac{1.03°C}{1.86 \frac{°C \cdot kg}{mol}} = 0.554 \text{ mol/kg} = 0.554 \, m$$

$$\Pi = MRT; \quad M = \frac{\Pi}{RT} = \frac{(12.16 \text{ atm})}{\left(0.082\,06 \frac{L \cdot atm}{K \cdot mol}\right)(298 \text{ K})} = 0.497 \text{ M}$$

Assume 1.000 L = 1000 mL of solution.

mass of solution = (1000 mL)(1.063 g/mL) = 1063 g

$$\text{mass of } H_2O \text{ in 1000 mL of solution} = \frac{1000 \text{ g } H_2O}{0.554 \text{ mol of solute}} \times 0.497 \text{ mol} = 897 \text{ g } H_2O$$

mass of solute = total mass – mass of H_2O = 1063 g – 897 g = 166 g solute

$$\text{molar mass} = \frac{166 \text{ g}}{0.497 \text{ mol}} = 334 \text{ g/mol}$$

11.126 (a) NaCl, 58.44 amu; $CaCl_2$, 110.98 amu; H_2O, 18.02 amu

$$\text{mol NaCl} = 100.0 \text{ g NaCl} \times \frac{1 \text{ mol NaCl}}{58.44 \text{ g NaCl}} = 1.711 \text{ mol NaCl}$$

$$\text{mol } CaCl_2 = 100.0 \text{ g } CaCl_2 \times \frac{1 \text{ mol } CaCl_2}{110.98 \text{ g } CaCl_2} = 0.9011 \text{ mol } CaCl_2$$

mass of solution = (1000 mL)(1.15 g/mL) = 1150 g
mass of H_2O in solution = mass of solution − mass NaCl − mass $CaCl_2$
$$= 1150 \text{ g} - 100.0 \text{ g} - 100.0 \text{ g} = 950 \text{ g}$$
$$= 950 \text{ g} \times \frac{1 \text{ kg}}{1000 \text{ g}} = 0.950 \text{ kg}$$

$$\Delta T_b = K_b \cdot (m_{NaCl} \cdot i + m_{CaCl_2} \cdot i)$$

$$\Delta T_b = \left(0.51 \frac{°C \cdot kg}{mol}\right)\left(\frac{(1.711 \text{ mol NaCl} \cdot 2) + (0.9011 \text{ mol } CaCl_2 \cdot 3)}{0.950 \text{ kg}}\right) = 3.3°C$$

solution boiling point = 100.0°C + ΔT_b = 100.0°C + 3.3°C = 103.3°C

(b) mol H_2O = 950 g H_2O × $\dfrac{1 \text{ mol } H_2O}{18.02 \text{ g } H_2O}$ = 52.7 mol H_2O

$$P_{Solution} = P° \cdot X_{H_2O}$$

$$P_{Solution} = P° \cdot \left(\frac{52.7 \text{ mol } H_2O}{(52.7 \text{ mol } H_2O) + (1.711 \text{ mol NaCl} \cdot 2) + (0.9011 \text{ mol } CaCl_2 \cdot 3)}\right)$$

$$P_{Solution} = (23.8 \text{ mm Hg})(0.896) = 21.3 \text{ mm Hg}$$

11.128 (a) KI, 166.00 amu
Assume you have 1.000 L of 1.24 M solution.
mass of solution = (1000 mL)(1.15 g/mL) = 1150 g
mass of KI in solution = 1.24 mol KI × $\dfrac{166.00 \text{ g KI}}{1 \text{ mol KI}}$ = 206 g KI
mass of H_2O in solution = mass of solution − mass KI = 1150 g − 206 g = 944 g
$$= 944 \text{ g} \times \frac{1 \text{ kg}}{1000 \text{ g}} = 0.944 \text{ kg}$$

molality = $\dfrac{1.24 \text{ mol KI}}{0.944 \text{ kg } H_2O}$ = 1.31 mol/kg = 1.31 *m*

(b) For KI, i = 2 assuming complete dissociation.

$$\Delta T_f = K_f \cdot m \cdot i = \left(1.86 \frac{°C \cdot kg}{mol}\right)(1.31 \text{ } m)(2) = 4.87°C$$

Solution freezing point = 0.00°C − ΔT_f = 0.00°C − 4.87°C = − 4.87°C

(c) i = $\dfrac{\Delta T_f}{K_f \cdot m}$ = $\dfrac{4.46°C}{\left(1.86 \dfrac{°C \cdot kg}{mol}\right)(1.31 \text{ mol/kg})}$ = 1.83

Because the calculated i is only 1.83 and not 2, the percent dissociation for KI is 83%.

11.130 NaCl, 58.44 amu; $C_{12}H_{22}O_{11}$, 342.3 amu
Let X = mass NaCl and Y = mass $C_{12}H_{22}O_{11}$, then X + Y = 100.0 g.
500.0 g = 0.5000 kg
Solution freezing point = −2.25°C = 0.00°C − ΔT_f, = ΔT_f = 0.00°C + 2.25°C = 2.25°C

$$\Delta T_f = K_f \cdot (m_{NaCl} \cdot i + m_{C_{12}H_{22}O_{11}})$$

$$\Delta T_b = \left(1.86 \frac{°C \cdot kg}{mol}\right)\left(\frac{(mol\ NaCl \cdot 2) + (mol\ C_{12}H_{22}O_{11})}{0.5000\ kg}\right) = 2.25°C$$

$$mol\ NaCl = X\ g\ NaCl \times \frac{1\ mol\ NaCl}{58.44\ g\ NaCl} = X/58.44\ mol$$

$$mol\ C_{12}H_{22}O_{11} = Y\ g\ C_{12}H_{22}O_{11} \times \frac{1\ mol\ C_{12}H_{22}O_{11}}{342.3\ g\ C_{12}H_{22}O_{11}} = Y/342.3\ mol$$

$$\Delta T_b = \left(1.86 \frac{°C \cdot kg}{mol}\right)\left(\frac{((X/58.44) \cdot 2\ mol) + ((Y/342.3)\ mol)}{0.5000\ kg}\right) = 2.25°C$$

$$X = 100 - Y$$

$$\left(1.86 \frac{°C \cdot kg}{mol}\right)\left(\frac{\{[(100-Y)/58.44] \cdot 2\ mol]\} + [(Y/342.3)\ mol]}{0.5000\ kg}\right) = 2.25°C$$

$$\left(\frac{[(200/58.44) - (2Y/58.44) + (Y/342.3)]\ mol}{0.5000\ kg}\right) = \frac{2.25\,°C}{\left(1.86\dfrac{°C \cdot kg}{mol}\right)} = 1.21\ mol/kg$$

$$\left(\frac{[(3.42) - (0.0313Y)]\ mol}{0.5000\ kg}\right) = 1.21\ mol/kg$$

$$[(3.42) - (0.0313Y)] = (0.5000\ kg)(1.21) = 0.605$$

$$-0.0313\ Y = 0.605 - 3.42 = -2.81$$

$$Y = (-2.81)/(-0.0313) = 89.9\ g\ of\ C_{12}H_{22}O_{11}$$

$$X = 100.0\ g - Y = 100.0\ g - 89.9\ g = 10.1\ g\ of\ NaCl$$

Multi-Concept Problems

11.132 (a) 20.00 mL = 0.02000 L

mol NaOH = (0.02000 L)(2.00 mol/L) = 0.0400 mol NaOH

$$mol\ CO_2 = 0.0400\ mol\ NaOH \times \frac{1\ mol\ CO_2}{2\ mol\ NaOH} = 0.0200\ mol\ CO_2$$

$$mol\ C = 0.0200\ mol\ CO_2 \times \frac{1\ mol\ C}{1\ mol\ CO_2} = 0.0200\ mol\ C$$

$$mass\ C = 0.0200\ mol\ C \times \frac{12.011\ g\ C}{1\ mol\ C} = 0.240\ g\ C$$

mass H = mass of compound – mass of C = 0.270 g – 0.240 g = 0.030 g H

$$mol\ H = 0.030\ g\ H \times \frac{1\ mol\ H}{1.008\ g\ H} = 0.030\ mol\ H$$

The mole ratio of C and H in the molecule is $C_{0.0200}\ H_{0.030}$.

$C_{0.0200}\ H_{0.030}$, divide both subscripts by the smaller of the two, 0.0200.

$C_{0.0200\ /\ 0.0200}\ H_{0.030\ /\ 0.0200}$

$C_1H_{1.5}$, multiply both subscripts by 2.

Chapter 11 – Solutions and Their Properties

$C_{(2 \times 1)} H_{(2 \times 1.5)}$
C_2H_3 (27.05 amu) is the empirical formula.

(b) $\Delta T_f = K_f \cdot m$; $m = \dfrac{\Delta T_f}{K_f} = \dfrac{(179.8°C - 177.9°C)}{37.7 \dfrac{°C \cdot kg}{mol}} = 0.050$ mol/kg $= 0.050\ m$

$50.0\ g \times \dfrac{1\ kg}{1000\ g} = 0.0500\ kg$

mol solute $= (0.050\ mol/kg)(0.0500\ kg) = 0.0025$ mol

molar mass $= \dfrac{0.270\ g}{0.0025\ mol} = 108$ g/mol; molecular mass $= 108$ amu

(c) To find the molecular formula, first divide the molecular mass by the mass of the empirical formula unit.

$\dfrac{108}{27} = 4$

Multiply the subscripts in the empirical formula by the result of this division, 4.
$C_{(4 \times 2)} H_{(4 \times 3)}$
C_8H_{12} is the molecular formula of the compound.

11.134   AgCl, 143.32 amu
Solution freezing point $= -4.42°C = 0.00°C - \Delta T_f$; $\Delta T_f = 0.00°C + 4.42°C = 4.42°C$
$\Delta T_f = K_f \cdot m$

total ion $m = \dfrac{\Delta T_f}{K_f} = \dfrac{4.42°C}{1.86 \dfrac{°C \cdot kg}{mol}} = 2.376$ mol/kg $= 2.376\ m$

$150.0\ g \times \dfrac{1\ kg}{1000\ g} = 0.1500\ kg$

total mol of ions $= (2.376\ mol/kg)(0.1500\ kg) = 0.3564$ mol of ions
An excess of $AgNO_3$ reacts with all Cl^- to produce 27.575 g AgCl.

total mol $Cl^- = 27.575\ g\ AgCl \times \dfrac{1\ mol\ AgCl}{143.32\ g\ AgCl} \times \dfrac{1\ mol\ Cl^-}{1\ mol\ AgCl} = 0.1924$ mol Cl^-

Let P = mol XCl and Q = mol YCl_2.
0.3564 mol ions $= 2 \times$ mol XCl $+ 3 \times$ mol $YCl_2 = (2 \times P) + (3 \times Q)$
0.1924 mol $Cl^- =$ mol XCl $+ 2 \times$ mol $YCl_2 = P + (2 \times Q)$
$P = 0.1924 - (2 \times Q)$
$0.3564 = 2 \times [0.1924 - (2 \times Q)] + (3 \times Q) = 0.3848 - (4 \times Q) + (3 \times Q)$
$Q = 0.3848 - 0.3564 = 0.0284$ mol YCl_2
$P = 0.1924 - (2 \times Q) = 0.1924 - (2 \times 0.0284) = 0.1356$ mol XCl

mass Cl in XCl $= 0.1356\ mol\ XCl \times \dfrac{1\ mol\ Cl}{1\ mol\ XCl} \times \dfrac{35.453\ g\ Cl}{1\ mol\ Cl} = 4.81$ g Cl

mass Cl in YCl_2 = 0.0284 mol YCl_2 x $\dfrac{2 \text{ mol Cl}}{1 \text{ mol } YCl_2}$ x $\dfrac{35.453 \text{ g Cl}}{1 \text{ mol Cl}}$ = 2.01 g Cl

total mass of XCl and YCl_2 = 8.900 g

mass of X + Y = total mass – mass Cl = 8.900 g – 4.81 g – 2.01 g = 2.08 g

X is an alkali metal and there are 0.1356 mol of X in XCl.

If X = Li, then mass of X = (0.1356 mol)(6.941 g/mol) = 0.941 g

If X = Na, then mass of X = (0.1356 mol)(22.99 g/mol) = 3.12 g but this is not possible because 3.12 g is greater than the total mass of X + Y. Therefore, X is Li.

mass of Y = 2.08 – mass of X = 2.08 g – 0.941 g = 1.14 g

Y is an alkaline earth metal and there are 0.0284 mol of Y in YCl_2.

molar mass of Y = 1.14 g/0.0284 mol = 40.1 g/mol. Therefore, Y is Ca.

mass LiCl = 0.1356 mol LiCl x $\dfrac{42.39 \text{ g LiCl}}{1 \text{ mol LiCl}}$ = 5.75 g LiCl

mass $CaCl_2$ = 0.0284 mol $CaCl_2$ x $\dfrac{110.98 \text{ g } CaCl_2}{1 \text{ mol } CaCl_2}$ = 3.15 g $CaCl_2$

Chemical Kinetics

12.1 $3\text{ I}^-(\text{aq}) + H_3AsO_4(\text{aq}) + 2\text{ H}^+(\text{aq}) \rightarrow I_3^-(\text{aq}) + H_3AsO_3(\text{aq}) + H_2O(\text{l})$

(a) $-\dfrac{\Delta[I^-]}{\Delta t} = 4.8 \times 10^{-4}$ M/s

$$\frac{\Delta[I_3^-]}{\Delta t} = \frac{1}{3}\left(-\frac{\Delta[I^-]}{\Delta t}\right) = \left(\frac{1}{3}\right)(4.8 \times 10^{-4} \text{ M/s}) = 1.6 \times 10^{-4} \text{ M/s}$$

(b) $-\dfrac{\Delta[H^+]}{\Delta t} = 2\left(\dfrac{\Delta[I_3^-]}{\Delta t}\right) = (2)(1.6 \times 10^{-4} \text{ M/s}) = 3.2 \times 10^{-4} \text{ M/s}$

12.2 $2\text{ N}_2O_5(\text{g}) \rightarrow 4\text{ NO}_2(\text{g}) + O_2(\text{g})$

time	$[N_2O_5]$	$[O_2]$
200 s	0.0142 M	0.0029 M
300 s	0.0120 M	0.0040 M

Rate of decomposition of $N_2O_5 = -\dfrac{\Delta[N_2O_5]}{\Delta t} = -\dfrac{0.0120 \text{ M} - 0.0142 \text{ M}}{300 \text{ s} - 200 \text{ s}} = 2.2 \times 10^{-5}$ M/s

Rate of formation of $O_2 = \dfrac{\Delta[O_2]}{\Delta t} = \dfrac{0.0040 \text{ M} - 0.0029 \text{ M}}{300 \text{ s} - 200 \text{ s}} = 1.1 \times 10^{-5}$ M/s

12.3 Rate $= k[BrO_3^-][Br^-][H^+]^2$
1st order in BrO_3^-, 1st order in Br^-, 2nd order in H^+, 4th order overall
Rate $= k[H_2][I_2]$, 1st order in H_2, 1st order in I_2, 2nd order overall
Rate $= k[CH_3CHO]^{3/2}$, 3/2 order in CH_3CHO, 3/2 order overall

12.4 $H_2O_2(\text{aq}) + 3\text{ I}^-(\text{aq}) + 2\text{ H}^+(\text{aq}) \rightarrow I_3^-(\text{aq}) + 2\text{ H}_2O(\text{l})$

Rate $= \dfrac{\Delta[I_3^-]}{\Delta t} = k[H_2O_2]^m[I^-]^n$

(a) $\dfrac{Rate_3}{Rate_1} = \dfrac{2.30 \times 10^{-4} \text{ M/s}}{1.15 \times 10^{-4} \text{ M/s}} = 2 \qquad \dfrac{[H_2O_2]_3}{[H_2O_2]_1} = \dfrac{0.200 \text{ M}}{0.100 \text{ M}} = 2$

Because both ratios are the same, m = 1.

$\dfrac{Rate_2}{Rate_1} = \dfrac{2.30 \times 10^{-4} \text{ M/s}}{1.15 \times 10^{-4} \text{ M/s}} = 2 \qquad \dfrac{[I^-]_2}{[I^-]_1} = \dfrac{0.200 \text{ M}}{0.100 \text{ M}} = 2$

Because both ratios are the same, n = 1.
The rate law is: Rate $= k[H_2O_2][I^-]$

(b) $k = \dfrac{\text{Rate}}{[H_2O_2][I^-]}$

Using data from Experiment 1: $k = \dfrac{1.15 \times 10^{-4}\,M/s}{(0.100\,M)(0.100\,M)} = 1.15 \times 10^{-2}\,/(M \cdot s)$

(c) $\text{Rate} = k[H_2O_2][I^-] = [1.15 \times 10^{-2}/(M \cdot s)](0.300\,M)(0.400\,M) = 1.38 \times 10^{-3}\,M/s$

12.5

Rate Law	Units of k
$\text{Rate} = k[(CH_3)_3CBr]$	$1/s$
$\text{Rate} = k[Br_2]$	$1/s$
$\text{Rate} = k[BrO_3^-][Br^-][H^+]^2$	$1/(M^3 \cdot s)$
$\text{Rate} = k[H_2][I_2]$	$1/(M \cdot s)$
$\text{Rate} = [CH_3CHO]^{3/2}$	$1/(M^{1/2} \cdot s)$

12.6   (a) The reactions in vessels (a) and (b) have the same rate, the same number of B molecules, but different numbers of A molecules. Therefore, the rate does not depend on A and its reaction order is zero. The same conclusion can be drawn from the reactions in vessels (c) and (d).

The rate for the reaction in vessel (c) is four times the rate for the reaction in vessel (a). Vessel (c) has twice as many B molecules than does vessel (a). Because the rate quadruples when the concentration of B doubles, the reaction order for B is two.

(b) $\text{rate} = k[B]^2$

12.7   (a) $\ln \dfrac{[Co(NH_3)_5Br^{2+}]_t}{[Co(NH_3)_5Br^{2+}]_0} = -kt$

$k = 6.3 \times 10^{-6}/s;$ $t = 10.0\,h \times \dfrac{3600\,s}{1\,h} = 36{,}000\,s$

$\ln[Co(NH_3)_5Br^{2+}]_t = -kt + \ln[Co(NH_3)_5Br^{2+}]_0$

$\ln[Co(NH_3)_5Br^{2+}]_t = -(6.3 \times 10^{-6}/s)(36{,}000\,s) + \ln(0.100)$

$\ln[Co(NH_3)_5Br^{2+}]_t = -2.5294;$ After 10.0 h, $[Co(NH_3)_5Br^{2+}] = e^{-2.5294} = 0.080\,M$

(b) $[Co(NH_3)_5Br^{2+}]_0 = 0.100\,M$

If 75% of the $Co(NH_3)_5Br^{2+}$ reacts then 25% remains.

$[Co(NH_3)_5Br^{2+}]_t = (0.25)(0.100\,M) = 0.025\,M$

$\ln \dfrac{[Co(NH_3)_5Br^{2+}]_t}{[Co(NH_3)_5Br^{2+}]_0} = -kt;\; t = \dfrac{\ln \dfrac{[Co(NH_3)_5Br^{2+}]_t}{[Co(NH_3)_5Br^{2+}]_0}}{-k}$

$t = \dfrac{\ln\left(\dfrac{0.025}{0.100}\right)}{-(6.3 \times 10^{-6}/s)} = 2.2 \times 10^5\,s;\quad t = 2.2 \times 10^5\,s \times \dfrac{1\,h}{3600\,s} = 61\,h$

12.8

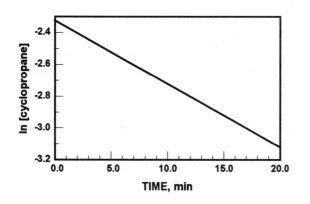

Slope = – 0.03989/min = – 6.6 x 10^{-4}/s and k = – slope

A plot of ln[cyclopropane] versus time is linear, indicating that the data fit the equation for a first-order reaction. k = 6.6 x 10^{-4}/s (0.040/min)

12.9 (a) k = 1.8 x 10^{-5}/s

$$t_{1/2} = \frac{0.693}{k} = \frac{0.693}{1.8 \times 10^{-5}/s} = 38,500 \text{ s}; \qquad t_{1/2} = 38,500 \text{ s} \times \frac{1 \text{ h}}{3600 \text{ s}} = 11 \text{ h}$$

(b) 0.30 M $\xrightarrow{t_{1/2}}$ 0.15 M $\xrightarrow{t_{1/2}}$ 0.075 M $\xrightarrow{t_{1/2}}$ 0.0375 M $\xrightarrow{t_{1/2}}$ 0.019 M

(c) Because 25% of the initial concentration corresponds to 1/4 or $(1/2)^2$ of the initial concentration, the time required is two half-lives: t = $2t_{1/2}$ = 2(11 h) = 22 h

12.10 After one half-life, there would be four A molecules remaining. After two half-lives, there would be two A molecules remaining. This is represented by the drawing at t = 10 min. 10 min is equal to two half-lives, therefore, $t_{1/2}$ = 5 min for this reaction. After 15 min (three half-lives) only one A molecule would remain.

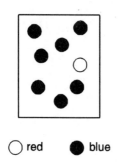

○ red   ● blue

12.11

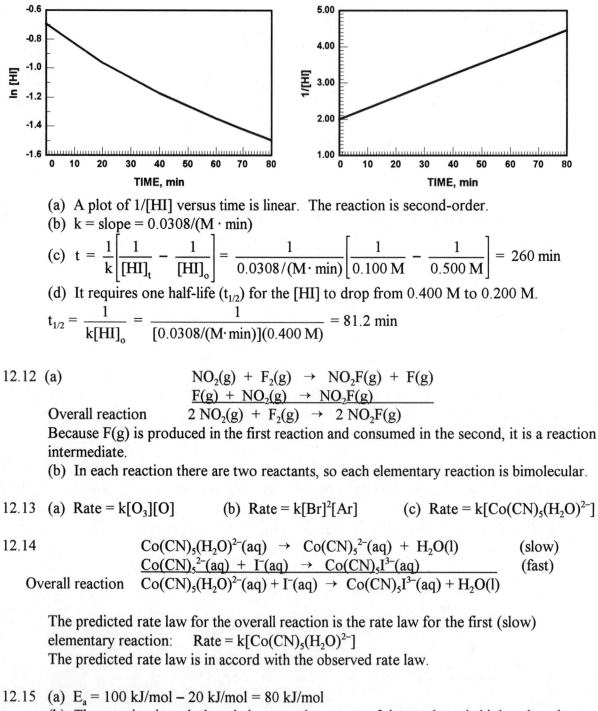

(a) A plot of 1/[HI] versus time is linear. The reaction is second-order.

(b) $k = \text{slope} = 0.0308/(M \cdot min)$

(c) $t = \dfrac{1}{k}\left[\dfrac{1}{[HI]_t} - \dfrac{1}{[HI]_o}\right] = \dfrac{1}{0.0308/(M \cdot min)}\left[\dfrac{1}{0.100\ M} - \dfrac{1}{0.500\ M}\right] = 260\ min$

(d) It requires one half-life $(t_{1/2})$ for the [HI] to drop from 0.400 M to 0.200 M.

$t_{1/2} = \dfrac{1}{k[HI]_o} = \dfrac{1}{[0.0308/(M \cdot min)](0.400\ M)} = 81.2\ min$

12.12 (a)

$$NO_2(g) + F_2(g) \rightarrow NO_2F(g) + F(g)$$
$$\underline{F(g) + NO_2(g) \rightarrow NO_2F(g)}$$

Overall reaction $\quad 2\ NO_2(g) + F_2(g) \rightarrow 2\ NO_2F(g)$

Because F(g) is produced in the first reaction and consumed in the second, it is a reaction intermediate.

(b) In each reaction there are two reactants, so each elementary reaction is bimolecular.

12.13 (a) Rate = $k[O_3][O]$ (b) Rate = $k[Br]^2[Ar]$ (c) Rate = $k[Co(CN)_5(H_2O)^{2-}]$

12.14

$$Co(CN)_5(H_2O)^{2-}(aq) \rightarrow Co(CN)_5^{2-}(aq) + H_2O(l) \qquad \text{(slow)}$$
$$\underline{Co(CN)_5^{2-}(aq) + I^-(aq) \rightarrow Co(CN)_5I^{3-}(aq)} \qquad \text{(fast)}$$

Overall reaction $\quad Co(CN)_5(H_2O)^{2-}(aq) + I^-(aq) \rightarrow Co(CN)_5I^{3-}(aq) + H_2O(l)$

The predicted rate law for the overall reaction is the rate law for the first (slow) elementary reaction: Rate = $k[Co(CN)_5(H_2O)^{2-}]$

The predicted rate law is in accord with the observed rate law.

12.15 (a) $E_a = 100\ kJ/mol - 20\ kJ/mol = 80\ kJ/mol$

(b) The reaction is endothermic because the energy of the products is higher than the energy of the reactants.

(c) A---C
     | |
    B---D

12.16 (a) $\ln\left(\dfrac{k_2}{k_1}\right) = \left(\dfrac{-E_a}{R}\right)\left(\dfrac{1}{T_2} - \dfrac{1}{T_1}\right)$

$k_1 = 3.7 \times 10^{-5}/s,\ T_1 = 25°C = 298\ K$

$k_2 = 1.7 \times 10^{-3}/s,\ T_2 = 55°C = 328\ K$

$E_a = -\dfrac{[\ln k_2 - \ln k_1]R}{\left(\dfrac{1}{T_2} - \dfrac{1}{T_1}\right)}$

$E_a = -\dfrac{[\ln(1.7 \times 10^{-3}) - \ln(3.7 \times 10^{-5})][8.314 \times 10^{-3}\ kJ/(K\cdot mol)]}{\left(\dfrac{1}{328\ K} - \dfrac{1}{298\ K}\right)} = 104\ kJ/mol$

(b) $k_1 = 3.7 \times 10^{-5}/s,\ T_1 = 25°C = 298\ K$

solve for $k_2,\ T_2 = 35°C = 308\ K$

$\ln k_2 = \left(\dfrac{-E_a}{R}\right)\left(\dfrac{1}{T_2} - \dfrac{1}{T_1}\right) + \ln k_1$

$\ln k_2 = \left(\dfrac{-104\ kJ/mol}{8.314 \times 10^{-3}\ kJ/(K\cdot mol)}\right)\left(\dfrac{1}{308\ K} - \dfrac{1}{298\ K}\right) + \ln(3.7 \times 10^{-5})$

$\ln k_2 = -8.84;\ \ k_2 = e^{-8.84} = 1.4 \times 10^{-4}/s$

12.17 Assume that concentration is proportional to the number of each molecule in a box.
(a) From boxes (1) and (2), the concentration of A doubles, B and C_2 remain the same and the rate does not change. This means the reaction is zeroth-order in A.
From boxes (1) and (3), the concentration of C_2 doubles, A and B remain the same and the rate doubles. This means the reaction is first-order in C_2.
From boxes (1) and (4), the concentration of B triples, A and C_2 remain the same and the rate triples. This means the reaction is first-order in B.
(b) Rate = $k\ [B][C_2]$
(c)

$$\begin{array}{llll} B + C_2 & \rightarrow & BC_2 & \text{(slow)} \\ A + BC_2 & \rightarrow & AC + BC & \\ \underline{A + BC} & \rightarrow & \underline{AC + B} & \\ 2A + C_2 & \rightarrow & 2AC & \text{(overall)} \end{array}$$

(d) B doesn't appear in the overall reaction because it is consumed in the first step and regenerated in the third step. B is therefore a catalyst. BC_2 and BC are intermediates because they are formed in one step and then consumed in a subsequent step in the reaction.

12.18 Nitroglycerin contains three nitro groups per molecule. Because the bonds in nitro groups are relatively weak (about 200 kJ/mol) and because the explosion products (CO_2, N_2, H_2O, and O_2) are extremely stable, a great deal of energy is released (very exothermic) during an explosion.

12.19 Secondary explosives are generally less sensitive to heat and shock than primary explosives. This would indicate that secondary explosives should have a higher activation energy than primary explosives.

12.20 $C_5H_8N_4O_{12}(s) \rightarrow 4\ CO_2(g) + 4\ H_2O(g) + 2\ N_2(g) + C(s)$
$\Delta H°_{rxn} = [4\ \Delta H°_f\ (CO_2) + 4\ \Delta H°_f\ (H_2O)] - \Delta H°_f\ (C_5H_8N_4O_{12})$
$\Delta H°_{rxn} = [(4\ mol)(-393.5\ kJ/mol) + (4\ mol)(-241.8\ kJ/mol)] - [(1\ mol)(537\ kJ/mol)]$
$\Delta H°_{rxn} = -3078\ kJ$

12.21 $C_5H_8N_4O_{12}(s) \rightarrow 4\ CO_2(g) + 4\ H_2O(g) + 2\ N_2(g) + C(s)$
$C_5H_8N_4O_{12}$, 316.14 amu; 1.54 kg = 1.54 x 10^3 g; 800°C = 1073 K
From the reaction, 1 mole of PETN produces 10 moles of gas.

$$mol\ gas = 1.54\ x\ 10^3\ g\ PETN\ x\ \frac{1\ mol\ PETN}{316.14\ g\ PETN}\ x\ \frac{10\ mol\ gas}{1\ mol\ PETN} = 48.7\ mol$$

$PV = nRT$

$$V = \frac{nRT}{P} = \frac{(48.7\ mol)\left(0.082\ 06\ \frac{L\cdot atm}{K\cdot mol}\right)(1073\ K)}{0.975\ atm} = 4.40\ x\ 10^3\ L$$

Understanding Key Concepts

12.22 (a) Because Rate = k[A][B], the rate is proportional to the product of the number of A molecules and the number of B molecules. The relative rates of the reaction in vessels (a) - (d) are 2 : 1 : 4 : 2.
(b) Because the same reaction takes place in each vessel, the k's are all the same.

12.24 (a) For the first-order reaction, half of the A molecules are converted to B molecules each minute.

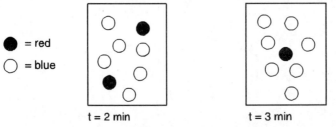

(b) Because half of the A molecules are converted to B molecules in 1 min, the half-life is 1 min.

12.26 (a) Because the half-life is inversely proportional to the concentration of A molecules, the reaction is second-order in A.
(b) Rate = $k[A]^2$
(c) The second box represents the passing of one half-life, and the third box represents the passing of a second half-life for a second-order reaction. A relative value of k can be calculated.

$$k = \frac{1}{t_{1/2}[A]} = \frac{1}{(1)(16)} = 0.0625$$

t$_{1/2}$ in going from box 3 to box 4 is: $t_{1/2} = \dfrac{1}{k[A]} = \dfrac{1}{(0.0625)(4)} = 4$ min

(For fourth box, t = 7 min)

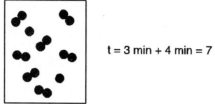

t = 3 min + 4 min = 7 min

12.28 (a) BC + D → B + CD
(b) 1. B–C + D (reactants), A (catalyst); 2. B---C---A (transition state), D (reactant);
3. A–C (intermediate), B (product), D (reactant); 4. A---C---D (transition state),
B (product); 5. A (catalyst), C–D + B (products)
(c) The first step is rate determining because the first maximum in the potential energy
curve is greater than the second (relative) maximum; Rate = k[A][BC]
(d) Endothermic

Additional Problems
Reaction Rates

12.30 M/s or $\dfrac{\text{mol}}{\text{L·s}}$

12.32 (a) Rate = $\dfrac{-\Delta[\text{cyclopropane}]}{\Delta t} = -\dfrac{0.080\,\text{M} - 0.098\,\text{M}}{5.0\,\text{min} - 0.0\,\text{min}} = 3.6 \times 10^{-3}$ M/min

Rate = $3.6 \times 10^{-3}\ \dfrac{\text{M}}{\text{min}} \times \dfrac{1\,\text{min}}{60\,\text{s}} = 6.0 \times 10^{-5}$ M/s

(b) Rate = $\dfrac{-\Delta[\text{cyclopropane}]}{\Delta t} = -\dfrac{0.044\,\text{M} - 0.054\,\text{M}}{20.0\,\text{min} - 15.0\,\text{min}} = 2.0 \times 10^{-3}$ M/min

Rate = $2.0 \times 10^{-3}\ \dfrac{\text{M}}{\text{min}} \times \dfrac{1\,\text{min}}{60\,\text{s}} = 3.3 \times 10^{-5}$ M/s

12.34

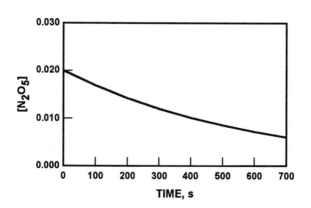

(a) The instantaneous rate of decomposition of N_2O_5 at $t = 200$ s is determined from the slope of the curve at $t = 200$ s.

$$\text{Rate} = -\frac{\Delta[N_2O_5]}{\Delta t} = -\text{slope} = -\frac{(1.20 \times 10^{-2}\text{ M}) - (1.69 \times 10^{-2}\text{ M})}{300\text{ s} - 100\text{ s}} = 2.4 \times 10^{-5}\text{ M/s}$$

(b) The initial rate of decomposition of N_2O_5 is determined from the slope of the curve at $t = 0$ s. This is equivalent to the slope of the curve from 0 s to 100 s because in this time interval the curve is almost linear.

$$\text{Initial rate} = -\frac{\Delta[N_2O_5]}{\Delta t} = -\text{slope} = -\frac{(1.69 \times 10^{-2}\text{ M}) - (2.00 \times 10^{-2}\text{ M})}{100\text{ s} - 0\text{ s}} = 3.1 \times 10^{-5}\text{ M/s}$$

12.36 (a) $-\dfrac{\Delta[H_2]}{\Delta t} = -3\dfrac{\Delta[N_2]}{\Delta t}$; The rate of consumption of H_2 is 3 times faster.

(b) $\dfrac{\Delta[NH_3]}{\Delta t} = -2\dfrac{\Delta[N_2]}{\Delta t}$; The rate of formation of NH_3 is 2 times faster.

12.38 $N_2(g) + 3 H_2(g) \rightarrow 2 NH_3(g)$; $-\dfrac{\Delta[N_2]}{\Delta t} = -\dfrac{1}{3}\dfrac{\Delta[H_2]}{\Delta t} = \dfrac{1}{2}\dfrac{\Delta[NH_3]}{\Delta t}$

Rate Laws

12.40 Rate = $k[NO]^2[Br_2]$; 2nd order in NO; 1st order in Br_2; 3rd order overall

12.42 Rate = $k[H_2][ICl]$; units for k are $\dfrac{\text{L}}{\text{mol}\cdot\text{s}}$ or $1/(\text{M}\cdot\text{s})$

12.44 (a) Rate = $k[CH_3Br][OH^-]$
(b) Because the reaction is first-order in OH^-, if the $[OH^-]$ is decreased by a factor of 5, the rate will also decrease by a factor of 5.
(c) Because the reaction is first-order in each reactant, if both reactant concentrations are doubled, the rate will increase by a factor of $2 \times 2 = 4$.

12.46 (a) Rate = $k[CH_3COCH_3]^m$

$$m = \frac{\ln\left(\dfrac{\text{Rate}_2}{\text{Rate}_1}\right)}{\ln\left(\dfrac{[CH_3COCH_3]_2}{[CH_3COCH_3]_1}\right)} = \frac{\ln\left(\dfrac{7.8 \times 10^{-5}}{5.2 \times 10^{-5}}\right)}{\ln\left(\dfrac{9.0 \times 10^{-3}}{6.0 \times 10^{-3}}\right)} = 1; \quad \text{Rate} = k[CH_3COCH_3]$$

(b) From Experiment 1: $k = \dfrac{\text{Rate}}{[CH_3COCH_3]} = \dfrac{5.2 \times 10^{-5}\text{ M/s}}{6.0 \times 10^{-3}\text{ M}} = 8.7 \times 10^{-3}/\text{s}$

(c) Rate = $k[CH_3COCH_3] = (8.7 \times 10^{-3}/\text{s})(1.8 \times 10^{-3}\text{M}) = 1.6 \times 10^{-5}\text{ M/s}$

12.48 (a) $\text{Rate} = k[NH_4^+]^m[NO_2^-]^n$

$$m = \frac{\ln\left(\dfrac{\text{Rate}_2}{\text{Rate}_1}\right)}{\ln\left(\dfrac{[NH_4^+]_2}{[NH_4^+]_1}\right)} = \frac{\ln\left(\dfrac{3.6 \times 10^{-6}}{7.2 \times 10^{-6}}\right)}{\ln\left(\dfrac{0.12}{0.24}\right)} = 1; \quad n = \frac{\ln\left(\dfrac{\text{Rate}_3}{\text{Rate}_2}\right)}{\ln\left(\dfrac{[NO_2^-]_3}{[NO_2^-]_2}\right)} = \frac{\ln\left(\dfrac{5.4 \times 10^{-6}}{3.6 \times 10^{-6}}\right)}{\ln\left(\dfrac{0.15}{0.10}\right)} = 1$$

$\text{Rate} = k[NH_4^+][NO_2^-]$

(b) From Experiment 1: $k = \dfrac{\text{Rate}}{[NH_4^+][NO_2^-]} = \dfrac{7.2 \times 10^{-6} \text{ M/s}}{(0.24 \text{ M})(0.10 \text{ M})} = 3.0 \times 10^{-4}/(\text{M} \cdot \text{s})$

(c) $\text{Rate} = k[NH_4^+][NO_2^-] = [3.0 \times 10^{-4}/(\text{M} \cdot \text{s})](0.39 \text{ M})(0.052 \text{ M}) = 6.1 \times 10^{-6} \text{ M/s}$

Integrated Rate Law; Half-Life

12.50 $\ln\dfrac{[C_3H_6]_t}{[C_3H_6]_0} = -kt, \; k = 6.7 \times 10^{-4}/s$

(a) $t = 30 \text{ min} \times \dfrac{60 \text{ s}}{1 \text{ min}} = 1800 \text{ s}$

$\ln[C_3H_6]_t = -kt + \ln[C_3H_6]_0 = -(6.7 \times 10^{-4}/s)(1800 \text{ s}) + \ln(0.0500) = -4.202$
$[C_3H_6]_t = e^{-4.202} = 0.015 \text{ M}$

(b) $t = \dfrac{\ln\dfrac{[C_3H_6]_t}{[C_3H_6]_0}}{-k} = \dfrac{\ln\left(\dfrac{0.0100}{0.0500}\right)}{-(6.7 \times 10^{-4}/s)} = 2402 \text{ s}; \quad t = 2402 \text{ s} \times \dfrac{1 \text{ min}}{60 \text{ s}} = 40 \text{ min}$

(c) $[C_3H_6]_0 = 0.0500 \text{ M}$;   If 25% of the C_3H_6 reacts then 75% remains.
$[C_3H_6]_t = (0.75)(0.0500 \text{ M}) = 0.0375 \text{ M}$.

$t = \dfrac{\ln\dfrac{[C_3H_6]_t}{[C_3H_6]_0}}{-k} = \dfrac{\ln\left(\dfrac{0.0375}{0.0500}\right)}{-(6.7 \times 10^{-4}/s)} = 429 \text{ s}; \quad t = 429 \text{ s} \times \dfrac{1 \text{ min}}{60 \text{ s}} = 7.2 \text{ min}$

2.52 $t_{1/2} = \dfrac{0.693}{k} = \dfrac{0.693}{6.7 \times 10^{-4}/s} = 1034 \text{ s} = 17 \text{ min}$

$t = \dfrac{\ln\dfrac{[C_3H_6]_t}{[C_3H_6]_0}}{-k} = \dfrac{\ln\dfrac{(0.0625)(0.0500)}{(0.0500)}}{-6.7 \times 10^{-4}/s} = 4140 \text{ s}$

$t = 4140 \text{ s} \times \dfrac{1 \text{ min}}{60 \text{ s}} = 69 \text{ min}$

This is also 4 half-lives. $100 \xrightarrow{t_{1/2}} 50 \xrightarrow{t_{1/2}} 25 \xrightarrow{t_{1/2}} 12.5 \xrightarrow{t_{1/2}} 6.25$

12.54 $t_{1/2} = 8.0$ h

$$0.60 \text{ M} \overset{t_{1/2}}{\rightarrow} 0.30 \text{ M} \overset{t_{1/2}}{\rightarrow} 0.15 \text{ M} \quad \text{requires 2 half-lives so it will take 16.0 h.}$$

12.56 $kt = \dfrac{1}{[C_4H_6]_t} - \dfrac{1}{[C_4H_6]_0}$, $\qquad k = 4.0 \times 10^{-2}/(\text{M} \cdot \text{s})$

(a) $t = 1.00 \text{ h} \times \dfrac{60 \text{ min}}{1 \text{ hr}} \times \dfrac{60 \text{ s}}{1 \text{ min}} = 3600 \text{ s}$

$$\dfrac{1}{[C_4H_6]_t} = kt + \dfrac{1}{[C_4H_6]_0} = (4.0 \times 10^{-2}/(\text{M} \cdot \text{s}))(3600 \text{ s}) + \dfrac{1}{0.0200 \text{ M}}$$

$$\dfrac{1}{[C_4H_6]_t} = 194/\text{M} \quad \text{and} \quad [C_4H_6] = 5.2 \times 10^{-3} \text{ M}$$

(b) $t = \dfrac{1}{k}\left[\dfrac{1}{[C_4H_6]_t} - \dfrac{1}{[C_4H_6]_0} \right]$

$$t = \dfrac{1}{4.0 \times 10^{-2}/(\text{M} \cdot \text{s})}\left[\dfrac{1}{(0.0020 \text{ M})} - \dfrac{1}{(0.0200 \text{ M})} \right] = 11{,}250 \text{ s}$$

$$t = 11{,}250 \text{ s} \times \dfrac{1 \text{ min}}{60 \text{ s}} \times \dfrac{1 \text{ hr}}{60 \text{ min}} = 3.1 \text{ h}$$

12.58 $t_{1/2} = \dfrac{1}{k[C_4H_6]_o} = \dfrac{1}{[4.0 \times 10^{-2}/(\text{M}\cdot\text{s})](0.0200 \text{ M})} = 1250 \text{ s} = 21 \text{ min}$

$$t = t_{1/2} = \dfrac{1}{k[C_4H_6]_o} = \dfrac{1}{[4.0 \times 10^{-2}/(\text{M}\cdot\text{s})](0.0100 \text{ M})} = 2500 \text{ s} = 42 \text{ min}$$

12.60

time (min)	$[N_2O]$	$\ln[N_2O]$	$1/[N_2O]$
0	0.250	−1.386	4.00
60	0.218	−1.523	4.59
90	0.204	−1.590	4.90
120	0.190	−1.661	5.26
180	0.166	−1.796	6.02

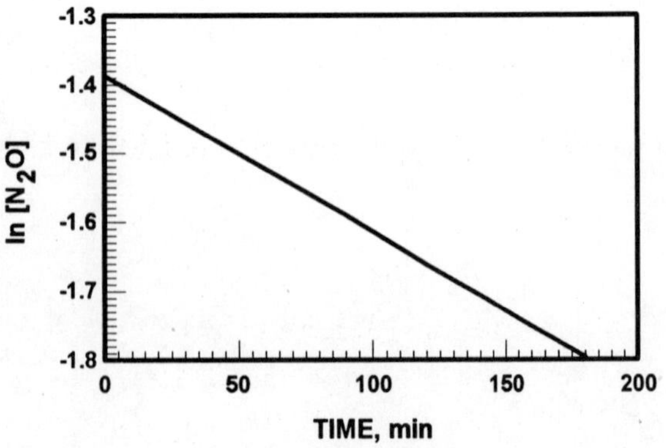

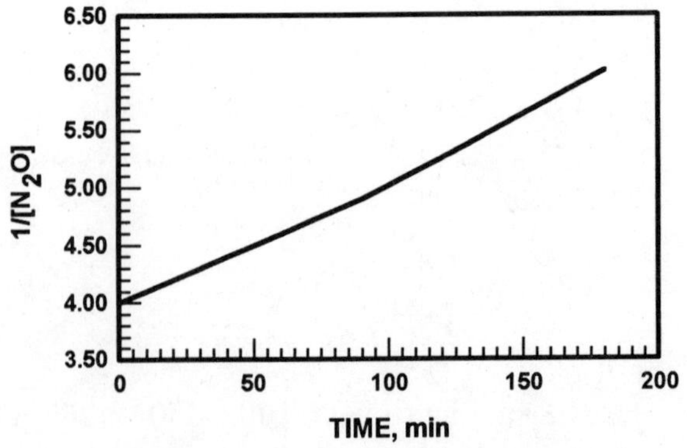

A plot of ln [N_2O] versus time is linear. The reaction is first-order in N_2O.

$k = -$ (slope) $= -(-2.28 \times 10^{-3}/min) = 2.28 \times 10^{-3}/min$

$k = 2.28 \times 10^{-3}/min \times \dfrac{1\ min}{60\ s} = 3.79 \times 10^{-5}/s$

12.62 $k = \dfrac{0.693}{t_{1/2}} = \dfrac{0.693}{248\ s} = 2.79 \times 10^{-3}/s$

12.64 (a) The units for the rate constant, k, indicate the reaction is zeroth-order.
(b) For a zeroth-order reaction, $[A]_t - [A]_o = -kt$

$t = 30\ min \times \dfrac{60\ s}{1\ min} = 1800\ s$

$[A]_t = -kt + [A]_o = -(3.6 \times 10^{-5}\ M/s)(1800\ s) + 0.096\ M = 0.031\ M$

(c) Let $[A]_t = [A]_o/2$

$t_{1/2} = \dfrac{[A]_o/2 - [A]_o}{-k} = \dfrac{0.096/2\ M - 0.096\ M}{-3.6 \times 10^{-5}\ M/s} = 1333\ s$

$t_{1/2} = 1333\ s \times \dfrac{1\ min}{60\ s} = 22\ min$

Reaction Mechanisms

12.66 An elementary reaction is a description of an individual molecular event that involves the breaking and/or making of chemical bonds. By contrast, the overall reaction describes only the stoichiometry of the overall process but provides no information about how the reaction occurs.

12.68 There is no relationship between the coefficients in a balanced chemical equation for an overall reaction and the exponents in the rate law unless the overall reaction occurs in a single elementary step, in which case the coefficients in the balanced equation are the exponents in the rate law.

12.70 (a)

$$H_2(g) + ICl(g) \rightarrow HI(g) + HCl(g)$$
$$\underline{HI(g) + ICl(g) \rightarrow I_2(g) + HCl(g)}$$

Overall reaction $H_2(g) + 2\ ICl(g) \rightarrow I_2(g) + 2\ HCl(g)$

(b) Because HI(g) is produced in the first step and consumed in the second step, it is a reaction intermediate.
(c) In each reaction there are two reactant molecules, so each elementary reaction is bimolecular.

12.72 (a) bimolecular, Rate $= k[O_3][Cl]$ (b) unimolecular, Rate $= k[NO_2]$
(c) bimolecular, Rate $= k[ClO][O]$ (d) termolecular, Rate $= k[Cl]^2[N_2]$

12.74 (a) $\qquad$ $NO_2Cl(g) \rightarrow NO_2(g) + Cl(g)$

$\underline{Cl(g) + NO_2Cl(g) \rightarrow NO_2(g) + Cl_2(g)}$

Overall reaction $\quad 2\,NO_2Cl(g) \rightarrow 2\,NO_2(g) + Cl_2(g)$

(b) 1. unimolecular; 2. bimolecular

(c) Rate = $k[NO_2Cl]$

12.76 $NO_2(g) + F_2(g) \rightarrow NO_2F(g) + F(g)$ $\qquad$ (slow)

$\quad\;\; F(g) + NO_2(g) \rightarrow NO_2F(g)$ $\qquad\qquad$ (fast)

The Arrhenius Equation

12.78 Very few collisions involve a collision energy greater than or equal to the activation energy, and only a fraction of those have the proper orientation for reaction.

12.80 Plot ln k versus 1/T to determine the activation energy, E_a

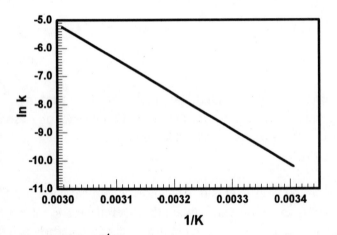

Slope = -1.25×10^4 K

$E_a = -(R)(slope) = -(8.314 \times 10^{-3}\ kJ/(K \cdot mol))(-1.25 \times 10^4\ K) = 104\ kJ/mol$

12.82 (a) $\ln\left(\dfrac{k_2}{k_1}\right) = \left(\dfrac{-E_a}{R}\right)\left(\dfrac{1}{T_2} - \dfrac{1}{T_1}\right)$

$k_1 = 1.3/(M \cdot s),\ T_1 = 700\ K$

$k_2 = 23.0/(M \cdot s),\ T_2 = 800\ K$

$E_a = -\dfrac{[\ln k_2 - \ln k_1](R)}{\left(\dfrac{1}{T_2} - \dfrac{1}{T_1}\right)}$

$$E_a = -\frac{[\ln(23.0) - \ln(1.3)][8.314 \times 10^{-3} \text{ kJ/(K} \cdot \text{mol)}]}{\left(\dfrac{1}{800 \text{ K}} - \dfrac{1}{700 \text{ K}}\right)} = 134 \text{ kJ/mol}$$

(b) $k_1 = 1.3/(M \cdot s)$, $T_1 = 700$ K
    solve for k_2, $T_2 = 750$ K

$$\ln k_2 = \left(\frac{-E_a}{R}\right)\left(\frac{1}{T_2} - \frac{1}{T_1}\right) + \ln k_1$$

$$\ln k_2 = \left(\frac{-133.8 \text{ kJ/mol}}{8.314 \times 10^{-3} \text{ kJ/(K} \cdot \text{mol)}}\right)\left(\frac{1}{750 \text{ K}} - \frac{1}{700 \text{ K}}\right) + \ln(1.3) = 1.795$$

$$k_2 = e^{1.795} = 6.0/(M \cdot s)$$

12.84 $\ln\left(\dfrac{k_2}{k_1}\right) = \left(\dfrac{-E_a}{R}\right)\left(\dfrac{1}{T_2} - \dfrac{1}{T_1}\right)$

assume $k_1 = 1.0/(M \cdot s)$ at $T_1 = 25°C = 298$ K
assume $k_2 = 15/(M \cdot s)$ ad $T_2 = 50°C = 323$ K

$$E_a = -\frac{[\ln k_2 - \ln k_1](R)}{\left(\dfrac{1}{T_2} - \dfrac{1}{T_1}\right)}$$

$$E_a = -\frac{[\ln(15) - \ln(1.0)][8.314 \times 10^{-3} \text{ kJ/(K} \cdot \text{mol)}]}{\left(\dfrac{1}{323 \text{ K}} - \dfrac{1}{298 \text{ K}}\right)} = 87 \text{ kJ/mol}$$

12.86

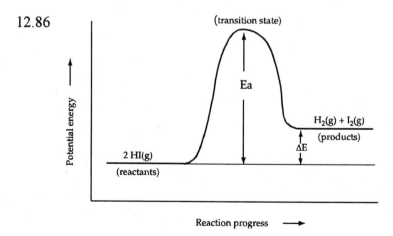

Catalysis

12.88 A catalyst does participate in the reaction, but it is not consumed because it reacts in one step of the reaction and is regenerated in a subsequent step.

12.90 A catalyst increases the rate of a reaction by changing the reaction mechanism and lowering the activation energy.

12.92 (a) $O_3(g) + O(g) \rightarrow 2\ O_2(g)$ (b) Cl acts as a catalyst.
(c) ClO is a reaction intermediate.
(d) A catalyst reacts in one step and is regenerated in a subsequent step. A reaction intermediate is produced in one step and consumed in another.

12.94 (a) $NH_2NO_2(aq) + OH^-(aq) \rightarrow NHNO_2^-(aq) + H_2O(l)$
$\underline{NHNO_2^-(aq) \rightarrow N_2O(g) + OH^-(aq)}$
Overall reaction $NH_2NO_2(aq) \rightarrow N_2O(g) + H_2O(l)$
(b) OH^- acts as a catalyst because it is used in the first step and regenerated in the second. $NHNO_2^-$ is a reaction intermediate because it is produced in the first step and consumed in the second.
(c) The rate will decrease because added acid decreases the concentration of OH^-, which appears in the rate law since it is a catalyst.

General Problems

12.96 $2\ AB_2 \rightarrow A_2 + 2\ B_2$
(a) Measure the change in the concentration of AB_2 as a function of time.
(b) and (c) If a plot of $[AB_2]$ versus time is linear, the reaction is zeroth-order and $k = -$ slope. If a plot of $\ln [AB_2]$ versus time is linear, the reaction is first-order and $k = -$ slope. If a plot of $1/[AB_2]$ versus time is linear, the reaction is second-order and $k =$ slope.

12.98 (a) Rate $= k[B_2][C]$
(b) $B_2 + C \rightarrow CB + B$ (slow)
   $CB + A \rightarrow AB + C$ (fast)
(c) C is a catalyst. C does not appear in the chemical equation because it is consumed in the first step and regenerated in the second step.

12.100 The first maximum represents the potential energy of the transition state for the first step. The second maximum represents the potential energy of the transition state for the second step. The saddle point between the two maxima represents the potential energy of the intermediate products.

12.102 (a) The reaction rate will increase with an increase in temperature at constant volume.
(b) The reaction rate will decrease with an increase in volume at constant temperature because reactant concentrations will decrease.

(c) The reaction rate will increase with the addition of a catalyst.

(d) Addition of an inert gas at constant volume will not affect the reaction rate.

12.104 (a) Rate $= k[C_2H_4Br_2]^m[I^-]^n$

$$m = \frac{\ln\left(\dfrac{Rate_2}{Rate_1}\right)}{\ln\left(\dfrac{[C_2H_4Br_2]_2}{[C_2H_4Br_2]_1}\right)} = \frac{\ln\left(\dfrac{1.74 \times 10^{-4}}{6.45 \times 10^{-5}}\right)}{\ln\left(\dfrac{0.343}{0.127}\right)} = 1$$

$$n = \frac{\ln\left(\dfrac{Rate_3 \cdot [C_2H_4Br_2]_2}{Rate_2 \cdot [C_2H_4Br_2]_3}\right)}{\ln\left(\dfrac{[I^-]_3}{[I^-]_2}\right)} = \frac{\ln\left(\dfrac{(1.26 \times 10^{-4})(0.343)}{(1.74 \times 10^{-4})(0.203)}\right)}{\ln\left(\dfrac{0.125}{0.102}\right)} = 1$$

Rate $= k[C_2H_4Br_2][I^-]$

(b) From Experiment 1:

$$k = \frac{Rate}{[C_2H_4Br_2][I^-]} = \frac{6.45 \times 10^{-5}\ M/s}{(0.127\ M)(0.102\ M)} = 4.98 \times 10^{-3}/(M \cdot s)$$

(c) Rate $= k[C_2H_4Br_2][I^-] = [4.98 \times 10^{-3}(M \cdot s)](0.150\ M)(0.150\ M) = 1.12 \times 10^{-4}\ M/s$

12.106 For $E_a = 50$ kJ/mol

$$f = e^{-E_a/RT} = \exp\left\{\frac{-50\ kJ/mol}{[8.314 \times 10^{-3}\ kJ/(K \cdot mol)](300\ K)}\right\} = 2.0 \times 10^{-9}$$

For $E_a = 100$ kJ/mol

$$f = e^{-E_a/RT} = \exp\left\{\frac{-100\ kJ/mol}{[8.314 \times 10^{-3}\ kJ/(K \cdot mol)](300\ K)}\right\} = 3.9 \times 10^{-18}$$

12.108 (a) $2\ NO(g) + Br_2(g) \rightarrow 2\ NOBr(g)$

(b) Since $NOBr_2$ is generated in the first step and consumed in the second step, $NOBr_2$ is a reaction intermediate.

(c) Rate $= k[NO][Br_2]$

(d) It can't be the first step. It must be the second step.

12.110 (a)

$$2\ NO(g) \rightleftharpoons N_2O_2(g) \qquad \text{(fast)}$$
$$N_2O_2(g) + H_2(g) \rightarrow N_2O(g) + H_2O(g) \qquad \text{(slow)}$$
$$\underline{N_2O(g) + H_2(g) \rightarrow N_2(g) + H_2O(g)} \qquad \text{(fast)}$$

Overall reaction $\qquad 2\ NO(g) + 2\ H_2(g) \rightarrow N_2(g) + 2\ H_2O(g)$

(b) N_2O_2 and N_2O are reaction intermediates because they are produced in one step of the reaction and used up in a subsequent step.

(c) Rate $= k_2[N_2O_2][H_2]$

(d) Because the forward and reverse rates in step 1 are equal, $k_1[NO]^2 = k_{-1}[N_2O_2]$. Solving for $[N_2O_2]$ and substituting into the rate law for the second step gives

$$\text{Rate} = k_2[N_2O_2][H_2] = \frac{k_1 k_2}{k_{-1}}[NO]^2[H_2]$$

Because the rate law for the overall reaction is equal to the rate law for the rate-determining step, the rate law for the overall reaction is

$$\text{Rate} = k[NO]^2[H_2] \quad \text{where } k = \frac{k_1 k_2}{k_{-1}}$$

12.112 (a) $\text{Rate}_f = k_f[A]$ and $\text{Rate}_r = k_r[B]$

(b)

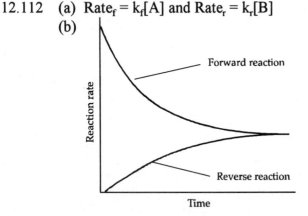

(c) When $\text{Rate}_f = \text{Rate}_r$, $k_f[A] = k_r[B]$, and $\dfrac{[B]}{[A]} = \dfrac{k_f}{k_r} = \dfrac{(3.0 \times 10^{-3})}{(1.0 \times 10^{-3})} = 3$

12.114 $k = \dfrac{0.693}{t_{1/2}} = \dfrac{0.693}{5730 \text{ y}} = 1.21 \times 10^{-4}/\text{y}$

$$t = \frac{\ln \dfrac{(^{14}C)_t}{(^{14}C)_o}}{-k} = \frac{\ln \dfrac{(2.3)}{(15.3)}}{-1.21 \times 10^{-4}/\text{y}} = 1.6 \times 10^4 \text{ y}$$

12.116 $X \rightarrow$ products is a first-order reaction

$$t = 60 \text{ min} \times \frac{60 \text{ s}}{1 \text{ min}} = 3600 \text{ s}$$

$$\ln \frac{[X]_t}{[X]_o} = -kt; \qquad k = \frac{\ln \dfrac{[X]_t}{[X]_o}}{-t}$$

At 25°C, calculate k_1:

$$k_1 = \frac{\ln\left(\dfrac{0.600\ \text{M}}{1.000\ \text{M}}\right)}{-3600\ \text{s}} = 1.42 \times 10^{-4}\ \text{s}^{-1}$$

At 35°C, calculate k_2:

$$k_2 = \frac{\ln\left(\dfrac{0.200\ \text{M}}{0.600\ \text{M}}\right)}{-3600\ \text{s}} = 3.05 \times 10^{-4}\ \text{s}^{-1}$$

At an unknown temperature calculate k_3.

$$k_3 = \frac{\ln\left(\dfrac{0.010\ \text{M}}{0.200\ \text{M}}\right)}{-3600\ \text{s}} = 8.32 \times 10^{-4}\ \text{s}^{-1}$$

$T_1 = 25°C = 25 + 273 = 298\ \text{K}$
$T_2 = 35°C = 35 + 273 = 308\ \text{K}$

Calculate E_a using k_1 and k_2.

$$\ln\left(\frac{k_2}{k_1}\right) = \left(\frac{-E_a}{R}\right)\left(\frac{1}{T_2} - \frac{1}{T_1}\right)$$

$$E_a = -\frac{[\ln k_2 - \ln k_1](R)}{\left(\dfrac{1}{T_2} - \dfrac{1}{T_1}\right)}$$

$$E_a = -\frac{[\ln(3.05 \times 10^{-4}) - \ln(1.42 \times 10^{-4})][8.314 \times 10^{-3}\ \text{kJ/(K}\cdot\text{mol})]}{\left(\dfrac{1}{308\ \text{K}} - \dfrac{1}{298\ \text{K}}\right)} \overset{\bullet}{=} 58.3\ \text{kJ/mol}$$

Use E_a, k_1, and k_3 to calculate T_3.

$$\frac{1}{T_3} = \frac{\ln\left(\dfrac{k_3}{k_1}\right)}{\left(\dfrac{-E_a}{R}\right)} + \frac{1}{T_1} = \frac{\ln\left(\dfrac{8.32 \times 10^{-4}}{1.42 \times 10^{-4}}\right)}{\left(\dfrac{-58.3\ \text{kJ/mol}}{8.314 \times 10^{-3}\ \text{kJ/(K}\cdot\text{mol})}\right)} + \frac{1}{298\ \text{K}} = 0.003104/\text{K}$$

$$T_3 = \frac{1}{0.003104/\text{K}} = 322\ \text{K} = 322 - 273 = 49°C$$

At 3:00 p.m. raise the temperature to 49°C to finish the reaction by 4:00 p.m.

12.118 (a) When equal volumes of two solutions are mixed, both concemtrations are cut in half.
$[H_3O^+]_o = [OH^-]_o = 1.0\ \text{M}$
When 99.999% of the acid is neutralized,

$$[H_3O^+] = [OH^-] = 1.0\ \text{M} - (1.0\ \text{M} \times 0.99999) = 1.0 \times 10^{-5}\ \text{M}$$

Using the 2nd order integrated rate law:

$$kt = \frac{1}{[H_3O^+]_t} - \frac{1}{[H_3O^+]_o}; \qquad t = \frac{1}{k}\left[\frac{1}{[H_3O^+]_t} - \frac{1}{[H_3O^+]_o}\right]$$

$$t = \frac{1}{(1.3 \times 10^{11} \, M^{-1} s^{-1})} \left[\frac{1}{(1.0 \times 10^{-5} \, M)} - \frac{1}{(1.0 \, M)} \right] = 7.7 \times 10^{-7} \, s$$

(b) The rate of an acid-base neutralization reaction would be limited by the speed of mixing, which is much slower than the intrinsic rate of the reaction itself.

12.120 Looking at the two experiments at 600 K, when the NO_2 concentration is doubled, the rate increased by a factor of 4. Therefore, the reaction is 2nd order.

Rate = $k \, [NO_2]^2$

Calculate k_1 at 600 K: $\quad k_1 = $ Rate/$[NO_2]^2 = 5.4 \times 10^{-7} \, M \, s^{-1}/(0.0010 \, M)^2 = 0.54 \, M^{-1} \, s^{-1}$

Calculate k_2 at 700 K: $\quad k_2 = $ Rate/$[NO_2]^2 = 5.2 \times 10^{-6} \, M \, s^{-1}/(0.0020 \, M)^2 = 13 \, M^{-1} \, s^{-1}$

Calculate E_a using k_1 and k_2.

$$\ln\left(\frac{k_2}{k_1}\right) = \left(\frac{-E_a}{R}\right)\left(\frac{1}{T_2} - \frac{1}{T_1}\right)$$

$$E_a = -\frac{[\ln k_2 - \ln k_1](R)}{\left(\dfrac{1}{T_2} - \dfrac{1}{T_1}\right)}$$

$$E_a = -\frac{[\ln(13) - \ln(0.54)][8.314 \times 10^{-3} \, kJ/(K \cdot mol)]}{\left(\dfrac{1}{700 \, K} - \dfrac{1}{600 \, K}\right)} = 111 \, kJ/mol$$

Calculate k_3 at 650 K using E_a and k_1.

Solve for k_3.

$$\ln k_3 = \frac{-E_a}{R}\left(\frac{1}{T_3} - \frac{1}{T_1}\right) + \ln k_1$$

$$\ln k_3 = \frac{-111 \, kJ/mol}{[8.314 \times 10^{-3} \, kJ/(K \cdot mol)]}\left(\frac{1}{650 \, K} - \frac{1}{600 \, K}\right) + \ln(0.54) = 1.0955$$

$$k_3 = e^{1.0955} = 3.0 \, M^{-1} \, s^{-1}$$

$$k_3 t = \frac{1}{[NO_2]_t} - \frac{1}{[NO_2]_o}; \qquad t = \frac{1}{k_3}\left[\frac{1}{[NO_2]_t} - \frac{1}{[NO_2]_o}\right]$$

$$t = \frac{1}{(3.0 \, M^{-1} s^{-1})}\left[\frac{1}{(0.0010 \, M)} - \frac{1}{(0.0050 \, M)}\right] = 2.7 \times 10^2 \, s$$

12.122 $A \rightarrow C$ is a first-order reaction.

The reaction is complete at 200 s when the absorbance of C reaches 1.200.

Because there is a one to one stoichiometry between A and C, the concentration of A must be proportional to 1.200 – absorbance of C. Any two data points can be used to find k. Let $[A]_o \propto 1.200$ and at 100 s, $[A]_t \propto 1.200 - 1.188 = 0.012$

$$\ln \frac{[A]_t}{[A]_o} = -kt; \qquad k = \frac{\ln\dfrac{[A]_t}{[A]_o}}{-t}; \qquad k = \frac{\ln\left(\dfrac{0.012 \, M}{1.200 \, M}\right)}{-100 \, s} = 0.0461 \, s^{-1}$$

$$t_{1/2} = \frac{0.693}{k} = \frac{0.693}{0.0461 \text{ s}^{-1}} = 15 \text{ s}$$

12.124   For radioactive decay, $\ln \dfrac{N}{N_o} = -kt$

For ^{235}U, $k_1 = \dfrac{0.693}{t_{1/2}} = \dfrac{0.693}{7.1 \times 10^8 \text{ y}} = 9.76 \times 10^{-10} \text{ y}^{-1}$

For ^{238}U, $k_2 = \dfrac{0.693}{t_{1/2}} = \dfrac{0.693}{4.51 \times 10^9 \text{ y}} = 1.54 \times 10^{-10} \text{ y}^{-1}$

For ^{235}U, $\ln \dfrac{N_1}{N_{o1}} = -k_1 t$ and $\ln \dfrac{N_1}{N_{o1}} + k_1 t = 0$

For ^{238}U, $\ln \dfrac{N_2}{N_{o2}} = -k_2 t$ and $\ln \dfrac{N_2}{N_{o2}} + k_2 t = 0$

Set the two equations that are equal to zero equal to each other and solve for t.

$$\ln \frac{N_1}{N_{o1}} + k_1 t = \ln \frac{N_2}{N_{o2}} + k_2 t$$

$$\ln \frac{N_1}{N_{o1}} - \ln \frac{N_2}{N_{o2}} = k_2 t - k_1 t = (k_2 - k_1)t$$

$$\ln \frac{\left(\dfrac{N_1}{N_{o1}}\right)}{\left(\dfrac{N_2}{N_{o2}}\right)} = (k_2 - k_1)t, \text{ now } N_{o1} = N_{o2}, \text{ so } \ln \frac{N_1}{N_2} = (k_2 - k_1)t$$

$\dfrac{N_1}{N_2} = 7.25 \times 10^{-3}$, so $\ln(7.25 \times 10^{-3}) = (1.54 \times 10^{-10} \text{ y}^{-1} - 9.76 \times 10^{-10} \text{ y}^{-1})t$

$$t = \frac{-4.93}{-8.22 \times 10^{-10} \text{ y}^{-1}} = 6.0 \times 10^9 \text{ y}$$

The age of the elements is 6.0×10^9 y (6 billion years).

Multi-Concept Problems

12.126 $2 \text{ HI(g)} \rightarrow \text{H}_2\text{(g)} + \text{I}_2\text{(g)}$

(a) mass HI $= 1.50 \text{ L} \times \dfrac{1000 \text{ mL}}{1 \text{ L}} \times \dfrac{0.0101 \text{ g}}{1 \text{ mL}} = 15.15 \text{ g HI}$

$15.15 \text{ g HI} \times \dfrac{1 \text{ mol HI}}{127.91 \text{ g HI}} = 0.118 \text{ mol HI}$

$[\text{HI}] = \dfrac{0.118 \text{ mol}}{1.50 \text{ L}} = 0.0787 \text{ mol/L}$

$$-\frac{\Delta[HI]}{\Delta t} = k[HI]^2 = (0.031/(M \cdot min))(0.0787\ M)^2 = 1.92 \times 10^{-4}\ M/min$$

$$2\ HI(g) \rightarrow H_2(g) + I_2(g)$$

$$\frac{\Delta[I_2]}{\Delta t} = \frac{1}{2}\left(-\frac{\Delta[HI]}{\Delta t}\right) = \frac{1.92 \times 10^{-4}\ M/min}{2} = 9.60 \times 10^{-5}\ M/min$$

$$(9.60 \times 10^{-5}\ M/min)(1.50\ L)(6.022 \times 10^{23}\ molecules/mol) = 8.7 \times 10^{19}\ molecules/min$$

(b) Rate = $k[HI]^2$

$$\frac{1}{[HI]_t} = kt + \frac{1}{[HI]_o} = (0.031/(M \cdot min))\left(8.00\ h \times \frac{60.0\ min}{1\ h}\right) + \frac{1}{0.0787\ M} = 27.59/M$$

$$[HI]_t = \frac{1}{27.59/M} = 0.0362\ M$$

From stoichiometry, $[H_2]_t = 1/2\ ([HI]_o - [HI]_t) = 1/2\ (0.0787\ M - 0.0362\ M) = 0.0212\ M$

$410°C = 683\ K$

$PV = nRT$

$$P_{H_2} = \left(\frac{n}{V}\right)RT = (0.0212\ mol/L)\left(0.082\ 06\ \frac{L \cdot atm}{K \cdot mol}\right)(683\ K) = 1.2\ atm$$

12.128 (a) N_2O_5, 108.01 amu

$$[N_2O_5]_o = \frac{\left(2.70\ g\ N_2O_5 \times \dfrac{1\ mol\ N_2O_5}{108.01\ g\ N_2O_5}\right)}{2.00\ L} = 0.0125\ mol/L$$

$$\ln[N_2O_5]_t = -kt + \ln[N_2O_5]_o = -(1.7 \times 10^{-3}\ s^{-1})\left(13.0\ min \times \frac{60.0\ s}{1\ min}\right) + \ln(0.0125) = -5.71$$

$$[N_2O_5]_t = e^{-5.71} = 3.31 \times 10^{-3}\ mol/L$$

After 13.0 min, mol $N_2O_5 = (3.31 \times 10^{-3}\ mol/L)(2.00\ L) = 6.62 \times 10^{-3}\ mol\ N_2O_5$

	$N_2O_5(g)$ $\rightarrow$	$2\ NO_2(g)$ +	$1/2\ O_2(g)$
before reaction (mol)	0.0250	0	0
change (mol)	−x	+2x	+1/2x
after reaction (mol)	0.0250 − x	2x	1/2x

After 13.0 min, mol $N_2O_5 = 6.62 \times 10^{-3} = 0.0250 - x$

$x = 0.0184\ mol$

After 13.0 min, $n_{total} = n_{N_2O_5} + n_{NO_2} + n_{O_2} = (6.62 \times 10^{-3}) + 2(0.0184) + 1/2(0.0184)$

$n_{total} = 0.0526\ mol$

$55°C = 328\ K$

$PV = nRT$

$$P_{total} = \frac{nRT}{V} = \frac{(0.0526\ mol)\left(0.082\ 06\ \dfrac{L \cdot atm}{K \cdot mol}\right)(328\ K)}{2.00\ L} = 0.71\ atm$$

(b) $N_2O_5(g) \rightarrow 2 NO_2(g) + 1/2 O_2(g)$

$\Delta H°_{rxn} = 2 \Delta H°_f(NO_2) - \Delta H°_f(N_2O_5)$

$\Delta H°_{rxn} = (2 \text{ mol})(33.2 \text{ kJ/mol}) - (1 \text{ mol})(11 \text{ kJ/mol}) = 55.4 \text{ kJ} = 5.54 \times 10^4 \text{ J}$

initial rate $= k[N_2O_5]_0 = (1.7 \times 10^{-3} \text{ s}^{-1})(0.0125 \text{ mol/L}) = 2.125 \times 10^{-5} \text{ mol/(L} \cdot \text{s})$

initial rate absorbing heat $= [2.125 \times 10^{-5} \text{ mol/(L} \cdot \text{s})](2.00 \text{ L})(5.54 \times 10^4 \text{ J/mol}) = 2.4 \text{ J/s}$

(c)

$$\ln [N_2O_5]_t = -kt + \ln [N_2O_5]_0 = -(1.7 \times 10^{-3} \text{ s}^{-1})\left(10.0 \text{ min} \times \frac{60.0 \text{ s}}{1 \text{ min}} \right) + \ln (0.0125) = -5.40$$

$[N_2O_5]_t = e^{-5.40} = 4.52 \times 10^{-3} \text{ mol/L}$

After 10.0 min, mol $N_2O_5 = (4.52 \times 10^{-3} \text{ mol/L})(2.00 \text{ L}) = 9.03 \times 10^{-3} \text{ mol } N_2O_5$

	$N_2O_5(g)$	$\rightarrow$	$2 NO_2(g)$	+	$1/2 O_2(g)$
before reaction (mol)	0.0250		0		0
change (mol)	–x		+2x		+1/2x
after reaction (mol)	0.0250 – x		2x		1/2x

After 10.0 min, mol $N_2O_5 = 9.03 \times 10^{-3} = 0.0250 - x$

x = 0.0160 mol

heat absorbed = (0.0160 mol)(55.4 kJ/mol) = 0.89 kJ

12.130 H_2O_2, 34.01 amu

mass $H_2O_2 = (0.500 \text{ L})(1000 \text{ mL/1 L})(1.00 \text{ g/ 1 mL})(0.0300) = 15.0 \text{ g } H_2O_2$

$$\text{mol } H_2O_2 = 15.0 \text{ g } H_2O_2 \times \frac{1 \text{ mol } H_2O_2}{34.01 \text{ g } H_2O_2} = 0.441 \text{ } H_2O_2$$

$$[H_2O_2]_0 = \frac{0.441 \text{ mol}}{0.500 \text{L}} = 0.882 \text{ mol /L}$$

$$k = \frac{0.693}{t_{1/2}} = \frac{0.693}{10.7 \text{ h}} = 6.48 \times 10^{-2}/\text{h}$$

$\ln [H_2O_2]_t = -kt + \ln [H_2O_2]_0$

$\ln [H_2O_2]_t = -(6.48 \times 10^{-2}/\text{h})(4.02 \text{ h}) + \ln (0.882)$

$\ln [H_2O_2]_t = -0.386; \quad [H_2O_2]_t = e^{-0.386} = 0.680 \text{ mol/L}$

mol $H_2O_2 = (0.680 \text{ mol/L})(0.500 \text{ L}) = 0.340 \text{ mol}$

	$2 H_2O_2(aq)$	$\rightarrow$	$2 H_2O(l)$	+	$O_2(g)$
before reaction (mol)	0.441		0		0
change (mol)	– 2x		+2x		+x
after reaction	0.441 – 2x		2x		x

After 4.02 h, mol $H_2O_2 = 0.340 \text{ mol} = 0.441 - 2x$; solve for x.

2x = 0.101

x = 0.0505 mol = mol O_2

$$P = 738 \text{ mm Hg} \times \frac{1.00 \text{ atm}}{760 \text{ mm Hg}} = 0.971 \text{ atm}$$

$PV = nRT$

$$V = \frac{nRT}{P} = \frac{(0.0505 \text{ mol})\left(0.082\ 06\ \dfrac{\text{L} \cdot \text{atm}}{\text{K} \cdot \text{mol}}\right)(293 \text{ K})}{0.971 \text{ atm}} = 1.25 \text{ L}$$

$P\Delta V = (0.971 \text{ atm})(1.25 \text{ L}) = 1.21 \text{ L} \cdot \text{atm}$

$$w = -P\Delta V = -1.21 \text{ L} \cdot \text{atm} = (-1.21 \text{ L} \cdot \text{atm})\left(101\ \frac{\text{J}}{\text{L} \cdot \text{atm}}\right) = -122 \text{ J}$$

Chemical Equilibrium

13.1 (a) $K_c = \dfrac{[SO_3]^2}{[SO_2]^2[O_2]}$ (b) $K_c = \dfrac{[SO_2]^2[O_2]}{[SO_3]^2}$

13.2 (a) $K_c = \dfrac{[SO_3]^2}{[SO_2]^2[O_2]} = \dfrac{(5.0 \times 10^{-2})^2}{(3.0 \times 10^{-3})^2(3.5 \times 10^{-3})} = 7.9 \times 10^4$

(b) $K_c = \dfrac{[SO_2]^2[O_2]}{[SO_3]^2} = \dfrac{(3.0 \times 10^{-3})^2(3.5 \times 10^{-3})}{(5.0 \times 10^{-2})^2} = 1.3 \times 10^{-5}$

13.3 (a) $K_c = \dfrac{[H^+][C_3H_5O_3^-]}{[C_3H_6O_3]}$

(b) $K_c = \dfrac{[(0.100)(0.0365)]^2}{[0.100 - (0.100)(0.0365)]} = 1.38 \times 10^{-4}$

13.4 From (1), $K_c = \dfrac{[AB][B]}{[A][B_2]} = \dfrac{(1)(2)}{(1)(2)} = 1$

For a mixture to be at equilibrium, $\dfrac{[AB][B]}{[A][B_2]}$ must be equal to 1.

For (2), $\dfrac{[AB][B]}{[A][B_2]} = \dfrac{(2)(1)}{(2)(1)} = 1$. This mixture is at equilibrium.

For (3), $\dfrac{[AB][B]}{[A][B_2]} = \dfrac{(1)(1)}{(4)(2)} = 0.125$. This mixture is not at equilibrium.

For (4), $\dfrac{[AB][B]}{[A][B_2]} = \dfrac{(2)(1)}{(4)(1)} = 0.5$. This mixture is not at equilibrium.

13.5 $K_p = \dfrac{(P_{CO_2})(P_{H_2})}{(P_{CO})(P_{H_2O})} = \dfrac{(6.12)(20.3)}{(1.31)(10.0)} = 9.48$

13.6 $2\,NO(g) + O_2 \rightleftarrows 2\,NO_2(g);$ $\Delta n = 2 - 3 = -1$
$K_p = K_c(RT)^{\Delta n},$ $K_c = K_p(1/RT)^{\Delta n}$
at 500 K: $K_p = (6.9 \times 10^5)[(0.082\ 06)(500)]^{-1} = 1.7 \times 10^4$

at 1000 K: $K_c = (1.3 \times 10^{-2})\left(\dfrac{1}{(0.082\ 06)(1000)}\right)^{-1} = 1.1$

13.7 (a) $K_c = \dfrac{[H_2]^3}{[H_2O]^3}$, $\quad K_p = \dfrac{(P_{H_2})^3}{(P_{H_2O})^3}$, $\quad \Delta n = (3) - (3) = 0$ and $K_p = K_c$

(b) $K_c = [H_2]^2[O_2]$, $\quad K_p = (P_{H_2})^2(P_{O_2})$, $\quad \Delta n = (3) - (0) = 3$ and $K_p = K_c(RT)^3$

(c) $K_c = \dfrac{[HCl]^4}{[SiCl_4][H_2]^2}$, $\quad K_p = \dfrac{(P_{HCl})^4}{(P_{SiCl_4})(P_{H_2})^2}$, $\quad \Delta n = (4) - (3) = 1$ and $K_p = K_c(RT)$

(d) $K_c = \dfrac{1}{[Hg_2^{2+}][Cl^-]^2}$

13.8 $K_c = 1.2 \times 10^{-42}$. Since K_c is very small, the equilibrium mixture contains mostly H_2 molecules. H is in periodic group 1A. A very small value of K_c is consistent with strong bonding between 2 H atoms, each with one valence electron.

13.9 The container volume of 5.0 L must be included to calculate molar concentrations.

(a) $Q_c = \dfrac{[NO_2]_t^2}{[NO]_t^2[O_2]_t} = \dfrac{(0.80 \text{ mol}/5.0 \text{ L})^2}{(0.060 \text{ mol}/5.0 \text{ L})^2(1.0 \text{ mol}/5.0 \text{ L})} = 890$

Because $Q_c < K_c$, the reaction is not at equilibrium. The reaction will proceed to the right to reach equilibrium.

(b) $Q_c = \dfrac{[NO_2]_t^2}{[NO]_t^2[O_2]_t} = \dfrac{(4.0 \text{ mol}/5.0 \text{ L})^2}{(5.0 \times 10^{-3} \text{ mol}/5.0 \text{ L})^2(0.20 \text{ mol}/5.0 \text{ L})} = 1.6 \times 10^7$

Because $Q_c > K_c$, the reaction is not at equilibrium. The reaction will proceed to the left to reach equilibrium.

13.10 $K_c = \dfrac{[AB]^2}{[A_2][B_2]} = 4$; For a mixture to be at equilibrium, $\dfrac{[AB]^2}{[A_2][B_2]}$ must be equal to 4.

For (1), $Q_c = \dfrac{[AB]^2}{[A_2][B_2]} = \dfrac{(6)^2}{(1)(1)} = 36$, $Q_c > K_c$

For (2), $Q_c = \dfrac{[AB]^2}{[A_2][B_2]} = \dfrac{(4)^2}{(2)(2)} = 4$, $Q_c = K_c$

For (3), $Q_c = \dfrac{[AB]^2}{[A_2][B_2]} = \dfrac{(2)^2}{(3)(3)} = 0.44$, $Q_c < K_c$

(a) (2)   (b) (1), reverse; (3), forward

13.11 $K_c = \dfrac{[H]^2}{[H_2]} = 1.2 \times 10^{-42}$

(a) $[H] = \sqrt{K_c[H_2]} = \sqrt{(1.2 \times 10^{-42})(0.10)} = 3.5 \times 10^{-22}$ M

(b) H atoms = $(3.5 \times 10^{-22}$ mol/L$)(1.0$ L$)(6.022 \times 10^{23}$ atoms/mol$) = 210$ H atoms

H_2 molecules = $(0.10$ mol/L$)(1.0$ L$)(6.022 \times 10^{23}$ molecules/mol$) = 6.0 \times 10^{22}$ H_2 molecules

13.12

	$CO(g)$	+	$H_2O(g)$	$\rightleftharpoons$	$CO_2(g)$	+	$H_2(g)$
initial (M)	0.150		0.150		0		0
change (M)	$-x$		$-x$		$+x$		$+x$
equil (M)	$0.150 - x$		$0.150 - x$		x		x

$$K_c = 4.24 = \frac{[CO_2][H_2]}{[CO][H_2O]} = \frac{x^2}{(0.150 - x)^2}$$

Take the square root of both sides and solve for x.

$$\sqrt{4.24} = \sqrt{\frac{x^2}{(0.150-x)^2}} ; \quad 2.06 = \frac{x}{0.150 - x} ; \quad x = 0.101$$

At equilibrium, $[CO_2] = [H_2] = x = 0.101$ M

$[CO] = [H_2O] = 0.150 - x = 0.150 - 0.101 = 0.049$ M

13.13

	$N_2O_4(g)$	$\rightleftharpoons$	$2\,NO_2(g)$
initial (M)	0.0500		0
change (M)	$-x$		$+2x$
equil (M)	$0.0500 - x$		$2x$

$$K_c = 4.64 \times 10^{-3} = \frac{[NO_2]^2}{[N_2O_4]} = \frac{(2x)^2}{(0.0500 - x)}$$

$$4x^2 + (4.64 \times 10^{-3})x - (2.32 \times 10^{-4}) = 0$$

Use the quadratic formula to solve for x.

$$x = \frac{-(4.64 \times 10^{-3}) \pm \sqrt{(4.64 \times 10^{-3})^2 - 4(4)(-2.32 \times 10^{-4})}}{2(4)} = \frac{-0.00464 \pm 0.06110}{8}$$

$x = -0.008\ 22$ and $0.007\ 06$

Discard the negative solution ($-0.008\ 22$) because it will lead to negative concentrations and that is impossible.

$[N_2O_4] = 0.0500 - x = 0.0500 - 0.007\ 06 = 0.0429$ M

$[NO_2] = 2x = 2(0.007\ 06) = 0.0141$ M

13.14 $N_2O_4(g)$ $\rightleftharpoons$ $2\,NO_2(g)$

$$Q_c = \frac{[NO_2]_t^2}{[N_2O_4]_t} = \frac{(0.0300 \text{ mol/L})^2}{(0.0200 \text{ mol/L})} = 0.0450; \quad Q_c > K_c$$

The reaction will approach equilibrium by going from right to left.

	$N_2O_4(g)$	$\rightleftharpoons$	$2\,NO_2(g)$
initial (M)	0.0200		0.0300
change (M)	$+x$		$-2x$
equil (M)	$0.0200 + x$		$0.0300 - 2x$

$$K_c = 4.64 \times 10^{-3} = \frac{[NO_2]^2}{[N_2O_4]} = \frac{(0.0300 - 2x)^2}{(0.0200 + x)}$$

$$4x^2 - 0.1246x + (8.072 \times 10^{-4}) = 0$$

Use the quadratic formula to solve for x.

$$x = \frac{-(-0.1246) \pm \sqrt{(-0.1246)^2 - 4(4)(8.072 \times 10^{-4})}}{2(4)} = \frac{0.1246 \pm 0.05109}{8}$$

x = 0.0220 and 0.009 19

Discard the larger solution (0.0220) because it will lead to a negative concentration of NO_2, and that is impossible.

$[N_2O_4] = 0.0200 + x = 0.0200 + 0.009\ 19 = 0.0292$ M

$[NO_2] = 0.0300 - 2x = 0.0300 - 2(0.009\ 19) = 0.0116$ M

13.15 $K_p = \dfrac{(P_{CO})(P_{H_2})}{(P_{H_2O})} = 2.44$, $Q_p = \dfrac{(1.00)(1.40)}{(1.20)} = 1.17$, $Q_p < K_p$ and the reaction goes to

the right to reach equilibrium.

	C(s)	+	H_2O(g)	$\rightleftarrows$	CO(g)	+	H_2(g)
initial (atm)			1.20		1.00		1.40
change (atm)			−x		+x		+x
equil (atm)			1.20 − x		1.00 + x		1.40 + x

$$K_p = \frac{(P_{CO})(P_{H_2})}{(P_{H_2O})} = 2.44 = \frac{(1.00 + x)(1.40 + x)}{(1.20 - x)}$$

$$x^2 + 4.84x - 1.53 = 0$$

Use the quadratic formula to solve for x.

$$x = \frac{-(4.84) \pm \sqrt{(4.84)^2 - 4(1)(-1.53)}}{2(1)} = \frac{-4.84 \pm 5.44}{2}$$

x = −5.14 and 0.300

Discard the negative solution (−5.14) because it will lead to negative partial pressures and that is impossible.

$P_{H_2O} = 1.20 - x = 1.20 - 0.300 = 0.90$ atm

$P_{CO} = 1.00 + x = 1.00 + 0.300 = 1.30$ atm

$P_{H_2} = 1.40 + x = 1.40 + 0.300 = 1.70$ atm

13.16 (a) CO(reactant) added, H_2 concentration increases.
(b) CO_2 (product) added, H_2 concentration decreases.
(c) H_2O (reactant) removed, H_2 concentration decreases.
(d) CO_2 (product) removed, H_2 concentration increases.

At equilibrium, $Q_c = K_c = \dfrac{[CO_2][H_2]}{[CO][H_2O]}$. If some CO_2 is removed from the

equilibrium mixture, the numerator in Q_c is decreased, which means that $Q_c < K_c$ and the reaction will shift to the right, increasing the H_2 concentration.

13.17   (a) Because there are 2 mol of gas on both sides of the balanced equation, the composition of the equilibrium mixture is unaffected by a change in pressure. The number of moles of reaction products remains the same.
(b) Because there are 2 mol of gas on the left side and 1 mol of gas on the right side of the balanced equation, the stress of an increase in pressure is relieved by a shift in the reaction to the side with fewer moles of gas (in this case, to products). The number of moles of reaction products increases.
(c) Because there is 1 mol of gas on the left side and 2 mol of gas on the right side of the balanced equation, the stress of an increase in pressure is relieved by a shift in the reaction to the side with fewer moles of gas (in this case, to reactants). The number of moles of reaction product decreases.

13.18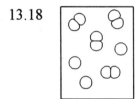

13.19   Le Châtelier's principle predicts that a stress of added heat will be relieved by net reaction in the direction that absorbs the heat. Since the reaction is endothermic, the equilibrium will shift from left to right (K_c will increase) with an increase in temperature. Therefore, the equilibrium mixture will contain more of the offending NO, the higher the temperature.

13.20   The reaction is exothermic. As the temperature is increased the reaction shifts from right to left. The amount of ethyl acetate decreases.

$$K_c = \frac{[CH_3CO_2C_2H_5][H_2O]}{[CH_3CO_2H][C_2H_5OH]}$$

As the temperature is decreased, the reaction shifts from left to right. The product concentrations increase, and the reactant concentrations decrease. This corresponds to an increase in K_c.

13.21   There are more AB(g) molecules at the higher temperature. The equilibrium shifted to the right at the higher temperature, which means the reaction is endothermic.

13.22   (a) A catalyst does not affect the equilibrium composition. The amount of CO remains the same.
(b) The reaction is exothermic. An increase in temperature shifts the reaction toward reactants. The amount of CO increases.
(c) Because there are 3 mol of gas on the left side and 2 mol of gas on the right side of the balanced equation, the stress of an increase in pressure is relieved by a shift in the reaction to the side with fewer moles of gas (in this case, to products). The amount of CO decreases.

(d) An increase in pressure as a result of the addition of an inert gas (with no volume change) does not affect the equilibrium composition. The amount of CO remains the same.
(e) Adding O_2 increases the O_2 concentration and shifts the reaction toward products. The amount of CO decreases.

13.23 (a) Because K_c is so large, k_f is larger than k_r.

(b) $K_c = \dfrac{k_f}{k_r}$; $\qquad k_r = \dfrac{k_f}{K_c} = \dfrac{8.5 \times 10^6 \, M^{-1} s^{-1}}{3.4 \times 10^{34}} = 2.5 \times 10^{-28} \, M^{-1} \, s^{-1}$

(c) Because the reaction is exothermic, E_a (forward) is less than E_a (reverse). Consequently, as the temperature decreases, k_r decreases more than k_f decreases, and therefore $K_c = \dfrac{k_f}{k_r}$ increases.

13.24 $Hb + O_2 \rightleftharpoons Hb(O_2)$
If CO binds to Hb, Hb is removed from the reaction and the reaction will shift to the left resulting in O_2 being released from $Hb(O_2)$. This will decrease the effectiveness of Hb for carrying O_2.

13.25 The equilibrium shifts to the left because at the higher altitude the concentration of O_2 is decreased.

13.26 There are 26 π electrons.

13.27 The partial pressure of O_2 in the atmosphere is 0.2095 atm.
$PV = nRT$

$$n = \frac{PV}{RT} = \frac{(0.2095 \text{ atm})(0.500 \text{ L})}{\left(0.082 \ 06 \ \dfrac{\text{L} \cdot \text{atm}}{\text{K} \cdot \text{mol}}\right)(298 \text{ K})} = 4.28 \times 10^{-3} \text{ mol } O_2$$

$$4.28 \times 10^{-3} \text{ mol } O_2 \times \frac{6.022 \times 10^{23} \, O_2 \text{ molecules}}{1 \text{ mol } O_2} = 2.58 \times 10^{21} \, O_2 \text{ molecules}$$

Understanding Key Concepts

13.28 (a) (1) and (3) because the number of A and B's are the same in the third and fourth box.

(b) $K_c = \dfrac{[B]}{[A]} = \dfrac{6}{4} = 1.5$

(c) Because the same number of molecules appear on both sides of the equation, the volume terms in K_c all cancel. Therefore, we can calculate K_c without including the volume.

13.30 (a) Only reaction (3), $K_c = \dfrac{[A][AB]}{[A_2][B]} = \dfrac{(2)(4)}{(2)(2)} = 2$, is at equilibrium.

(b) $Q_c = \dfrac{[A][AB]}{[A_2][B]} = \dfrac{(3)(5)}{(1)(1)} = 15$ for reaction (1). Because $Q_c > K_c$, the reaction will go

in the reverse direction to reach equilibrium.

$Q_c = \dfrac{[A][AB]}{[A_2][B]} = \dfrac{(1)(3)}{(3)(3)} = 1/3$ for reaction (2). Because $Q_c < K_c$, the reaction will go in

the forward direction to reach equilibrium.

13.32 When the stopcock is opened, the reaction will go in the reverse direction because there will be initially an excess of AB molecules.

13.34 (a) AB $\rightarrow$ A + B

(b) The reaction is endothermic because a stress of added heat (higher temperature) shifts the AB $\rightleftharpoons$ A + B equilibrium to the right.

(c) If the volume is increased, the pressure is decreased. The stress of decreased pressure will be relieved by a shift in the equilibrium from left to right, thus increasing the number of A atoms.

13.36 (a)   (b)   (c)

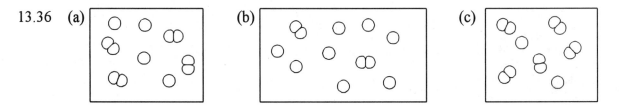

Additional Problems
Equilibrium Expressions and Equilibrium Constants

13.38 (a) $K_c = \dfrac{[CO][H_2]^3}{[CH_4][H_2O]}$ (b) $K_c = \dfrac{[ClF_3]^2}{[F_2]^3[Cl_2]}$ (c) $K_c = \dfrac{[HF]^2}{[H_2][F_2]}$

13.40 (a) $K_p = \dfrac{(P_{CO})(P_{H_2})^3}{(P_{CH_4})(P_{H_2O})}$, $\Delta n = 2$ and $K_p = K_c(RT)^2$

(b) $K_p = \dfrac{(P_{ClF_3})^2}{(P_{F_2})^3(P_{Cl_2})}$, $\Delta n = -2$ and $K_p = K_c(RT)^{-2}$

(c) $K_p = \dfrac{(P_{HF})^2}{(P_{H_2})(P_{F_2})}$, $\Delta n = 0$ and $K_p = K_c$

13.42 $K_c = \dfrac{[C_2H_5OC_2H_5][H_2O]}{[C_2H_5OH]^2}$

13.44 $K_c = \dfrac{[Isocitrate]}{[Citrate]}$

13.46 The two reactions are the reverse of each other.

$K_c(reverse) = \dfrac{1}{K_c(forward)} = \dfrac{1}{7.5 \times 10^{-9}} = 1.3 \times 10^8$

13.48 $K_c = \dfrac{[PCl_3][Cl_2]}{[PCl_5]} = \dfrac{(1.5 \times 10^{-2})(3.2 \times 10^{-2})}{(8.3 \times 10^{-3})} = 0.058$

13.50 The container volume of 2.00 L must be included to calculate molar concentrations.
Initial [HI] = 9.30 x 10⁻³ mol/2.00 L = 4.65 x 10⁻³ M = 0.004 65 M

$$H_2(g) + I_2(g) \rightleftharpoons 2\,HI(g)$$

initial (M)	0	0	0.004 65
change (M)	+x	+x	–2x
equil (M)	x	x	0.004 65 – 2x

x = [H₂] = [I₂] = 6.29 x 10⁻⁴ M = 0.000 629 M
[HI] = 0.004 65 – 2x = 0.004 65 – 2(0.000 629) = 0.003 39 M

$K_c = \dfrac{[HI]^2}{[H_2][I_2]} = \dfrac{(0.003\ 39)^2}{(0.000\ 629)^2} = 29.0$

13.52 (a) $K_c = \dfrac{[CH_3CO_2C_2H_5][H_2O]}{[CH_3CO_2H][C_2H_5OH]}$

(b) $$CH_3CO_2H(soln) + C_2H_5OH(soln) \rightleftharpoons CH_3CO_2C_2H_5(soln) + H_2O(soln)$$

initial (mol)	1.00	1.00	0	0
change (mol)	–x	–x	+x	+x
equil (mol)	1.00 – x	1.00 – x	x	x

x = 0.65 mol; 1.00 – x = 0.35 mol; $K_c = \dfrac{(0.65)^2}{(0.35)^2} = 3.4$

Because there are the same number of molecules on both sides of the equation, the volume terms in K_c cancel. Therefore, we can calculate K_c without including the volume.

13.54 Δn = 1 and $K_p = K_c(RT) = (0.575)(0.082\ 06)(500) = 23.6$

13.56 $K_p = P_{H_2O} = 0.0313$ atm; Δn = 1

$K_c = K_p\left(\dfrac{1}{RT}\right) = (0.0313)\left(\dfrac{1}{(0.082\ 06)(298)}\right) = 1.28 \times 10^{-3}$

13.58 (a) $K_c = \dfrac{[CO_2]^3}{[CO]^3}$, $K_p = \dfrac{(P_{CO_2})^3}{(P_{CO})^3}$ (b) $K_c = \dfrac{1}{[O_2]^3}$, $K_p = \dfrac{1}{(P_{O_2})^3}$

(c) $K_c = [SO_3]$, $K_p = P_{SO_3}$ (d) $K_c = [Ba^{2+}][SO_4^{2-}]$

Using the Equilibrium Constant

13.60 (a) Because K_c is very large, the equilibrium mixture contains mostly product.
(b) Because K_c is very small, the equilibrium mixture contains mostly reactants.

13.62 (a) Because K_c is very small, the equilibrium mixture contains mostly reactant.
(b) Because K_c is very large, the equilibrium mixture contains mostly product.
(c) Because $K_c = 1.8$, the equilibrium mixture contains an appreciable concentration of both reactants and products.

13.64 $K_c = 1.2 \times 10^{82}$ is very large. When equilibrium is reached, very little if ar y ethanol will remain because the reaction goes to completion.

13.66 The container volume of 10.0 L must be included to calculate molar concentrations.

$$Q_c = \frac{[CS_2]_t[H_2]_t^4}{[CH_4]_t[H_2S]_t^2} = \frac{(3.0 \text{ mol}/10.0 \text{ L})(3.0 \text{ mol}/10.0 \text{ L})^4}{(2.0 \text{ mol}/10.0 \text{ L})(4.0 \text{ mol}/10.0 \text{ L})^2} = 7.6 \times 10^{-2}; \quad K_c = 2.5 \times 10^{-3}$$

The reaction is not at equilibrium because $Q_c > K_c$. The reaction will proceed from right to left to reach equilibrium.

13.68 $K_c = \dfrac{[NH_3]^2}{[N_2][H_2]^3} = 0.29$;   At equilibrium, $[N_2] = 0.036$ M and $[H_2] = 0.15$ M

$$[NH_3] = \sqrt{[N_2] \times [H_2]^3 \times K_c} = \sqrt{(0.036)(0.15)^3(0.29)} = 5.9 \times 10^{-3} \text{ M}$$

13.70

	$N_2(g)$	+	$O_2(g)$	$\rightleftarrows$	$2\,NO(g)$
initial (M)	1.40		1.40		0
change (M)	$-x$		$-x$		$+2x$
equil (M)	$1.40 - x$		$1.40 - x$		$2x$

$$K_c = 1.7 \times 10^{-3} = \frac{[NO]^2}{[N_2][O_2]} = \frac{(2x)^2}{(1.40 - x)^2}$$

Take the square root of both sides and solve for x.

$$\sqrt{1.7 \times 10^{-3}} = \sqrt{\frac{(2x)^2}{(1.40-x)^2}}; \quad 4.1 \times 10^{-2} = \frac{2x}{1.40 - x}; \quad x = 2.8 \times 10^{-2}$$

At equilibrium, $[NO] = 2x = 2(2.8 \times 10^{-2}) = 0.056$ M
$[N_2] = [O_2] = 1.40 - x = 1.40 - (2.8 \times 10^{-2}) = 1.37$ M

Chapter 13 – Chemical Equilibrium

13.72

$$PCl_5(g) \rightleftharpoons PCl_3(g) + Cl_2(g)$$

initial (M)	0.160	0	0
change (M)	−x	+x	+x
equil (M)	0.160 − x	x	x

$$K_c = \frac{[PCl_3][Cl_2]}{[PCl_5]} = 5.8 \times 10^{-2} = \frac{x^2}{0.160 - x}$$

$$x^2 + (5.8 \times 10^{-2})x - 0.00928 = 0$$

Use the quadratic formula to solve for x.

$$x = \frac{(-5.8 \times 10^{-2}) \pm \sqrt{(5.8 \times 10^{-2})^2 - 4(1)(-0.00928)}}{2(1)} = \frac{(-5.8 \times 10^{-2}) \pm 0.20}{2}$$

x = 0.071 and −0.129

Discard the negative solution (−0.129) because it gives negative concentrations of PCl_3 and Cl_2 and that is impossible.

$[PCl_3] = [Cl_2] = x = 0.071$ M; $[PCl_5] = 0.160 - x = 0.160 - 0.071 = 0.089$ M

13.74   (a) $K_c = \dfrac{[CH_3CO_2C_2H_5][H_2O]}{[CH_3CO_2H][C_2H_5OH]} = 3.4 = \dfrac{(x)(12.0)}{(4.0)(6.0)}$;   x = 6.8 moles $CH_3CO_2C_2H_5$

Note that the volume cancels because the same number of molecules appear on both sides of the chemical equation.

(b) $$CH_3CO_2H(soln) + C_2H_5OH(soln) \rightleftharpoons CH_3CO_2C_2H_5(soln) + H_2O(soln)$$

initial (mol)	1.00	10.00	0	0
change (mol)	−x	−x	+x	+x
equil (mol)	1.00 − x	10.00 − x	x	x

$$K_c = 3.4 = \frac{x^2}{(1.00 - x)(10.00 - x)}$$

$$2.4x^2 - 37.4x + 34 = 0$$

Use the quadratic formula to solve for x.

$$x = \frac{-(-37.4) \pm \sqrt{(-37.4)^2 - 4(2.4)(34)}}{2(2.4)} = \frac{37.4 \pm 32.75}{4.8}$$

x = 0.969 and 14.6

Discard the larger solution (14.6) because it leads to negative concentrations and that is impossible.

mol CH_3CO_2H = 1.00 − x = 1.00 − 0.969 = 0.03 mol

mol C_2H_5OH = 10.00 − x = 10.00 − 0.969 = 9.03 mol

mol $CH_3CO_2C_2H_5$ = mol H_2O = x = 0.97 mol

13.76

$$ClF_3(g) \rightleftharpoons ClF(g) + F_2(g)$$

initial (atm)	1.47	0	0
change (atm)	−x	+x	+x
equil (atm)	1.47 − x	x	x

$$K_p = \frac{(P_{ClF})(P_{F_2})}{(P_{ClF_3})} = 0.140 = \frac{(x)(x)}{1.47 - x}$$

$x^2 + 0.140x - 0.2058 = 0$

Use the quadratic formula to solve for x.

$$x = \frac{-(0.140) \pm \sqrt{(0.140)^2 - (4)(1)(-0.2058)}}{2(1)}$$

$$x = \frac{-0.140 \pm 0.918}{2}$$

$x = 0.389$ and -0.529

Discard the negative solution (–0.529) because it gives negative partial pressures and that is impossible.

$P_{ClF} = P_{F_2} = x = 0.389$ atm

$P_{ClF_3} = 1.47 - x = 1.47 - 0.389 = 1.08$ atm

Le Châtelier's Principle

13.78   (a) Cl^- (reactant) added, $AgCl(s)$ increases
(b) Ag^+ (reactant) added, $AgCl(s)$ increases
(c) Ag^+ (reactant) removed, $AgCl(s)$ decreases
(d) Cl^- (reactant) removed, $AgCl(s)$ decreases

Disturbing the equilibrium by decreasing [Cl^-] increases Q_c $\left(Q_c = \dfrac{1}{[Ag^+]_t[Cl^-]_t} \right)$ to a

value greater than K_c. To reach a new state of equilibrium, Q_c must decrease, which means that the denominator must increase; that is, the reaction must go from right to left, thus decreasing the amount of solid AgCl.

13.80   (a) Because there are 2 mol of gas on the left side and 3 mol of gas on the right side of the balanced equation, the stress of an increase in pressure is relieved by a shift in the reaction to the side with fewer moles of gas (in this case, to reactants). The number of moles of reaction products decreases.
(b) Because there are 2 mol of gas on both sides of the balanced equation, the composition of the equilibrium mixture is unaffected by a change in pressure. The number of moles of reaction product remains the same.
(c) Because there are 2 mol of gas on the left side and 1 mol of gas on the right side of the balanced equation, the stress of an increase in pressure is relieved by a shift in the reaction to the side with fewer moles of gas (in this case, to products). The number of moles of reaction products increases.

13.82 $CO(g) + H_2O(g) \rightleftharpoons CO_2(g) + H_2(g)$ $\qquad\qquad \Delta H° = -41.2$ kJ
The reaction is exothermic. [H_2] decreases when the temperature is increased.
As the temperature is decreased, the reaction shifts to the right. [CO_2] and [H_2] increase, [CO] and [H_2O] decrease, and K_c increases.

13.84 (a) HCl is a source of Cl^- (product), the reaction shifts left, the equilibrium $[CoCl_4^{2-}]$ increases.
(b) $Co(NO_3)_2$ is a source of $Co(H_2O)_6^{2+}$ (product), the reaction shifts left, the equilibrium $[CoCl_4^{2-}]$ increases.
(c) All concentrations will initially decrease and the reaction will shift to the right, the equilibrium $[CoCl_4^{2-}]$ decreases.
(d) For an exothermic reaction, the reaction shifts to the left when the temperature is increased, the equilibrium $[CoCl_4^{2-}]$ increases.

13.86 (a) The reaction is exothermic. The amount of CH_3OH (product) decreases as the temperature increases.
(b) When the volume decreases, the reaction shifts to the side with fewer gas molecules. The amount of CH_3OH increases.
(c) Addition of an inert gas (He) does not affect the equilibrium composition. There is no change.
(d) Addition of CO (reactant) shifts the reaction toward product. The amount of CH_3OH increases.
(e) Addition or removal of a catalyst does not affect the equilibrium composition. There is no change.

Chemical Equilibrium and Chemical Kinetics

13.88 A + B ⇌ C
$rate_f = k_f[A][B]$ and $rate_r = k_r[C]$; at equilibrium, $rate_f = rate_r$

$$k_f[A][B] = k_r[C]; \qquad \frac{k_f}{k_r} = \frac{[C]}{[A][B]} = K_c$$

13.90 $K_c = \dfrac{k_f}{k_r} = \dfrac{0.13}{6.2 \times 10^{-4}} = 210$

13.92 k_r increases more than k_f, this means that E_a (reverse) is greater than E_a (forward). The reaction is exothermic when E_a (reverse) > E_a (forward).

General Problems

13.94 (a) $[N_2O_4] = \dfrac{0.500 \text{ mol}}{4.00 \text{ L}} = 0.125$ M

	$N_2O_4(g)$	⇌	$2 NO_2(g)$
initial (M)	0.125		0
change (M)	$-(0.793)(0.125)$		$+(2)(0.793)(0.125)$
equil (M)	$0.125 - (0.793)(0.125)$		$(2)(0.793)(0.125)$

At equilibrium, $[N_2O_4] = 0.125 - (0.793)(0.125) = 0.0259$ M
$[NO_2] = (2)(0.793)(0.125) = 0.198$ M

$$K_c = \frac{[NO_2]^2}{[N_2O_4]} = \frac{(0.198)^2}{(0.0259)} = 1.51$$

$\Delta n = 2 - 1 = 1$ and $K_p = K_c(RT)^{\Delta n}$; $\quad K_p = K_c(RT) = (1.51)(0.082\ 06)(400) = 49.6$

(b)

13.96 $\quad K_c = \dfrac{[NH_3]^2}{[N_2][H_2]^3} = 0.291$

At equilibrium, $[N_2] = 1.0 \times 10^{-3}$ M and $[H_2] = 2.0 \times 10^{-3}$ M

$[NH_3] = \sqrt{[N_2] \times [H_2]^3 \times K_c} = \sqrt{(1.0 \times 10^{-3})(2.0 \times 10^{-3})^3(0.291)} = 1.5 \times 10^{-6}$ M

13.98 $\quad 2\ HI(g) \rightleftharpoons H_2(g) + I_2(g)$

Calculate K_c. $\quad K_c = \dfrac{[H_2][I_2]}{[HI]^2} = \dfrac{(0.13)(0.70)}{(2.1)^2} = 0.0206$

$[HI] = \dfrac{0.20\ \text{mol}}{0.5000\ \text{L}} = 0.40$ M

	$2\ HI(g)$	$\rightleftharpoons$	$H_2(g)$	$+$	$I_2(g)$
initial (M)	0.40		0		0
change (M)	$-2x$		$+x$		$+x$
equil (M)	$0.40 - 2x$		x		x

$$K_c = 0.0206 = \frac{[H_2][I_2]}{[HI]^2} = \frac{x^2}{(0.40 - 2x)^2}$$

Take the square root of both sides, and solve for x.

$$\sqrt{0.0206} = \sqrt{\frac{x^2}{(0.40 - 2x)^2}}; \quad 0.144 = \frac{x}{0.40 - 2x}; \quad x = 0.045$$

At equilibrium, $[H_2] = [I_2] = x = 0.045$ M; $\quad [HI] = 0.40 - 2x = 0.40 - 2(0.045) = 0.31$ M

13.100 $\quad [H_2O] = \dfrac{6.00\ \text{mol}}{5.00\ \text{L}} = 1.20$ M

	$C(s)$	$+$	$H_2O(g)$	$\rightleftharpoons$	$CO(g)$	$+$	$H_2(g)$
initial (M)			1.20		0		0
change (M)			$-x$		$+x$		$+x$
equil (M)			$1.20 - x$		x		x

$$K_c = \frac{[CO][H_2]}{[H_2O]} = 3.0 \times 10^{-2} = \frac{x^2}{1.20 - x}$$

$x^2 + (3.0 \times 10^{-2})x - 0.036 = 0$

Use the quadratic formula to solve for x.

$$x = \frac{-(0.030) \pm \sqrt{(0.030)^2 - 4(-0.036)}}{2(1)} = \frac{-0.030 \pm 0.381}{2}$$

$x = 0.176$ and -0.206

Discard negative solution (-0.206) because it leads to negative concentrations and that is impossible.

$[CO] = [H_2] = x = 0.18$ M; $\quad [H_2O] = 1.20 - x = 1.20 - 0.18 = 1.02$ M

13.102 A decrease in volume (a) and the addition of reactants (c) will affect the composition of the equilibrium mixture, but leave the value of K_c unchanged.
A change in temperature (b) affects the value of K_c.
Addition of a catalyst (d) or an inert gas (e) affects neither the composition of the equilibrium mixture nor the value of K_c.

13.104 2 monomer $\rightleftarrows$ dimer
   (a) In benzene, $K_c = 1.51 \times 10^2$

	2 monomer	$\rightleftarrows$	dimer
initial (M)	0.100		0
change (M)	$-2x$		$+x$
equil (M)	$0.100 - 2x$		x

$$K_c = \frac{[dimer]}{[monomer]^2} = 1.51 \times 10^2 = \frac{x}{(0.100 - 2x)^2}$$

$604x^2 - 61.4x + 1.51 = 0$

Use the quadratic formula to solve for x.

$$x = \frac{-(-61.4) \pm \sqrt{(-61.4)^2 - (4)(604)(1.51)}}{2(604)} = \frac{61.4 \pm 11.04}{1208}$$

$x = 0.0600$ and 0.0417

Discard the larger solution (0.0600) because it gives a negative concentration of the monomer and that is impossible.

$[monomer] = 0.100 - 2x = 0.100 - 2(0.0417) = 0.017$ M; $\quad$ $[dimer] = x = 0.0417$ M

$$\frac{[dimer]}{[monomer]} = \frac{0.0417 \, M}{0.017 \, M} = 2.5$$

   (b) In H_2O, $K_c = 3.7 \times 10^{-2}$

	2 monomer	$\rightleftarrows$	dimer
initial (M)	0.100		0
change (M)	$-2x$		$+x$
equil (M)	$0.100 - 2x$		x

$$K_c = \frac{[dimer]}{[monomer]^2} = 3.7 \times 10^{-2} = \frac{x}{(0.100 - 2x)^2}$$

$0.148x^2 - 1.0148x + 0.000\,37 = 0$

Use the quadratic formula to solve for x.

$$x = \frac{-(-1.0148) \pm \sqrt{(-1.0148)^2 - (4)(0.148)(0.00037)}}{2(0.148)} = \frac{1.0148 \pm 1.0147}{0.296}$$

$x = 6.86$ and 3.7×10^{-4}

Discard the larger solution (6.86) because it gives a negative concentration of the monomer and that is impossible.

[monomer] = $0.100 - 2x = 0.100 - 2(3.7 \times 10^{-4}) = 0.099$ M; [dimer] = $x = 3.7 \times 10^{-4}$ M

$$\frac{[\text{dimer}]}{[\text{monomer}]} = \frac{3.7 \times 10^{-4} \text{ M}}{0.099 \text{ M}} = 0.0038$$

(c) K_c for the water solution is so much smaller than K_c for the benzene solution because H_2O can hydrogen bond with acetic acid, thus preventing acetic acid dimer formation. Benzene cannot hydrogen bond with acetic acid.

13.106 (a) $[PCl_5] = 1.000$ mol/5.000 L = 0.2000 M

	$PCl_5(g)$	$\rightleftharpoons$	$PCl_3(g)$	+	$Cl_2(g)$
initial (M)	0.2000		0		0
change (M)	−(0.2000)(0.7850)		+(0.2000)(0.7850)		+(0.2000)(0.7850)
equil (M)	0.0430		0.1570		0.1570

$$K_c = \frac{[PCl_3][Cl_2]}{[PCl_5]} = \frac{(0.1570)(0.1570)}{(0.0430)} = 0.573$$

$\Delta n = 1$ and $K_p = K_c(RT) = (0.573)(0.082\ 06)(500) = 23.5$

(b) $Q_c = \dfrac{[PCl_3][Cl_2]}{[PCl_5]} = \dfrac{(0.150)(0.600)}{(0.500)} = 0.18$

Because $Q_c < K_c$, the reaction proceeds to the right to reach equilibrium.

	$PCl_5(g)$	$\rightleftharpoons$	$PCl_3(g)$	+	$Cl_2(g)$
initial (M)	0.500		0.150		0.600
change (M)	− x		+x		+x
equil (M)	0.500 − x		0.150 + x		0.600 + x

$$K_c = \frac{[PCl_3][Cl_2]}{[PCl_5]} = 0.573 = \frac{(0.150 + x)(0.600 + x)}{(0.500 - x)}; \text{ solve for x.}$$

$x^2 + 1.323x - 0.1965 = 0$

$$x = \frac{-(1.323) \pm \sqrt{(1.323)^2 - (4)(1)(-0.1965)}}{2(1)} = \frac{-1.323 \pm 1.593}{2}$$

$x = -1.458$ and 0.135

Discard the negative solution (−1.458) because it will lead to negative concentrations and that is impossible.

$[PCl_5] = 0.500 - x = 0.500 - 0.135 = 0.365$ M
$[PCl_3] = 0.150 + x = 0.150 + 0.135 = 0.285$ M
$[Cl_2] = 0.600 + x = 0.600 + 0.135 = 0.735$ M

13.108 (a) $K_c = \dfrac{[C_2H_6][C_2H_4]}{[C_4H_{10}]}$ $\qquad$ $K_p = \dfrac{(P_{C_2H_6})(P_{C_2H_4})}{P_{C_4H_{10}}}$

(b) $K_p = 12$; $\Delta n = 1$; $\quad K_c = K_p\left(\dfrac{1}{RT}\right) = (12)\left(\dfrac{1}{(0.082\,06)(773)}\right) = 0.19$

(c)

	$C_4H_{10}(g)$	$\rightleftarrows$	$C_2H_6(g)$	+	$C_2H_4(g)$
initial (atm)	50		0		0
change (atm)	$-x$		$+x$		$+x$
equil (atm)	$50-x$		x		x

$K_p = 12 = \dfrac{x^2}{50-x};\qquad x^2 + 12x - 600 = 0$

Use the quadratic formula to solve for x.

$x = \dfrac{(-12) \pm \sqrt{(12)^2 - 4(1)(-600)}}{2(1)} = \dfrac{-12 \pm 50.44}{2}$

$x = -31.22$ and 19.22

Discard the negative solution (–31.22) because it leads to negative concentrations and that is impossible.

% C_4H_{10} converted $= \dfrac{19.22}{50} \times 100\% = 38\%$

$P_{total} = P_{C_4H_{10}} + P_{C_2H_6} + P_{C_2H_4} = (50-x) + x + x = (50 - 19) + 19 + 19 = 69$ atm

(d) A decrease in volume would decrease the % conversion of C_4H_{10}.

13.110 (a) $K_p = 3.45$; $\Delta n = 1$; $\quad K_c = K_p\left(\dfrac{1}{RT}\right) = (3.45)\left(\dfrac{1}{(0.082\,06)(500)}\right) = 0.0840$

(b) $[(CH_3)_3CCl] = 1.00$ mol/5.00 L $= 0.200$ M

	$(CH_3)_3CCl(g)$	$\rightleftarrows$	$(CH_3)_2C{=}CH_2(g)$	+	$HCl(g)$
initial (M)	0.200		0		0
change (M)	$-x$		$+x$		$+x$
equil (M)	$0.200 - x$		x		x

$K_c = 0.0840 = \dfrac{x^2}{0.200 - x};\qquad x^2 + 0.0840x - 0.0168 = 0$

Use the quadratic formula to solve for x.

$x = \dfrac{(-0.0840) \pm \sqrt{(0.0840)^2 - 4(1)(-0.0168)}}{2(1)} = \dfrac{-0.0840 \pm 0.272}{2}$

$x = -0.178$ and 0.094

Discard the negative solution (–0.178) because it leads to negative concentrations and that is impossible.

$[(CH_3)_2C{=}CCH_2] = [HCl] = x = 0.094$ M

$[(CH_3)_3CCl] = 0.200 - x = 0.200 - 0.094 = 0.106$ M

(c) $K_p = 3.45$

$$(CH_3)_3CCl(g) \rightleftharpoons (CH_3)_2C{=}CH_2(g) + HCl(g)$$

	$(CH_3)_3CCl(g)$	$(CH_3)_2C{=}CH_2(g)$	$HCl(g)$
initial (atm)	0	0.400	0.600
change (atm)	+x	−x	−x
equil (atm)	x	0.400 − x	0.600 − x

$$K_p = 3.45 = \frac{(0.400 - x)(0.600 - x)}{x}$$

$x^2 - 4.45x + 0.240 = 0$

Use the quadratic formula to solve for x.

$$x = \frac{-(-4.45) \pm \sqrt{(-4.45)^2 - 4(1)(0.240)}}{2(1)} = \frac{4.45 \pm 4.34}{2}$$

$x = 0.055$ and 4.40

Discard the larger solution (4.40) because it leads to a negative partial pressures and that is impossible.

$P_{t\text{–butyl chloride}} = x = 0.055$ atm; $\quad P_{isobutylene} = 0.400 - x = 0.400 - 0.055 = 0.345$ atm

$P_{HCl} = 0.600 - x = 0.600 - 0.055 = 0.545$ atm

13.112 The activation energy (E_a) is positive, and for an exothermic reaction, $E_{a,r} > E_{a,f}$.

$$k_f = A_f\, e^{-E_{a,f}/RT}, \quad k_r = A_r\, e^{-E_{a,r}/RT}$$

$$K_c = \frac{k_f}{k_r} = \frac{A_f e^{-E_{a,f}/RT}}{A_r e^{-E_{a,r}/RT}} = \frac{A_f}{A_r} e^{(E_{a,r}-E_{a,f})/RT}$$

$(E_{a,r} - E_{a,f})$ is positive, so the exponent is always positive. As the temperature increases, the exponent, $(E_{a,r} - E_{a,f})/RT$, decreases and the value for K_c decreases as well.

13.114 (a) $PV = nRT$, $\quad n_{total} = \dfrac{PV}{RT} = \dfrac{(0.588 \text{ atm})(1.00 \text{ L})}{\left(0.082\ 06\ \dfrac{\text{L} \cdot \text{atm}}{\text{K} \cdot \text{mol}}\right)(300 \text{ K})} = 0.0239$ mol

$$2\ NOBr(g) \rightleftharpoons 2\ NO(g) + Br_2(g)$$

	$2\ NOBr(g)$	$2\ NO(g)$	$Br_2(g)$
initial (mol)	0.0200	0	0
change (mol)	−2x	+2x	+x
equil (mol)	0.0200 − 2x	2x	x

$n_{total} = 0.0239$ mol $= (0.0200 - 2x) + 2x + x = 0.0200 + x$

$x = 0.0239 - 0.0200 = 0.0039$ mol

Because the volume is 1.00 L, the molarity equals the number of moles.

$[NOBr] = 0.0200 - 2x = 0.0200 - 2(0.0039) = 0.0122$ M

$[NO] = 2x = 2(0.0039) = 0.0078$ M

$[Br_2] = x = 0.0039$ M

$$K_c = \frac{[NO]^2[Br_2]}{[NOBr]^2} = \frac{(0.0078)^2(0.0039)}{(0.0122)^2} = 1.6 \times 10^{-3}$$

(b) $\Delta n = (3) - (2) = 1$, $K_p = K_c(RT) = (1.6 \times 10^{-3})(0.082\ 06)(300) = 0.039$

13.116 (a) $W(s) + 4\ Br(g) \rightleftharpoons WBr_4(g)$

$$K_p = \frac{P_{WBr_4}}{(P_{Br})^4} = 100, \quad P_{WBr_4} = (P_{Br})^4(100) = (0.010\ atm)^4(100) = 1.0 \times 10^{-6}\ atm$$

(b) Because K_p is smaller at the higher temperature, the reaction has shifted toward reactants at the higher temperature, which means the reaction is exothermic.

(c) At 2800 K, $Q_p = \frac{(1.0 \times 10^{-6})}{(0.010)^4} = 100$, $Q_p > K_p$ so the reaction will go from products to reactants, depositing tungsten back onto the filament.

13.118 $2\ NO_2(g) \rightleftharpoons N_2O_4(g)$

$\Delta n = (1) - (2) = -1$ and $K_p = K_c(RT)^{-1} = (216)[(0.082\ 06)(298)]^{-1} = 8.83$

$$K_p = \frac{P_{N_2O_4}}{(P_{NO_2})^2} = 8.83$$

Let $X = P_{N_2O_4}$ and $Y = P_{NO_2}$.

$P_{total} = 1.50\ atm = X + Y$ and $\dfrac{X}{Y^2} = 8.83$. Use these two equations to solve for X and Y.

$X = 1.50 - Y$

$$\frac{1.50 - Y}{Y^2} = 8.83$$

$8.83Y^2 + Y - 1.50 = 0$

Use the quadratic formula to solve for Y.

$$Y = \frac{-(1) \pm \sqrt{(1)^2 - 4(8.83)(-1.50)}}{2(8.83)} = \frac{-1 \pm 7.35}{17.7}$$

$Y = -0.472$ and 0.359

Discard the negative solution (-0.472) because it leads to a negative partial pressure and that is impossible.

$Y = P_{NO_2} = 0.359\ atm$

$X = P_{N_2O_4} = 1.50\ atm - Y = 1.50\ atm - 0.359\ atm = 1.14\ atm$

13.120

	$N_2(g)$	$+$	$3\ H_2$	$\rightleftharpoons$	$2\ NH_3$
initial (mol)	0		0		X
change (mol)	$+y$		$+3y$		$-2y$
equil (mol)	y		$3y$		$X - 2y$

$y = 0.200\ mol$

Because the volume is 1.00 L, the molarity equals the number of moles.

$[N_2] = y = 0.200$ M; $[H_2] = 3y = 3(0.200) = 0.600$ M

$$K_c = \frac{[NH_3]^2}{[N_2][H_2]^3} = \frac{[NH_3]^2}{(0.200)(0.600)^3} = 4.20, \text{ solve for } [NH_3]_{eq}$$

$$[NH_3]_{eq}^2 = [N_2][H_2]^3(4.20) = (0.200)(0.600)^3(4.20)$$

$$[NH_3]_{eq} = \sqrt{[N_2][H_2]^3(4.20)} = \sqrt{(0.200)(0.600)^3(4.20)} = 0.426 \text{ M}$$

$[NH_3]_{eq} = 0.426$ M $= X - 2(0.200) = [NH_3]_o - 2(0.200)$

$[NH_3]_o = 0.426 + 2(0.200) = 0.826$ M

0.826 mol of NH_3 were placed in the 1.00 L reaction vessel.

Multi-Concept Problems

13.122 (a) CO_2, 44.01 amu; CO, 28.01 amu

$$79.2 \text{ g } CO_2 \times \frac{1 \text{ mol } CO_2}{44.01 \text{ g } CO_2} = 1.80 \text{ mol } CO_2$$

	$CO_2(g)$	+	C(s)	⇌	2 CO(g)
initial (mol)	1.80				0
change (mol)	−x				+2x
equil (mol)	1.80 − x				2x

total mass of gas in flask $= (16.3 \text{ g/L})(5.00 \text{ L}) = 81.5$ g

$81.5 = (1.80 - x)(44.01) + (2x)(28.01)$

$81.5 = 79.22 - 44.01x + 56.02x;$ $2.28 = 12.01x;$ $x = 2.28/12.01 = 0.19$

$n_{CO_2} = 1.80 - x = 1.80 - 0.19 = 1.61 \text{ mol } CO_2;$ $n_{CO} = 2x = 2(0.19) = 0.38 \text{ mol CO}$

$$P_{CO_2} = \frac{nRT}{V} = \frac{(1.61 \text{ mol})\left(0.082\,06 \dfrac{L \cdot atm}{K \cdot mol}\right)(1000 \text{ K})}{5.0 \text{ L}} = 26.4 \text{ atm}$$

$$P_{CO} = \frac{nRT}{V} = \frac{(0.38 \text{ mol})\left(0.082\,06 \dfrac{L \cdot atm}{K \cdot mol}\right)(1000 \text{ K})}{5.0 \text{ L}} = 6.24 \text{ atm}$$

$$K_p = \frac{(P_{CO})^2}{(P_{CO_2})} = \frac{(6.24)^2}{(26.4)} = 1.47$$

(b) At 1100K, the total mass of gas in flask $= (16.9 \text{ g/L})(5.00 \text{ L}) = 84.5$ g

$84.5 = (1.80 - x)(44.01) + (2x)(28.01)$

$84.5 = 79.22 - 44.01x + 56.02x;$ $5.28 = 12.01x;$ $x = 5.28/12.01 = 0.44$

$n_{CO_2} = 1.80 - x = 1.80 - 0.44 = 1.36 \text{ mol } CO_2;$ $n_{CO} = 2x = 2(0.44) = 0.88 \text{ mol CO}$

$$P_{CO_2} = \frac{nRT}{V} = \frac{(1.36 \text{ mol})\left(0.082\,06 \dfrac{L \cdot atm}{K \cdot mol}\right)(1100 \text{ K})}{5.0 \text{ L}} = 24.6 \text{ atm}$$

$$P_{CO} = \frac{nRT}{V} = \frac{(0.88\ \text{mol})\left(0.082\ 06\ \frac{\text{L} \cdot \text{atm}}{\text{K} \cdot \text{mol}}\right)(1100\ \text{K})}{5.0\ \text{L}} = 15.9\ \text{atm}$$

$$K_p = \frac{(P_{CO})^2}{(P_{CO_2})} = \frac{(15.9)^2}{(24.6)} = 10.3$$

(c) In agreement with Le Châtelier's principle, the reaction is endothermic because K_p increases with increasing temperature.

13.124 (a) N_2O_4, 92.01 amu

$$14.58\ \text{g}\ N_2O_4 \times \frac{1\ \text{mol}\ N_2O_4}{92.01\ \text{g}\ N_2O_4} = 0.1585\ \text{mol}\ N_2O_4$$

$PV = nRT$

$$P_{N_2O_4} = \frac{nRT}{V} = \frac{(0.1585\ \text{mol})\left(0.082\ 06\ \frac{\text{L} \cdot \text{atm}}{\text{K} \cdot \text{mol}}\right)(400\ \text{K})}{1.000\ \text{L}} = 5.20\ \text{atm}$$

	$N_2O_4(g)$	$\rightleftharpoons$	$2\ NO_2(g)$
initial (atm)	5.20		0
change (atm)	$-x$		$+2x$
equil (atm)	$5.20 - x$		$2x$

$P_{total} = P_{N_2O_4} + P_{NO_2} = (5.20 - x) + (2x) = 9.15\ \text{atm}$

$5.20 + x = 9.15\ \text{atm}$

$x = 3.95\ \text{atm}$

$P_{N_2O_4} = 5.20 - x = 5.20 - 3.95 = 1.25\ \text{atm}$

$P_{NO_2} = 2x = 2(3.95) = 7.90\ \text{atm}$

$$K_p = \frac{(P_{NO_2})^2}{(P_{N_2O_4})} = \frac{(7.90)^2}{(1.25)} = 49.9$$

$$\Delta n = 1\ \text{and}\ K_c = K_p\left(\frac{1}{RT}\right) = \frac{(49.9)}{(0.082\ 06)(400)} = 1.52$$

(b) $\Delta H°_{rxn} = [2\ \Delta H°_f(NO_2)] - \Delta H°_f(N_2O_4)$

$\Delta H°_{rxn} = [(2\ \text{mol})(33.2\ \text{kJ/mol})] - [(1\ \text{mol})(9.16\ \text{kJ/mol})] = 57.2\ \text{kJ}$

$PV = nRT$

$$\text{moles}\ N_2O_4\ \text{reacted} = n = \frac{PV}{RT} = \frac{(3.95\ \text{atm})(1.000\ \text{L})}{\left(0.082\ 06\ \frac{\text{L} \cdot \text{atm}}{\text{K} \cdot \text{mol}}\right)(400\ \text{K})} = 0.1203\ \text{mol}\ N_2O_4$$

$q = (57.24\ \text{kJ/mol}\ N_2O_4)(0.1203\ \text{mol}\ N_2O_4) = 6.89\ \text{kJ}$

13.126 The atmosphere is 21% (0.21) O_2; $P_{O_2} = (0.21)\left(720 \text{ mm Hg} \times \dfrac{1 \text{ atm}}{760 \text{ mm Hg}}\right) = 0.199 \text{ atm}$

$$2 \, O_3(g) \rightleftharpoons 3 \, O_2(g)$$

$$K_p = \frac{(P_{O_2})^3}{(P_{O_3})^2}; \qquad P_{O_3} = \sqrt{\frac{(P_{O_2})^3}{K_p}} = \sqrt{\frac{(0.199)^3}{1.3 \times 10^{57}}} = 2.46 \times 10^{-30} \text{ atm}$$

$$\text{vol} = 10 \times 10^6 \text{ m}^3 \times \left(\frac{100 \text{ cm}}{1 \text{ m}}\right)^3 \times \frac{1 \text{ L}}{1000 \text{ cm}^3} = 1.0 \times 10^{10} \text{ L}$$

$$n_{O_3} = \frac{PV}{RT} = \frac{(2.46 \times 10^{-30} \text{ atm})(1.0 \times 10^{10} \text{ L})}{\left(0.082 \, 06 \, \dfrac{\text{L} \cdot \text{atm}}{\text{K} \cdot \text{mol}}\right)(298 \text{ K})} = 1.0 \times 10^{-21} \text{ mol } O_3$$

$$O_3 \text{ molecules} = 1.0 \times 10^{-21} \text{ mol } O_3 \times \frac{6.022 \times 10^{23} \, O_3 \text{ molecules}}{1 \text{ mol } O_3} = 6.0 \times 10^2 \, O_3 \text{ molecules}$$

13.128 $PCl_5(g) \rightleftharpoons PCl_3(g) + Cl_2(g)$

$\Delta n = (2) - (1) = 1$ and at 700 K, $K_p = K_c(RT) = (46.9)(0.082\,06)(700) = 2694$

(a) Because K_p is larger at the higher temperature, the reaction has shifted toward products at the higher temperature, which means the reaction is endothermic. Because the reaction involves breaking two P–Cl bonds and forming just one Cl–Cl bond, it should be endothermic.

(b) PCl_5, 208.24 amu

$$\text{mol } PCl_5 = 1.25 \text{ g } PCl_5 \times \frac{1 \text{ mol } PCl_5}{208.24 \text{ g } PCl_5} = 6.00 \times 10^{-3} \text{ mol}$$

$$PV = nRT, \quad P_{PCl_5} = \frac{nRT}{V} = \frac{(6.00 \times 10^{-3} \text{ mol})\left(0.082 \, 06 \, \dfrac{\text{L} \cdot \text{atm}}{\text{K} \cdot \text{mol}}\right)(700 \text{ K})}{0.500 \text{ L}} = 0.689 \text{ atm}$$

Because K_p is so large, first assume the reaction goes to completion and then allow for a small back reaction.

	$PCl_5(g)$	$\rightleftharpoons$	$PCl_3(g)$	$+$	$Cl_2(g)$
before rxn (atm)	0.689		0		0
change (atm)	−0.689		+0.689		+0.689
after rxn (atm)	0		0.689		0.689
change (atm)	+x		−x		−x
equil (atm)	x		0.689 − x		0.689 − x

$$K_p = \frac{(P_{PCl_3})(P_{Cl_2})}{P_{PCl_5}} = 2694 = \frac{(0.689 - x)^2}{x} \approx \frac{(0.689)^2}{x}$$

$$x = P_{PCl_5} = \frac{(0.689)^2}{2694} = 1.76 \times 10^{-4} \text{ atm}$$

$$P_{\text{total}} = P_{PCl_5} + P_{PCl_3} + P_{Cl_2}$$

$$P_{total} = x + (0.689 - x) + (0.689 - x) = 0.689 + 0.689 - 1.76 \times 10^{-4} = 1.38 \text{ atm}$$

$$\% \text{ dissociation} = \frac{(P_{PCl_5})_o - (P_{PCl_5})}{(P_{PCl_5})_o} \times 100\% = \frac{0.689 - (1.76 \times 10^{-4})}{0.689} \times 100\% = 99.97\%$$

(c)

The molecular geometry is trigonal bipyramidal. There is no dipole moment because of a symmetrical distribution of Cl's around the central P.

The molecular geometry is trigonal pyramidal. There is a dipole moment because of the lone pair of electrons on the P and an unsymmetrical distribution of Cl's around the central P.

14 Hydrogen, Oxygen, and Water

14.1 $PV = nRT$; $PV = \dfrac{g}{molar\ mass}RT$

$$d_{H_2} = \frac{g}{V} = \frac{P(molar\ mass)}{RT} = \frac{(1.00\ atm)(2.016\ g/mol)}{\left(0.08206\ \dfrac{L \cdot atm}{K \cdot mol}\right)(298\ K)} = 0.0824\ g/L$$

$1\ L = 1000\ mL = 1000\ cm^3$
$d_{H_2} = 0.0824\ g/1000\ cm^3 = 8.24 \times 10^{-5}\ g/cm^3$

$$\frac{d_{air}}{d_{H_2}} = \frac{1.185 \times 10^{-3}\ g/cm^3}{8.24 \times 10^{-5}\ g/cm^3} = 14.4;\ \ \text{Air is 14 times more dense than } H_2.$$

14.2   For every 100.0 g, there are:
61.4 g O, 22.9 g C, 10.0 g H, 2.6 g N, and 3.1 g other

$$22.9\ g\ C \times \frac{1\ mol\ C}{12.011\ g\ C} = 1.907\ mol\ C$$

$$10.0\ g\ H \times \frac{1\ mol\ H}{1.008\ g\ H} = 9.921\ mol\ H$$

Assume the sample contains 1.907 mol ^{13}C and 9.921 mol D.

$$mass\ ^{13}C = 1.907\ mol\ ^{13}C \times \frac{13.0034\ g\ ^{13}C}{1\ mol\ ^{13}C} = 24.8\ g\ ^{13}C$$

$$mass\ D = 9.921\ mol\ D \times \frac{2.0141\ g\ D}{1\ mol\ D} = 20.0\ g\ D$$

(a) Total mass if all H is D is:
61.4 g O + 22.9 C + 20.0 g D + 2.6 g N + 3.1 g other = 110.0 g

$$mass\ \%\ D = \frac{20.0\ g\ D}{110.0\ g} \times 100\% = 18.2\%\ D$$

(b) Total mass if all C is ^{13}C is:
61.4 g O + 24.8 g ^{13}C + 10.0 g H + 2.6 g N + 3.1 g other = 101.9 g

$$mass\ \%\ ^{13}C = \frac{24.8\ g\ ^{13}C}{101.9\ g} \times 100\% = 24.3\%\ ^{13}C$$

(c) The isotope effect for H is larger than that for C because D is two times the mass of ^{1}H while ^{13}C is only about 8% heavier than ^{12}C.

14.3 $2\ Ga(s) + 6\ H^+(aq) \rightarrow 3\ H_2(g) + 2\ Ga^{3+}(aq)$

14.4   (a) $SrH_2(s) + 2\ H_2O(l) \rightarrow 2\ H_2(g) + Sr^{2+}(aq) + 2\ OH^-(aq)$
        (b) $KH(s) + H_2O(l) \rightarrow H_2(g) + K^+(aq) + OH^-(aq)$

14.5 $CaH_2(s) + 2 H_2O(l) \rightarrow 2 H_2(g) + Ca^{2+}(aq) + 2 OH^-(aq)$
CaH_2, 42.09 amu; 25°C = 298 K

$$PV = nRT; \quad n_{H_2} = \frac{PV}{RT} = \frac{(1.00 \text{ atm})(2.0 \times 10^5 \text{ L})}{\left(0.082\ 06 \dfrac{L \cdot atm}{K \cdot mol}\right)(298 \text{ K})} = 8.18 \times 10^3 \text{ mol H}_2$$

$$8.18 \times 10^3 \text{ mol H}_2 \times \frac{1 \text{ mol CaH}_2}{2 \text{ mol H}_2} \times \frac{42.09 \text{ g CaH}_2}{1 \text{ mol CaH}_2} \times \frac{1 \text{ kg}}{1000 \text{ g}} = 1.7 \times 10^2 \text{ kg CaH}_2$$

14.6   (a) (1) ZrH_x, interstitial   (2) PH_3, covalent   (3) HBr, covalent   (4) LiH, ionic
(b) (1) and (4) are likely to be solids at 25°C. (2) and (3) are likely to be gases at 25°C.
Covalent hydrides, like (2) and (3), form discrete molecules and have only relatively weak
intermolecular forces, resulting in gases. (4) is an ionic metal hydride with strong ion-ion
forces holding the 3-dimensional lattice together in the solid state. (1) is an interstitial
hydride with the metal atoms in a solid crystal lattice and H's occupying holes.
(c) $LiH(s) + H_2O(l) \rightarrow H_2(g) + Li^+(aq) + OH^-(aq)$

14.7   Assume 12.0 g of Pd with a volume of 1.0 cm³.
$V_{H_2} = 935 \text{ cm}^3 = 935 \text{ mL} = 0.935 \text{ L}$

$$PV = nRT; \quad n_{H_2} = \frac{PV}{RT} = \frac{(1.00 \text{ atm})(0.935 \text{ L})}{\left(0.082\ 06 \dfrac{L \cdot atm}{K \cdot mol}\right)(273 \text{ K})} = 0.0417 \text{ mol H}_2$$

$n_H = 2 n_{H_2} = 0.0834 \text{ mol H}$

$$12.0 \text{ g Pd} \times \frac{1 \text{ mol Pd}}{106.42 \text{ g Pd}} = 0.113 \text{ mol Pd}$$

$Pd_{0.113}H_{0.0834}$
$Pd_{0.113/0.113}H_{0.0834/0.113}$
$PdH_{0.74}$
$g H = (0.0834 \text{ mol H})(1.008 \text{ g/mol}) = 0.0841 \text{ g H}$

$d_H = 0.0841 \text{ g/cm}^3; \quad M_H = \dfrac{0.0834 \text{ mol}}{0.001 \text{ L}} = 83.4 \text{ M}$

14.8 $2 KMnO_4(s) \rightarrow K_2MnO_4(s) + MnO_2(s) + O_2(g)$
$KMnO_4$, 158.03 amu; 25°C = 298 K

$$\text{mol O}_2 = 0.200 \text{ g KMnO}_4 \times \frac{1 \text{ mol KMnO}_4}{158.03 \text{ g KMnO}_4} \times \frac{1 \text{ mol O}_2}{2 \text{ mol KMnO}_4} = 6.33 \times 10^{-4} \text{ mol O}_2$$

$$PV = nRT; \quad V = \frac{nRT}{P} = \frac{(6.33 \times 10^{-4} \text{ mol})\left(0.082\ 06 \dfrac{L \cdot atm}{K \cdot mol}\right)(298 \text{ K})}{1.00 \text{ atm}} = 0.0155 \text{ L}$$

$$V = 0.0155 \text{ L} \times \frac{1000 \text{ mL}}{1 \text{ L}} = 15.5 \text{ mL O}_2$$

14.9   A is Li; B is Ga; C is C
(a) Li_2O, Ga_2O_3, CO_2
(b) Li_2O is the most ionic. CO_2 is the most covalent.
(c) CO_2 is the most acidic. Li_2O is the most basic.
(d) Ga_2O_3 is amphoteric and can react with both $H^+(aq)$ and $OH^-(aq)$.

14.10   (a) $Li_2O(s) + H_2O(l) \rightarrow 2\,Li^+(aq) + 2\,OH^-(aq)$
(b) $SO_3(l) + H_2O(l) \rightarrow H^+(aq) + HSO_4^-(aq)$
(c) $Cr_2O_3(s) + 6\,H^+(aq) \rightarrow 2\,Cr^{3+}(aq) + 3\,H_2O(l)$
(d) $Cr_2O_3(s) + 2\,OH^-(aq) + 3\,H_2O(l) \rightarrow 2\,Cr(OH)_4^-(aq)$

14.11   (a) Rb_2O_2 Rb +1, O −1, peroxide   (b) CaO   Ca +2, O −2, oxide
(c) CsO_2 Cs +1, O −1/2, superoxide   (d) SrO_2 Sr +2, O −1, peroxide
(e) CO_2 C +4, O −2, oxide

14.12   (a) $Rb_2O_2(s) + H_2O(l) \rightarrow 2\,Rb^+(aq) + HO_2^-(aq) + OH^-(aq)$
(b) $CaO(s) + H_2O(l) \rightarrow Ca^{2+}(aq) + 2\,OH^-(aq)$
(c) $2\,CsO_2(s) + H_2O(l) \rightarrow O_2(g) + 2\,Cs^+(aq) + HO_2^-(aq) + OH^-(aq)$
(d) $SrO_2(s) + H_2O(l) \rightarrow Sr^{2+}(aq) + HO_2^-(aq) + OH^-(aq)$
(e) $CO_2(g) + H_2O(l) \rightarrow H^+(aq) + HCO_3^-(aq)$

14.13

σ^*_{2p} ___

π^*_{2p} ↑↓   ↑

π_{2p} ↑↓   ↑↓

σ_{2p} ↑↓

σ^*_{2s} ↑↓

σ_{2s} ↑↓

O_2^-

O_2^- is paramagnetic with one unpaired electron.

$$\text{Bond order} = \frac{\left(\begin{array}{c}\text{number of}\\\text{bonding electrons}\end{array}\right) - \left(\begin{array}{c}\text{number of}\\\text{antibonding electrons}\end{array}\right)}{2}$$

$$O_2^- \text{ bond order} = \frac{8-5}{2} = 1.5$$

14.14   H—Ö—Ö—H The electron dot structure indicates a single bond (see text Table 14.2) which is consistent with an O–O bond length of 148 pm.

14.15 $PbS(s) + 4\,H_2O_2(aq) \rightarrow PbSO_4(s) + 4\,H_2O(l)$

14.16   (a) $2\,Li(s) + 2\,H_2O(l) \rightarrow H_2(g) + 2\,Li^+(aq) + 2\,OH^-(aq)$
(b) $Sr(s) + 2\,H_2O(l) \rightarrow H_2(g) + Sr^{2+}(aq) + 2\,OH^-(aq)$
(c) $Br_2(l) + H_2O(l) \rightleftharpoons HOBr(aq) + H^+(aq) + Br^-(aq)$

14.17   mass of H_2O = 5.62 g – 3.10 g = 2.52 g H_2O

$$2.52 \text{ g } H_2O \times \frac{1 \text{ mol } H_2O}{18.02 \text{ g } H_2O} = 0.140 \text{ mol } H_2O$$

$$3.10 \text{ g NiSO}_4 \times \frac{1 \text{ mol NiSO}_4}{154.8 \text{ g NiSO}_4} = 0.0200 \text{ mol NiSO}_4$$

$$\text{number of } H_2O\text{'s in hydrate} = \frac{n_{H_2O}}{n_{NiSO_4}} = \frac{0.140 \text{ mol}}{0.0200 \text{ mol}} = 7$$

Hydrate formula is $NiSO_4 \cdot 7\, H_2O$

14.18   Hydrogen can be stored as a solid in the form of solid interstitial hydrides or in the recently discovered tube-shaped molecules called carbon nanotubes.

14.19 $H_2(g) + 1/2\, O_2(g) \rightarrow H_2O(g)$ $\Delta H° = -242$ kJ

$$\text{mol } H_2 = 1.45 \times 10^6 \text{ L} \times \frac{0.088 \text{ g}}{1 \text{ L}} \times \frac{1 \text{ mol } H_2}{2.016 \text{ g } H_2} = 6.33 \times 10^4 \text{ mol } H_2$$

$$q = 6.33 \times 10^4 \text{ mol } H_2 \times \frac{242 \text{ kJ}}{1 \text{ mol } H_2} = 1.5 \times 10^7 \text{ kJ}$$

$$\text{mass } O_2 = 6.33 \times 10^4 \text{ mol } H_2 \times \frac{0.5 \text{ mol } O_2}{1 \text{ mol } H_2} \times \frac{32.00 \text{ g } O_2}{1 \text{ mol } O_2} \times \frac{1 \text{ kg}}{1000 \text{ g}} = 1.0 \times 10^3 \text{ kg } O_2$$

Understanding Key Concepts

14.20   (a)   (1) covalent   (2) ionic   (3) covalent   (4) interstitial
          (b)   (1) H, +1; other element, –3
                (2) H, –1; other element, +1
                (3) H, +1; other element, –2

14.22   (a) Because of the unpaired electron, the compound is a superoxide. The chemical formula is KO_2.
         (b) Because of the unpaired electron, the compound is paramagnetic and attracted by a magnetic field.
         (c) O_2 has a bond order of 2; O_2^- has a bond order of 1.5. The O–O bond length in O_2^- is longer and the bond energy is smaller than in O_2.
         (d) The solution is basic because of the following reaction that produces OH^-:
         $2\, KO_2(s) + H_2O(l) \rightarrow O_2(g) + 2\, K^+(aq) + HO_2^-(aq) + OH^-(aq)$

14.24   (a) (1) –2, +2;   (2) –2, +1;   (3) –2, +5
         (b) (1) three-dimensional;   (2) molecular;   (3) molecular
         (c) (1) solid;   (2) gas or liquid;   (3) gas or liquid
         (d) (2) hydrogen;   (3) nitrogen

14.26   (a) The ionic hydride (4) has the highest melting point.
         (b) (1), (2), and (3) are covalent hydrides. (1) and (2) can hydrogen bond, (3) cannot. Consequently, (3) has the lowest boiling point.
         (c) (1), water, and (4), the ionic hydride react together to form $H_2(g)$.

Additional Problems
Chemistry of Hydrogen

14.28   Quantitative differences in properties that arise from the differences in the masses of the isotopes are known as isotope effects.
         Examples: H_2 and D_2 have different melting and boiling points.
                       H_2O and D_2O have different dissociation constants.

14.30 $$\frac{\text{mass }^2D - \text{mass }^1H}{\text{mass }^1H} \times 100\% = \frac{2.0141 \text{ amu} - 1.0078 \text{ amu}}{1.0078 \text{ amu}} \times 100\% = 98.85\%$$

$$\frac{\text{mass }^3H - \text{mass }^2H}{\text{mass }^2H} \times 100\% = \frac{3.0160 \text{ amu} - 2.0141 \text{ amu}}{2.0141 \text{ amu}} \times 100\% = 49.74\%$$

        The differences in properties will be larger for H_2O and D_2O rather than for D_2O and T_2O because of the larger relative difference in mass for H and D versus D and T. This is supported by the data in Table 14.1.

14.32   There are 18 kinds of H_2O

$H_2{}^{16}O$	$H_2{}^{17}O$	$H_2{}^{18}O$
$D_2{}^{16}O$	$D_2{}^{17}O$	$D_2{}^{18}O$
$T_2{}^{16}O$	$T_2{}^{17}O$	$T_2{}^{18}O$
$HD^{16}O$	$HD^{17}O$	$HD^{18}O$
$HT^{16}O$	$HT^{17}O$	$HT^{18}O$
$DT^{16}O$	$DT^{17}O$	$DT^{18}O$

14.34   (a) $Zn(s) + 2 H^+(aq) \rightarrow H_2(g) + Zn^{2+}(aq)$
         (b) at 1000°C, $H_2O(g) + C(s) \rightarrow CO(g) + H_2(g)$
         (c) at 1100°C with a Ni catalyst, $H_2O(g) + CH_4(g) \rightarrow CO(g) + 3 H_2(g)$
         (d) There are a number of possibilities. (b) and (c) above are two; electrolysis is another:
             $2 H_2O(l) \rightarrow 2 H_2(g) + O_2(g)$

14.36   The steam-hydrocarbon reforming process is the most important industrial preparation of hydrogen.

$$CH_4(g) + H_2O(g) \xrightarrow[\text{Ni catalyst}]{1100°C} CO(g) + 3 H_2(g)$$

$$CO(g) + H_2O(g) \xrightarrow{400°C} CO_2(g) + H_2(g)$$
$$CO_2(g) + 2\,OH^-(aq) \rightarrow CO_3^{2-}(aq) + H_2O(l)$$

14.38 (a) LiH, 7.95 amu; CaH$_2$, 42.09 amu

$$LiH(s) + H_2O(l) \rightarrow H_2(g) + Li^+(aq) + OH^-(aq)$$
$$CaH_2(s) + 2\,H_2O(l) \rightarrow 2\,H_2(g) + Ca^{2+}(aq) + 2\,OH^-(aq)$$

You obtain $\dfrac{1\text{ mol }H_2}{7.95\text{ g LiH}} = 0.126$ mol H$_2$/g LiH

and $\dfrac{2\text{ mol }H_2}{42.09\text{ g CaH}_2} = 0.0475$ mol H$_2$/g CaH$_2$;   Therefore, LiH gives more H$_2$.

(b) 25°C = 298 K

$$PV = nRT; \quad n_{H_2} = \frac{PV}{RT} = \frac{(150\text{ atm})(100\text{ L})}{\left(0.082\,06\,\dfrac{L\cdot atm}{K\cdot mol}\right)(298K)} = 613.4\text{ mol }H_2$$

mass CaH$_2$ = 613.4 mol H$_2$ x $\dfrac{1\text{ mol CaH}_2}{2\text{ mol }H_2}$ x $\dfrac{42.09\text{ g CaH}_2}{1\text{ mol CaH}_2}$ x $\dfrac{1\text{ kg}}{1000\text{ g}}$ = 12.9 kg CaH$_2$

14.40 (a) MgH$_2$, H$^-$ (b) PH$_3$, covalent   (c) KH, H$^-$ (d) HBr, covalent

14.42 H$_2$S   covalent hydride, gas, weak acid in H$_2$O

NaH   ionic hydride, solid (salt like), reacts with H$_2$O to produce H$_2$

PdH$_x$ metallic (interstitial) hydride, solid, stores hydrogen

14.44 (a) CH$_4$, covalent bonding   (b) NaH, ionic bonding

14.46 (a) H—S̈e—H , bent   (b) H—Äs—H, trigonal pyramidal
$$\qquad\qquad\qquad\qquad\qquad\qquad\quad |$$
$$\qquad\qquad\qquad\qquad\qquad\qquad\quad H$$

(c)   H   , tetrahedral
$$\qquad\qquad | $$
$$\quad H—Si—H$$
$$\qquad\qquad |$$
$$\qquad\qquad H$$

14.48 A nonstoichiometric compound is a compound whose atomic composition cannot be expressed as a ratio of small whole numbers. An example is PdH$_x$. The lack of stoichiometry results from the hydrogen occupying holes in the solid state structure.

14.50 (a) TiH$_2$, 49.90 amu; Assume 1.0 cm^3 of TiH$_2$ which has a mass of 3.9 g.

3.9 g TiH$_2$ x $\dfrac{1\text{ mol TiH}_2}{49.90\text{ g TiH}_2}$ = 0.078 mol TiH$_2$

0.078 mol TiH$_2$ x $\dfrac{2\text{ mol H}}{1\text{ mol TiH}_2}$ = 0.156 mol H

$$0.156 \text{ mol H} \times \frac{1.008 \text{ g H}}{1 \text{ mol H}} = 0.157 \text{ g H}$$

$d_H = 0.16 \text{ g/cm}^3$; the density of H in TiH_2 is about 2.25 times the density of liquid H_2.

(b)

$$PV = nRT; \quad V = \frac{nRT}{P} = \frac{\left(0.16 \text{ g} \times \frac{1 \text{ mol}}{2.016 \text{ g}}\right)\left(0.082 \ 06 \ \frac{L \cdot atm}{K \cdot mol}\right)(273 \text{ K})}{1.00 \text{ atm}} = 1.8 \text{ L } H_2$$

$$1.8 \text{ L} = 1.8 \times 10^3 \text{ mL} = 1.8 \times 10^3 \text{ cm}^3$$

Chemistry of Oxygen

14.52 (a) O_2 is obtained in industry by the fractional distillation of liquid air.
(b) In the laboratory, O_2 is prepared by the thermal decomposition of $KClO_3(s)$.

$$2 \text{ KClO}_3(s) \ \overset{heat}{\underset{MnO_2}{\rightarrow}} \ 2 \text{ KCl}(s) + 3 \text{ O}_2(g)$$

14.54 $$2 \text{ H}_2O_2(aq) \overset{catalyst}{\rightarrow} 2 \text{ H}_2O(l) + O_2(g)$$
H_2O_2, 34.01 amu; $25°C = 298$ K

$$\text{mol } O_2 = 20.4 \text{ g } H_2O_2 \times \frac{1 \text{ mol } H_2O_2}{34.01 \text{ g } H_2O_2} \times \frac{1 \text{ mol } O_2}{2 \text{ mol } H_2O_2} = 0.300 \text{ mol } O_2$$

$$PV = nRT; \quad V = \frac{nRT}{P} = \frac{(0.300 \text{ mol})\left(0.082 \ 06 \ \frac{L \cdot atm}{K \cdot mol}\right)(298 \text{ K})}{1.00 \text{ atm}} = 7.34 \text{ L } O_2$$

14.56 (a) $4 \text{ Li}(s) + O_2(g) \rightarrow 2 \text{ Li}_2O(s)$
(b) $P_4(s) + 5 O_2(g) \rightarrow P_4O_{10}(s)$
(c) $4 \text{ Al}(s) + 3 O_2(g) \rightarrow 2 \text{ Al}_2O_3(s)$
(d) $Si(s) + O_2(g) \rightarrow SiO_2(s)$

14.58 $:\ddot{O}::\ddot{O}:$ The electron dot structure shows an O=O double bond. It also shows all electrons paired. This is not consistent with the fact that O_2 is paramagnetic.

14.60 $Li_2O < BeO < B_2O_3 < CO_2 < N_2O_5$ (see Figure 14.6)

14.62 $N_2O_5 < Al_2O_3 < K_2O < Cs_2O$ (see Figure 14.6)

14.64 (a) CrO_3 (higher Cr oxidation state)   (b) N_2O_5 (higher N oxidation state)
(c) SO_3 (higher S oxidation)

14.66 (a) $Cl_2O_7(l) + H_2O(l) \rightarrow 2 \text{ H}^+(aq) + 2 \text{ ClO}_4^-(aq)$
(b) $K_2O(s) + H_2O(l) \rightarrow 2 \text{ K}^+(aq) + 2 \text{ OH}^-(aq)$
(c) $SO_3(l) + H_2O(l) \rightarrow H^+(aq) + HSO_4^-(aq)$

14.68 (a) $ZnO(s) + 2 H^+(aq) \rightarrow Zn^{2+}(aq) + H_2O(l)$
(b) $ZnO(s) + 2 OH^-(aq) + H_2O(l) \rightarrow Zn(OH)_4^{2-}(aq)$

14.70 A peroxide has oxygen in the –1 oxidation state, for example, H_2O_2. A superoxide has oxygen in the –1/2 oxidation state, for example, KO_2.

14.72 (a) BaO_2 (b) CaO (c) CsO_2 (d) Li_2O (e) Na_2O_2

14.74

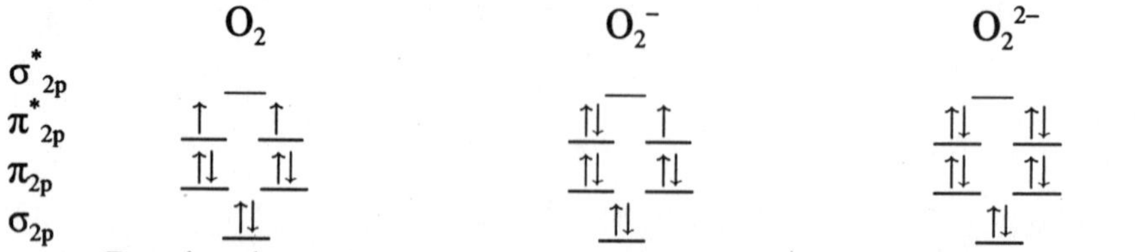

$$O_2 \qquad\qquad O_2^- \qquad\qquad O_2^{2-}$$

σ^*_{2p}

π^*_{2p} ↑ ↑   ↑↓ ↑   ↑↓ ↑↓

π_{2p} ↑↓ ↑↓   ↑↓ ↑↓   ↑↓ ↑↓

σ_{2p} ↑↓   ↑↓   ↑↓

     Bond order = 2   Bond order = 1.5   Bond order = 1

(a) The O–O bond length increases because the bond order decreases. The bond order decreases because of the increased occupancy of antibonding orbitals.
(b) O_2^- has 1 unpaired electron and is paramagnetic. O_2^{2-} has no unpaired electrons and is diamagnetic.

14.76 (a) $H_2O_2(aq) + 2 H^+(aq) + 2 I^-(aq) \rightarrow I_2(aq) + 2 H_2O(l)$
(b) $3 H_2O_2(aq) + 8 H^+(aq) + Cr_2O_7^{2-}(aq) \rightarrow 2 Cr^{3+}(aq) + 3 O_2(g) + 7 H_2O(l)$

14.78

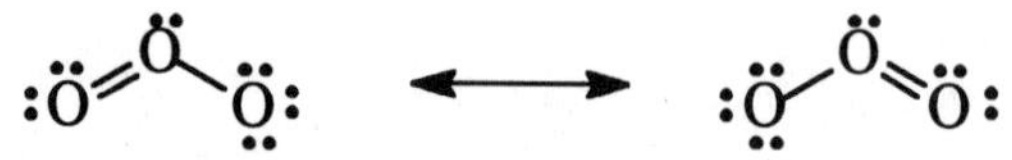

Ozone has two resonance structures, consistent with two equivalent O–O bond lengths.

14.80 $3 O_2(g) \xrightarrow{\text{electric discharge}} 2 O_3(g)$

Chemistry of Water

14.82 (a) $2 F_2(g) + 2 H_2O(l) \rightarrow O_2(g) + 4 HF(aq)$
(b) $Cl_2(g) + H_2O(l) \rightleftharpoons HOCl(aq) + H^+(aq) + Cl^-(aq)$
(c) $I_2(s) + H_2O(l) \rightarrow HOI(aq) + H^+(aq) + I^-(aq)$
(d) $Ba(s) + 2 H_2O(l) \rightarrow H_2(g) + Ba^{2+}(aq) + 2 OH^-(aq)$

14.84 $AlCl_3 \cdot 6 H_2O$

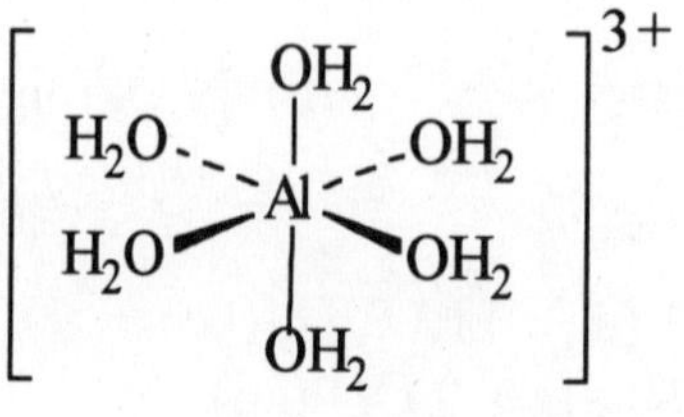

14.86 $CaSO_4 \cdot \frac{1}{2} H_2O$, 145.15 amu; H_2O, 18.02 amu

Assume one mole of $CaSO_4 \cdot \frac{1}{2} H_2O$

$$\text{mass \% } H_2O = \frac{\text{mass } H_2O}{\text{mass hydrate}} \times 100\% = \frac{\frac{1}{2}(18.02 \text{ g})}{145.15 \text{ g}} \times 100\% = 6.21\%$$

14.88 $CaSO_4 \cdot \frac{1}{2} H_2O$, 145.15 amu; H_2O, 18.02 amu

mass of H_2O lost = 3.44 g – 2.90 g = 0.54 g H_2O

$$2.90 \text{ g } CaSO_4 \cdot \frac{1}{2} H_2O \times \frac{1 \text{ mol}}{145.15 \text{ g}} = 0.020 \text{ mol } CaSO_4 \cdot \frac{1}{2} H_2O$$

$$0.54 \text{ g } H_2O \times \frac{1 \text{ mol}}{18.02 \text{ g}} = 0.030 \text{ mol}$$

$$\text{number of } H_2O\text{'s lost} = \frac{0.030 \text{ mol}}{0.020 \text{ mol}} = 1.5 \text{ } H_2O \text{ per } CaSO_4 \cdot \frac{1}{2} H_2O \text{ formed}$$

The mineral gypsum is $CaSO_4 \cdot 2 H_2O$; x = 2

14.90 Convert mi^3 to cm^3; $(1.0 \text{ mi}^3)\left(\frac{1609 \text{ m}}{1 \text{ mi}}\right)^3\left(\frac{100 \text{ cm}}{1 \text{ m}}\right)^3 = 4.2 \times 10^{15} \text{ cm}^3$

mass of sea water = volume x density = $(4.2 \times 10^{15} \text{ cm}^3)(1.025 \text{ g/cm}^3) = 4.3 \times 10^{15}$ g

mass of salts = $(0.035)(4.3 \times 10^{15} \text{ g})\left(\frac{1 \text{ kg}}{1000 \text{ g}}\right) = 1.5 \times 10^{11}$ kg

General Problems

14.92 React H_2O with a reducing agent to produce H_2. Ca or Al could be used.

14.94 Butadiene, C_4H_6, 54.09 amu; 2.7 kg = 2700 g

$$\text{moles } H_2 = 2700 \text{ g } C_4H_6 \times \frac{1 \text{ mol } C_4H_6}{54.09 \text{ g } C_4H_6} \times \frac{2 \text{ mol } H_2}{1 \text{ mol } C_4H_6} = 99.8 \text{ mol } H_2$$

At STP, P = 1.00 atm and T = 273 K

$$PV = nRT; \quad V = \frac{nRT}{P} = \frac{(99.8 \text{ mol})\left(0.082\ 06 \frac{L \cdot atm}{K \cdot mol}\right)(273 \text{ K})}{1.00 \text{ atm}} = 2.2 \times 10^3 \text{ L of } H_2$$

14.96 (a) B_2O_3, diboron trioxide (b) H_2O_2, hydrogen peroxide (c) SrH_2, strontium hydride (d) CsO_2, cesium superoxide (e) $HClO_4$, perchloric acid (f) BaO_2, barium peroxide

14.98 (a) 6; $^{16}O_2$, $^{17}O_2$, $^{18}O_2$, $^{16}O^{17}O$, $^{16}O^{18}O$, $^{17}O^{18}O$

(b) 18

$^{16}O_3$	$^{16}O_2{}^{17}O$	$^{18}O_3$
$^{17}O_2{}^{16}O$	$^{17}O_3$	$^{16}O_2{}^{18}O$
$^{18}O_2{}^{16}O$	$^{18}O_2{}^{17}O$	$^{17}O_2{}^{18}O$
$^{16}O^{17}O^{16}O$	$^{17}O^{18}O^{17}O$	$^{16}O^{17}O^{18}O$
$^{16}O^{18}O^{16}O$	$^{17}O^{16}O^{17}O$	$^{18}O^{16}O^{18}O$
$^{17}O^{18}O^{16}O$	$^{18}O^{16}O^{17}O$	$^{18}O^{17}O^{18}O$

14.100 (a) $2\,H_2(g) + O_2(g) \rightarrow 2\,H_2O(l)$
(b) $O_3(g) + 2\,I^-(aq) + H_2O(l) \rightarrow O_2(g) + I_2(aq) + 2\,OH^-(aq)$
(c) $H_2O_2(aq) + 2\,H^+(aq) + 2\,Br^-(aq) \rightarrow 2\,H_2O(l) + Br_2(aq)$
(d) $2\,Na(l) + H_2(g) \rightarrow 2\,NaH(s)$
(e) $2\,Na(s) + 2\,H_2O(l) \rightarrow H_2(g) + 2\,Na^+(aq) + 2\,OH^-(aq)$

14.102 K is oxidized by water. F_2 is reduced by water.
Cl_2 and Br_2 disproportionate when treated with water.

14.104 (a) $CO(g) + 2\,H_2(g) \rightarrow CH_3OH(l)$
$\Delta H^\circ = \Delta H^\circ_f(CH_3OH) - \Delta H^\circ_f(CO)$
$\Delta H^\circ = (1\ mol)(-238.7\ kJ/mol) - (1\ mol)(-110.5\ kJ/mol) = -128.2\ kJ$
(b) $CO(g) + H_2O(g) \rightarrow CO_2(g) + H_2(g)$
$\Delta H^\circ = \Delta H^\circ_f(CO_2) - [\Delta H^\circ_f(CO) + \Delta H^\circ_f(H_2O)]$
$\Delta H^\circ = (1\ mol)(-393.5\ kJ/mol) - [(1\ mol)(-110.5\ kJ/mol)$
$+ (1\ mol)(-241.8\ kJ/mol)] = -41.2\ kJ$
(c) $2\,KClO_3(s) \rightarrow 2\,KCl(s) + 3\,O_2(g)$
$\Delta H^\circ = 2\,\Delta H^\circ_f(KCl) - 2\,\Delta H^\circ_f(KClO_3)$
$\Delta H^\circ = (2\ mol)(-436.7\ kJ/mol) - (2\ mol)(-397.7\ kJ/mol) = -78.0\ kJ$
(d) $6\,CO_2(g) + 6\,H_2O(l) \rightarrow 6\,O_2(g) + C_6H_{12}O_6(s)$
$\Delta H^\circ = \Delta H^\circ_f(C_6H_{12}O_6) - [6\,\Delta H^\circ_f(CO_2) + 6\,\Delta H^\circ_f(H_2O)]$
$\Delta H^\circ = (1\ mol)(-1260\ kJ/mol) - [(6\ mol)(-393.5\ kJ/mol)$
$+ (6\ mol)(-285.8\ kJ/mol)] = 2816\ kJ$

Multi-Concept Problems

14.106 $2\,O_3(g) \rightarrow 3\,O_2(g) \qquad \Delta H^\circ = -258\ kJ$
$PV = nRT$

$$n_{O_3} = \frac{PV}{RT} = \frac{\left(63.6\ mm\ Hg \times \dfrac{1.00\ atm}{760\ mm\ Hg}\right)(1.000\ L)}{\left(0.082\ 06\ \dfrac{L \cdot atm}{K \cdot mol}\right)(293\ K)} = 3.48 \times 10^{-3}\ mol\ O_3$$

$$q = 3.48 \times 10^{-3}\ mol\ O_3 \times \frac{285\ kJ}{2\ mol\ O_3} \times \frac{1000\ J}{1\ kJ} = 496\ J \text{ are liberated.}$$

14.108 $MH_2(s) + 2\,HCl(aq) \rightarrow 2\,H_2(g) + M^{2+}(aq) + 2\,Cl^-(aq)$

$PV = nRT$

$$n_{H_2} = \frac{PV}{RT} = \frac{\left(750\ \text{mm Hg} \times \dfrac{1.00\ \text{atm}}{760\ \text{mm Hg}}\right)(1.000\ \text{L})}{\left(0.082\ 06\ \dfrac{\text{L} \cdot \text{atm}}{\text{K} \cdot \text{mol}}\right)(293\ \text{K})} = 0.0410\ \text{mol}\ H_2$$

$$0.0410\ \text{mol}\ H_2 \times \frac{1\ \text{mol}\ MH_2}{2\ \text{mol}\ H_2} = 0.0205\ \text{mol}\ MH_2$$

$$\text{molar mass} = \frac{1.84\ \text{g}}{0.0205\ \text{mol}} = 89.8\ \text{g/mol}$$

mass M = 89.8 – mass of 2 H = 89.8 – 2(1.008) = 87.7 g/mol;   M = Sr; SrH_2

14.110 N_2O_5, 108.01 amu

$N_2O_5(g) + H_2O(l) \rightarrow 2\,HNO_3(aq)$

$2\,HNO_3(aq) + Zn(s) \rightarrow H_2(g) + Zn(NO_3)_2(aq)$

$$5.4\ \text{g}\ N_2O_5 \times \frac{1\ \text{mol}\ N_2O_5}{108.01\ \text{g}\ N_2O_5} \times \frac{2\ \text{mol}\ HNO_3}{1\ \text{mol}\ N_2O_5} \times \frac{1\ \text{mol}\ H_2}{2\ \text{mol}\ HNO_3} = 0.050\ \text{mol}\ H_2$$

$$PV = nRT; \quad P_{H_2} = \frac{nRT}{V} = \frac{(0.050\ \text{mol})\left(0.082\ 06\ \dfrac{\text{L} \cdot \text{atm}}{\text{K} \cdot \text{mol}}\right)(298\ \text{K})}{0.500\ \text{L}} = 2.45\ \text{atm}$$

$$P_{H_2} = 2.45\ \text{atm} \times \frac{760\ \text{mm Hg}}{1.00\ \text{atm}} = 1862\ \text{mm Hg}$$

(a) In H there is 0.0156 atom % D.

To get P_{HD} multiply P_{H_2} by the atom % D and then by 2 because H_2 is diatomic.

$$P_{H_2} = (1862\ \text{mm Hg})(0.000156)(2) = 0.58\ \text{mm Hg}$$

(b) $PV = nRT$

$$n_{HD} = \frac{PV}{RT} = \frac{\left(0.58\ \text{mm Hg} \times \dfrac{1.00\ \text{atm}}{760\ \text{mm Hg}}\right)(0.5000\ \text{L})}{\left(0.082\ 06\ \dfrac{\text{L} \cdot \text{atm}}{\text{K} \cdot \text{mol}}\right)(298\ \text{K})} = 1.56 \times 10^{-5}\ \text{mol}\ HD$$

$$1.56 \times 10^{-5}\ \text{mol}\ HD \times \frac{6.022 \times 10^{23}\ \text{HD molecules}}{1\ \text{mol}\ HD} = 9.4 \times 10^{18}\ \text{HD molecules}$$

(c) $P_{D_2} = (1862 \text{ mm Hg})(0.000156)^2 = 4.53 \times 10^{-5} \text{ mm Hg}$

$PV = nRT$

$$n_{D_2} = \frac{PV}{RT} = \frac{\left(4.53 \times 10^{-5} \text{ mm Hg} \times \dfrac{1.00 \text{ atm}}{760 \text{ mm Hg}}\right)(0.5000 \text{ L})}{\left(0.082 \ 06 \ \dfrac{\text{L} \cdot \text{atm}}{\text{K} \cdot \text{mol}}\right)(298 \text{ K})} = 1.22 \times 10^{-9} \text{ mol D}_2$$

$$1.22 \times 10^{-9} \text{ mol D}_2 \times \frac{6.022 \times 10^{23} \text{ D}_2 \text{ molecules}}{1 \text{ mol D}_2} = 7.3 \times 10^{14} \text{ D}_2 \text{ molecules}$$

14.112

$$\ln P_2 = \ln P_1 + \frac{\Delta H_{vap}}{R}\left(\frac{1}{T_1} - \frac{1}{T_2}\right)$$

$$\Delta H_{vap} = \frac{(\ln P_2 - \ln P_1)(R)}{\left(\dfrac{1}{T_1} - \dfrac{1}{T_2}\right)}$$

$P_1 = 75 \text{ mm Hg};$ $T_1 = 89°C = 89 + 273 = 362 \text{ K}$
$P_2 = 319.2 \text{ mm Hg};$ $T_2 = 125°C = 125 + 273 = 398 \text{ K}$

$$\Delta H_{vap} = \frac{[\ln(319.2) - \ln(75.0)](8.314 \times 10^{-3} \text{ kJ/(K} \cdot \text{mol)}}{\left(\dfrac{1}{362 \text{ K}} - \dfrac{1}{398 \text{ K}}\right)} = 48.2 \text{ kJ/mol}$$

Now calculate the normal boiling point.
$P_1 = 75 \text{ mm Hg};$ $T_1 = 89°C = 89 + 273 = 362 \text{ K}$
$P_2 = 760 \text{ mm Hg};$ $T_2 = ?$

$$\ln P_2 = \ln P_1 + \frac{\Delta H_{vap}}{R}\left(\frac{1}{T_1} - \frac{1}{T_2}\right)$$

$$(\ln P_2 - \ln P_1)\left(\frac{R}{\Delta H_{vap}}\right) = \frac{1}{T_1} - \frac{1}{T_2}$$

Solve for T_2 (the boiling point for H_2O_2 at 760 mm Hg).

$$\frac{1}{T_1} - (\ln P_2 - \ln P_1)\left(\frac{R}{\Delta H_{vap}}\right) = \frac{1}{T_2}$$

$$\frac{1}{362 \text{ K}} - [\ln(760) - \ln(75.0)]\left(\frac{8.314 \times 10^{-3} \text{ kJ/(K} \cdot \text{mol)}}{48.2 \text{ kJ/mol}}\right) = \frac{1}{T_2} = 0.002363/\text{K}$$

$T_2 = 1/0.002363/\text{K} = 423 \text{ K} = 423 - 273 = 150°C$ (boiling point for H_2O_2)
The calculated boiling point is the same as the one listed in section 14.11.

Aqueous Equilibria: Acids and Bases

15.1　(a) $H_2SO_4(aq) + H_2O(l) \rightleftharpoons H_3O^+(aq) + HSO_4^-(aq)$
 conjugate base

(b) $HSO_4^-(aq) + H_2O(l) \rightleftharpoons H_3O^+(aq) + SO_4^{2-}(aq)$
 conjugate base

(c) $H_3O^+(aq) + H_2O(l) \rightleftharpoons H_3O^+(aq) + H_2O(l)$
 conjugate base

(d) $NH_4^+(aq) + H_2O(l) \rightleftharpoons H_3O^+(aq) + NH_3(aq)$
 conjugate base

15.2　(a) $HCO_3^-(aq) + H_2O(l) \rightleftharpoons H_2CO_3(aq) + OH^-(aq)$
 conjugate acid

(b) $CO_3^{2-}(aq) + H_2O(l) \rightleftharpoons HCO_3^-(aq) + OH^-(aq)$
$\phantom{(b) CO_3^{2-}(aq) + H_2O(l) \rightleftharpoons }$ conjugate acid

(c) $OH^-(aq) + H_2O(l) \rightleftharpoons H_2O(l)(aq) + OH^-(aq)$
 conjugate acid

(d) $H_2PO_4^-(aq) + H_2O(l) \rightleftharpoons H_3PO_4(aq) + OH^-(aq)$
 conjugate acid

15.3　$HCl(aq) + NH_3(aq) \rightleftharpoons NH_4^+(aq) + Cl^-(aq)$
acid$$base$$acid$$base

conjugate acid-base pairs

15.4　(a) $HF(aq) + NO_3^-(aq) \rightleftharpoons HNO_3(aq) + F^-(aq)$
HNO_3 is a stronger acid than HF, and F^- is a stronger base than NO_3^- (see Table 15.1).
Because proton transfer occurs from the stronger acid to the stronger base, the reaction
proceeds from right to left.

(b) $NH_4^+(aq) + CO_3^{2-}(aq) \rightleftharpoons HCO_3^-(aq) + NH_3(aq)$
NH_4^+ is a stronger acid than HCO_3^-, and CO_3^{2-} is a stronger base than NH_3 (see Table
15.1). Because proton transfer occurs from the stronger acid to the stronger base, the
reaction proceeds from left to right.

15.5　(a) Both HX and HY have the same initial concentration. HY is more dissociated than
HX. Therefore, HY is the stronger acid.
(b) The conjugate base (X^-) of the weaker acid (HX) is the stronger base.
(c) $HX + Y^- \rightleftharpoons HY + X^-$
Proton transfer occurs from the stronger acid to the stronger base. The reaction proceeds
to the left.

15.6 $[H_3O^+] = \dfrac{K_w}{[OH^-]} = \dfrac{1.0 \times 10^{-14}}{5.0 \times 10^{-6}} = 2.0 \times 10^{-9} \text{ M}$

Because $[OH^-] > [H_3O^+]$, the solution is basic.

15.7 $K_w = [H_3O^+][OH^-]$;   In a neutral solution, $[H_3O^+] = [OH^-]$

At 50°C, $[H_3O^+] = [OH^-] = \sqrt{K_w} = \sqrt{5.5 \times 10^{-14}} = 2.3 \times 10^{-7} \text{ M}$

15.8   (a) $[H_3O^+] = \dfrac{K_w}{[OH^-]} = \dfrac{1.0 \times 10^{-14}}{1.58 \times 10^{-6}} = 6.3 \times 10^{-9} \text{ M}$

pH $= -\log[H_3O^+] = -\log(6.3 \times 10^{-9}) = 8.20$

(b) pH $= -\log[H_3O^+] = -\log(6.0 \times 10^{-5}) = 4.22$

15.9   (a) $[H_3O^+] = 10^{-pH} = 10^{-7.40} = 4.0 \times 10^{-8} \text{ M}$

$[OH^-] = \dfrac{K_w}{[H_3O^+]} = \dfrac{1.0 \times 10^{-14}}{4.0 \times 10^{-8}} = 2.5 \times 10^{-7} \text{ M}$

(b) $[H_3O^+] = 10^{-pH} = 10^{-2.8} = 2 \times 10^{-3} \text{ M}$

$[OH^-] = \dfrac{K_w}{[H_3O^+]} = \dfrac{1.0 \times 10^{-14}}{2 \times 10^{-3}} = 5 \times 10^{-12} \text{ M}$

15.10   (a) Because $HClO_4$ is a strong acid, $[H_3O^+] = 0.050$ M.

pH $= -\log[H_3O^+] = -\log(0.050) = 1.30$

(b) Because HCl is a strong acid, $[H_3O^+] = 6.0$ M.

pH $= -\log[H_3O^+] = -\log(6.0) = -0.78$

(c) Because KOH is a strong base, $[OH^-] = 4.0$ M.

$[H_3O^+] = \dfrac{K_w}{[OH^-]} = \dfrac{1.0 \times 10^{-14}}{4.0} = 2.5 \times 10^{-15} \text{ M}$

pH $= -\log[H_3O^+] = -\log(2.5 \times 10^{-15}) = 14.60$

(d) Because $Ba(OH)_2$ is a strong base, $[OH^-] = 2(0.010 \text{ M}) = 0.020$ M.

$[H_3O^+] = \dfrac{K_w}{[OH^-]} = \dfrac{1.0 \times 10^{-14}}{0.020} = 5.0 \times 10^{-13} \text{ M}$

pH $= -\log[H_3O^+] = -\log(5.0 \times 10^{-13}) = 12.30$

15.11 $BaO(s) + H_2O(l) \rightarrow Ba(OH)_2(aq)$

BaO, 153.33 amu

$0.25 \text{ g BaO} \times \dfrac{1 \text{ mol BaO}}{153.33 \text{ g BaO}} \times \dfrac{1 \text{ mol } Ba(OH)_2}{1 \text{ mol BaO}} \times \dfrac{2 \text{ mol } OH^-}{1 \text{ mol } Ba(OH)_2} = 3.26 \times 10^{-3} \text{ mol } OH^-$

$[OH^-] = \dfrac{3.26 \times 10^{-3} \text{ mol } OH^-}{0.500 \text{ L}} = 6.52 \times 10^{-3} \text{ M}$

$$[H_3O^+] = \frac{K_w}{[OH^-]} = \frac{1.0 \times 10^{-14}}{6.52 \times 10^{-3}} = 1.53 \times 10^{-12} \text{ M}$$

$$pH = -\log[H_3O^+] = -\log(1.53 \times 10^{-12}) = 11.81$$

15.12

$$HOCl(aq) + H_2O(l) \rightleftarrows H_3O^+(aq) + OCl^-(aq)$$

initial (M)	0.10	~0	0
change (M)	−x	+x	+x
equil (M)	0.10 − x	x	x

$x = [H_3O^+] = 10^{-pH} = 10^{-4.23} = 5.9 \times 10^{-5}$ M

$[OCl^-] = x = 5.9 \times 10^{-5}$ M; $[HOCl] = 0.10 - x = (0.10 - 5.9 \times 10^{-5})$ M

$$K_a = \frac{[H_3O^+][OCl^-]}{[HOCl]} = \frac{(5.9 \times 10^{-5})(5.9 \times 10^{-5})}{(0.10 - 5.9 \times 10^{-5})} = 3.5 \times 10^{-8}$$

This value of K_a agrees with the value in Table 15.2.

15.13 (a) HZ is completely dissociated. HX and HY are at the same concentration and HX is more dissociated than HY. The strongest acid is HZ, the weakest is HY.

K_a (HY) $<$ K_a (HX) $<$ K_a (HZ)

(b) HZ

(c) HY has the highest pH; HX has the lowest pH (highest $[H_3O^+]$).

15.14 (a)

$$CH_3CO_2H(aq) + H_2O(l) \rightleftarrows H_3O^+(aq) + CH_3CO_2^-(aq)$$

initial (M)	1.00	~0	0
change (M)	−x	+x	+x
equil (M)	1.00 − x	x	x

$$K_a = \frac{[H_3O^+][CH_3CO_2^-]}{[CH_3CO_2H]} = 1.8 \times 10^{-5} = \frac{x^2}{1.00 - x} \approx \frac{x^2}{1.00}$$

Solve for x. $x = [H_3O^+] = 4.2 \times 10^{-3}$ M

$pH = -\log[H_3O^+] = -\log(4.2 \times 10^{-3}) = 2.38$

$[CH_3CO_2^-] = x = 4.2 \times 10^{-3}$ M; $[CH_3CO_2H] = 1.00 - x = 1.00$ M

$$[OH^-] = \frac{K_w}{[H_3O^+]} = \frac{1.0 \times 10^{-14}}{4.2 \times 10^{-3}} = 2.4 \times 10^{-12} \text{ M}$$

(b)

$$CH_3CO_2H(aq) + H_2O(l) \rightleftarrows H_3O^+(aq) + CH_3CO_2^-(aq)$$

initial (M)	0.0100	~0	0
change (M)	−x	+x	+x
equil (M)	0.0100 − x	x	x

$$K_a = \frac{[H_3O^+][CH_3CO_2^-]}{[CH_3CO_2H]} = 1.8 \times 10^{-5} = \frac{x^2}{0.0100 - x}$$

$x^2 + (1.8 \times 10^{-5})x - (1.8 \times 10^{-7}) = 0$

Use the quadratic formula to solve for x.

$$x = \frac{-(1.8 \times 10^{-5}) \pm \sqrt{(1.8 \times 10^{-5})^2 - 4(-1.8 \times 10^{-7})}}{2(1)} = \frac{(-1.8 \times 10^{-5}) \pm (8.5 \times 10^{-4})}{2}$$

$x = 4.2 \times 10^{-4}$ and -4.3×10^{-4}

Of the two solutions for x, only the positive value of x has physical meaning because x is the $[H_3O^+]$.

$x = [H_3O^+] = 4.2 \times 10^{-4}$ M

$pH = -\log[H_3O^+] = -\log(4.2 \times 10^{-4}) = 3.38$

$[CH_3CO_2^-] = x = 4.2 \times 10^{-4}$ M

$[CH_3CO_2H] = 0.0100 - x = 0.0100 - (4.2 \times 10^{-4}) = 0.0096$ M

$$[OH^-] = \frac{K_w}{[H_3O^+]} = \frac{1.0 \times 10^{-14}}{4.2 \times 10^{-4}} = 2.4 \times 10^{-11} \text{ M}$$

15.15 $C_6H_8O_6$, 176.13 amu; 250 mg = 0.250 g; 250 mL = 0.250 L

$$[C_6H_8O_6] = \frac{\left(0.250 \text{ g} \times \dfrac{1 \text{ mol}}{176.13 \text{ g}}\right)}{0.250 \text{ L}} = 5.68 \times 10^{-3} \text{ M}$$

$$\begin{array}{lccc}
 & C_6H_8O_6(aq) + H_2O(l) & \rightleftharpoons & H_3O^+(aq) + C_6H_7O_6^-(aq) \\
\text{initial (M)} & 5.68 \times 10^{-3} & & \sim 0 \qquad\qquad 0 \\
\text{change (M)} & -x & & +x \qquad\qquad +x \\
\text{equil (M)} & (5.68 \times 10^{-3}) - x & & x \qquad\qquad\; x
\end{array}$$

$$K_a = \frac{[H_3O^+][C_6H_7O_6^-]}{[C_6H_8O_6]} = 8.0 \times 10^{-5} = \frac{x^2}{(5.68 \times 10^{-3}) - x}$$

$x^2 + (8.0 \times 10^{-5})x - (4.54 \times 10^{-7}) = 0$

Use the quadratic formula to solve for x.

$$x = \frac{-(8.0 \times 10^{-5}) \pm \sqrt{(8.0 \times 10^{-5})^2 - (4)(-4.54 \times 10^{-7})}}{2(1)} = \frac{(-8.0 \times 10^{-5}) \pm 0.001\ 35}{2}$$

$x = 6.35 \times 10^{-4}$ and -7.15×10^{-4}

Of the two solutions for x, only the positive value of x has physical meaning because x is the $[H_3O^+]$.

$x = [H_3O^+] = 6.35 \times 10^{-4}$ M

$pH = -\log[H_3O^+] = -\log(6.35 \times 10^{-4}) = 3.20$

15.16 (a) From Example 15.9:

$[H_3O^+] = [HF]_{diss} = 4.0 \times 10^{-3}$ M

$$\% \text{ dissociation} = \frac{[HF]_{diss}}{[HF]_{initial}} \times 100\% = \frac{4.0 \times 10^{-3} \text{ M}}{0.050 \text{ M}} \times 100\% = 8.0\% \text{ dissociation}$$

(b)

$$\begin{array}{lccc}
 & HF(aq) + H_2O(l) & \rightleftharpoons & H_3O^+(aq) + F^-(aq) \\
\text{initial (M)} & 0.50 & & \sim 0 \qquad\qquad 0 \\
\text{change (M)} & -x & & +x \qquad\qquad +x \\
\text{equil (M)} & 0.50 - x & & x \qquad\qquad\; x
\end{array}$$

$$K_a = \frac{[H_3O^+][F^-]}{[HF]} = 3.5 \times 10^{-4} = \frac{x^2}{0.50 - x}$$

$x^2 + (3.5 \times 10^{-4})x - (1.75 \times 10^{-4}) = 0$

Use the quadratic formula to solve for x.

$$x = \frac{-(3.5 \times 10^{-4}) \pm \sqrt{(3.5 \times 10^{-4})^2 - 4(1)(-1.75 \times 10^{-4})}}{2(1)} = \frac{(-3.5 \times 10^{-4}) \pm 0.0265}{2}$$

x = 0.0131 and −0.0134

Of the two solutions for x, only the positive value of x has physical meaning, because x is the $[H_3O^+]$.

$[H_3O^+] = [HF]_{diss} = 0.013$ M

% dissociation $= \dfrac{[HF]_{diss}}{[HF]_{initial}} \times 100\% = \dfrac{0.013\ M}{0.50\ M} \times 100\% = 2.6\%$ dissociation

15.17

$$H_2SO_3(aq) + H_2O(l) \rightleftharpoons H_3O^+(aq) + HSO_3^-(aq)$$

initial (M)	0.10	~0	0
change (M)	−x	+x	+x
equil (M)	0.10 − x	x	x

$$K_{a1} = \frac{[H_3O^+][HSO_3^-]}{[H_2SO_3]} = 1.5 \times 10^{-2} = \frac{x^2}{0.10 - x}$$

$x^2 + 0.015x - 0.0015 = 0$

Use the quadratic formula to solve for x.

$$x = \frac{-(0.015) \pm \sqrt{(0.015)^2 - (4)(-0.0015)}}{2(1)} = \frac{-0.015 \pm 0.079}{2}$$

x = 0.032 and −0.047

Of the two solutions for x, only the positive value of x has physical meaning since x is the $[H_3O^+]$.

x = $[H_3O^+]$ = $[HSO_3^-]$ = 0.032 M; $[H_2SO_3]$ = 0.10 − x = 0.10 − 0.032 = 0.07 M

The second dissociation of H_2SO_3 produces a negligible amount of H_3O^+ compared with that from the first dissociation.

$$HSO_3^-(aq) + H_2O(l) \rightleftharpoons H_3O^+(aq) + SO_3^{2-}(aq)$$

$$K_{a2} = \frac{[H_3O^+][SO_3^{2-}]}{[HSO_3^-]} = 6.3 \times 10^{-8} = \frac{(0.032)[SO_3^{2-}]}{(0.032)}$$

$[SO_3^{2-}] = K_{a2} = 6.3 \times 10^{-8}$ M

$$[OH^-] = \frac{K_w}{[H_3O^+]} = \frac{1.0 \times 10^{-14}}{0.032} = 3.1 \times 10^{-13}\ M$$

pH = −log$[H_3O^+]$ = −log(0.032) = 1.49

15.18   From the complete dissociation of the first proton, $[H_3O^+] = [HSO_4^-] = 0.50$ M.
For the dissociation of the second proton, the following equilibrium must be considered:

$$HSO_4^-(aq) \;+\; H_2O(l) \;\rightleftharpoons\; H_3O^+(aq) \;+\; SO_4^{2-}(aq)$$

initial (M)	0.50	0.50	0
change (M)	$-x$	$+x$	$+x$
equil (M)	$0.50 - x$	$0.50 + x$	x

$$K_{a2} = \frac{[H_3O^+][SO_4^{2-}]}{[HSO_4^-]} = 1.2 \times 10^{-2} = \frac{(0.50 + x)(x)}{0.50 - x}$$

$x^2 + 0.512x - 0.0060 = 0$

Use the quadratic formula to solve for x.

$$x = \frac{-(0.512) \pm \sqrt{(0.512)^2 - 4(1)(-0.0060)}}{2(1)} = \frac{-0.512 \pm 0.535}{2}$$

$x = 0.011$ and -0.524

Of the two solutions for x, only the positive value of x has physical meaning, since x is the $[SO_4^{2-}]$.

$[H_2SO_4] = 0$ M; $[HSO_4^-] = 0.50 - x = 0.49$ M; $[SO_4^{2-}] = x = 0.011$ M

$[H_3O^+] = 0.50 + x = 0.51$ M

pH $= -\log[H_3O^+] = -\log(0.51) = 0.29$

$$[OH^-] = \frac{K_w}{[H_3O^+]} = \frac{1.0 \times 10^{-14}}{0.51} = 2.0 \times 10^{-14} \text{ M}$$

15.19

$$NH_3(aq) \;+\; H_2O(l) \;\rightleftharpoons\; NH_4^+(aq) \;+\; OH^-(aq)$$

initial (M)	0.40	0	~0
change (M)	$-x$	$+x$	$+x$
equil (M)	$0.40 - x$	x	x

$$K_b = \frac{[NH_4^+][OH^-]}{[NH_3]} = 1.8 \times 10^{-5} = \frac{x^2}{0.40 - x} \approx \frac{x^2}{0.40}$$

Solve for x. $x = [OH^-] = 2.7 \times 10^{-3}$ M

$[NH_4^+] = x = 2.7 \times 10^{-3}$ M; $[NH_3] = 0.40 - x = 0.40$ M

$$[H_3O^+] = \frac{K_w}{[OH^-]} = \frac{1.0 \times 10^{-14}}{2.7 \times 10^{-3}} = 3.7 \times 10^{-12} \text{ M}$$

pH $= -\log[H_3O^+] = -\log(3.7 \times 10^{-12}) = 11.43$

15.20 $C_{21}H_{22}N_2O_2$, 334.42 amu; 16 mg = 0.016 g

$$\text{molarity} = \frac{\left(0.016 \text{ g} \times \dfrac{1 \text{ mol}}{334.42 \text{ g}}\right)}{0.100 \text{ L}} = 4.8 \times 10^{-4} \text{ M}$$

$$C_{21}H_{22}N_2O_2(aq) \;+\; H_2O(l) \;\rightleftharpoons\; C_{21}H_{23}N_2O_2^+(aq) \;+\; OH^-(aq)$$

initial (M)	4.8×10^{-4}	0	~0
change (M)	$-x$	$+x$	$+x$
equil (M)	$(4.8 \times 10^{-4}) - x$	x	x

$$K_b = \frac{[C_{21}H_{23}N_2O_2^+][OH^-]}{[C_{21}H_{22}N_2O_2]} = 1.8 \times 10^{-6} = \frac{x^2}{(4.8 \times 10^{-4}) - x}$$

$$x^2 + (1.8 \times 10^{-6})x - (8.6 \times 10^{-10}) = 0$$

Use the quadratic formula to solve for x.

$$x = \frac{-(1.8 \times 10^{-6}) \pm \sqrt{(1.8 \times 10^{-6})^2 - (4)(-8.6 \times 10^{-10})}}{2(1)} = \frac{(-1.8 \times 10^{-6}) \pm (5.87 \times 10^{-5})}{2}$$

$x = 2.84 \times 10^{-5}$ and -3.02×10^{-5}

Of the two solutions for x, only the positive value of x has physical meaning, because x is the $[OH^-]$.

$[OH^-] = 2.84 \times 10^{-5} \ M$

$$[H_3O^+] = \frac{K_w}{[OH^-]} = \frac{1.0 \times 10^{-14}}{2.84 \times 10^{-5}} = 3.52 \times 10^{-10} \ M$$

$pH = -\log[H_3O^+] = -\log(3.52 \times 10^{-10}) = 9.45$

15.21　(a) $K_a = \dfrac{K_w}{K_b \text{ for } C_5H_{11}N} = \dfrac{1.0 \times 10^{-14}}{1.3 \times 10^{-3}} = 7.7 \times 10^{-12}$

(b) $K_b = \dfrac{K_w}{K_a \text{ for } HOCl} = \dfrac{1.0 \times 10^{-14}}{3.5 \times 10^{-8}} = 2.9 \times 10^{-7}$

15.22　(a) 0.25 M NH_4Br

NH_4^+ is an acidic cation. Br^- is a neutral anion. The salt solution is acidic.

For NH_4^+, $K_a = \dfrac{K_w}{K_b \text{ for } NH_3} = \dfrac{1.0 \times 10^{-14}}{1.8 \times 10^{-5}} = 5.6 \times 10^{-10}$

	$NH_4^+(aq)$	$+ \ H_2O(l)$	$\rightleftharpoons$	$H_3O^+(aq)$	$+ \ NH_3(aq)$
initial (M)	0.25			~0	0
change (M)	−x			+x	+x
equil (M)	0.25 − x			x	x

$$K_a = \frac{[H_3O^+][NH_3]}{[NH_4^+]} = 5.6 \times 10^{-10} = \frac{x^2}{0.25 - x} \approx \frac{x^2}{0.25}$$

Solve for x. $x = [H_3O^+] = 1.2 \times 10^{-5} \ M$

$pH = -\log[H_3O^+] = -\log(1.2 \times 10^{-5}) = 4.92$

(b) 0.40 M $ZnCl_2$

Zn^{2+} is an acidic cation. Cl^- is a neutral anion. The salt solution is acidic.

	$Zn(H_2O)_6^{2+}(aq)$	$+ \ H_2O(l)$	$\rightleftharpoons$	$H_3O^+(aq)$	$+ \ Zn(H_2O)_5(OH)^+(aq)$
initial (M)	0.40			~0	0
change (M)	−x			+x	+x
equil(M)	0.40 − x			x	x

$$K_a = \frac{[H_3O^+][Zn(H_2O)_5(OH)^+]}{[Zn(H_2O)_6^{2+}]} = 2.5 \times 10^{-10} = \frac{x^2}{0.40 - x} \approx \frac{x^2}{0.40}$$

Solve for x. $x = [H_3O^+] = 1.0 \times 10^{-5}$ M

$pH = -\log[H_3O^+] = -\log(1.0 \times 10^{-5}) = 5.00$

15.23   For NO_2^-, $K_b = \dfrac{K_w}{K_a \text{ for } HNO_2} = \dfrac{1.0 \times 10^{-14}}{4.6 \times 10^{-4}} = 2.2 \times 10^{-11}$

	$NO_2^-(aq)$	$+$	$H_2O(l)$	$\rightleftharpoons$	$HNO_2(aq)$	$+$	$OH^-(aq)$
initial (M)	0.20				0		~0
change (M)	$-x$				$+x$		$+x$
equil (M)	$0.20 - x$				x		x

$$K_b = \frac{[HNO_2][OH^-]}{[NO_2^-]} = 2.2 \times 10^{-11} = \frac{x^2}{0.20 - x} \approx \frac{x^2}{0.20}$$

Solve for x. $x = [OH^-] = 2.1 \times 10^{-6}$ M

$$[H_3O^+] = \frac{K_w}{[OH^-]} = \frac{1.0 \times 10^{-14}}{2.1 \times 10^{-6}} = 4.8 \times 10^{-9} \text{ M}$$

$pH = -\log[H_3O^+] = -\log(4.8 \times 10^{-9}) = 8.32$

15.24   For NH_4^+, $K_a = \dfrac{K_w}{K_b \text{ for } NH_3} = \dfrac{1.0 \times 10^{-14}}{1.8 \times 10^{-5}} = 5.6 \times 10^{-10}$

For CN^-, $K_b = \dfrac{K_w}{K_a \text{ for } HCN} = \dfrac{1.0 \times 10^{-14}}{4.9 \times 10^{-10}} = 2.0 \times 10^{-5}$

Because $K_b > K_a$, the solution is basic.

15.25   (a) KBr: K^+, neutral cation; Br^-, neutral anion; solution is neutral

(b) $NaNO_2$: Na^+, neutral cation; NO_2^-, basic anion; solution is basic

(c) NH_4Br: NH_4^+, acidic cation; Br^-, neutral anion; solution is acidic

(d) $ZnCl_2$: Zn^{2+}, acidic cation; Cl^-, neutral anion; solution is acidic

(e) NH_4F

For NH_4^+, $K_a = \dfrac{K_w}{K_b \text{ for } NH_3} = \dfrac{1.0 \times 10^{-14}}{1.8 \times 10^{-5}} = 5.6 \times 10^{-10}$

For F^-, $K_b = \dfrac{K_w}{K_a \text{ for } HF} = \dfrac{1.0 \times 10^{-14}}{3.5 \times 10^{-4}} = 2.9 \times 10^{-11}$

Because $K_a > K_b$, the solution is acidic.

15.26   (a) H_2Se is a stronger acid than H_2S because Se is below S in the 6A group and the H–Se bond is weaker than the H–S bond.
(b) HI is a stronger acid than H_2Te because I is to the right of Te in the same row of the periodic table, I is more electronegative than Te, and the H–I bond is more polar.
(c) HNO_3 is a stronger acid than HNO_2 because acid strength increases with increasing oxidation number of N. The oxidation number for N is +5 in HNO_3 and +3 in HNO_2.
(d) H_2SO_3 is a stronger acid than H_2SeO_3 because acid strength increases with increasing electronegativity of the central atom. S is more electronegative than Se.

15.27   (a) Lewis acid, $AlCl_3$; Lewis base, Cl^- (b) Lewis acid, Ag^+; Lewis base, NH_3
(c) Lewis acid, SO_2; Lewis base, OH^- (d) Lewis acid, Cr^{3+}; Lewis base, H_2O

15.28

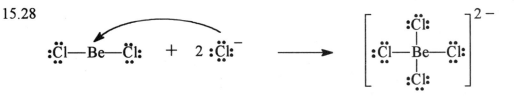

15.29   Lewis acids include not only H^+ but also other cations and neutral molecules having vacant valence orbitals that can accept a share in a pair of electrons donated by a Lewis base. The O^{2-} from CaO is the Lewis base and SO_2 is the Lewis acid.

$$:\ddot{O}-\overset{\cdot\cdot}{\underset{\cdot\cdot}{S}}-\ddot{O}: \; + \; :\ddot{O}:^{2-} \longrightarrow \; :SO_3^{2-}$$

15.30 NO_2, 46.01 amu

$$mol\ NO_2 = 5.47\ mg \times \frac{1.00 \times 10^{-3}\ g}{1\ mg} \times \frac{1\ mol\ NO_2}{46.01\ g\ NO_2} = 1.19 \times 10^{-4}\ mol\ NO_2$$

$$3\ NO_2(g) \; + \; H_2O(l) \; \rightarrow \; 2\ HNO_3(aq) \; + \; NO(g)$$

$$mol\ HNO_3 = 1.19 \times 10^{-4}\ mol\ NO_2 \times \frac{2\ mol\ HNO_3}{3\ mol\ NO_2} = 7.93 \times 10^{-5}\ mol\ HNO_3$$

$$[HNO_3] = \frac{7.93 \times 10^{-5}\ mol\ HNO_3}{1.00\ L} = 7.93 \times 10^{-5}\ M$$

HNO_3 is a strong acid and is completely dissociated therefore $[H_3O^+] = 7.93 \times 10^{-5}\ M$
$pH = -log[H_3O^+] = -log(7.93 \times 10^{-5}\ M) = 4.101$

Understanding Key Concepts

15.32   (a) X^-, Y^-, Z^- (b) HX < HZ < HY   (c) HY   (d) HX   (e) (2/10) x 100% = 20%

15.34   For H_2SO_4, there is complete dissociation of only the first H^+. At equilibrium, there should be H_3O^+, HSO_4^-, and a small amount of SO_4^{2-}. This is best represented by (b).

15.36 (a) $Y^- < Z^- < X^-$

(b) The weakest base, Y^-, has the strongest conjugate acid.

(c) The numbers of HA molecules and OH^- ions are equal because the reaction of A^- with water has a 1:1 stoichiometry: $A^- + H_2O \rightleftharpoons HA + OH^-$

15.38 (a) Brønsted-Lowry acids: NH_4^+, $H_2PO_4^-$;

Brønsted-Lowry bases: SO_3^{2-}, OCl^-, $H_2PO_4^-$

(b) Lewis acids: Fe^{3+}, BCl_3

Lewis bases: SO_3^{2-}, OCl^-, $H_2PO_4^-$

Additional Problems
Acid–Base Concepts

15.40 NH_3, CN^-, and NO_2^-

15.42 (a) SO_4^{2-} (b) HSO_3^- (c) HPO_4^{2-} (d) NH_3 (e) OH^- (f) NH_2^-

15.44 (a) $CH_3CO_2H(aq) + NH_3(aq) \rightleftharpoons NH_4^+(aq) + CH_3CO_2^-(aq)$

      acid   base ——— acid   base

(b) $CO_3^{2-}(aq) + H_3O^+(aq) \rightleftharpoons H_2O(l) + HCO_3^-(aq)$

      base   acid ——— base   acid

(c) $HSO_3^-(aq) + H_2O(l) \rightleftharpoons H_3O^+(aq) + SO_3^{2-}(aq)$

      acid   base ——— acid   base

(d) $HSO_3^-(aq) + H_2O(l) \rightleftharpoons H_2SO_3(aq) + OH^-(aq)$

      base   acid   acid   base

15.46 From data in Table 15.1:   Strong acids: HNO_3 and H_2SO_4;   Strong bases: H^- and O^{2-}

15.48 (a) left, HCO_3^- is the stronger base   (b) left, F^- is the stronger base

(c) right, NH_3 is the stronger base   (d) right, CN^- is the stronger base

Dissociation of Water; pH

15.50 If $[H_3O^+] > 1.0 \times 10^{-7}$ M, solution is acidic.

If $[H_3O^+] < 1.0 \times 10^{-7}$ M, solution is basic.

If $[H_3O^+] = [OH^-] = 1.0 \times 10^{-7}$ M, solution is neutral.

If $[OH^-] > 1.0 \times 10^{-7}$ M, solution is basic

If $[OH^-] < 1.0 \times 10^{-7}$ M, solution is acidic.

(a) $[OH^-] = \dfrac{K_w}{[H_3O^+]} = \dfrac{1.0 \times 10^{-14}}{3.4 \times 10^{-9}} = 2.9 \times 10^{-6}$ M, basic

(b) $[H_3O^+] = \dfrac{K_w}{[OH^-]} = \dfrac{1.0 \times 10^{-14}}{0.010} = 1.0 \times 10^{-12}$ M, basic

(c) $[H_3O^+] = \dfrac{K_w}{[OH^-]} = \dfrac{1.0 \times 10^{-14}}{1.0 \times 10^{-10}} = 1.0 \times 10^{-4}$ M, acidic

(d) $[OH^-] = \dfrac{K_w}{[H_3O^+]} = \dfrac{1.0 \times 10^{-14}}{1.0 \times 10^{-7}} = 1.0 \times 10^{-7}$ M, neutral

(e) $[OH^-] = \dfrac{K_w}{[H_3O^+]} = \dfrac{1.0 \times 10^{-14}}{8.6 \times 10^{-5}} = 1.2 \times 10^{-10}$ M, acidic

15.52 (a) $pH = -\log[H_3O^+] = -\log(2.0 \times 10^{-5}) = 4.70$

(b) $[H_3O^+] = \dfrac{K_w}{[OH^-]} = \dfrac{1.0 \times 10^{-14}}{4 \times 10^{-3}} = 2.5 \times 10^{-12}$ M

$pH = -\log[H_3O^+] = -\log(2.5 \times 10^{-12}) = 11.6$

(c) $pH = -\log[H_3O^+] = -\log(3.56 \times 10^{-9}) = 8.449$

(d) $pH = -\log[H_3O^+] = -\log(10^{-3}) = 3$

(e) $[H_3O^+] = \dfrac{K_w}{[OH^-]} = \dfrac{1.0 \times 10^{-14}}{12} = 8.3 \times 10^{-16}$ M

$pH = -\log[H_3O^+] = -\log(8.3 \times 10^{-16}) = 15.08$

15.54 $[H_3O^+] = 10^{-pH}$;   (a) 8×10^{-5} M   (b) 1.5×10^{-11} M   (c) 1.0 M
     (d) 5.6×10^{-15} M   (e) 10 M   (f) 5.78×10^{-6} M

15.56 $\Delta pH = \log(\Delta[H_3O^+])$;   (a) $\Delta pH = \log(1000) = 3$
(b) $\Delta pH = \log(1.0 \times 10^5) = 5.00$ (c) $\Delta pH = \log(2.0) = 0.30$

15.58 (a) $pH = -\log[H_3O^+] = -\log(10^{-2}) = 2$
(b) $pH = -\log[H_3O^+] = -\log(4 \times 10^{-8}) = 7.4$

(c) $[H_3O^+] = \dfrac{K_w}{[OH^-]} = \dfrac{1.0 \times 10^{-14}}{8 \times 10^{-8}} = 1.25 \times 10^{-7}$ M

$pH = -\log[H_3O^+] = -\log(1.25 \times 10^{-7}) = 6.9$

(d) $[H_3O^+] = \dfrac{K_w}{[OH^-]} = \dfrac{1.0 \times 10^{-14}}{6 \times 10^{-10}} = 1.7 \times 10^{-5}$ M

$pH = -\log[H_3O^+] = -\log(1.7 \times 10^{-5}) = 4.8$

$[H_3O^+] = \dfrac{K_w}{[OH^-]} = \dfrac{1.0 \times 10^{-14}}{2 \times 10^{-6}} = 5 \times 10^{-9}$ M

$pH = -\log[H_3O^+] = -\log(5 \times 10^{-9}) = 8.3$
$pH = 4.8$ to 8.3

(e) $[H_3O^+] = \dfrac{K_w}{[OH^-]} = \dfrac{1.0 \times 10^{-14}}{2 \times 10^{-7}} = 5 \times 10^{-8}$ M

$pH = -\log[H_3O^+] = -\log(5 \times 10^{-8}) = 7.3$

Strong Acids and Strong Bases

15.60 (a) $[H_3O^+] = 0.40$ M; $pH = -\log[H_3O^+] = -\log(0.40) = 0.40$

(b) $[OH^-] = 3.7 \times 10^{-4}$ M

$[H_3O^+] = \dfrac{K_w}{[OH^-]} = \dfrac{1.0 \times 10^{-14}}{3.7 \times 10^{-4}} = 2.7 \times 10^{-11}$ M

$pH = -\log[H_3O^+] = -\log(2.7 \times 10^{-11}) = 10.57$

(c) $[OH^-] = 2(5.0 \times 10^{-5}$ M$) = 1.0 \times 10^{-4}$ M

$[H_3O^+] = \dfrac{K_w}{[OH^-]} = \dfrac{1.0 \times 10^{-14}}{1.0 \times 10^{-4}} = 1.0 \times 10^{-10}$ M

$pH = -\log[H_3O^+] = -\log(1.0 \times 10^{-10}) = 10.00$

15.62 (a) LiOH, 23.95 amu; 250 mL = 0.250 L

$$\text{molarity of LiOH(aq)} = \dfrac{\left(4.8 \text{ g} \times \dfrac{1 \text{ mol}}{23.95 \text{ g}}\right)}{0.250 \text{ L}} = 0.80 \text{ M}$$

LiOH is a strong base; therefore $[OH^-] = 0.80$ M.

$[H_3O^+] = \dfrac{K_w}{[OH^-]} = \dfrac{1.0 \times 10^{-14}}{0.80} = 1.25 \times 10^{-14}$ M

$pH = -\log[H_3O^+] = -\log(1.25 \times 10^{-14}) = 13.90$

(b) HCl, 36.46 amu

$$\text{molarity of HCl(aq)} = \dfrac{\left(0.93 \text{ g} \times \dfrac{1 \text{ mol}}{36.46 \text{ g}}\right)}{0.40 \text{ L}} = 0.064 \text{ M}$$

HCl is a strong acid; therefore $[H_3O^+] = 0.064$ M

$pH = -\log[H_3O^+] = -\log(0.064) = 1.19$

(c) $M_f \cdot V_f = M_i \cdot V_i$

$M_f = \dfrac{M_i \cdot V_i}{V_f} = \dfrac{(0.10 \text{ M})(50 \text{ mL})}{(1000 \text{ mL})} = 5.0 \times 10^{-3}$ M

$pH = -\log[H_3O^+] = -\log(5.0 \times 10^{-3}) = 2.30$

(d) For HCl, $M_f = \dfrac{M_i \cdot V_i}{V_f} = \dfrac{(2.0 \times 10^{-3} \text{ M})(100 \text{ mL})}{(500 \text{ mL})} = 4.0 \times 10^{-4}$ M

For HClO$_4$, $M_f = \dfrac{M_i \cdot V_i}{V_f} = \dfrac{(1.0 \times 10^{-3} \text{ M})(400 \text{ mL})}{(500 \text{ mL})} = 8.0 \times 10^{-4}$ M

$[H_3O^+] = (4.0 \times 10^{-4}$ M$) + (8.0 \times 10^{-4}$ M$) = 1.2 \times 10^{-3}$ M

$pH = -\log[H_3O^+] = -\log(1.2 \times 10^{-3}) = 2.92$

Weak Acids

15.64 (a) $HClO_2(aq) + H_2O(l) \rightleftharpoons H_3O^+(aq) + ClO_2^-(aq)$; $K_a = \dfrac{[H_3O^+][ClO_2^-]}{[HClO_2]}$

(b) $HOBr(aq) + H_2O(l) \rightleftharpoons H_3O^+(aq) + OBr^-(aq)$; $K_a = \dfrac{[H_3O^+][OBr^-]}{[HOBr]}$

(c) $HCO_2H(aq) + H_2O(l) \rightleftharpoons H_3O^+(aq) + HCO_2^-(aq)$; $K_a = \dfrac{[H_3O^+][HCO_2^-]}{[HCO_2H]}$

15.66 (a) The larger the K_a, the stronger the acid.
$C_6H_5OH < HOCl < CH_3CO_2H < HNO_3$
(b) The larger the K_a, the larger the percent dissociation for the same concentration.
$HNO_3 > CH_3CO_2H > HOCl > C_6H_5OH$
1 M HNO_3, $[H_3O^+] = 1$ M

1 M CH_3CO_2H, $[H_3O^+] = \sqrt{[HA] \times K_a} = \sqrt{(1\,M)(1.8 \times 10^{-5})} = 4 \times 10^{-3}$ M

1 M $HOCl$, $[H_3O^+] = \sqrt{[HA] \times K_a} = \sqrt{(1\,M)(3.5 \times 10^{-8})} = 2 \times 10^{-4}$ M

1 M C_6H_5OH, $[H_3O^+] = \sqrt{[HA] \times K_a} = \sqrt{(1\,M)(1.3 \times 10^{-10})} = 1 \times 10^{-5}$ M

15.68
	$HOBr(aq)$	+ $H_2O(l)$	$\rightleftharpoons$	$H_3O^+(aq)$	+ $OBr^-(aq)$
initial (M)	0.040			~0	0
change (M)	$-x$			$+x$	$+x$
equil (M)	$0.040 - x$			x	x

$x = [H_3O^+] = 10^{-pH} = 10^{-5.05} = 8.9 \times 10^{-6}$ M

$K_a = \dfrac{[H_3O^+][OBr^-]}{[HOBr]} = \dfrac{x^2}{0.040 - x} = \dfrac{(8.9 \times 10^{-6})^2}{0.040 - (8.9 \times 10^{-6})} = 2.0 \times 10^{-9}$

15.70
	$C_6H_5OH(aq)$	+ $H_2O(l)$	$\rightleftharpoons$	$H_3O^+(aq)$	+ $C_6H_5O^-(aq)$
initial (M)	0.10			~0	0
change (M)	$-x$			$+x$	$+x$
equil (M)	$0.10 - x$			x	x

$K_a = \dfrac{[H_3O^+][C_6H_5O^-]}{[C_6H_5OH]} = 1.3 \times 10^{-10} = \dfrac{x^2}{0.10 - x} \approx \dfrac{x^2}{0.10}$

Solve for x. $x = 3.6 \times 10^{-6}$ M $= [H_3O^+] = [C_6H_5O^-]$
$[C_6H_5OH] = 0.10 - x = 0.10$ M
$pH = -\log[H_3O^+] = -\log(3.6 \times 10^{-6}) = 5.44$

$[OH^-] = \dfrac{K_w}{[H_3O^+]} = \dfrac{1.0 \times 10^{-14}}{3.6 \times 10^{-6}} = 2.8 \times 10^{-9}$ M

% dissociation $= \dfrac{[C_6H_5OH]_{diss}}{[C_6H_5OH]_{initial}} \times 100\% = \dfrac{3.6 \times 10^{-6}\,M}{0.10\,M} \times 100\% = 0.0036\%$

15.72

$$HNO_2(aq) + H_2O(l) \rightleftharpoons H_3O^+(aq) + NO_2^-(aq)$$

initial (M)	1.5	~0	0
change (M)	−x	+x	+x
equil (M)	1.5 − x	x	x

$$K_a = \frac{[H_3O^+][NO_2^-]}{[HNO_2]} = 4.5 \times 10^{-4} = \frac{x^2}{1.5-x} \approx \frac{x^2}{1.5}$$

Solve for x. $x = 0.026 \ M = [H_3O^+]$

$pH = -\log[H_3O^+] = -\log(0.026) = 1.59$

$$\% \text{ dissociation} = \frac{[HNO_2]_{diss}}{[HNO_2]_{initial}} \times 100\% = \frac{0.026 \ M}{1.5 \ M} \times 100\% = 1.7\%$$

Polyprotic Acids

15.74 $H_2SeO_4(aq) + H_2O(l) \rightleftharpoons H_3O^+(aq) + HSeO_4^-(aq); \quad K_{a1} = \dfrac{[H_3O^+][HSeO_4^-]}{[H_2SeO_4]}$

$HSeO_4^-(aq) + H_2O(l) \rightleftharpoons H_3O^+(aq) + SeO_4^{2-}(aq); \quad K_{a2} = \dfrac{[H_3O^+][SeO_4^{2-}]}{[HSeO_4^-]}$

15.76

$$H_2CO_3(aq) + H_2O(l) \rightleftharpoons H_3O^+(aq) + HCO_3^-(aq)$$

initial (M)	0.010	~0	0
change (M)	−x	+x	+x
equil (M)	0.010 − x	x	x

$$K_{a1} = \frac{[H_3O^+][HCO_3^-]}{[H_2CO_3]} = 4.3 \times 10^{-7} = \frac{x^2}{0.010-x} \approx \frac{x^2}{0.010}$$

Solve for x. $x = 6.6 \times 10^{-5}$

$[H_3O^+] = [HCO_3^-] = x = 6.6 \times 10^{-5} \ M; \qquad [H_2CO_3] = 0.010 - x = 0.010 \ M$

The second dissociation of H_2CO_3 produces a negligible amount of H_3O^+ compared with that from the first dissociation.

$$HCO_3^-(aq) + H_2O(l) \rightleftharpoons H_3O^+(aq) + CO_3^{2-}(aq)$$

$$K_{a2} = \frac{[H_3O^+][CO_3^{2-}]}{[HCO_3^-]} = 5.6 \times 10^{-11} = \frac{(6.6 \times 10^{-5})[CO_3^{2-}]}{(6.6 \times 10^{-5})}$$

$[CO_3^{2-}] = K_{a2} = 5.6 \times 10^{-11} \ M$

$$[OH^-] = \frac{K_w}{[H_3O^+]} = \frac{1.0 \times 10^{-14}}{6.6 \times 10^{-5}} = 1.5 \times 10^{-10} \ M$$

$pH = -\log[H_3O^+] = -\log(6.6 \times 10^{-5}) = 4.18$

15.78 For the dissociation of the first proton, the following equilibrium must be considered:

$$H_2C_2O_4(aq) + H_2O(l) \rightleftharpoons H_3O^+(aq) + HC_2O_4^-(aq)$$

initial (M)	0.20	~0	0
change (M)	−x	+x	+x
equil (M)	0.20 − x	x	x

$$K_{a1} = \frac{[H_3O^+][HC_2O_4^-]}{[H_2C_2O_4]} = 5.9 \times 10^{-2} = \frac{x^2}{0.20 - x}$$

$x^2 + 0.059x - 0.0118 = 0$

Use the quadratic formula to solve for x.

$$x = \frac{-(0.059) \pm \sqrt{(0.059)^2 - 4(1)(-0.0118)}}{2(1)} = \frac{-0.059 \pm 0.225}{2}$$

x = 0.083 and −0.142

Of the two solutions for x, only the positive value of x has physical meaning, because x is the $[H_3O^+]$.

$[H_3O^+] = [HC_2O_4^-] = 0.083$ M

For the dissociation of the second proton, the following equilibrium must be considered:

$$HC_2O_4^-(aq) + H_2O(l) \rightleftharpoons H_3O^+(aq) + C_2O_4^{2-}(aq)$$

initial (M)	0.083	0.083	0
change (M)	−x	+x	+x
equil (M)	0.083 − x	0.083 + x	x

$$K_{a2} = \frac{[H_3O^+][C_2O_4^{2-}]}{[HC_2O_4^-]} = 6.4 \times 10^{-5} = \frac{(0.083 + x)(x)}{0.083 - x} \approx \frac{(0.083)(x)}{0.083} = x$$

$[H_3O^+] = 0.083 + x = 0.083$ M

$pH = -\log[H_3O^+] = -\log(0.083) = 1.08$

$[C_2O_4^{2-}] = x = 6.4 \times 10^{-5}$ M

Weak Bases; Relation Between K_a and K_b

15.80 (a) $(CH_3)_2NH(aq) + H_2O(l) \rightleftharpoons (CH_3)_2NH_2^+(aq) + OH^-(aq)$; $K_b = \dfrac{[(CH_3)_2NH_2^+][OH^-]}{[(CH_3)_2NH]}$

(b) $C_6H_5NH_2(aq) + H_2O(l) \rightleftharpoons C_6H_5NH_3^+(aq) + OH^-(aq)$; $K_b = \dfrac{[C_6H_5NH_3^+][OH^-]}{[C_6H_5NH_2]}$

(c) $CN^-(aq) + H_2O(l) \rightleftharpoons HCN(aq) + OH^-(aq)$; $K_b = \dfrac{[HCN][OH^-]}{[CN^-]}$

15.82 $[H_3O^+] = 10^{-pH} = 10^{-9.5} = 3.16 \times 10^{-10}$ M

$$[OH^-] = \frac{K_w}{[H_3O^+]} = \frac{1.0 \times 10^{-14}}{3.16 \times 10^{-10}} = 3.16 \times 10^{-5} \text{ M}$$

$$C_{17}H_{19}NO_3(aq) + H_2O(l) \rightleftarrows C_{17}H_{20}NO_3^+(aq) + OH^-(aq)$$

initial (M)	7.0×10^{-4}	0	~0
change (M)	$-x$	$+x$	$+x$
equil (M)	$(7.0 \times 10^{-4}) - x$	x	x

$x = [OH^-] = 3.16 \times 10^{-5}$ M

$$K_b = \frac{[C_{17}H_{20}NO_3^+][OH^-]}{[C_{17}H_{19}NO_3]} = \frac{x^2}{(7.0 \times 10^{-4}) - x} =$$

$$\frac{(3.16 \times 10^{-5})^2}{(7.0 \times 10^{-4}) - (3.16 \times 10^{-5})} = 1.49 \times 10^{-6} = 1 \times 10^{-6}$$

15.84 (a)
$$CH_3NH_2(aq) + H_2O(l) \rightleftarrows CH_3NH_3^+(aq) + OH^-(aq)$$

initial (M)	0.24	0	~0
change (M)	$-x$	$+x$	$+x$
equil (M)	$0.24 - x$	x	x

$$K_b = \frac{[CH_3NH_3^+][OH^-]}{[CH_3NH_2]} = 3.7 \times 10^{-4} = \frac{x^2}{0.24 - x}$$

$x^2 + (3.7 \times 10^{-4})x - (8.9 \times 10^{-5}) = 0$

Use the quadratic formula to solve for x.

$$x = \frac{-(3.7 \times 10^{-4}) \pm \sqrt{(3.7 \times 10^{-4})^2 - (4)(-8.9 \times 10^{-5})}}{2(1)} = \frac{(-3.7 \times 10^{-4}) \pm 0.0189}{2}$$

$x = 0.0093$ and -0.0096

Of the two solutions for x, only the positive value of x has physical meaning because x is the $[OH^-]$.

$[OH^-] = x = 0.0093$ M

$$[H_3O^+] = \frac{K_w}{[OH^-]} = \frac{1.0 \times 10^{-14}}{0.0093} = 1.1 \times 10^{-12}$$ M

$pH = -\log[H_3O^+] = -\log(1.1 \times 10^{-12}) = 11.96$

(b)
$$C_5H_5N(aq) + H_2O(l) \rightleftarrows C_5H_5NH^+(aq) + OH^-(aq)$$

initial (M)	0.040	0	~0
change (M)	$-x$	$+x$	$+x$
equil (M)	$0.040 - x$	x	x

$$K_b = \frac{[C_5H_5NH^+][OH^-]}{[C_5H_5N]} = 1.8 \times 10^{-9} = \frac{x^2}{0.040 - x} \approx \frac{x^2}{0.040}$$

Solve for x. $x = [OH^-] = 8.5 \times 10^{-6}$ M

$$[H_3O^+] = \frac{K_w}{[OH^-]} = \frac{1.0 \times 10^{-14}}{8.5 \times 10^{-6}} = 1.2 \times 10^{-9}$$ M

$pH = -\log[H_3O^+] = -\log(1.2 \times 10^{-9}) = 8.92$

(c) $\qquad$ $NH_2OH(aq) + H_2O(l) \rightleftharpoons NH_3OH^+(aq) + OH^-(aq)$

initial (M)	0.075	0	~0
change (M)	–x	+x	+x
equil (M)	0.075 – x	x	x

$$K_b = \frac{[NH_3OH^+][OH^-]}{[NH_2OH]} = 9.1 \times 10^{-9} = \frac{x^2}{0.075-x} \approx \frac{x^2}{0.075}$$

Solve for x. $x = [OH^-] = 2.6 \times 10^{-5}$ M

$$[H_3O^+] = \frac{K_w}{[OH^-]} = \frac{1.0 \times 10^{-14}}{2.6 \times 10^{-5}} = 3.8 \times 10^{-10} \text{ M}$$

$$pH = -\log[H_3O^+] = -\log(3.8 \times 10^{-10}) = 9.42$$

15.86 (a) $K_a = \dfrac{K_w}{K_b \text{ for } C_3H_7NH_2} = \dfrac{1.0 \times 10^{-14}}{5.1 \times 10^{-4}} = 2.0 \times 10^{-11}$

(b) $K_a = \dfrac{K_w}{K_b \text{ for } NH_2OH} = \dfrac{1.0 \times 10^{-14}}{9.1 \times 10^{-9}} = 1.1 \times 10^{-6}$

(c) $K_a = \dfrac{K_w}{K_b \text{ for } C_6H_5NH_2} = \dfrac{1.0 \times 10^{-14}}{4.3 \times 10^{-10}} = 2.3 \times 10^{-5}$

(d) $K_a = \dfrac{K_w}{K_b \text{ for } C_5H_5N} = \dfrac{1.0 \times 10^{-14}}{1.8 \times 10^{-9}} = 5.6 \times 10^{-6}$

Acid–Base Properties of Salts

15.88 (a) $CH_3NH_3^+(aq) + H_2O(l) \rightleftharpoons H_3O^+(aq) + CH_3NH_2(aq)$
      acid   base —— acid   base

(b) $Cr(H_2O)_6^{3+}(aq) + H_2O(l) \rightleftharpoons H_3O^+(aq) + Cr(H_2O)_5(OH)^{2+}(aq)$
      acid   base —— acid   base

(c) $CH_3CO_2^-(aq) + H_2O(l) \rightleftharpoons CH_3CO_2H(aq) + OH^-(aq)$
      base   acid   acid   base

(d) $PO_4^{3-}(aq) + H_2O(l) \rightleftharpoons HPO_4^{2-}(aq) + OH^-(aq)$
      base   acid   acid   base

15.90 (a) F^- (conjugate base of a weak acid), basic solution

(b) Br^- (anion of a strong acid), neutral solution

(c) NH_4^+ (conjugate acid of a weak base), acidic solution

(d) $K(H_2O)_6^+$ (neutral cation), neutral solution

(e) SO_3^{2-} (conjugate base of a weak acid), basic solution

(f) $Cr(H_2O)_6^{3+}$ (acidic cation), acidic solution

15.92 (a) $(C_2H_5NH_3)NO_3$: $C_2H_5NH_3^+$, acidic cation; NO_3^-, neutral anion

$C_2H_5NH_2$, $K_b = 6.4 \times 10^{-4}$

$$C_2H_5NH_3^+, \quad K_a = \frac{K_w}{K_b \text{ for } C_2H_5NH_2} = \frac{1.0 \times 10^{-14}}{6.4 \times 10^{-4}} = 1.56 \times 10^{-11}$$

$$C_2H_5NH_3^+(aq) \; + \; H_2O(l) \; \rightleftharpoons \; H_3O^+(aq) \; + \; C_2H_5NH_2(aq)$$

initial (M)	0.10	~0	0
change (M)	−x	+x	+x
equil (M)	0.10 − x	x	x

$$K_a = \frac{[H_3O^+][C_2H_5NH_2]}{[C_2H_5NH_3^+]} = 1.56 \times 10^{-11} = \frac{x^2}{0.10 - x} \approx \frac{x^2}{0.10}$$

Solve for x. $x = 1.25 \times 10^{-6}$ M $= 1.2 \times 10^{-6}$ M $= [H_3O^+] = [C_2H_5NH_2]$

pH $= -\log[H_3O^+] = -\log(1.25 \times 10^{-6}) = 5.90$

$[C_2H_5NH_3^+] = 0.10 - x = 0.10$ M; $\quad [NO_3^-] = 0.10$ M

$$[OH^-] = \frac{K_w}{[H_3O^+]} = \frac{1.0 \times 10^{-14}}{1.25 \times 10^{-6} \text{ M}} = 8.0 \times 10^{-9}$$

(b) $Na(CH_3CO_2)$: Na^+, neutral cation; $CH_3CO_2^-$, basic anion

CH_3CO_2H, $K_a = 1.8 \times 10^{-5}$

$$CH_3CO_2^-, \quad K_b = \frac{K_w}{K_a \text{ for } CH_3CO_2H} = \frac{1.0 \times 10^{-14}}{1.8 \times 10^{-5}} = 5.6 \times 10^{-10}$$

$$CH_3CO_2^-(aq) \; + \; H_2O(aq) \; \rightleftharpoons \; CH_3CO_2H(aq) \; + \; OH^-(aq)$$

initial (M)	0.10	0	~0
change (M)	−x	+x	+x
equil (M)	0.10 − x	x	x

$$K_b = \frac{[CH_3CO_2H][OH^-]}{[CH_3CO_2^-]} = 5.6 \times 10^{-10} = \frac{x^2}{0.10 - x} \approx \frac{x^2}{0.10}$$

Solve for x. $x = 7.5 \times 10^{-6}$ M $= [CH_3CO_2H] = [OH^-]$

$[CH_3CO_2^-] = 0.10 - x = 0.10$ M; $\quad [Na^+] = 0.10$ M

$$[H_3O^+] = \frac{K_w}{[OH^-]} = \frac{1.0 \times 10^{-14}}{7.5 \times 10^{-6}} = 1.3 \times 10^{-9} \text{ M}$$

pH $= -\log[H_3O^+] = -\log(1.3 \times 10^{-9}) = 8.89$

(c) $NaNO_3$: Na^+, neutral cation; NO_3^-, neutral anion

$[Na^+] = [NO_3^-] = 0.10$ M

$[H_3O^+] = [OH^-] = 1.0 \times 10^{-7}$ M; pH $= 7.00$

Factors That Affect Acid Strength

15.94 (a) PH_3 < H_2S < HCl; electronegativity increases from P to Cl
(b) NH_3 < PH_3 < AsH_3; X–H bond strength decreases from N to As (down a group)
(c) HBrO < $HBrO_2$ < $HBrO_3$; acid strength increases with the number of O atoms

15.96 (a) HCl; The strength of a binary acid H_nA increases as A moves from left to right and from top to bottom in the periodic table.
(b) $HClO_3$; The strength of an oxoacid increases with increasing electronegativity and increasing oxidation state of the central atom.
(c) HBr; The strength of a binary acid H_nA increases as A moves from left to right and from top to bottom in the periodic table.

15.98 (a) H_2Te, weaker X–H bond
(b) H_3PO_4, P has higher electronegativity
(c) $H_2PO_4^-$, lower negative charge
(d) NH_4^+, higher positive charge and N is more electronegative than C

Lewis Acids and Bases

15.100 (a) Lewis acid, SiF_4; Lewis base, F^- (b) Lewis acid, Zn^{2+}; Lewis base, NH_3
(c) Lewis acid, $HgCl_2$; Lewis base, Cl^- (d) Lewis acid, CO_2; Lewis base, H_2O

15.102 (a) $2 : \ddot{F} :^- \; + \; SiF_4 \longrightarrow SiF_6^{2-}$

(b) $4 \; \ddot{N}H_3 \; + \; Zn^{2+} \longrightarrow Zn(NH_3)_4^{2+}$

(c) $2 : \ddot{C}l :^- \; + \; HgCl_2 \longrightarrow HgCl_4^{2-}$

(d) $H_2\ddot{O} : \; + \; CO_2 \longrightarrow H_2CO_3$

15.104 (a) CN^-, Lewis base   (b) H^+, Lewis acid   (c) H_2O, Lewis base
(d) Fe^{3+}, Lewis acid   (e) OH^-, Lewis base   (f) CO_2, Lewis acid
(g) $P(CH_3)_3$, Lewis base   (h) $B(CH_3)_3$, Lewis acid

General Problems

15.106 In aqueous solution:
(1) H_2S acts as an acid only.
(2) HS^- can act as both an acid and a base.
(3) S^{2-} can act as a base only.
(4) H_2O can act as both an acid and a base.
(5) H_3O^+ acts as an acid only.
(6) OH^- acts as a base only.

15.108

H_3O^+ can hydrogen bond with additional H_2O molecules.

15.110 $HCO_3^-(aq) + Al(H_2O)_6^{3+}(aq) \rightarrow H_2O(l) + CO_2(g) + Al(H_2O)_5(OH)^{2+}(aq)$

15.112 H_2O, 18.02 amu

at 0°C, $[H_2O] = \dfrac{\left(0.9998 \text{ g} \times \dfrac{1 \text{ mol}}{18.02 \text{ g}}\right)}{0.001 \text{ L}} = 55.48 \text{ M}$

$K_w = [H_3O^+][OH^-]$, for a neutral solution $[H_3O^+] = [OH^-]$

$[H_3O^+] = \sqrt{K_w} = \sqrt{1.14 \times 10^{-15}} = 3.376 \times 10^{-8} \text{ M}$

$pH = -\log[H_3O^+] = -\log(3.376 \times 10^{-8}) = 7.472$

fraction dissociated $= \dfrac{[H_2O]_{diss}}{[H_2O]_{initial}} = \dfrac{3.376 \times 10^{-8} \text{ M}}{55.48 \text{ M}} = 6.09 \times 10^{-10}$

% dissociation $= \dfrac{[H_2O]_{diss}}{[H_2O]_{initial}} \times 100\% = \dfrac{3.376 \times 10^{-8} \text{ M}}{55.48 \text{ M}} \times 100\% = 6.09 \times 10^{-8}\%$

15.114 For $C_{10}H_{14}N_2H^+$, $K_{a1} = \dfrac{K_w}{K_{b1} \text{ for } C_{10}H_{14}N_2} = \dfrac{1.0 \times 10^{-14}}{1.0 \times 10^{-6}} = 1.0 \times 10^{-8}$

For $C_{10}H_{14}N_2H_2^{2+}$, $K_{a2} = \dfrac{K_w}{K_{b2} \text{ for } C_{10}H_{14}N_2H^+} = \dfrac{1.0 \times 10^{-14}}{1.3 \times 10^{-11}} = 7.7 \times 10^{-4}$

15.116 (a) $A^-(aq) + H_2O(l) \rightleftharpoons HA(aq) + OH^-(aq)$; basic

(b) $M(H_2O)_6^{3+}(aq) + H_2O(l) \rightleftharpoons H_3O^+(aq) + M(H_2O)_5(OH)^{2+}(aq)$; acidic

(c) $2 H_2O(l) \rightleftharpoons H_3O^+(aq) + OH^-(aq)$; neutral

(d) $M(H_2O)_6^{3+}(aq) + A^-(aq) \rightleftharpoons HA(aq) + M(H_2O)_5(OH)^{2+}(aq)$;
acidic because K_a for $M(H_2O)_6^{3+}$ (10^{-4}) is greater than K_b for A^- (10^{-9}).

15.118

	$HIO_3(aq)$	$+\ H_2O(l) \rightleftharpoons$	$H_3O^+(aq)$	$+\ IO_3^-(aq)$
initial (M)	0.0500		~0	0
change (M)	−x		+x	+x
equil (M)	0.0500 − x		x	x

$K_a = \dfrac{[H_3O^+][IO_3^-]}{[HIO_3]} = 1.7 \times 10^{-1} = \dfrac{x^2}{0.0500 - x}$

$x^2 + 0.17x - 0.0085 = 0$

Use the quadratic formula to solve for x.

$$x = \frac{-(0.17) \pm \sqrt{(0.17)^2 - (4)(1)(-0.0085)}}{2(1)} = \frac{(-0.17) \pm 0.251}{2}$$

$x = -0.210$ and 0.0405

Of the two solutions for x, only the positive value of x has physical meaning because x is the $[H_3O^+]$.

$x = [H_3O^+] = 0.0405\ M = 0.040\ M$

$pH = -\log[H_3O^+] = -\log(0.040) = 1.39$

$[HIO_3] = 0.0500 - x = 0.0500 - 0.040 = 0.010\ M$

$[IO_3^-] = x = 0.040\ M$

$$[OH^-] = \frac{K_w}{[H_3O^+]} = \frac{1.0 \times 10^{-14}}{0.040} = 2.5 \times 10^{-13}\ M$$

15.120 For $H_2C_2O_4$, $K_{a1} = 5.9 \times 10^{-2}$ and $K_{a2} = 6.4 \times 10^{-5}$.

For $C_2O_4^{2-}$, $K_{b1} = \dfrac{K_w}{K_{a2}} = \dfrac{1.0 \times 10^{-14}}{6.4 \times 10^{-5}} = 1.6 \times 10^{-10}$

For $HC_2O_4^-$, $K_{b2} = \dfrac{K_w}{K_{a1}} = \dfrac{1.0 \times 10^{-14}}{5.9 \times 10^{-2}} = 1.7 \times 10^{-13}$

	$C_2O_4^{2-}(aq)$	$+$	$H_2O(l)$	$\rightleftharpoons$	$HC_2O_4^-(aq)$	$+$	$OH^-(aq)$
initial (M)	0.100				0		~0
change (M)	$-x$				$+x$		$+x$
equil (M	$0.100 - x$				x		x

$$K_{b1} = \frac{[HC_2O_4^-][OH^-]}{[C_2O_4^{2-}]} = 1.6 \times 10^{-10} = \frac{x^2}{0.100 - x} \approx \frac{x^2}{0.100}$$

$[OH^-] = x = \sqrt{(1.6 \times 10^{-10})(0.100)} = 4.0 \times 10^{-6}\ M$

$$[H_3O^+] = \frac{K_w}{[OH^-]} = \frac{1.0 \times 10^{-14}}{4.0 \times 10^{-6}} = 2.5 \times 10^{-9}\ M$$

The second dissociation produces a negligible additional amount of OH^-.

$pH = -\log[H_3O^+] = -\log(2.5 \times 10^{-9}) = 8.60$

15.122 (a) NH_4F; For NH_4^+, $K_a = 5.6 \times 10^{-10}$ and for F^-, $K_b = 2.9 \times 10^{-11}$
Because $K_a > K_b$, the salt solution is acidic.
(b) $(NH_4)_2SO_3$; For NH_4^+, $K_a = 5.6 \times 10^{-10}$ and for SO_3^{2-}, $K_b = 1.6 \times 10^{-7}$
Because $K_b > K_a$, the salt solution is basic.

15.124 Fraction dissociated $= \dfrac{[HA]_{diss}}{[HA]_{initial}}$

For a weak acid, $[HA]_{diss} = [H_3O^+] = [A^-]$

$$K_a = \frac{[H_3O^+][A^-]}{[HA]} = \frac{[H_3O^+]^2}{[HA]}; \qquad [H_3O^+] = \sqrt{K_a[HA]}$$

$$\text{Fraction dissociated} = \frac{[HA]_{diss}}{[HA]} = \frac{[H_3O^+]}{[HA]} = \frac{\sqrt{K_a[HA]}}{[HA]} = \sqrt{\frac{K_a}{[HA]}}$$

When the concentration of HA that dissociates is negligible compared with its initial concentration, the equilibrium concentration, [HA], equals the initial concentration, $[HA]_{initial}$.

$$\% \text{ dissociation} = \sqrt{\frac{K_a}{[HA]_{initial}}} \times 100\%$$

15.126 Both reactions occur together.
Let x = $[H_3O^+]$ from CH_3CO_2H and y = $[H_3O^+]$ from $C_6H_5CO_2H$
The following two equilibria must be considered:

$$CH_3CO_2H(aq) + H_2O(l) \rightleftharpoons H_3O^+(aq) + CH_3CO_2^-(aq)$$

initial (M)	0.10	y	0
change (M)	−x	+x	+x
equil (M)	0.10 − x	x + y	x

$$C_6H_5CO_2H(aq) + H_2O(l) \rightleftharpoons H_3O^+(aq) + C_6H_5CO_2^-(aq)$$

initial (M)	0.10	x	0
change (M)	−y	+y	+y
equil (M)	0.10 − y	x + y	y

$$K_a(\text{for } CH_3CO_2H) = \frac{[H_3O^+][CH_3CO_2^-]}{[CH_3CO_2H]} = 1.8 \times 10^{-5} = \frac{(x+y)(x)}{0.10-x} \approx \frac{(x+y)(x)}{0.10}$$

$$1.8 \times 10^{-6} = (x+y)(x)$$

$$K_a(\text{for } C_6H_5CO_2H) = \frac{[H_3O^+][C_6H_5CO_2^-]}{[C_6H_5CO_2H]} = 6.5 \times 10^{-5} = \frac{(x+y)(y)}{0.10-y} \approx \frac{(x+y)(y)}{0.10}$$

$$6.5 \times 10^{-6} = (x+y)(y)$$

$$1.8 \times 10^{-6} = (x+y)(x)$$
$$6.5 \times 10^{-6} = (x+y)(y)$$

These two equations must be solved simultaneously for x and y. Divide the first equation by the second.

$$\frac{x}{y} = \frac{1.8 \times 10^{-6}}{6.5 \times 10^{-6}}; \quad x = 0.277y$$

$6.5 \times 10^{-6} = (x+y)(y)$; substitute x = 0.277y into this equation and solve for y.
$6.5 \times 10^{-6} = (0.277y + y)(y) = 1.277y^2$
y = 0.002 256
x = 0.277y = (0.277)(0.002 256) = 0.000 624 9
$[H_3O^+]$ = (x + y) = (0.000 624 9 + 0.002 256) = 0.002 881 M
pH = $-\log[H_3O^+]$ = $-\log(0.002\ 881)$ = 2.54

15.128 $2 NO_2(g) + H_2O(l) \rightarrow HNO_3(aq) + HNO_2(aq)$

$$\text{mol } HNO_3 = 0.0500 \text{ mol } NO_2 \times \frac{1 \text{ mol } HNO_3}{2 \text{ mol } NO_2} = 0.0250 \text{ mol } HNO_3$$

$$\text{mol } HNO_2 = 0.0500 \text{ mol } NO_2 \times \frac{1 \text{ mol } HNO_2}{2 \text{ mol } NO_2} = 0.0250 \text{ mol } HNO_2$$

Because the volume is 1.00 L, mol and molarity are the same.
HNO_3 is a strong acid and completely dissociated. From HNO_3, $[NO_3^-] = [H_3O^+] = 0.0250$ M

	$HNO_2(aq)$	$+ H_2O(l)$	$\rightleftharpoons$	$H_3O^+(aq)$	$+ NO_2^-(aq)$
initial (M)	0.0250			0.0250	0
change (M)	$-x$			$+x$	$+x$
equil (M)	$0.0250 - x$			$0.0250 + x$	x

$$K_a = \frac{[H_3O^+][NO_2^-]}{[HNO_2]} = 4.5 \times 10^{-4} = \frac{(0.0250 + x)x}{0.0250 - x}$$

$x^2 + 0.02545x - 1.125 \times 10^{-5} = 0$
Solve for x using the quadratic formula.

$$x = \frac{-(0.02545) \pm \sqrt{(0.02545)^2 - (4)(1)(-1.125 \times 10^{-5})}}{2(1)} = \frac{(-0.02545) \pm (0.02632)}{2}$$

$x = -0.0259$ and 4.35×10^{-4}
Of the two solutions for x, only the positive value of x has physical meaning, because x is the $[NO_2^-]$.
$[H_3O^+] = (0.0250) + x = (0.0250) + (4.35 \times 10^{-4}) = 0.0254$ M
$pH = -\log[H_3O^+] = -\log(0.02543) = 1.59$

$$[OH^-] = \frac{K_w}{[H_3O^+]} = \frac{1.0 \times 10^{-14}}{0.0254} = 3.9 \times 10^{-13} \text{ M}$$

$[NO_3^-] = 0.0250$ M
$[HNO_2] = 0.0250 - x = 0.0250 - 4.35 \times 10^{-4} = 0.0246$ M
$[NO_2^-] = x = 4.3 \times 10^{-4}$ M

Multi-Concept Problems

15.130 H_3PO_4, 98.00 amu
Assume 1.000 L of solution.

$$\text{Mass of solution} = 1.000 \text{ L} \times \frac{1000 \text{ mL}}{1 \text{ L}} \times \frac{1.0353 \text{ g}}{1 \text{ mL}} = 1035.3 \text{ g}$$

Mass $H_3PO_4 = (0.070)(1035.3 \text{ g}) = 72.47 \text{ g } H_3PO_4$

$$\text{mol } H_3PO_4 = 72.47 \text{ g } H_3PO_4 \times \frac{1 \text{ mol } H_3PO_4}{98.00 \text{ g } H_3PO_4} = 0.740 \text{ mol } H_3PO_4$$

$$[H_3PO_4] = \frac{0.740 \text{ mol } H_3PO_4}{1.000 \text{ L}} = 0.740 \text{ M}$$

For the dissociation of the first proton, the following equilibrium must be considered:

$$H_3PO_4(aq) + H_2O(l) \rightleftharpoons H_3O^+(aq) + H_2PO_4^-(aq)$$

initial (M)	0.740	~0	0
change (M)	–x	+x	+x
equil (M)	0.740 – x	x	x

$$K_{a1} = \frac{[H_3O^+][H_2PO_4^-]}{[H_3PO_4]} = 7.5 \times 10^{-3} = \frac{x^2}{0.740 - x}$$

$$x^2 + (7.5 \times 10^{-3})x - (5.55 \times 10^{-3}) = 0$$

Solve for x using the quadratic formula.

$$x = \frac{-(7.5 \times 10^{-3}) \pm \sqrt{(7.5 \times 10^{-3})^2 - (4)(1)(-5.55 \times 10^{-3})}}{2(1)} = \frac{(-7.5 \times 10^{-3}) \pm 0.149}{2}$$

$x = 0.0708$ and -0.0783

Of the two solutions for x, only the positive value of x has physical meaning, because x is the $[H_3O^+]$.

$x = 0.0708$ M $= [H_2PO_4^-] = [H_3O^+]$

For the dissociation of the second proton, the following equilibrium must be considered:

$$H_2PO_4^-(aq) + H_2O(l) \rightleftharpoons H_3O^+(aq) + HPO_4^{2-}(aq)$$

initial (M)	0.0708	0.0708	0
change (M)	–y	+y	+y
equil (M)	0.0708 – y	0.0708 + y	y

$$K_{a2} = \frac{[H_3O^+][HPO_4^{2-}]}{[H_2PO_4^-]} = 6.2 \times 10^{-8} = \frac{(0.0708 + y)(y)}{0.0708 - y} \approx \frac{(0.0708)(y)}{0.0708} = y$$

$y = 6.2 \times 10^{-8}$ M $= [HPO_4^{2-}]$

For the dissociation of the third proton, the following equilibrium must be considered:

$$HPO_4^{2-}(aq) + H_2O(l) \rightleftharpoons H_3O^+(aq) + PO_4^{3-}(aq)$$

initial (M)	6.2×10^{-8}	0.0708	0
change (M)	–z	+z	+z
equil (M)	$(6.2 \times 10^{-8}) - z$	0.0708 + z	z

$$K_{a3} = \frac{[H_3O^+][PO_4^{3-}]}{[HPO_4^{2-}]} = 4.8 \times 10^{-13} = \frac{(0.0708 + z)(z)}{(6.2 \times 10^{-8}) - z} \approx \frac{(0.0708)(z)}{6.2 \times 10^{-8}}$$

$z = 4.2 \times 10^{-19}$ M $= [PO_4^{3-}]$

$[H_3PO_4] = 0.740 - x = 0.740 - 0.0708 = 0.67$ M

$[H_2PO_4^-] = [H_3O^+] = 0.0708$ M $= 0.071$ M

$[HPO_4^{2-}] = 6.2 \times 10^{-8}$ M; $\qquad [PO_4^{3-}] = 4.2 \times 10^{-19}$ M

$$[OH^-] = \frac{K_w}{[H_3O^+]} = \frac{1.0 \times 10^{-14}}{0.0708} = 1.4 \times 10^{-13} \text{ M}$$

$$pH = -\log[H_3O^+] = -\log(0.0708) = 1.15$$

15.132 $[H_3O^+] = 10^{-pH} = 10^{-9.07} = 8.51 \times 10^{-10}$ M

$[H_3O^+][OH^-] = K_w; \quad [OH^-] = \dfrac{K_w}{[H_3O^+]} = \dfrac{1.0 \times 10^{-14}}{8.51 \times 10^{-10}} = 1.18 \times 10^{-5}$ M = x below.

$K_a = 1.8 \times 10^{-5}$ for CH_3CO_2H and $K_b = \dfrac{K_w}{K_a} = \dfrac{1.0 \times 10^{-14}}{1.8 \times 10^{-5}} = 5.56 \times 10^{-10}$

Use the equilibrium associated with a weak base to solve for $[CH_3CO_2^-] = y$ below.

$$CH_3CO_2^-(aq) + H_2O(l) \rightleftharpoons CH_3CO_2H(aq) + OH^-(aq)$$

initial (M)	y	~0	0
change (M)	−x	+x	+x
equil (M)	y − x	x	x

$K_b = \dfrac{[CH_3CO_2H][OH^-]}{[CH_3CO_2^-]} = 5.56 \times 10^{-10} = \dfrac{x^2}{y - x} = \dfrac{(1.18 \times 10^{-5})^2}{[y - (1.18 \times 10^{-5})]}$

Solve for y.
$(5.56 \times 10^{-10})[y - (1.18 \times 10^{-5})] = (1.18 \times 10^{-5})^2$
$(5.56 \times 10^{-10})y - 6.56 \times 10^{-15} = 1.39 \times 10^{-10}$
$(5.56 \times 10^{-10})y = 1.39 \times 10^{-10}$
$y = (1.39 \times 10^{-10})/(5.56 \times 10^{-10}) = [CH_3CO_2^-] = 0.25$ M

In 1.00 L of solution, the mass of CH_3CO_2Na solute

$$= (0.25 \text{ mol/L})\left(\dfrac{82.035 \text{ g } CH_3CO_2Na}{1 \text{ mol } CH_3CO_2Na}\right) = 20.5 \text{ g}$$

$$\text{mass of solution} = (1000 \text{ mL})\left(\dfrac{1.0085 \text{ g}}{1 \text{ mL}}\right) = 1008.5 \text{ g}$$

mass of solvent = 1008.5 g − 20.5 g = 988 g = 0.988 kg

$$m = \dfrac{0.25 \text{ mol } CH_3CO_2Na}{0.988 \text{ kg}} = 0.25 \ m$$

Because CH_3CO_2Na is a strong electrolyte, the ionic compound is completely dissociated and $[CH_3CO_2^-] = [Na^+]$. The contribution of CH_3CO_2H and OH^- to the total molality of the solution is negligible.
$\Delta T_f = K_f \cdot (2 \cdot m) = (1.86 \text{ °C/m})(2)(0.25 \ m) = 0.93°C$
Solution freezing point = 0.00°C − ΔT_f = 0.00°C − 0.93°C = −0.93°C

15.134 Na_3PO_4, 163.94 amu

$$3.28 \text{ g } Na_3PO_4 \times \dfrac{1 \text{ mol } Na_3PO_4}{163.94 \text{ g } Na_3PO_4} = 0.0200 \text{ mol} = 20.0 \text{ mmol } Na_3PO_4$$

300.0 mL x 0.180 mmol/mL = 54.0 mmol HCl

$$H_3O^+(aq) + PO_4^{3-}(aq) \rightleftharpoons HPO_4^{2-}(aq) + H_2O(l)$$

before (mmol)	54.0	20.0	0
change (mmol)	−20.0	−20.0	+20.0
after (mmol)	34.0	0	20.0

$$H_3O^+(aq) + HPO_4^{2-}(aq) \rightleftharpoons H_2PO_4^-(aq) + H_2O(l)$$

before (mmol)	34.0	20.0	0
change (mmol)	−20.0	−20.0	+20.0
after (mmol)	14.0	0	20.0

$$H_3O^+(aq) + H_2PO_4^-(aq) \rightleftharpoons H_3PO_4(aq) + H_2O(l)$$

before (mmol)	14.0	20.0	0
change (mmol)	−14.0	−14.0	+14.0
after (mmol)	0	6.0	14.0

$$[H_3PO_4] = \frac{14.0 \text{ mmol}}{300.0 \text{ mL}} = 0.047 \text{ M}; \quad [H_2PO_4^-] = \frac{6.0 \text{ mmol}}{300.0 \text{ mL}} = 0.020 \text{ M}$$

$$H_3PO_4(aq) + H_2O(l) \rightleftharpoons H_3O^+(aq) + H_2PO_4^-(aq)$$

initial (M)	0.047	~0	0.020
change (M)	−x	+x	+x
equil (M)	0.047 − x	x	0.020 + x

$$K_a = \frac{[H_3O^+][H_2PO_4^{2-}]}{[H_3PO_4]} = 7.5 \times 10^{-3} = \frac{x(0.020 + x)}{(0.047 - x)}$$

$$x^2 + 0.0275x - (3.525 \times 10^{-4}) = 0$$

Solve for x using the quadratic formula.

$$x = \frac{-(0.0275) \pm \sqrt{(0.0275)^2 - (4)(-3.525 \times 10^{-4})}}{2(1)} = \frac{-0.0275 \pm 0.0465}{2}$$

x = 0.009 52 and −0.0370

Of the two solutions for x, only the positive value of x has physical meaning, because x is the $[H_3O^+]$.

$$pH = -\log[H_3O^+] = -\log(0.009\ 52) = 2.02$$

15.136 (a) $PV = nRT$, $n = \dfrac{PV}{RT} = \dfrac{(0.601 \text{ atm})(1.000 \text{ L})}{\left(0.082\ 06 \dfrac{L \cdot atm}{K \cdot mol}\right)(293.1 \text{ K})} = 0.0250$ mol HF

$$50.0 \text{ mL} \times \frac{1.00 \text{ L}}{1000 \text{ mL}} = 0.0500 \text{ L}$$

$$[HF] = \frac{0.0250 \text{ mol HF}}{0.0500 \text{ L}} = 0.500 \text{ M}$$

$$HF(aq) + H_2O(l) \rightleftharpoons H_3O^+(aq) + F^-(aq)$$

initial (M)	0.500	~0	0
change (M)	−x	+x	+x
equil (M)	0.500 − x	x	x

$$K_a = \frac{[H_3O^+][F^-]}{[HF]} = 3.5 \times 10^{-4} = \frac{x^2}{0.500 - x}$$

$$x^2 + (3.5 \times 10^{-4})x - (1.75 \times 10^{-4}) = 0$$

Solve for x using the quadratic formula.

$$x = \frac{-(3.5 \times 10^{-4}) \pm \sqrt{(3.5 \times 10^{-4})^2 - (4)(1)(-1.75 \times 10^{-4})}}{2(1)} = \frac{(-3.5 \times 10^{-4}) \pm 0.0265}{2}$$

x = −0.0134 and 0.0131

Of the two solutions for x, only the positive value of x has physical meaning, because x is the $[H_3O^+]$.

$pH = -\log[H_3O^+] = -\log(0.0131) = 1.883 = 1.88$

(b) % dissociation $= \dfrac{0.0131 \, M}{0.500} \times 100\% = 2.62\% = 2.6\%$

New % dissociation = (3)(2.62 %) = 7.86 %

Let X equal the concentration of HF dissociated and Y the new volume (in liters) that would triple the % dissociation.

$$K_a = \frac{X^2}{(0.0250/Y) - X} = 3.5 \times 10^{-4}$$

% dissociation $= \dfrac{X}{(0.0250/Y)} \times 100\% = 7.86\%$ and $\dfrac{X}{(0.0250/Y)} = 0.0786$

$X = 1.965 \times 10^{-3}/Y$

Substitute X into the K_a equation.

$$\frac{(1.965 \times 10^{-3}/Y)^2}{(0.0250/Y) - (1.965 \times 10^{-3}/Y)} = 3.5 \times 10^{-4}$$

$$\frac{3.861 \times 10^{-6}/Y^2}{0.0230/Y} = 3.5 \times 10^{-4}$$

$$\frac{3.861 \times 10^{-6}/Y}{0.0230} = 3.5 \times 10^{-4}$$

$$\frac{3.861 \times 10^{-6}}{8.05 \times 10^{-6}} = Y = 0.48 \, L$$

The result in Problem 15.124 can't be used here because the concentration of HF that dissociates can't be neglected compared with the initial HF concentration.

15.138 (a) Rate = $k[OCl^-]^x[NH_3]^y[OH^-]^z$

From experiments 1 & 2, the $[OCl^-]$ doubles and the rate doubles, therefore x = 1.

From experiments 2 & 3, the $[NH_3]$ triples and the rate triples, therefore y = 1.

From experiments 3 & 4, the $[OH^-]$ goes up by a factor of 10 and the rate goes down by a factor of 10, therefore z = −1.

$$\text{Rate} = k \frac{[OCl^-][NH_3]}{[OH^-]}$$

$[H_3O^+] = 10^{-pH} = 10^{-12} = 1 \times 10^{-12} \, M$

$[OH^-] = \dfrac{K_w}{[H_3O^+]} = \dfrac{1.0 \times 10^{-14}}{1 \times 10^{-12}} = 0.01 \, M$

Using experiment 1: $k = \dfrac{(\text{Rate})[\text{OH}^-]}{[\text{OCl}^-][\text{NH}_3]} = \dfrac{(0.017\,\text{M/s})(0.01\,\text{M})}{(0.001\,\text{M})(0.01\,\text{M})} = 17\,\text{s}^{-1}$

(b) $K_1 = K_b(\text{OCl}^-) = \dfrac{K_w}{K_a(\text{HOCl})} = \dfrac{1.0 \times 10^{-14}}{3.5 \times 10^{-8}} = 2.9 \times 10^{-7}$

For second step, Rate $= k_2\,[\text{HOCl}][\text{NH}_3]$

$$\text{Rate} = K_1\,k_2\,\dfrac{\dfrac{[\text{HOCl}][\text{NH}_3]}{[\text{HOCl}][\text{OH}^-]}}{[\text{OCl}^-]} = K_1\,k_2\,\dfrac{[\text{HOCl}][\text{NH}_3][\text{OCl}^-]}{[\text{HOCl}][\text{OH}^-]} = K_1\,k_2\,\dfrac{[\text{OCl}^-][\text{NH}_3]}{[\text{OH}^-]}$$

$K_1\,k_2 = k = 17\,\text{s}^{-1}$

$k_2 = \dfrac{17\,\text{s}^{-1}}{2.9 \times 10^{-7}\,\text{M}} = 5.9 \times 10^7\,\text{M}^{-1}\text{s}^{-1}$

16 Applications of Aqueous Equilibria

16.1   (a) $HNO_2(aq) + OH^-(aq) \rightleftharpoons NO_2^-(aq) + H_2O(l)$; NO_2^- (basic anion), pH > 7.00

(b) $H_3O^+(aq) + NH_3(aq) \rightleftharpoons NH_4^+(aq) + H_2O(l)$; NH_4^+ (acidic cation), pH < 7.00

(c) $OH^-(aq) + H_3O^+(aq) \rightleftharpoons 2\,H_2O(l)$;   pH = 7.00

16.2   (a) $HF(aq) + OH^-(aq) \rightleftharpoons H_2O(l) + F^-(aq)$

$$K_n = \frac{K_a}{K_w} = \frac{3.5 \times 10^{-4}}{1.0 \times 10^{-14}} = 3.5 \times 10^{10}$$

(b) $H_3O^+(aq) + OH^-(aq) \rightleftharpoons 2\,H_2O(l)$

$$K_n = \frac{1}{K_w} = \frac{1}{1.0 \times 10^{-14}} = 1.0 \times 10^{14}$$

(c) $HF(aq) + NH_3(aq) \rightleftharpoons NH_4^+(aq) + F^-(aq)$

$$K_n = \frac{K_a K_b}{K_w} = \frac{(3.5 \times 10^{-4})(1.8 \times 10^{-5})}{1.0 \times 10^{-14}} = 6.3 \times 10^5$$

The tendency to proceed to completion is determined by the magnitude of K_n. The larger the value of K_n, the further does the reaction proceed to completion.

The tendency to proceed to completion is: reaction (c) < reaction (a) < reaction (b)

16.3

	$HCN(aq)$ +	$H_2O(l)$ $\rightleftharpoons$	$H_3O^+(aq)$ +	$CN^-(aq)$
initial (M)	0.025		~0	0.010
change (M)	$-x$		$+x$	$+x$
equil (M)	$0.025 - x$		x	$0.010 + x$

$$K_a = \frac{[H_3O^+][CN^-]}{[HCN]} = 4.9 \times 10^{-10} = \frac{x(0.010 + x)}{0.025 - x} \approx \frac{x(0.010)}{0.025}$$

Solve for x. $x = 1.23 \times 10^{-9}$ M $= 1.2 \times 10^{-9}$ M $= [H_3O^+]$

$pH = -\log[H_3O^+] = -\log(1.23 \times 10^{-9}) = 8.91$

$$[OH^-] = \frac{K_w}{[H_3O^+]} = \frac{1.0 \times 10^{-14}}{1.23 \times 10^{-9}} = 8.2 \times 10^{-6} \text{ M}$$

$[Na^+] = [CN^-] = 0.010$ M; $[HCN] = 0.025$ M

$$\% \text{ dissociation} = \frac{[HCN]_{diss}}{[HCN]_{initial}} \times 100\% = \frac{1.23 \times 10^{-9} \text{ M}}{0.025 \text{ M}} \times 100\% = 4.9 \times 10^{-6} \%$$

16.4   From $NH_4Cl(s)$, $[NH_4^+]_{initial} = \dfrac{0.10\ mol}{0.500\ L} = 0.20\ M$

$$NH_3(aq) + H_2O(l) \rightleftharpoons NH_4^+(aq) + OH^-(aq)$$

	NH_3	NH_4^+	OH^-
initial (M)	0.40	0.20	~0
change (M)	−x	+x	+x
equil (M)	0.40 − x	0.20 + x	x

$K_b = \dfrac{[NH_4^+][OH^-]}{[NH_3]} = 1.8 \times 10^{-5} = \dfrac{(0.20+x)(x)}{(0.40-x)} \approx \dfrac{(0.20)(x)}{(0.40)}$

Solve for x. $x = [OH^-] = 3.6 \times 10^{-5}\ M$

$[H_3O^+] = \dfrac{K_w}{[OH^-]} = \dfrac{1.0 \times 10^{-14}}{3.6 \times 10^{-5}} = 2.8 \times 10^{-10}\ M$

$pH = -\log[H_3O^+] = -\log(2.8 \times 10^{-10}) = 9.55$

16.5   Each solution contains the same number of B molecules. The presence of BH^+ from BHCl lowers the percent dissociation of B. Solution (2) contains no BH^+, therefore it has the largest percent dissociation. BH^+ is the conjugate acid of B. Solution (1) has the largest amount of BH^+ and it would be the most acidic solution and have the lowest pH.

16.6   (a) (1) and (3). Both pictures show equal concentrations of HA and A^-.
(b) (3). It contains a higher concentration of HA and A^-.

16.7
$$HF(aq) + H_2O(l) \rightleftharpoons H_3O^+(aq) + F^-(aq)$$

	HF	H_3O^+	F^-
initial (M)	0.25	~0	0.50
change (M)	−x	+x	+x
equil (M)	0.25 − x	x	0.50 + x

$K_a = \dfrac{[H_3O^+][F^-]}{[HF]} = 3.5 \times 10^{-4} = \dfrac{x(0.50+x)}{0.25-x} \approx \dfrac{x(0.50)}{0.25}$

Solve for x. $x = 1.75 \times 10^{-4}\ M = [H_3O^+]$

For the buffer, $pH = -\log[H_3O^+] = -\log(1.75 \times 10^{-4}) = 3.76$

(a) mol HF = 0.025 mol; mol F^- = 0.050 mol; vol = 0.100 L

$$F^-(aq) + H_3O^+(aq) \xrightarrow{100\%} HF(aq) + H_2O(l)$$

	F^-	H_3O^+	HF
before (mol)	0.050	0.002	0.025
change (mol)	−0.002	−0.002	+0.002
after (mol)	0.048	0	0.027

$[H_3O^+] = K_a \dfrac{[HF]}{[F^-]} = (3.5 \times 10^{-4})\left(\dfrac{0.27}{0.48}\right) = 1.97 \times 10^{-4}\ M$

$pH = -\log[H_3O^+] = -\log(1.97 \times 10^{-4}) = 3.71$

(b) mol HF = 0.025 mol; mol F⁻ = 0.050 mol; vol = 0.100 L

$$\overset{100\%}{HF(aq) \; + \; OH^-(aq) \; \to \; F^-(aq) \; + \; H_2O(l)}$$

	HF	OH⁻	F⁻
before (mol)	0.025	0.004	0.050
change (mol)	−0.004	−0.004	+0.004
after (mol)	0.021	0	0.054

$$[H_3O^+] = K_a \frac{[HF]}{[F^-]} = (3.5 \times 10^{-4})\left(\frac{0.21}{0.54}\right) = 1.36 \times 10^{-4} \text{ M}$$

$$pH = -\log[H_3O^+] = -\log(1.36 \times 10^{-4}) = 3.87$$

16.8

$$HF(aq) \; + \; H_2O(l) \; \rightleftharpoons \; H_3O^+(aq) \; + \; F^-(aq)$$

	HF		H₃O⁺	F⁻
initial (M)	0.050		~0	0.100
change (M)	−x		+x	+x
equil (M)	0.050 − x		x	0.100 + x

$$K_a = \frac{[H_3O^+][F^-]}{[HF]} = 3.5 \times 10^{-4} = \frac{x(0.100 + x)}{0.050 - x} \approx \frac{x(0.100)}{0.050}$$

Solve for x. x = [H₃O⁺] = 1.75 × 10⁻⁴ M

pH = −log[H₃O⁺] = −log(1.75 × 10⁻⁴) = 3.76

mol HF = 0.050 mol/L × 0.100 L = 0.0050 mol HF

mol F⁻ = 0.100 mol/L × 0.100 L = 0.0100 mol F⁻

mol HNO₃ = mol H₃O⁺ = 0.002 mol

$$\text{Neutralization reaction:} \quad \overset{100\%}{F^-(aq) \; + \; H_3O^+(aq) \; \to \; HF(aq) \; + \; H_2O(l)}$$

	F⁻	H₃O⁺	HF
before reaction (mol)	0.0100	0.002	0.0050
change (mol)	−0.002	−0.002	+0.002
after reaction (mol)	0.008	0	0.007

$$[HF] = \frac{0.007 \text{ mol}}{0.100 \text{ L}} = 0.07 \text{ M}; \qquad [F^-] = \frac{0.008 \text{ mol}}{0.100 \text{ L}} = 0.08 \text{ M}$$

$$[H_3O^+] = K_a \frac{[HF]}{[F^-]} = (3.5 \times 10^{-4})\frac{(0.07)}{(0.08)} = 3 \times 10^{-4} \text{ M}$$

pH = −log[H₃O⁺] = −log(3 × 10⁻⁴) = 3.5

This solution has less buffering capacity than the solution in Problem 16.7 because it contains less HF and F⁻ per 100 mL. Note that the change in pH is greater than that in Problem 16.7.

16.9 When equal volumes of two solutions are mixed together, the concentration of each solution is cut in half.

$$pH = pK_a + \log \frac{[base]}{[acid]} = pK_a + \log \frac{[CO_3^{2-}]}{[HCO_3^-]}$$

For HCO₃⁻, $K_a = 5.6 \times 10^{-11}$, $pK_a = -\log K_a = -\log(5.6 \times 10^{-11}) = 10.25$

$$pH = 10.25 + \log\left(\frac{0.050}{0.10}\right) = 10.25 - 0.30 = 9.95$$

16.10 $\quad pH = pK_a + \log\dfrac{[base]}{[acid]} = pK_a + \log\dfrac{[CO_3^{2-}]}{[HCO_3^-]}$

For HCO_3^-, $K_a = 5.6 \times 10^{-11}$, $pK_a = -\log K_a = -\log(5.6 \times 10^{-11}) = 10.25$

$10.40 = 10.25 + \log\dfrac{[CO_3^{2-}]}{[HCO_3^-]}$; $\qquad \log\dfrac{[CO_3^{2-}]}{[HCO_3^-]} = 10.40 - 10.25 = 0.15$

$\dfrac{[CO_3^{2-}]}{[HCO_3^-]} = 10^{0.15} = 1.4$

To obtain a buffer solution with pH 10.40, make the Na_2CO_3 concentration 1.4 times the concentration of $NaHCO_3$.

16.11 Look for an acid with pK_a near the required pH of 7.50.
$K_a = 10^{-pH} = 10^{-7.50} = 3.2 \times 10^{-8}$
Suggested buffer system: $HOCl$ ($K_a = 3.5 \times 10^{-8}$) and $NaOCl$.

16.12 (a) serine is 66% dissociated at $pH = 9.15 + \log\left(\dfrac{66}{34}\right) = 9.44$

(b) serine is 5% dissociated at $pH = 9.15 + \log\left(\dfrac{5}{95}\right) = 7.87$

16.13 (a) mol HCl = mol H_3O^+ = 0.100 mol/L x 0.0400 L = 0.004 00 mol
mol NaOH = mol OH^- = 0.100 mol/L x 0.0350 L = 0.003 50 mol

Neutralization reaction:	$H_3O^+(aq)$	+	$OH^-(aq)$	$\rightarrow$	$2 H_2O(l)$
before reaction (mol)	0.004 00		0.003 50		
change (mol)	−0.003 50		−0.003 50		
after reaction (mol)	0.000 50		0		

$[H_3O^+] = \dfrac{0.000\ 50\ mol}{(0.0400\ L\ +\ 0.0350\ L)} = 6.7 \times 10^{-3}$ M

$pH = -\log[H_3O^+] = -\log(6.7 \times 10^{-3}) = 2.17$

(b) mol HCl = mol H_3O^+ = 0.100 mol/L x 0.0400 L = 0.004 00 mol
mol NaOH = mol OH^- = 0.100 mol/L x 0.0450 L = 0.004 50 mol

Neutralization reaction:	$H_3O^+(aq)$	+	$OH^-(aq)$	$\rightarrow$	$2 H_2O(l)$
before reaction (mol)	0.004 00		0.004 50		
change (mol)	−0.004 00		−0.004 00		
after reaction (mol)	0		0.000 50		

$[OH^-] = \dfrac{0.000\ 50\ mol}{(0.0400\ L\ +\ 0.0450\ L)} = 5.9 \times 10^{-3}$ M

$[H_3O^+] = \dfrac{K_w}{[OH^-]} = \dfrac{1.0 \times 10^{-14}}{5.9 \times 10^{-3}} = 1.7 \times 10^{-12}$ M

$pH = -\log[H_3O^+] = -\log(1.7 \times 10^{-12}) = 11.77$
The results obtained here are consistent with the pH data in Table 16.1

16.14 (a) mol NaOH = mol OH⁻ = 0.100 mol/L x 0.0400 L = 0.004 00 mol
mol HCl = mol H_3O^+ = 0.0500 mol/L x 0.0600 L = 0.003 00 mol
Neutralization reaction: $H_3O^+(aq) + OH^-(aq) \rightarrow 2 H_2O(l)$
before reaction (mol)   0.003 00   0.004 00
change (mol)   −0.003 00   −0.003 00
after reaction (mol)   0   0.001 00

$$[OH^-] = \frac{0.001\ 00\ mol}{(0.0400\ L\ +\ 0.0600\ L)} = 1.0 \times 10^{-2}\ M$$

$$[H_3O^+] = \frac{K_w}{[OH^-]} = \frac{1.0 \times 10^{-14}}{1.0 \times 10^{-2}} = 1.0 \times 10^{-12}\ M$$

pH = −log[H_3O^+] = −log(1.0 x 10⁻¹²) = 12.00
(b) mol NaOH = mol OH⁻ = 0.100 mol/L x 0.0400 L = 0.004 00 mol
mol HCl = mol H_3O^+ = 0.0500 mol/L x 0.0802 L = 0.004 01 mol
Neutralization reaction: $H_3O^+(aq) + OH^-(aq) \rightarrow 2 H_2O(l)$
before reaction (mol)   0.004 01   0.004 00
change (mol)   −0.004 00   −0.004 00
after reaction (mol)   0.000 01   0

$$[H_3O^+] = \frac{0.000\ 01\ mol}{(0.0400\ L\ +\ 0.0802\ L)} = 8.3 \times 10^{-5}\ M$$

pH = −log[H_3O^+] = −log(8.3 x 10⁻⁵) = 4.08
(c) mol NaOH = mol OH⁻ = 0.100 mol/L x 0.0400 L = 0.004 00 mol
mol HCl = mol H_3O^+ = 0.0500 mol/L x 0.1000 L = 0.005 00 mol
Neutralization reaction: $H_3O^+(aq) + OH^-(aq) \rightarrow 2 H_2O(l)$
before reaction (mol)   0.005 00   0.004 00
change (mol)   −0.004 00   −0.004 00
after reaction (mol)   0.001 00   0

$$[H_3O^+] = \frac{0.001\ 00\ mol}{(0.0400\ L\ +\ 0.1000\ L)} = 7.1 \times 10^{-3}\ M$$

pH = −log[H_3O^+] = −log(7.1 x 10⁻³) = 2.15

16.15 (a) (3), only HA present   (b) (1), HA and A⁻ present
(c) (4), only A⁻ present   (d) (2), A⁻ and OH⁻ present

16.16 mol NaOH required = $\left(\dfrac{0.016\ mol\ HOCl}{L}\right)(0.100\ L)\left(\dfrac{1\ mol\ NaOH}{1\ mol\ HOCl}\right) = 0.0016\ mol$

vol NaOH required = $(0.0016\ mol)\left(\dfrac{1\ L}{0.0400\ mol}\right) = 0.040\ L = 40\ mL$

40 mL of 0.0400 M NaOH are required to reach the equivalence point.
(a) mmol HOCl = 0.016 mmol/mL x 100.0 mL = 1.6 mmol
mmol NaOH = mmol OH⁻ = 0.0400 mmol/mL x 10.0 mL = 0.400 mmol
Neutralization reaction: $HOCl(aq) + OH^-(aq) \rightarrow OCl^-(aq) + H_2O(l)$
before reaction (mmol)   1.6   0.400   0
change (mmol)   −0.400   −0.400   +0.400
after reaction (mmol)   1.2   0   0.400

$$[\text{HOCl}] = \frac{1.2 \text{ mmol}}{(100.0 \text{ mL} + 10.0 \text{ mL})} = 1.09 \times 10^{-2} \text{ M}$$

$$[\text{OCl}^-] = \frac{0.400 \text{ mmol}}{(100.0 \text{ mL} + 10.0 \text{ mL})} = 3.64 \times 10^{-3} \text{ M}$$

$$\text{HOCl(aq)} + \text{H}_2\text{O(l)} \rightleftharpoons \text{H}_3\text{O}^+\text{(aq)} + \text{OCl}^-\text{(aq)}$$

initial (M)	0.0109	~0	0.003 64
change (M)	−x	+x	+x
equil (M)	0.0109 − x	x	0.003 64 + x

$$K_a = \frac{[\text{H}_3\text{O}^+][\text{OCl}^-]}{[\text{HOCl}]} = 3.5 \times 10^{-8} = \frac{x(0.003\ 64 + x)}{0.0109 - x} \approx \frac{x(0.003\ 64)}{0.0109}$$

Solve for x. $x = [\text{H}_3\text{O}^+] = 1.05 \times 10^{-7} \text{ M}$

$\text{pH} = -\log[\text{H}_3\text{O}^+] = -\log(1.05 \times 10^{-7}) = 6.98$

(b) Halfway to the equivalence point, $[\text{OCl}^-] = [\text{HOCl}]$

$\text{pH} = \text{p}K_a = -\log K_a = -\log(3.5 \times 10^{-8}) = 7.46$

(c) At the equivalence point the solution contains the salt, NaOCl.

mol NaOCl = initial mol HOCl = 0.0016 mol = 1.6 mmol

$$[\text{OCl}^-] = \frac{1.6 \text{ mmol}}{(100.0 \text{ mL} + 40.0 \text{ mL})} = 1.1 \times 10^{-2} \text{ M}$$

$$\text{For OCl}^-, \ K_b = \frac{K_w}{K_a \text{ for HOCl}} = \frac{1.0 \times 10^{-14}}{3.5 \times 10^{-8}} = 2.9 \times 10^{-7}$$

$$\text{OCl}^-\text{(aq)} + \text{H}_2\text{O(l)} \rightleftharpoons \text{HOCl(aq)} + \text{OH}^-\text{(aq)}$$

initial (M)	0.011	0	~0
change (M)	−x	+x	+x
equil (M)	0.011 − x	x	x

$$K_b = \frac{[\text{HOCl}][\text{OH}^-]}{[\text{OCl}^-]} = 2.9 \times 10^{-7} = \frac{x^2}{0.011 - x} \approx \frac{x^2}{0.011}$$

Solve for x. $x = [\text{OH}^-] = 5.65 \times 10^{-5} \text{ M}$

$$[\text{H}_3\text{O}^+] = \frac{K_w}{[\text{OH}^-]} = \frac{1.0 \times 10^{-14}}{5.65 \times 10^{-5}} = 1.77 \times 10^{-10} = 1.8 \times 10^{-10} \text{ M}$$

$\text{pH} = -\log[\text{H}_3\text{O}^+] = -\log(1.77 \times 10^{-10}) = 9.75$

16.17 From Problem 16.16, pH = 9.75 at the equivalence point.
Use thymolphthalein (pH 9.4 − 10.6). Bromthymol blue is unacceptable because it changes color halfway to the equivalence point.

16.18 (a) mol NaOH required to reach first equivalence point

$$= \left(\frac{0.0800 \text{ mol H}_2\text{SO}_3}{\text{L}}\right)(0.0400 \text{ L})\left(\frac{1 \text{ mol NaOH}}{1 \text{ mol H}_2\text{SO}_3}\right) = 0.003\ 20 \text{ mol}$$

vol NaOH required to reach first equivalence point

$$= (0.003\ 20\ \text{mol})\left(\frac{1\ \text{L}}{0.160\ \text{mol}}\right) = 0.020\ \text{L} = 20.0\ \text{mL}$$

20.0 mL is enough NaOH solution to reach the first equivalence point for the titration of the diprotic acid, H_2SO_3.

For H_2SO_3,

$K_{a1} = 1.5 \times 10^{-2}$, $pK_{a1} = -\log K_{a1} = -\log(1.5 \times 10^{-2}) = 1.82$

$K_{a2} = 6.3 \times 10^{-8}$, $pK_{a2} = -\log K_{a2} = -\log(6.3 \times 10^{-8}) = 7.20$

At the first equivalence point, $pH = \dfrac{pK_{a1} + pK_{a2}}{2} = \dfrac{1.82 + 7.20}{2} = 4.51$

(b) mol NaOH required to reach second equivalence point

$$= \left(\frac{0.0800\ \text{mol}\ H_2SO_3}{\text{L}}\right)(0.0400\ \text{L})\left(\frac{2\ \text{mol NaOH}}{1\ \text{mol}\ H_2SO_3}\right) = 0.006\ 40\ \text{mol}$$

vol NaOH required to reach second equivalence point

$$= (0.006\ 40\ \text{mol})\left(\frac{1\ \text{L}}{0.160\ \text{mol}}\right) = 0.040\ \text{L} = 40.0\ \text{mL}$$

30.0 mL is enough NaOH solution to reach halfway to the second equivalent point. Halfway to the second equivalence point

$pH = pK_{a2} = -\log K_{a2} = -\log(6.3 \times 10^{-8}) = 7.20$

(c) mmol $HSO_3^- = 0.0800$ mmol/mL x 40.0 mL = 3.20 mmol

volume NaOH added after first equivalence point = 35.0 mL – 20.0 mL = 15.0 mL

mmol NaOH = mmol $OH^- = 0.160$ mmol/L x 15.0 mL = 2.40 mmol

Neutralization reaction:	$HSO_3^-(aq)$	$+ OH^-(aq)$	$\rightleftarrows SO_3^{2-}(aq)$	$+ H_2O(l)$
before reaction (mmol)	3.20	2.40	0	
change (mmol)	–2.40	–2.40	+2.40	
after reaction (mmol)	0.80	0	2.40	

$$[HSO_3^-] = \frac{0.80\ \text{mmol}}{(40.0\ \text{mL} + 35.0\ \text{mL})} = 0.0107\ \text{M}$$

$$[SO_3^{2-}] = \frac{2.40\ \text{mmol}}{(40.0\ \text{mL} + 35.0\ \text{mL})} = 0.0320\ \text{M}$$

	$HSO_3^-(aq)$	$+ H_2O(l)$	$\rightleftarrows H_3O^+(aq)$	$+ SO_3^{2-}(aq)$
initial (M)	0.0107		~0	0.0320
change (M)	–x		+x	+x
equil (M)	0.0107 – x		x	0.0320 + x

$$K_a = \frac{[H_3O^+][SO_3^{2-}]}{[HSO_3^-]} = 6.3 \times 10^{-8} = \frac{x(0.0320 + x)}{0.0107 - x} \approx \frac{x(0.0320)}{0.0107}$$

Solve for x. $x = [H_3O^+] = 2.1 \times 10^{-8}$ M

$pH = -\log[H_3O^+] = -\log(2.1 \times 10^{-8}) = 7.68$

16.19 Let H_2A^+ = valine cation

(a) mol NaOH required to reach first equivalence point

$$= \left(\frac{0.0250 \text{ mol } H_2A^+}{L} \right)(0.0400 \text{ L})\left(\frac{1 \text{ mol NaOH}}{1 \text{ mol } H_2A^+} \right) = 0.001\ 00 \text{ mol}$$

vol NaOH required to reach first equivalence point

$$= (0.001\ 00 \text{ mol})\left(\frac{1 \text{ L}}{0.100 \text{ mol}} \right) = 0.0100 \text{ L} = 10.0 \text{ mL}$$

10.0 mL is enough NaOH solution to reach the first equivalence point for the titration of the diprotic acid, H_2A^+.

For H_2A^+,

$K_{a1} = 4.8 \times 10^{-3}$, $pK_{a1} = -\log K_{a1} = -\log(4.8 \times 10^{-3}) = 2.32$

$K_{a2} = 2.4 \times 10^{-10}$, $pK_{a2} = -\log K_{a2} = -\log(2.4 \times 10^{-10}) = 9.62$

At the first equivalence point, $pH = \dfrac{pK_{a1} + pK_{a2}}{2} = \dfrac{2.32 + 9.62}{2} = 5.97$

(b) mol NaOH required to reach second equivalence point

$$= \left(\frac{0.0250 \text{ mol } H_2A^+}{L} \right)(0.0400 \text{ L})\left(\frac{2 \text{ mol NaOH}}{1 \text{ mol } H_2A^+} \right) = 0.002\ 00 \text{ mol}$$

vol NaOH required to reach second equivalence point

$$= (0.002\ 00 \text{ mol})\left(\frac{1 \text{ L}}{0.100 \text{ mol}} \right) = 0.0200 \text{ L} = 20.0 \text{ mL}$$

15.0 mL is enough NaOH solution to reach halfway to the second equivalent point.
Halfway to the second equivalence point

$pH = pK_{a2} = -\log K_{a2} = -\log(2.4 \times 10^{-10}) = 9.62$

(c) 20.0 mL is enough NaOH to reach the second equivalence point.
At the second equivalence point

mmol A^- = (0.0250 mmol/mL)(40.0 mL) = 1.00 mmol A^-

solution volume = 40.0 mL + 20.0 mL = 60.0 mL

$$[A^-] = \frac{1.00 \text{ mmol}}{60.0 \text{ mL}} = 0.0167 \text{ M}$$

	$A^-(aq)$	+	$H_2O(l)$	$\rightleftharpoons$	$HA(aq)$	+	$OH^-(aq)$
initial (M)	0.0167				0		~0
change (M)	$-x$				$+x$		$+x$
equil (M)	$0.0167 - x$				x		x

$$K_b = \frac{K_w}{K_a \text{ for HA}} = \frac{K_w}{K_{a2}} = \frac{1.0 \times 10^{-14}}{2.4 \times 10^{-10}} = 4.17 \times 10^{-5}$$

$$K_b = \frac{[HA][OH^-]}{[A^-]} = 4.17 \times 10^{-5} = \frac{x^2}{0.0167 - x}$$

$x^2 + (4.17 \times 10^{-5})x - (6.964 \times 10^{-7}) = 0$

Use the quadratic formula to solve for x.

$$x = \frac{-(4.17 \times 10^{-5}) \pm \sqrt{(4.17 \times 10^{-5})^2 - (4)(1)(-6.964 \times 10^{-7})}}{2(1)} = \frac{(-4.17 \times 10^{-5}) \pm (1.67 \times 10^{-3})}{2}$$

$x = 8.14 \times 10^{-4}$ and -8.56×10^{-4}

Of the two solutions for x, only the positive value has physical meaning because x is the $[OH^-]$.

$x = [OH^-] = 8.14 \times 10^{-4}$ M

$$[H_3O^+] = \frac{K_w}{[OH^-]} = \frac{1.0 \times 10^{-14}}{8.14 \times 10^{-4}} = 1.23 \times 10^{-11} \text{ M}$$

$pH = -\log[H_3O^+] = -\log(1.23 \times 10^{-11}) = 10.91$

16.20 (a) $K_{sp} = [Ag^+][Cl^-]$　　　　　　(b) $K_{sp} = [Pb^{2+}][I^-]^2$
　　　　(c) $K_{sp} = [Ca^{2+}]^3[PO_4^{3-}]^2$　　　　(d) $K_{sp} = [Cr^{3+}][OH^-]^3$

16.21 $K_{sp} = [Ca^{2+}]^3[PO_4^{3-}]^2 = (2.01 \times 10^{-8})^3(1.6 \times 10^{-5})^2 = 2.1 \times 10^{-33}$

16.22 $[Ba^{2+}] = [SO_4^{2-}] = 1.05 \times 10^{-5}$ M;　　$K_{sp} = [Ba^{2+}][SO_4^{2-}] = (1.05 \times 10^{-5})^2 = 1.10 \times 10^{-10}$

16.23 (a)　　　　　　$AgCl(s) \rightleftharpoons Ag^+(aq) + Cl^-(aq)$
equil (M)　　　　　　　　　x　　　　x
$K_{sp} = [Ag^+][Cl^-] = 1.8 \times 10^{-10} = (x)(x)$
molar solubility $= x = \sqrt{K_{sp}} = 1.3 \times 10^{-5}$ mol/L

AgCl, 143.32 amu

$$\text{solubility} = \frac{\left(1.3 \times 10^{-5} \text{ mol} \times \dfrac{143.32 \text{ g}}{1 \text{ mol}}\right)}{1 \text{ L}} = 0.0019 \text{ g/L}$$

(b)　　　　　　$Ag_2CrO_4(s) \rightleftharpoons 2\,Ag^+(aq) + CrO_4^{2-}(aq)$
equil (M)　　　　　　　　　2x　　　　x
$K_{sp} = [Ag^+]^2[CrO_4^{2-}] = 1.1 \times 10^{-12} = (2x)^2(x) = 4x^3$

$$\text{molar solubility} = x = \sqrt[3]{\frac{1.1 \times 10^{-12}}{4}} = 6.5 \times 10^{-5} \text{ mol/L}$$

Ag_2CrO_4, 331.73 amu

$$\text{solubility} = \frac{\left(6.5 \times 10^{-5} \text{ mol} \times \dfrac{331.73 \text{ g}}{1 \text{ mol}}\right)}{1 \text{ L}} = 0.022 \text{ g/L}$$

Ag_2CrO_4 has both the higher molar and gram solubility, despite its smaller value of K_{sp}.

16.24 Let the number of ions be proportional to its concentration.
　　　For AgX, $K_{sp} = [Ag^+][X^-] \propto (4)(4) = 16$
　　　For AgY, $K_{sp} = [Ag^+][Y^-] \propto (1)(9) = 9$
　　　For AgZ, $K_{sp} = [Ag^+][Z^-] \propto (3)(6) = 18$
　　　(a) AgZ　　　(b) AgY

16.25 $[Mg^{2+}]_0$ is from 0.10 M $MgCl_2$.

$$MgF_2(s) \rightleftharpoons Mg^{2+}(aq) + 2 F^-(aq)$$

initial (M)	0.10	0
change (M)	+x	+2x
equil (M)	0.10 + x	2x

$K_{sp} = 7.4 \times 10^{-11} = [Mg^{2+}][F^-]^2 = (0.10 + x)(2x)^2 \approx (0.10)(4x^2)$

$x = 1.4 \times 10^{-5}$, molar solubility = x = 1.4×10^{-5} M

16.26 Compounds that contain basic anions are more soluble in acidic solution than in pure water. AgCN, $Al(OH)_3$, and ZnS all contain basic anions.

16.27 $[Cu^{2+}] = (5.0 \times 10^{-3} \text{ mol})/(0.500 \text{ L}) = 0.010$ M

$$Cu^{2+}(aq) + 4 NH_3(aq) \rightleftharpoons Cu(NH_3)_4^{2+}(aq)$$

before reaction (M)	0.010	0.40	0
assume 100 % reaction (M)	–0.010	–4(0.010)	+0.010
after reaction (M)	0	0.36	0.010
assume small back reaction (M)	+x	+4x	–x
equil (M)	x	0.36 + 4x	0.010 – x

$$K_f = \frac{[Cu(NH_3)_4^{2+}]}{[Cu^{2+}][NH_3]^4} = 5.6 \times 10^{11} = \frac{(0.010 - x)}{(x)(0.36 + 4x)^4} \approx \frac{0.010}{x(0.36)^4}$$

Solve for x. x = $[Cu^{2+}] = 1.1 \times 10^{-12}$ M

16.28

$$AgBr(s) \rightleftharpoons Ag^+(aq) + Br^-(aq) \qquad K_{sp} = 5.4 \times 10^{-13}$$
$$Ag^+(aq) + 2 S_2O_3^{2-} \rightarrow Ag(S_2O_3)_2^{3-}(aq) \qquad K_f = 4.7 \times 10^{13}$$

dissolution $AgBr(s) + 2 S_2O_3^{2-}(aq) \rightleftharpoons Ag(S_2O_3)_2^{3-}(aq) + Br^-(aq)$
reaction

$K = (K_{sp})(K_f) = (5.4 \times 10^{-13})(4.7 \times 10^{13}) = 25.4$

$$AgBr(s) + 2 S_2O_3^{2-}(aq) \rightleftharpoons Ag(S_2O_3)_2^{3-}(aq) + Br^-(aq)$$

initial (M)	0.10	0	0
change (M)	–2x	x	x
equil (M)	0.10 – 2x	x	x

$$K = \frac{[Ag(S_2O_3)_2^{3-}][Br^-]}{[S_2O_3^{2-}]^2} = 25.4 = \frac{x^2}{(0.10 - 2x)^2}$$

Take the square root of both sides and solve for x.

$$\sqrt{25.4} = \sqrt{\frac{x^2}{(0.10 - 2x)^2}}; \quad 5.04 = \frac{x}{0.10 - 2x}; \quad x = \text{molar solubility} = 0.045 \text{ mol/L}$$

16.29 On mixing equal volumes of two solutions, the concentrations of both solutions are cut in half.

For $BaCO_3$, $K_{sp} = 2.6 \times 10^{-9}$

(a) IP = $[Ba^{2+}][CO_3^{2-}] = (1.5 \times 10^{-3})(1.0 \times 10^{-3}) = 1.5 \times 10^{-6}$

IP > K_{sp}; a precipitate of $BaCO_3$ will form.

(b) $IP = [Ba^{2+}][CO_3^{2-}] = (5.0 \times 10^{-6})(2.0 \times 10^{-5}) = 1.0 \times 10^{-10}$
$IP < K_{sp}$; no precipitate will form.

16.30 $pH = pK_a + \log \dfrac{[base]}{[acid]} = pK_a + \log \dfrac{[NH_3]}{[NH_4^+]}$

For NH_4^+, $K_a = 5.6 \times 10^{-10}$, $pK_a = -\log K_a = -\log(5.6 \times 10^{-10}) = 9.25$

$pH = 9.25 + \log \dfrac{(0.20)}{(0.20)} = 9.25$; $[H_3O^+] = 10^{-pH} = 10^{-9.25} = 5.6 \times 10^{-10}$ M

$[OH^-] = \dfrac{K_w}{[H_3O^+]} = \dfrac{1.0 \times 10^{-14}}{5.6 \times 10^{-10}} = 1.8 \times 10^{-5}$ M

$[Fe^{2+}] = [Mn^{2+}] = \dfrac{(25 \text{ mL})(1.0 \times 10^{-3} \text{ M})}{250 \text{ mL}} = 1.0 \times 10^{-4}$ M

For $Mn(OH)_2$, $K_{sp} = 2.1 \times 10^{-13}$
    $IP = [Mn^{2+}][OH^-]^2 = (1.0 \times 10^{-4})(1.8 \times 10^{-5})^2 = 3.2 \times 10^{-14}$
    $IP < K_{sp}$; no precipitate will form.
For $Fe(OH)_2$, $K_{sp} = 4.9 \times 10^{-17}$
    $IP = [Fe^{2+}][OH^-]^2 = (1.0 \times 10^{-4})(1.8 \times 10^{-5})^2 = 3.2 \times 10^{-14}$
    $IP > K_{sp}$; a precipitate of $Fe(OH)_2$ will form.

16.31 $MS(s) + 2 H_3O^+(aq) \rightleftharpoons M^{2+}(aq) + H_2S(aq) + 2 H_2O(l)$

$K_{spa} = \dfrac{[M^{2+}][H_2S]}{[H_3O^+]^2}$

For ZnS, $K_{spa} = 3 \times 10^{-2}$;   for CdS, $K_{spa} = 8 \times 10^{-7}$
$[Cd^{2+}] = [Zn^{2+}] = 0.005$ M
Because the two cation concentrations are equal, Q_c is the same for both.

$Q_c = \dfrac{[M^{2+}]_t[H_2S]_t}{[H_3O^+]_t^2} = \dfrac{(0.005)(0.10)}{(0.3)^2} = 6 \times 10^{-3}$

$Q_c > K_{spa}$ for CdS; CdS will precipitate. $Q_c < K_{spa}$ for ZnS; Zn^{2+} will remain in solution.

16.32 This protein has both acidic and basic sites. H_3PO_4–$H_2PO_4^-$ is an acidic buffer. It protonates the basic sites in the protein making them positive and the protein migrates towards the negative electrode. H_3BO_3–$H_2BO_3^-$ is a basic buffer. At basic pH's, the acidic sites in the protein are dissociated making them negative and the protein migrates towards the positive electrode.

16.33 To increase the rate at which the proteins migrates toward the negative electrode, increase the number of basic sites that are protonated by lowering the pH. Decrease the $[HPO_4^{2-}]/[H_2PO_4^-]$ ratio (less HPO_4^{2-}, more $H_2PO_4^-$) to lower the pH.

Understanding Key Concepts

16.34 A buffer solution contains a conjugate acid-base pair in about equal concentrations.
(a) (1), (3), and (4)
(b) (4) because it has the highest buffer concentration.

16.36 (4); only A^- and water should be present

16.38 (a) (i) (1), only B present   (ii) (4), equal amounts of B and BH^+ present
(iii) (3), only BH^+ present   (iv) (2), BH^+ and H_3O^+ present
(b) The pH is less than 7 because BH^+ is an acidic cation.

16.40 Let the number of ions be proportional to its concentration.
For Ag_2CrO_4, $K_{sp} = [Ag^+]^2[CrO_4^{2-}] \propto (4)^2(2) = 32$
For (2), $IP = [Ag^+]^2[CrO_4^{2-}] \propto (2)^2(4) = 16$
For (3), $IP = [Ag^+]^2[CrO_4^{2-}] \propto (6)^2(2) = 72$
For (4), $IP = [Ag^+]^2[CrO_4^{2-}] \propto (2)^2(6) = 24$
A precipitate will form when $IP > K_{sp}$. A precipitate will form only in (3).

Additional Problems
Neutralization Reactions

16.42 (a) $HI(aq) + NaOH(aq) \rightarrow H_2O(l) + NaI(aq)$
net ionic equation: $H_3O^+(aq) + OH^-(aq) \rightarrow 2\,H_2O(l)$
The solution at neutralization contains a neutral salt (NaI); pH = 7.00.
(b) $2\,HOCl(aq) + Ba(OH)_2(aq) \rightarrow 2\,H_2O(l) + Ba(OCl)_2(aq)$
net ionic equation: $HOCl(aq) + OH^-(aq) \rightarrow H_2O(l) + OCl^-(aq)$
The solution at neutralization contains a basic anion (OCl^-); pH > 7.00.
(c) $HNO_3(aq) + C_6H_5NH_2(aq) \rightarrow C_6H_5NH_3NO_3(aq)$
net ionic equation: $H_3O^+(aq) + C_6H_5NH_2(aq) \rightarrow H_2O(l) + C_6H_5NH_3^+(aq)$
The solution at neutralization contains an acidic cation ($C_6H_5NH_3^+$); pH < 7.00.
(d) $C_6H_5CO_2H(aq) + KOH(aq) \rightarrow H_2O(l) + C_6H_5CO_2K(aq)$
net ionic equation: $C_6H_5CO_2H(aq) + OH^-(aq) \rightarrow H_2O(l) + C_6H_5CO_2^-(aq)$
The solution at neutralization contains a basic anion ($C_6H_5CO_2^-$); pH > 7.00.

16.44 (a) Strong acid - strong base reaction $K_n = \dfrac{1}{K_w} = \dfrac{1}{1.0 \times 10^{-14}} = 1.0 \times 10^{14}$

(b) Weak acid - strong base reaction $K_n = \dfrac{K_a}{K_w} = \dfrac{3.5 \times 10^{-8}}{1.0 \times 10^{-14}} = 3.5 \times 10^6$

(c) Strong acid - weak base reaction $K_n = \dfrac{K_b}{K_w} = \dfrac{4.3 \times 10^{-10}}{1.0 \times 10^{-14}} = 4.3 \times 10^4$

(d) Weak acid - strong base reaction $K_n = \dfrac{K_a}{K_w} = \dfrac{6.5 \times 10^{-5}}{1.0 \times 10^{-14}} = 6.5 \times 10^9$

(c) < (b) < (d) < (a)

16.46 (a) After mixing, the solution contains the basic salt, NaF; pH > 7.00
(b) After mixing, the solution contains the neutral salt, NaCl; pH = 7.00
Solution (a) has the higher pH.

16.48 Weak acid - weak base reaction $K_n = \dfrac{K_a K_b}{K_w} = \dfrac{(1.3 \times 10^{-10})(1.8 \times 10^{-9})}{1.0 \times 10^{-14}} = 2.3 \times 10^{-5}$

K_n is small so the neutralization reaction does not proceed very far to completion.

The Common–Ion Effect

16.50 $HNO_2(aq) + H_2O(l) \rightleftharpoons H_3O^+(aq) + NO_2^-(aq)$
(a) $NaNO_2$ is a source of NO_2^- (reaction product). The equilibrium shifts towards reactants, and the percent dissociation of HNO_2 decreases.
(c) HCl is a source of H_3O^+ (reaction product). The equilibrium shifts towards reactants, and the percent dissociation of HNO_2 decreases.
(d) $Ba(NO_2)_2$ is a source of NO_2^- (reaction product). The equilibrium shifts towards reactants, and the percent dissociation of HNO_2 decreases.

16.52 (a) $HF(aq) + H_2O(l) \rightleftharpoons H_3O^+(aq) + F^-(aq)$
LiF is a source of F^- (reaction product). The equilibrium shifts toward reactants, and the $[H_3O^+]$ decreases. The pH increases.
(b) Because HI is a strong acid, addition of KI, a neutral salt, does not change the pH.
(c) $NH_3(aq) + H_2O(l) \rightleftharpoons NH_4^+(aq) + OH^-(aq)$
NH_4Cl is a source of NH_4^+ (reaction product). The equilibrium shifts toward reactants, and the $[OH^-]$ decreases. The pH decreases.

16.54 For 0.25 M HF and 0.10 M NaF

	$HF(aq) + H_2O(l) \rightleftharpoons$	$H_3O^+(aq) +$	$F^-(aq)$
initial (M)	0.25	~0	0.10
change (M)	−x	+x	+x
equil (M)	0.25 − x	x	0.10 + x

$K_a = \dfrac{[H_3O^+][F^-]}{[HF]} = 3.5 \times 10^{-4} = \dfrac{x(0.10 + x)}{0.25 - x} \approx \dfrac{x(0.10)}{0.25}$

Solve for x. x = $[H_3O^+]$ = 8.8 × 10⁻⁴ M
pH = −log$[H_3O^+]$ = −log(8.8 × 10⁻⁴) = 3.06

16.56   For 0.10 M HN_3:

$$HN_3(aq) \ + \ H_2O(l) \ \rightleftharpoons \ H_3O^+(aq) \ + \ N_3^-(aq)$$

initial (M)	0.10	~0	0
change (M)	−x	+x	+x
equil (M)	0.10 − x	x	x

$$K_a = \frac{[H_3O^+][N_3^-]}{[HN_3]} = 1.9 \times 10^{-5} = \frac{x^2}{0.10-x} \approx \frac{x^2}{0.10}$$

Solve for x. $x = 1.4 \times 10^{-3}$ M

$$\% \text{ dissociation} = \frac{[HN_3]_{diss}}{[HN_3]_{initial}} \times 100\% = \frac{1.4 \times 10^{-3} \text{ M}}{0.10 \text{ M}} \times 100\% = 1.4\%$$

For 0.10 M HN_3 in 0.10 M HCl:

$$HN_3(aq) \ + \ H_2O(l) \ \rightleftharpoons \ H_3O^+(aq) \ + \ N_3^-(aq)$$

initial (M)	0.10	0.10	0
change (M)	−x	+x	+x
equil (M)	0.10 − x	0.10 + x	x

$$K_a = \frac{[H_3O^+][N_3^-]}{[HN_3]} = 1.9 \times 10^{-5} = \frac{(0.10+x)(x)}{0.10-x} \approx \frac{(0.10)(x)}{0.10} = x$$

Solve for x. $x = 1.9 \times 10^{-5}$ M

$$\% \text{ dissociation} = \frac{[HN_3]_{diss}}{[HN_3]_{initial}} \times 100\% = \frac{1.9 \times 10^{-5} \text{ M}}{0.10 \text{ M}} \times 100\% = 0.019\%$$

The % dissociation is less because of the common ion (H_3O^+) effect.

Buffer Solutions

16.58   Solutions (a), (c) and (d) are buffer solutions. Neutralization reactions for (c) and (d) result in solutions with equal concentrations of HF and F^-.

16.60   Both solutions buffer at the same pH because in both cases the $[NO_2^-]/[HNO_2] = 1$. Solution (a), however, has a higher concentration of both HNO_2 and NO_2^-, and therefore it has the greater buffer capacity.

16.62   When blood absorbs acid, the equilibrium shifts to the left, decreasing the pH, but not by much because the $[HCO_3^-]/[H_2CO_3]$ ratio remains nearly constant. When blood absorbs base, the equilibrium shifts to the right, increasing the pH, but not by much because the $[HCO_3^-]/[H_2CO_3]$ ratio remains nearly constant.

16.64 $$pH = pK_a + \log\frac{[\text{base}]}{[\text{acid}]} = pK_a + \log\frac{[CN^-]}{[HCN]}$$

For HCN, $K_a = 4.9 \times 10^{-10}$, $pK_a = -\log K_a = -\log(4.9 \times 10^{-10}) = 9.31$

$$pH = 9.31 + \log\left(\frac{0.12}{0.20}\right) = 9.09$$

The pH of a buffer solution will not change on dilution because the acid and base concentrations will change by the same amount and their ratio will remain the same.

16.66 $pH = pK_a + \log\dfrac{[base]}{[acid]} = pK_a + \log\dfrac{[NH_3]}{[NH_4^+]}$

For NH_4^+, $K_a = 5.6 \times 10^{-10}$, $pK_a = -\log K_a = -\log(5.6 \times 10^{-10}) = 9.25$

For the buffer: $pH = 9.25 + \log\dfrac{(0.200)}{(0.200)} = 9.25$

(a) add 0.0050 mol NaOH, $[OH^-] = 0.0050$ mol/0.500 L = 0.010 M

	$NH_4^+(aq)$	+	$OH^-(aq)$	$\rightleftharpoons$	$NH_3(aq)$	+	$H_2O(l)$
before reaction (M)	0.200		0.010		0.200		
change (M)	−0.010		−0.010		+0.010		
after reaction (M)	0.200 − 0.010		0		0.200 + 0.010		

$pH = 9.25 + \log\dfrac{[NH_3]}{[NH_4^+]} = 9.25 + \log\dfrac{(0.200 + 0.010)}{(0.200 - 0.010)} = 9.29$

(b) add 0.020 mol HCl, $[H_3O^+] = 0.020$ mol/0.500 L = 0.040 M

	$NH_3(aq)$	+	$H_3O^+(aq)$	$\rightleftharpoons$	$NH_4^+(aq)$	+	$H_2O(l)$
before reaction (M)	0.200		0.040		0.200		
change (M)	−0.040		−0.040		+0.040		
after reaction (M)	0.200 − 0.040		0		0.200 + 0.040		

$pH = 9.25 + \log\dfrac{[NH_3]}{[NH_4^+]} = 9.25 + \log\dfrac{(0.200 - 0.040)}{(0.200 + 0.040)} = 9.07$

16.68

	Acid	K_a	$pK_a = -\log K_a$
(a)	H_3BO_3	5.8×10^{-10}	9.24
(b)	HCO_2H	1.8×10^{-4}	3.74
(c)	$HOCl$	3.5×10^{-8}	7.46

The stronger the acid (the larger the K_a), the smaller is the pK_a.

16.70 $pH = pK_a + \log\dfrac{[base]}{[acid]} = pK_a + \log\dfrac{[HCO_2^-]}{[HCO_2H]}$

For HCO_2H, $K_a = 1.8 \times 10^{-4}$; $pK_a = -\log K_a = -\log(1.8 \times 10^{-4}) = 3.74$

$pH = 3.74 + \log\dfrac{(0.50)}{(0.25)} = 4.04$

16.72 $pH = pK_a + \log\dfrac{[base]}{[acid]} = pK_a + \log\dfrac{[NH_3]}{[NH_4^+]}$

For NH_4^+, $K_a = 5.6 \times 10^{-10}$; $pK_a = -\log K_a = -\log(5.6 \times 10^{-10}) = 9.25$

$$9.80 = 9.25 + \log \frac{[NH_3]}{[NH_4^+]}; \quad 0.550 = \log \frac{[NH_3]}{[NH_4^+]}; \quad \frac{[NH_3]}{[NH_4^+]} = 10^{0.55} = 3.5$$

The volume of the 1.0 M NH_3 solution should be 3.5 times the volume of the 1.0 M NH_4Cl solution so that the mixture will buffer at pH 9.80.

16.74 H_3PO_4, $K_{a1} = 7.5 \times 10^{-3}$; $pK_{a1} = -\log K_{a1} = 2.12$

$H_2PO_4^-$, $K_{a2} = 6.2 \times 10^{-8}$; $pK_{a2} = -\log K_{a2} = 7.21$

HPO_4^{2-}, $K_{a3} = 4.8 \times 10^{-13}$; $pK_{a3} = -\log K_{a3} = 12.32$

The buffer system of choice for pH 7.00 is (b) $H_2PO_4^- - HPO_4^{2-}$ because the pK_a for $H_2PO_4^-$ (7.21) is closest to 7.00.

pH Titration Curves

16.76   (a) (0.060 L)(0.150 mol/L)(1000 mmol/mol) = 9.00 mmol HNO_3

(b) vol NaOH = $(9.00 \text{ mmol } HNO_3)\left(\dfrac{1 \text{ mmol NaOH}}{1 \text{ mmol } HNO_3}\right)\left(\dfrac{1 \text{ mL NaOH}}{0.450 \text{ mmol NaOH}}\right)$ = 20.0 mL NaOH

(c) At the equivalence point the solution contains the neutral salt $NaNO_3$. The pH is 7.00.

(d)

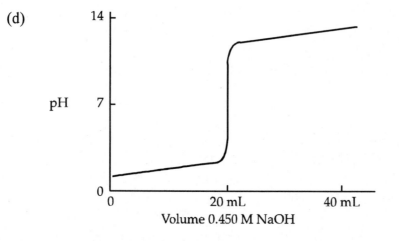

Volume 0.450 M NaOH

16.78   mmol OH^- = (20.0 mL)(0.150 mmol/mL) = 3.00 mmol

mmol acid present = mmol OH^- added = 3.00 mmol

$$[acid] = \frac{3.00 \text{ mmol}}{60.0 \text{ mL}} = 0.0500 \text{ M}$$

16.80 $HBr(aq) + NaOH(aq) \rightarrow Na^+(aq) + Br^-(aq) + H_2O(l)$

(a) $[H_3O^+] = 0.120$ M; pH = $-\log[H_3O^+]$ = $-\log (0.120) = 0.92$

(b) (50.0 mL)(0.120 mmol/mL) = 6.00 mmol HBr

(20.0 mL)(0.240 mmol/mL) = 4.80 mmol NaOH

6.00 mmol HBr − 4.80 mmol NaOH = 1.20 mmol HBr after neutralization

$$[H_3O^+] = \frac{1.20 \text{ mmol}}{(50.0 \text{ mL} + 20.0 \text{ mL})} = 0.0171 \text{ M}$$

$pH = -\log[H_3O^+] = -\log(0.0171) = 1.77$

(c) $(24.9\ \text{mL})(0.240\ \text{mmol/mL}) = 5.98\ \text{mmol NaOH}$

6.00 mmol HBr – 5.98 mmol NaOH = 0.02 mmol HBr after neutralization

$$[H_3O^+] = \frac{0.02\ \text{mmol}}{(50.0\ \text{mL}\ +\ 24.9\ \text{mL})} = 3 \times 10^{-4}\ \text{M}$$

$pH = -\log[H_3O^+] = -\log(3 \times 10^{-4}) = 3.5$

(d) The titration reaches the equivalence point when 25.0 mL of 0.240 M NaOH is added. At the equivalence point the solution contains the neutral salt NaBr. The pH is 7.00.

(e) $(25.1\ \text{mL})(0.240\ \text{mmol/mL}) = 6.024\ \text{mmol NaOH}$

6.024 mmol NaOH – 6.00 mmol HBr = 0.024 mmol NaOH after neutralization

$$[OH^-] = \frac{0.024\ \text{mmol}}{(50.0\ \text{mL}\ +\ 25.1\ \text{mL})} = 3.2 \times 10^{-4}\ \text{M}$$

$$[H_3O^+] = \frac{K_w}{[OH^-]} = \frac{1.0 \times 10^{-14}}{3.2 \times 10^{-4}} = 3.1 \times 10^{-11}\ \text{M}$$

$pH = -\log[H_3O^+] = -\log(3.1 \times 10^{-11}) = 10.5$

(f) $(40.0\ \text{mL})(0.240\ \text{mmol/mL}) = 9.60\ \text{mmol NaOH}$

9.60 mmol NaOH – 6.00 mmol HBr = 3.60 mmol NaOH after neutralization

$$[OH^-] = \frac{3.60\ \text{mmol}}{(50.0\ \text{mL}\ +\ 40.0\ \text{mL})} = 0.040\ \text{M}$$

$$[H_3O^+] = \frac{K_w}{[OH^-]} = \frac{1.0 \times 10^{-14}}{0.040} = 2.5 \times 10^{-13}\ \text{M}$$

$pH = -\log[H_3O^+] = -\log(2.5 \times 10^{-13}) = 12.60$

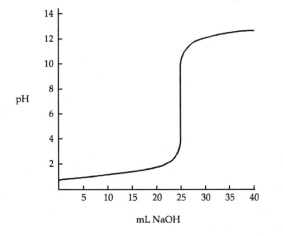

16.82 $\text{mmol HF} = (40.0\ \text{mL})(0.250\ \text{mmol/mL}) = 10.0\ \text{mmol}$

mmol NaOH required = mmol HF = 10.0 mmol

$$\text{mL NaOH required} = (10.0\ \text{mmol})\left(\frac{1.00\ \text{mL}}{0.200\ \text{mmol}}\right) = 50.0\ \text{mL}$$

50.0 mL of 0.200 M NaOH is required to reach the equivalence point.

For HF, $K_a = 3.5 \times 10^{-4}$; $pK_a = -\log K_a = -\log(3.5 \times 10^{-4}) = 3.46$

(a) mmol HF = 10.0 mmol

mmol NaOH = (0.200 mmol/mL)(10.0 mL) = 2.00 mmol

Neutralization reaction: $HF(aq) + OH^-(aq) \rightarrow F^-(aq) + H_2O(l)$

before reaction (mmol)	10.0	2.00	0
change (mmol)	−2.00	−2.00	+2.00
after reaction (mmol)	8.0	0	2.00

$$[HF] = \frac{8.0 \text{ mmol}}{(40.0 \text{ mL} + 10.0 \text{ mL})} = 0.16 \text{ M}; \quad [F^-] = \frac{2.00 \text{ mmol}}{(40.0 \text{ mL} + 10.0 \text{ mL})} = 0.0400 \text{ M}$$

$$HF(aq) + H_2O(l) \rightleftharpoons H_3O^+(aq) + F^-(aq)$$

initial (M)	0.16	~0	0.0400
change (M)	−x	+x	+x
equil (M)	0.16 − x	x	0.0400 + x

$$K_a = \frac{[H_3O^+][F^-]}{[HF]} = 3.5 \times 10^{-4} = \frac{x(0.0400 + x)}{0.16 - x} \approx \frac{x(0.0400)}{0.16}$$

Solve for x. $x = [H_3O^+] = 1.4 \times 10^{-3}$ M

$pH = -\log[H_3O^+] = -\log(1.4 \times 10^{-3}) = 2.85$

(b) Halfway to the equivalence point,

$pH = pK_a = -\log K_a = -\log(3.5 \times 10^{-4}) = 3.46$

(c) At the equivalence point only the salt NaF is in solution.

$$[F^-] = \frac{10.0 \text{ mmol}}{(40.0 \text{ mL} + 50.0 \text{ mL})} = 0.111 \text{ M}$$

$$F^-(aq) + H_2O(l) \rightleftharpoons HF(aq) + OH^-(aq)$$

initial (M)	0.111	0	~0
change (M)	−x	+x	+x
equil (M)	0.111 − x	x	x

$$\text{For } F^-, K_b = \frac{K_w}{K_a \text{ for HF}} = \frac{1.0 \times 10^{-14}}{3.5 \times 10^{-4}} = 2.9 \times 10^{-11}$$

$$K_b = \frac{[HF][OH^-]}{[F^-]} = 2.9 \times 10^{-11} = \frac{x^2}{0.111 - x} \approx \frac{x^2}{0.111}$$

Solve for x. $x = [OH^-] = 1.8 \times 10^{-6}$ M

$$[H_3O^+] = \frac{K_w}{[OH^-]} = \frac{1.0 \times 10^{-14}}{1.8 \times 10^{-6}} = 5.6 \times 10^{-9} \text{ M}$$

$pH = -\log[H_3O^+] = -\log(5.6 \times 10^{-9}) = 8.25$

(d) mmol HF = 10.0 mmol

mol NaOH = (0.200 mmol/mL)(80.0 mL) = 16.0 mmol

Neutralization reaction: $HF(aq) + OH^-(aq) \rightarrow F^-(aq) + H_2O(l)$

before reaction (mmol)	10.0	16.0	0
change (mmol)	−10.0	−10.0	+10.0
after reaction (mmol)	0	6.0	10.0

After the equivalence point, the pH of the solution is determined by the $[OH^-]$.

$$[OH^-] = \frac{6.0 \text{ mmol}}{(40.0 \text{ mL} + 80.0 \text{ mL})} = 5.0 \times 10^{-2} \text{ M}$$

$$[H_3O^+] = \frac{K_w}{[OH^-]} = \frac{1.0 \times 10^{-14}}{5.0 \times 10^{-2}} = 2.0 \times 10^{-13} \text{ M}$$

$$pH = -\log[H_3O^+] = -\log(2.0 \times 10^{-13}) = 12.70$$

16.84 For H_2A^+, $K_{a1} = 4.6 \times 10^{-3}$ and $K_{a2} = 2.0 \times 10^{-10}$
(a) (10.0 mL)(0.100 mmol/mL) = 1.00 mmol NaOH added = 1.00 mmol HA produced.
(50.0 mL)(0.100 mmol/mL) = 5.00 mmol H_2A^+
5.00 mmol H_2A^+ – 1.00 mmol NaOH = 4.00 mmol H_2A^+ after neutralization

$$[H_2A^+] = \frac{4.00 \text{ mmol}}{(50.0 \text{ mL} + 10.0 \text{ mL})} = 6.67 \times 10^{-2} \text{ M}$$

$$[HA] = \frac{1.00 \text{ mmol}}{(50.0 \text{ mL} + 10.0 \text{ mL})} = 1.67 \times 10^{-2} \text{ M}$$

$$pH = pK_{a1} + \log \frac{[HA]}{[H_2A^+]} = -\log(4.6 \times 10^{-3}) + \log\left(\frac{1.67 \times 10^{-2}}{6.67 \times 10^{-2}}\right) = 1.74$$

(b) Halfway to the first equivalence point, pH = pK_{a1} = 2.34

(c) At the first equivalence point, pH = $\dfrac{pK_{a1} + pK_{a2}}{2}$ = 6.02

(d) Halfway between the first and second equivalence points, pH = pK_{a2} = 9.70
(e) At the second equivalence point only the basic salt, NaA, is in solution.

$$K_b = \frac{K_w}{K_a \text{ for HA}} = \frac{K_w}{K_{a2}} = \frac{1.0 \times 10^{-14}}{2.0 \times 10^{-10}} = 5.0 \times 10^{-5}$$

mmol A^- = (50.0 mL)(0.100 mmol/mL) = 5.00 mmol

$$[A^-] = \frac{5.0 \text{ mmol}}{(50.0 \text{ mL} + 100.0 \text{ mL})} = 3.3 \times 10^{-2} \text{ M}$$

	A^-(aq)	+	H_2O(l)	⇌	HA(aq)	+	OH^-(aq)
initial (M)	0.033				0		~0
change (M)	–x				+x		+x
equil (M)	0.033 – x				x		x

$$K_b = \frac{[HA][OH^-]}{[A^-]} = 5.0 \times 10^{-5} = \frac{(x)(x)}{0.033 - x} \approx \frac{x^2}{0.033}$$

Solve for x.

$$x = [OH^-] = \sqrt{(5.0 \times 10^{-5})(0.033)} = 1.3 \times 10^{-3} \text{ M}$$

$$[H_3O^+] = \frac{K_w}{[OH^-]} = \frac{1.0 \times 10^{-14}}{1.3 \times 10^{-3}} = 7.7 \times 10^{-12} \text{ M}$$

$$pH = -\log[H_3O^+] = -\log(7.7 \times 10^{-12}) = 11.11$$

16.86 When equal volumes of acid and base react, all concentrations are cut in half.
(a) At the equivalence point only the salt $NaNO_2$ is in solution.
$[NO_2^-] = 0.050$ M

For NO_2^-, $K_b = \dfrac{K_w}{K_a \text{ for } HNO_2} = \dfrac{1.0 \times 10^{-14}}{4.5 \times 10^{-4}} = 2.2 \times 10^{-11}$

$$NO_2^-(aq) + H_2O(l) \rightleftharpoons HNO_2(aq) + OH^-(aq)$$

Initial (M)	0.050	0	~0
change (M)	−x	+x	+x
equil (M)	0.050 − x	x	x

$K_b = \dfrac{[HNO_2][OH^-]}{[NO_2^-]} = 2.2 \times 10^{-11} = \dfrac{(x)(x)}{0.050 - x} \approx \dfrac{x^2}{0.050}$

Solve for x. $x = [OH^-] = 1.1 \times 10^{-6}$ M

$[H_3O^+] = \dfrac{K_w}{[OH^-]} = \dfrac{1.0 \times 10^{-14}}{1.1 \times 10^{-6}} = 9.1 \times 10^{-9}$ M

pH = $-\log[H_3O^+] = -\log(9.1 \times 10^{-9}) = 8.04$
Phenol red would be a suitable indicator. (see Figure 15.4)
(b) The pH is 7.00 at the equivalence point for the titration of a strong acid (HI) with a strong base (NaOH).
Bromthymol blue or phenol red would be suitable indicators. (Any indicator that changes color in the pH range 4 – 10 is satisfactory for a strong acid – strong base titration.)
(c) At the equivalence point only the salt CH_3NH_3Cl is in solution.
$[CH_3NH_3^+] = 0.050$ M

For $CH_3NH_3^+$, $K_a = \dfrac{K_w}{K_b \text{ for } CH_3NH_2} = \dfrac{1.0 \times 10^{-14}}{3.7 \times 10^{-4}} = 2.7 \times 10^{-11}$

$$CH_3NH_3^+(aq) + H_2O(l) \rightleftharpoons H_3O^+(aq) + CH_3NH_2(aq)$$

initial (M)	0.050	~0	0
change (M)	−x	+x	+x
equil (M)	0.050 − x	x	x

$K_a = \dfrac{[H_3O^+][CH_3NH_2]}{[CH_3NH_3^+]} = 2.7 \times 10^{-11} = \dfrac{(x)(x)}{0.050 - x} \approx \dfrac{x^2}{0.050}$

Solve for x. $x = [H_3O^+] = 1.2 \times 10^{-6}$ M
pH = $-\log[H_3O^+] = -\log(1.2 \times 10^{-6}) = 5.92$
Chlorphenol red would be a suitable indicator.

Solubility Equilibria

16.88 (a) $Ag_2CO_3(s) \rightleftharpoons 2 Ag^+(aq) + CO_3^{2-}(aq)$ $K_{sp} = [Ag^+]^2[CO_3^{2-}]$
(b) $PbCrO_4(s) \rightleftharpoons Pb^{2+}(aq) + CrO_4^{2-}(aq)$ $K_{sp} = [Pb^{2+}][CrO_4^{2-}]$
(c) $Al(OH)_3(s) \rightleftharpoons Al^{3+}(aq) + 3 OH^-(aq)$ $K_{sp} = [Al^{3+}][OH^-]^3$
(d) $Hg_2Cl_2(s) \rightleftharpoons Hg_2^{2+}(aq) + 2 Cl^-(aq)$ $K_{sp} = [Hg_2^{2+}][Cl^-]^2$

16.90   (a) $K_{sp} = [Pb^{2+}][I^-]^2 = (5.0 \times 10^{-3})(1.3 \times 10^{-3})^2 = 8.4 \times 10^{-9}$

      (b) $[I^-] = \sqrt{\dfrac{K_{sp}}{[Pb^{2+}]}} = \sqrt{\dfrac{(8.4 \times 10^{-9})}{(2.5 \times 10^{-4})}} = 5.8 \times 10^{-3}$ M

      (c) $[Pb^{2+}] = \dfrac{K_{sp}}{[I^-]^2} = \dfrac{(8.4 \times 10^{-9})}{(2.5 \times 10^{-4})^2} = 0.13$ M

16.92 $Ag_2CO_3(s) \;\rightleftharpoons\; 2\,Ag^+(aq) \;+\; CO_3^{2-}(aq)$

     equil (M) $2x$ x

$[Ag^+] = 2x = 2.56 \times 10^{-4}$ M; $[CO_3^{2-}] = x = (2.56 \times 10^{-4} \text{ M})/2 = 1.28 \times 10^{-4}$ M

$K_{sp} = [Ag^+]^2[CO_3^{2-}] = (2.56 \times 10^{-4})^2(1.28 \times 10^{-4}) = 8.39 \times 10^{-12}$

16.94   (a) $BaCrO_4(s) \;\rightleftharpoons\; Ba^{2+}(aq) + CrO_4^{2-}(aq)$

       equil (M) x x

$K_{sp} = [Ba^{2+}][CrO_4^{2-}] = 1.2 \times 10^{-10} = (x)(x)$

molar solubility $= x = \sqrt{1.2 \times 10^{-10}} = 1.1 \times 10^{-5}$ M

      (b) $Mg(OH)_2(s) \;\rightleftharpoons\; Mg^{2+}(aq) + 2\,OH^-(aq)$

       equil (M) x $2x$

$K_{sp} = [Mg^{2+}][OH^-]^2 = 5.6 \times 10^{-12} = x(2x)^2 = 4x^3$

molar solubility $= x = \sqrt[3]{\dfrac{5.6 \times 10^{-12}}{4}} = 1.1 \times 10^{-4}$ M

      (c) $Ag_2SO_3(s) \;\rightleftharpoons\; 2\,Ag^+(aq) + SO_3^{2-}(aq)$

       equil (M) $2x$ x

$K_{sp} = [Ag^+]^2[SO_3^{2-}] = 1.5 \times 10^{-14} = (2x)^2x = 4x^3$

molar solubility $= x = \sqrt[3]{\dfrac{1.5 \times 10^{-14}}{4}} = 1.6 \times 10^{-5}$ M

Factors That Affect Solubility

16.96 $Ag_2CO_3(s) \;\rightleftharpoons\; 2\,Ag^+(aq) + CO_3^{2-}(aq)$

      (a) $AgNO_3$, source of Ag^+; equilibrium shifts left

      (b) HNO_3, source of H_3O^+, removes CO_3^{2-}; equilibrium shifts right

      (c) Na_2CO_3, source of CO_3^{2-}; equilibrium shifts left

      (d) NH_3, forms $Ag(NH_3)_2^+$; removes Ag^+; equilibrium shifts right

16.98   (a) $PbCrO_4(s) \;\rightleftharpoons\; Pb^{2+}(aq) + CrO_4^{2-}(aq)$

       equil (M) x x

$K_{sp} = [Pb^{2+}][CrO_4^{2-}] = 2.8 \times 10^{-13} = (x)(x)$

molar solubility $= x = \sqrt{2.8 \times 10^{-13}} = 5.3 \times 10^{-7}$ M

(b)
$$PbCrO_4(s) \rightleftharpoons Pb^{2+}(aq) + CrO_4^{2-}(aq)$$

	Pb^{2+}	CrO_4^{2-}
initial(M)	0	1.0×10^{-3}
equil (M)	x	$1.0 \times 10^{-3} + x$

$K_{sp} = [Pb^{2+}][CrO_4^{2-}] = 1.2 \times 10^{-10} = (x)(1.0 \times 10^{-3} + x) \approx (x)(1.0 \times 10^{-3})$

molar solubility $= x = \dfrac{2.8 \times 10^{-13}}{1 \times 10^{-3}} = 2.8 \times 10^{-10}$ M

16.100 (b), (c), and (d) are more soluble in acidic solution.

(a) $AgBr(s) \rightleftharpoons Ag^+(aq) + Br^-(aq)$

(b) $CaCO_3(s) + H_3O^+(aq) \rightleftharpoons Ca^{2+}(aq) + HCO_3^-(aq) + H_2O(l)$

(c) $Ni(OH)_2(s) + 2 H_3O^+(aq) \rightleftharpoons Ni^{2+}(aq) + 4 H_2O(l)$

(d) $Ca_3(PO_4)_2(s) + 2 H_3O^+(aq) \rightleftharpoons 3 Ca^{2+}(aq) + 2 HPO_4^{2-}(aq) + 2 H_2O(l)$

16.102 On mixing equal volumes of two solutions, the concentrations of both solutions are cut in half.

	$Ag^+(aq)$	+	$2 CN^-(aq)$	$\rightleftharpoons$	$Ag(CN)_2^-(aq)$
before reaction (M)	0.0010		0.10		0
assume 100% reaction	−0.0010		−2(0.0010)		0.0010
after reaction (M)	0		0.098		0.0010
assume small back rxn	+x		+2x		−x
equil (M)	x		0.098 + 2x		0.0010 − x

$K_f = 3.0 \times 10^{20} = \dfrac{[Ag(CN)_2^-]}{[Ag^+][CN^-]^2} = \dfrac{(0.0010 - x)}{x(0.098 + 2x)^2} \approx \dfrac{0.0010}{x(0.098)^2}$

Solve for x. $x = [Ag^+] = 3.5 \times 10^{-22}$ M

16.104 (a)

$$AgI(s) \rightleftharpoons Ag^+(aq) + I^-(aq) \qquad K_{sp} = 8.5 \times 10^{-17}$$
$$\underline{Ag^+(aq) + 2 CN^-(aq) \rightarrow Ag(CN)_2^-(aq)} \qquad K_f = 3.0 \times 10^{20}$$

dissolution rxn $AgI(s) + 2 CN^-(aq) \rightleftharpoons Ag(CN)_2^-(aq) + I^-(aq)$

$K = (K_{sp})(K_f) = (8.5 \times 10^{-17})(3.0 \times 10^{20}) = 2.6 \times 10^4$

(b)

$$Al(OH)_3(s) \rightleftharpoons Al^{3+}(aq) + 3 OH^-(aq) \qquad K_{sp} = 1.9 \times 10^{-33}$$
$$\underline{Al^{3+}(aq) + 4 OH^-(aq) \rightarrow Al(OH)_4^-(aq)} \qquad K_f = 3 \times 10^{33}$$

dissolution rxn $Al(OH)_3(s) + OH^-(aq) \rightleftharpoons Al(OH)_4^-(aq)$

$K = (K_{sp})(K_f) = (1.9 \times 10^{-33})(3 \times 10^{33}) = 6$

(c)

$$Zn(OH)_2(s) \rightleftharpoons Zn^{2+}(aq) + 2 OH^-(aq) \qquad K_{sp} = 4.1 \times 10^{-17}$$
$$\underline{Zn^{2+}(aq) + 4 NH_3(aq) \rightarrow Zn(NH_3)_4^{2+}(aq)} \qquad K_f = 7.8 \times 10^8$$

dissolution rxn $Zn(OH)_2(s) + 4 NH_3(aq) \rightleftharpoons Zn(NH_3)_4^{2+} + 2 OH^-(aq)$

$K = (K_{sp})(K_f) = (4.1 \times 10^{-17})(7.8 \times 10^8) = 3.2 \times 10^{-8}$

16.106 (a)

$$AgI(s) \rightleftarrows Ag^+(aq) + I^-(aq)$$

equil (M)   x   x

$K_{sp} = [Ag^+][I^-] = 8.5 \times 10^{-17} = (x)(x)$

molar solubility $= x = \sqrt{8.5 \times 10^{-17}} = 9.2 \times 10^{-9}$ M

(b)

$$AgI(s) + 2\,CN^-(aq) \rightleftarrows Ag(CN)_2^-(aq) + I^-(aq)$$

initial (M)   0.10   0   0
change (M)   −2x   +x   +x
equil (M)   0.10 − 2x   x   x

$K = (K_{sp})(K_f) = (8.5 \times 10^{-17})(3.0 \times 10^{20}) = 2.6 \times 10^4$

$$K = 2.6 \times 10^4 = \frac{[Ag(CN)_2^-][I^-]}{[CN^-]^2} = \frac{x^2}{(0.10 - 2x)^2}$$

Take the square root of both sides and solve for x.
molar solubility $= x = 0.050$ M

Precipitation; Qualitative Analysis

16.108 For $BaSO_4$, $K_{sp} = 1.1 \times 10^{-10}$
Total volume = 300 mL + 100 mL = 400 mL

$$[Ba^{2+}] = \frac{(4.0 \times 10^{-3}\ M)(100\ mL)}{(400\ mL)} = 1.0 \times 10^{-3}\ M$$

$$[SO_4^{2-}] = \frac{(6.0 \times 10^{-4}\ M)(300\ mL)}{(400\ mL)} = 4.5 \times 10^{-4}\ M$$

$IP = [Ba^{2+}]_t[SO_4^{2-}]_t = (1.0 \times 10^{-3})(4.5 \times 10^{-4}) = 4.5 \times 10^{-7}$
$IP > K_{sp}$; $BaSO_4(s)$ will precipitate.

16.110 $BaSO_4$, $K_{sp} = 1.1 \times 10^{-10}$; $Fe(OH)_3$, $K_{sp} = 2.6 \times 10^{-39}$
Total volume = 80 mL + 20 mL = 100 mL

$$[Ba^{2+}] = \frac{(1.0 \times 10^{-5}\ M)(80\ mL)}{(100\ mL)} = 8.0 \times 10^{-6}\ M$$

$[OH^-] = 2[Ba^{2+}] = 2(8.0 \times 10^{-6}) = 1.6 \times 10^{-5}\ M$

$$[Fe^{3+}] = \frac{2(1.0 \times 10^{-5}\ M)(20\ mL)}{(100\ mL)} = 4.0 \times 10^{-6}\ M$$

$$[SO_4^{2-}] = \frac{3(1.0 \times 10^{-5}\ M)(20\ mL)}{(100\ mL)} = 6.0 \times 10^{-6}\ M$$

For $BaSO_4$, $IP = [Ba^{2+}]_t[SO_4^{2-}]_t = (8.0 \times 10^{-6})(6.0 \times 10^{-6}) = 4.8 \times 10^{-11}$
$IP < K_{sp}$; $BaSO_4$ will not precipitate.
For $Fe(OH)_3$, $IP = [Fe^{3+}]_t[OH^-]_t^3 = (4.0 \times 10^{-6})(1.6 \times 10^{-5})^3 = 1.6 \times 10^{-20}$
$IP > K_{sp}$; $Fe(OH)_3(s)$ will precipitate.

16.112 pH = 10.80; $[H_3O^+] = 10^{-pH} = 10^{-10.80} = 1.6 \times 10^{-11}$ M

$$[OH^-] = \frac{K_w}{[H_3O^+]} = \frac{1.0 \times 10^{-14}}{1.6 \times 10^{-11}} = 6.2 \times 10^{-4} \text{ M}$$

For $Mg(OH)_2$, $K_{sp} = 5.6 \times 10^{-12}$
$IP = [Mg^{2+}]_t[OH^-]_t^2 = (2.5 \times 10^{-4})(6.2 \times 10^{-4})^2 = 9.6 \times 10^{-11}$
$IP > K_{sp}$; $Mg(OH)_2(s)$ will precipitate

16.114 $K_{spa} = \dfrac{[M^{2+}][H_2S]}{[H_3O^+]^2}$; FeS, $K_{spa} = 6 \times 10^2$; SnS, $K_{spa} = 1 \times 10^{-5}$

Fe^{2+} and Sn^{2+} can be separated by bubbling H_2S through an acidic solution containing the two cations because their K_{spa} values are so different.

For FeS and SnS, $Q_c = \dfrac{(0.01)(0.10)}{(0.3)^2} = 1.1 \times 10^{-2}$

For FeS, $Q_c < K_{spa}$, and no FeS will precipitate.
For SnS, $Q_c > K_{spa}$, and SnS will precipitate.

16.116 (a) add Cl^- to precipitate AgCl
　　　　(b) add CO_3^{2-} to precipitate $CaCO_3$
　　　　(c) add H_2S to precipitate MnS
　　　　(d) add NH_3 and NH_4Cl to precipitate $Cr(OH)_3$
　　　　(Need a buffer to control $[OH^-]$; excess OH^- produces the soluble $Cr(OH)_4^-$.)

General Problems

16.118 Prepare aqueous solutions of the three salts. Add a solution of $(NH_4)_2HPO_4$. If a white precipitate forms, the solution contains Mg^{2+}. Perform flame test on the other two solutions. A yellow flame test indicates Na^+. A violet flame test indicates K^+.

16.120 (a), solution contains H_2CO_3 and HCO_3^-
　　　　(b), solution contains HCO_3^- and CO_3^{2-}
　　　　(d), solution contains HCO_3^- and CO_3^{2-}

16.122 (a)

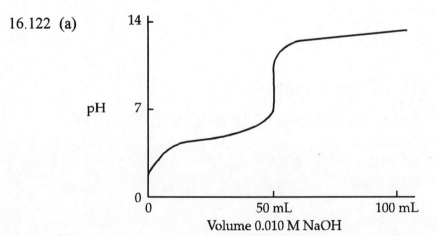

(b) mol NaOH required = $\left(\dfrac{0.010 \text{ mol HA}}{\text{L}}\right)(0.0500 \text{ L})\left(\dfrac{1 \text{ mol NaOH}}{1 \text{ mol HA}}\right) = 0.000\ 50 \text{ mol}$

vol NaOH required = $(0.000\ 50 \text{ mol})\left(\dfrac{1 \text{ L}}{0.010 \text{ mol}}\right) = 0.050 \text{ L} = 50 \text{ mL}$

(c) A basic salt is present at the equivalence point; pH > 7.00

(d) Halfway to the equivalence point, the pH = pK_a = 4.00

16.124 For NH_4^+, $K_a = \dfrac{K_w}{K_b \text{ for } NH_3} = \dfrac{1.0 \times 10^{-14}}{1.8 \times 10^{-5}} = 5.6 \times 10^{-10}$

$pK_a = -\log K_a = -\log(5.6 \times 10^{-10}) = 9.25$

$pH = pK_a + \log\dfrac{[NH_3]}{[NH_4^+]};\qquad 9.40 = 9.25 + \log\dfrac{[NH_3]}{[NH_4^+]}$

$\log\dfrac{[NH_3]}{[NH_4^+]} = 9.40 - 9.25 = 0.15;\qquad \dfrac{[NH_3]}{[NH_4^+]} = 10^{0.15} = 1.41$

Because the volume is the same for both NH_3 and NH_4^+, $\dfrac{\text{mol } NH_3}{\text{mol } NH_4^+} = 1.41$.

$\text{mol } NH_3 = (0.20 \text{ mol/L})(0.250 \text{ L}) = 0.050 \text{ mol } NH_3$

$\text{mol } NH_4^+ = \dfrac{\text{mol } NH_3}{1.41} = \dfrac{0.050}{1.41} = 0.035 \text{ mol } NH_4^+$

$\text{vol } NH_4^+ = (0.035 \text{ mol})\left(\dfrac{1 \text{ L}}{3.0 \text{ mol}}\right) = 0.012 \text{ L} = 12 \text{ mL}$

12 mL of 3.0 M NH_4Cl must be added to 250 mL of 0.20 M NH_3 to obtain a buffer solution having a pH = 9.40.

16.126 pH = 10.35; $[H_3O^+] = 10^{-pH} = 10^{-10.35} = 4.5 \times 10^{-11}$ M

$[OH^-] = \dfrac{K_w}{[H_3O^+]} = \dfrac{1.0 \times 10^{-14}}{4.5 \times 10^{-11}} = 2.2 \times 10^{-4}$ M

$[Mg^{2+}] = \dfrac{[OH^-]}{2} = \dfrac{2.2 \times 10^{-4}}{2} = 1.1 \times 10^{-4}$ M

$K_{sp} = [Mg^{2+}][OH^-]^2 = (1.1 \times 10^{-4})(2.2 \times 10^{-4})^2 = 5.3 \times 10^{-12}$

16.128 NaOH, 40.0 amu; $20 \text{ g} \times \dfrac{1 \text{ mol}}{40.0 \text{ g}} = 0.50 \text{ mol NaOH}$

$(0.500 \text{ L})(1.5 \text{ mol/L}) = 0.75 \text{ mol } NH_4Cl$

	$NH_4^+(aq)$	+	$OH^-(aq)$	$\rightleftharpoons$	$NH_3(aq)$	+	$H_2O(l)$
before reaction (mol)	0.75		0.50		0		
change (mol)	−0.50		−0.50		+0.50		
after reaction (mol)	0.25		0		0.50		

This reaction produces a buffer solution.

$[NH_4^+] = 0.25$ mol/0.500 L $= 0.50$ M; $[NH_3] = 0.50$ mol/0.500 L $= 1.0$ M

$$pH = pK_a + \log\frac{[base]}{[acid]} = pK_a + \log\frac{[NH_3]}{[NH_4^+]}$$

For NH_4^+, $K_a = \dfrac{K_w}{K_b \text{ for } NH_3} = \dfrac{1.0 \times 10^{-14}}{1.8 \times 10^{-5}} = 5.6 \times 10^{-10}$; $pK_a = -\log K_a = 9.25$

$$pH = 9.25 + \log\left(\frac{1.0}{0.5}\right) = 9.55$$

16.130 For NH_4^+, $K_a = \dfrac{K_w}{K_b \text{ for } NH_3} = \dfrac{1.0 \times 10^{-14}}{1.8 \times 10^{-5}} = 5.6 \times 10^{-10}$; $pK_a = -\log K_a = 9.25$

$$pH = pK_a + \log\frac{[NH_3]}{[NH_4^+]} = 9.25 + \log\frac{(0.50)}{(0.30)} = 9.47$$

$[H_3O^+] = 10^{-pH} = 10^{-9.47} = 3.4 \times 10^{-10}$ M

For MnS, $K_{spa} = \dfrac{[Mn^{2+}][H_2S]}{[H_3O^+]^2} = 3 \times 10^{10}$

molar solubility $= [Mn^{2+}] = \dfrac{K_{spa}[H_3O^+]^2}{[H_2S]} = \dfrac{(3 \times 10^{10})(3.4 \times 10^{-10})^2}{(0.10)} = 3.5 \times 10^{-8}$ M

MnS, 87.00 amu;   solubility $= (3.5 \times 10^{-8}$ mol/L$)(87.00$ g/mol$) = 3 \times 10^{-6}$ g/L

16.132 60.0 mL $= 0.0600$ L

mol $H_3PO_4 = 0.0600$ L $\times \dfrac{1.00 \text{ mol } H_3PO_4}{1.00 \text{ L}} = 0.0600$ mol H_3PO_4

mol LiOH $= 1.00$ L $\times \dfrac{0.100 \text{ mol LiOH}}{1.00 \text{ L}} = 0.100$ mol LiOH

	$H_3PO_4(aq)$	$+$ $OH^-(aq)$	$\rightarrow$ $H_2PO_4^-(aq)$	$+$ $H_2O(l)$
before reaction (mol)	0.0600	0.100	0	
change (mol)	−0.0600	−0.0600	+0.0600	
after reaction (mol)	0	0.040	0.0600	
	$H_2PO_4^-(aq)$	$+$ $OH^-(aq)$	$\rightarrow$ $HPO_4^{2-}(aq)$	$+$ $H_2O(l)$
before reaction (mol)	0.0600	0.040	0	
change (mol)	−0.040	−0.040	+0.040	
after reaction (mol)	0.020	0	0.040	

The resulting solution is a buffer because it contains the conjugate acid-base pair, $H_2PO_4^-$ and HPO_4^{2-}, at acceptable buffer concentrations.

For $H_2PO_4^-$, $K_{a2} = 6.2 \times 10^{-8}$ and $pK_{a2} = -\log K_{a2} = -\log (6.2 \times 10^{-8}) = 7.21$

$$pH = pK_{a2} + \log\frac{[HPO_4^{2-}]}{[H_2PO_4^-]} = 7.21 + \log\frac{(0.040 \text{ mol}/1.06 \text{ L})}{(0.020 \text{ mol}/1.06 \text{ L})}$$

$$pH = 7.21 + \log \frac{(0.040)}{(0.020)} = 7.21 + 0.30 = 7.51$$

16.134 For CH_3CO_2H, $K_a = 1.8 \times 10^{-5}$ and $pK_a = -\log K_a = -\log(1.8 \times 10^{-5}) = 4.74$
The mixture will be a buffer solution containing the conjugate acid-base pair, CH_3CO_2H and $CH_3CO_2^-$, having a pH near the pK_a of CH_3CO_2H.

$$pH = pK_a + \log \frac{[CH_3CO_2^-]}{[CH_3CO_2H]}$$

$$4.85 = 4.74 + \log \frac{[CH_3CO_2^-]}{[CH_3CO_2H]}; \quad 4.85 - 4.74 = \log \frac{[CH_3CO_2^-]}{[CH_3CO_2H]}$$

$$0.11 = \log \frac{[CH_3CO_2^-]}{[CH_3CO_2H]}; \quad \frac{[CH_3CO_2^-]}{[CH_3CO_2H]} = 10^{0.11} = 1.3$$

In the Henderson-Hasselbalch equation, moles can be used in place of concentrations because both components are in the same volume so the volume terms cancel.
20.0 mL = 0.0200 L
Let X equal the volume of 0.10 M CH_3CO_2H and Y equal the volume of 0.15 M $CH_3CO_2^-$.
Therefore, X + Y = 0.0200 L and

$$\frac{Y \times [CH_3CO_2^-]}{X \times [CH_3CO_2H]} = \frac{Y(0.15 \text{ mol/L})}{X(0.10 \text{ mol/L})} = 1.3$$

X = 0.0200 − Y

$$\frac{Y(0.15 \text{ mol/L})}{(0.020 - Y)(0.10 \text{ mol/L})} = 1.3$$

$$\frac{0.15Y}{0.0020 - 0.10Y} = 1.3$$

0.15Y = 1.3(0.0020 − 0.10Y)
0.15Y = 0.0026 − 0.13Y
0.15Y + 0.13Y = 0.0026
0.28Y = 0.0026
Y = 0.0026/0.28 = 0.0093 L
X = 0.0200 − Y = 0.0200 − 0.0093 = 0.0107 L
X = 0.0107 L = 10.7 mL and Y = 0.0093 L = 9.3 mL
You need to mix together 10.7 mL of 0.10 M CH_3CO_2H and 9.3 mL of 0.15 M $NaCH_3CO_2$ to prepare 20.0 mL of a solution with a pH of 4.85.

16.136 (a) HCl is a strong acid. HCN is a weak acid with $K_a = 4.9 \times 10^{-10}$. Before the titration, the $[H_3O^+] = 0.100$ M. The HCN contributes an insignificant amount of additional H_3O^+, so the $pH = -\log[H_3O^+] = -\log(0.100) = 1.00$
(b) 100.0 mL = 0.1000 L

$$\text{mol } H_3O^+ = 0.1000 \text{ L} \times \frac{0.100 \text{ mol HCl}}{1.00 \text{ L}} = 0.0100 \text{ mol } H_3O^+$$

add 75.0 mL of 0.100 M NaOH; 75.0 mL = 0.0750 L

$$\text{mol OH}^- = 0.0750 \text{ L} \times \frac{0.100 \text{ mol NaOH}}{1.00 \text{ L}} = 0.00750 \text{ mol OH}^-$$

	$H_3O^+(aq)$ +	$OH^-(aq)$	$\rightarrow$ $2 H_2O(l)$
before reaction (mol)	0.0100	0.0075	
change (mol)	−0.0075	−0.0075	
after reaction (mol)	0.0025	0	

$$[H_3O^+] = \frac{0.0025 \text{ mol } H_3O^+}{0.1000 \text{ L} + 0.0750 \text{ L}} = 0.0143 \text{ M}$$

$$pH = -\log[H_3O^+] = -\log(0.0143) = 1.84$$

(c) 100.0 mL of 0.100 M NaOH will completely neutralize all of the H_3O^+ from 100.0 mL of 0.100 M HCl. Only NaCl and HCN remain in the solution. NaCl is a neutral salt and does not affect the pH of the solution. [HCN] changes because of dilution. Because the solution volume is doubled, [HCN] is cut in half.

[HCN] = 0.100 M/2 = 0.0500 M

	$HCN(aq)$ +	$H_2O(l)$ $\rightleftharpoons$	$H_3O^+(aq)$ +	$CN^-(aq)$
initial (M)	0.0500		~0	0
change (M)	−x		+x	+x
equil (M)	0.0500 − x		x	x

$$K_a = \frac{[H_3O^+][CN^-]}{HCN} = 4.9 \times 10^{-10} = \frac{x^2}{0.0500 - x} \approx \frac{x^2}{0.0500}$$

$$[H_3O^+] = x = \sqrt{(0.0500)(4.9 \times 10^{-10})} = 4.95 \times 10^{-6} \text{ M}$$

$$pH = -\log[H_3O^+] = -\log(4.95 \times 10^{-6}) = 5.31$$

(d) Add an additional 25.0 mL of 0.100 M NaOH.

25.0 mL = 0.0250 L

$$\text{additional mol OH}^- = 0.0250 \text{ L} \times \frac{0.100 \text{ mol NaOH}}{1.00 \text{ L}} = 0.00250 \text{ mol OH}^-$$

$$\text{mol HCN} = 0.200 \text{ L} \times \frac{0.0500 \text{ mol HCN}}{1.00 \text{ L}} = 0.0100 \text{ mol HCN}$$

	$HCN(aq)$ +	$OH^-(aq)$	$\rightarrow$ $CN^-(aq)$ +	$H_2O(l)$
before reaction (mol)	0.0100	0.00250	0	
change (mol)	−0.00250	−0.00250	+0.00250	
after reaction (mol)	0.0075	0	0.00250	

The resulting solution is a buffer because it contains the conjugate acid-base pair, HCN and CN^-, at acceptable buffer concentrations.

For HCN, $K_a = 4.9 \times 10^{-10}$ and $pK_a = -\log K_a = -\log(4.9 \times 10^{-10}) = 9.31$

$$pH = pK_a + \log\frac{[CN^-]}{[HCN]} = 9.31 + \log\frac{(0.00250 \text{ mol}/0.2250 \text{ L})}{(0.0075 \text{ mol}/0.2250 \text{ L})}$$

$$pH = 9.31 + \log\frac{(0.00250)}{(0.0075)} = 9.31 - 0.48 = 8.83$$

16.138 (a) $\quad\quad\quad Zn(OH)_2(s) \rightleftharpoons Zn^{2+}(aq) + 2\ OH^-(aq)$

initial (M) $\quad\quad\quad\quad\quad\quad\quad 0 \quad\quad\quad\quad \sim 0$

equil (M) $\quad\quad\quad\quad\quad\quad\quad\ x \quad\quad\quad\quad\ 2x$

$K_{sp} = [Zn^{2+}][OH^-]^2 = 4.1 \times 10^{-17} = (x)(2x)^2 = 4x^3$

$$\text{molar solubility} = x = \sqrt[3]{\frac{4.1 \times 10^{-17}}{4}} = 2.2 \times 10^{-6}\ M$$

(b) $[OH^-] = 2x = 2(2.2 \times 10^{-6}\ M) = 4.4 \times 10^{-6}\ M$

$$[H_3O^+] = \frac{1.0 \times 10^{-14}}{4.4 \times 10^{-6}} = 2.3 \times 10^{-9}\ M$$

$pH = -\log[H_3O^+] = -\log(2.3 \times 10^{-9}) = 8.64$

(c) $\quad\quad\quad\ Zn(OH)_2(s) \rightleftharpoons Zn^{2+}(aq) + 2\ OH^-(aq) \quad\quad\quad K_{sp} = 4.1 \times 10^{-17}$

$\quad\quad\quad\quad\quad \underline{Zn^{2+}(aq) + 4\ OH^-(aq) \rightleftharpoons Zn(OH)_4^{2-}(aq)} \quad\quad K_f = 3 \times 10^{15}$

$\quad\quad\quad\quad\quad Zn(OH)_2(s) + 2\ OH^-(aq) \rightleftharpoons Zn(OH)_4^{2-}(aq) \quad K = K_{sp} \cdot K_f = 0.123$

initial (M) $\quad\quad\quad\quad\quad\quad 0.10 \quad\quad\quad\quad\quad\quad 0$

change (M) $\quad\quad\quad\quad\quad\ -2x \quad\quad\quad\quad\quad\ +x$

equil (M) $\quad\quad\quad\quad\quad\ 0.10 - 2x \quad\quad\quad\quad x$

$$K = \frac{[Zn(OH)_4^{2-}]}{[OH^-]^2} = 0.123 = \frac{x}{(0.10 - 2x)^2}$$

$0.492x^2 - 1.0492x + 0.00123 = 0$

Use the quadratic formula to solve for x.

$$x = \frac{-(-1.0492) \pm \sqrt{(-1.0492)^2 - (4)(0.492)(0.00123)}}{2(0.492)} = \frac{1.0492 \pm 1.0480}{0.984}$$

$x = 2.1$ and 1.2×10^{-3}

Of the two solutions for x, only 1.2×10^{-3} has physical meaning because the other solution leads to a negative $[OH^-]$.

molar solubility of $Zn(OH)_4^{2-}$ in 0.10 M NaOH $= x = 1.2 \times 10^{-3}\ M$

Multi-Concept Problems

16.140 (a) $HA^-(aq) + H_2O(l) \rightleftharpoons H_3O^+(aq) + A^{2-}(aq) \quad\quad\quad K_{a2} = 10^{-10}$

$\quad\quad\ HA^-(aq) + H_2O(l) \rightleftharpoons H_2A(aq) + OH^-(aq) \quad\quad\quad K_b = \dfrac{K_w}{K_{a1}} = 10^{-10}$

$\quad\quad\ 2\ HA^-(aq) \rightleftharpoons H_2A(aq) + A^{2-}(aq) \quad\quad\quad\quad\quad\quad K = \dfrac{K_{a2}}{K_{a1}} = 10^{-6}$

$\quad\quad\ 2\ H_2O(l) \rightleftharpoons H_3O^+(aq) + OH^-(aq) \quad\quad\quad\quad\quad K_w = 1.0 \times 10^{-14}$

The principal reaction of the four is the one with the largest K, and that is the third reaction.

(b) $K_{a1} = \dfrac{[H_3O^+][HA^-]}{[H_2A]}$ and $K_{a2} = \dfrac{[H_3O^+][A^{2-}]}{[HA^-]}$

$[H_3O^+] = \dfrac{K_{a1}[H_2A]}{[HA^-]}$ and $[H_3O^+] = \dfrac{K_{a2}[HA^-]}{[A^{2-}]}$

$\dfrac{K_{a1}[H_2A]}{[HA^-]} \times \dfrac{K_{a2}[HA^-]}{[A^{2-}]} = [H_3O^+]^2;\qquad \dfrac{K_{a1}K_{a2}[H_2A]}{[A^{2-}]} = [H_3O^+]^2$

Because the principal reaction is $2\,HA^-(aq) \rightleftharpoons H_2A(aq) + A^{2-}(aq)$, $[H_2A] = [A^{2-}]$.

$K_{a1}K_{a2} = [H_3O^+]^2$

$\log K_{a1} + \log K_{a2} = 2\log[H_3O^+]$

$\dfrac{\log K_{a1} + \log K_{a2}}{2} = \log[H_3O^+];\qquad \dfrac{-\log K_{a1} + (-\log K_{a2})}{2} = -\log[H_3O^+]$

$\dfrac{pK_{a1} + pK_{a2}}{2} = pH$

(c)
	$2\,HA^-(aq)$	$\rightleftharpoons$	$H_2A(aq)$	$+$	$A^{2-}(aq)$
initial (M)	1.0		0		0
change (M)	$-2x$		$+x$		$+x$
equil (M)	$1.0 - 2x$		x		x

$K = \dfrac{[H_2A][A^{2-}]}{[HA^-]^2} = 1 \times 10^{-6} = \dfrac{x^2}{(1.0-2x)^2}$

Take the square root of both sides and solve for x.

$x = [A^{2-}] = 1 \times 10^{-3}$ M

mol $A^{2-} = (1 \times 10^{-3}$ mol/L$)(0.0500$ L$) = 5 \times 10^{-5}$ mol A^{2-}

number of A^{2-} ions $= (5 \times 10^{-5}$ mol $A^{2-})(6.022 \times 10^{23}$ ions/mol$) = 3 \times 10^{19}$ A^{2-} ions

16.142 (a) The first equivalence point is reached when all the H_3O^+ from the HCl and the H_3O^+ form the first ionization of H_3PO_4 is consumed.

At the first equivalence point $pH = \dfrac{pK_{a1} + pK_{a2}}{2} = 4.66$

$[H_3O^+] = 10^{-pH} = 10^{(-4.66)} = 2.2 \times 10^{-5}$ M

$(88.0$ mL$)(0.100$ mmol/mL$) = 8.80$ mmol NaOH are used to get to the first equivalence point

(b) mmol $(HCl + H_3PO_4) =$ mmol NaOH $= 8.8$ mmol

mmol $H_3PO_4 = (126.4$ mL $- 88.0$ mL$)(0.100$ mmol/mL$) = 3.84$ mmol

mmol HCl $= (8.8 - 3.84) = 4.96$ mmol

$[HCl] = \dfrac{4.96\ \text{mmol}}{40.0\ \text{mL}} = 0.124$ M;$\qquad [H_3PO_4] = \dfrac{3.84\ \text{mmol}}{40.0\ \text{mL}} = 0.0960$ M

(c) 100% of the HCl is neutralized at the first equivalence point.

(d)
	$H_3PO_4(aq)$	$+$	$H_2O(l)$	$\rightleftharpoons$	$H_3O^+(aq)$	$+$	$H_2PO_4^-(aq)$
initial (M)	0.0960				0.124		0
change (M)	$-x$				$+x$		$+x$
equil (M)	$0.0960 - x$				$0.124 + x$		x

$$K_{a1} = \frac{[H_3O^+][H_2PO_4^-]}{[H_3PO_4]} = 7.5 \times 10^{-3} = \frac{(0.124 + x)(x)}{0.0960 - x}$$

$x^2 + 0.132x - (7.2 \times 10^{-4}) = 0$

Use the quadratic formula to solve for x.

$$x = \frac{-(0.132) \pm \sqrt{(0.132)^2 - 4(1)(-7.2 \times 10^{-4})}}{2(1)} = \frac{-0.132 \pm 0.142}{2}$$

x = –0.137 and 0.005

Of the two solutions for x, only the positive value of x has physical meaning because the other solution would give a negative $[H_3O^+]$.

$[H_3O^+] = 0.124 + x = 0.124 + 0.005 = 0.129$ M

pH = $-\log[H_3O^+]$ = $-\log(0.129)$ = 0.89

(e)

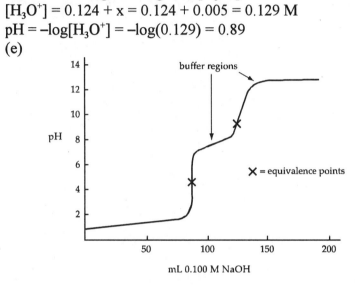

(f) Bromcresol green or methyl orange are suitable indicators for the first equivalence point. Thymolphthalein is a suitable indicator for the second equivalence point.

16.144 25°C = 25 + 273 = 298 K

$$\Pi = 2MRT; \quad M = \frac{\Pi}{2RT} = \frac{\left(74.4 \text{ mm Hg} \times \dfrac{1.00 \text{ atm}}{760 \text{ mm Hg}}\right)}{(2)\left(0.082\ 06\ \dfrac{L \cdot atm}{K \cdot mol}\right)(298 \text{ K})} = 0.00200 \text{ M}$$

$[M^+] = [X^-] = 0.00200$ M

$K_{sp} = [M^+][X^-] = (0.00200)^2 = 4.00 \times 10^{-6}$

16.146 (a) species present initially:

NH$_4^+$	CO$_3^{2-}$	H$_2$O
acid	base	acid or base

$2H_2O(l) \rightleftharpoons H_3O^+(aq) + OH^-(aq)$

$NH_4^+(aq) + H_2O(l) \rightleftharpoons NH_3(aq) + H_3O^+(aq)$

$CO_3^{2-}(aq) + H_2O(l) \rightleftharpoons HCO_3^-(aq) + OH^-(aq)$

NH_3, $K_b = 1.8 \times 10^{-5}$

NH_4^+, $K_a = 5.6 \times 10^{-10}$

CO_3^{2-}, $K_b = 1.8 \times 10^{-4}$

HCO_3^-, $K_a = 5.6 \times 10^{-11}$

In the mixture, proton transfer takes place from the stronger acid to the stronger base, so the principal reaction is $NH_4^+(aq) + CO_3^{2-}(aq) \rightleftharpoons HCO_3^-(aq) + NH_3(aq)$

(b)

$$NH_4^+(aq) + OH^-(aq) \rightleftharpoons NH_3(aq) + H_2O(l) \qquad K_1 = 1/K_b(NH_3)$$
$$\underline{CO_3^{2-}(aq) + H_2O(l) \rightleftharpoons HCO_3^-(aq) + OH^-(aq) \qquad K_2 = K_b(CO_3^{2-})}$$
$$NH_4^+(aq) + CO_3^{2-}(aq) \rightleftharpoons HCO_3^-(aq) + NH_3(aq) \qquad K = K_1 \cdot K_2$$

	NH_4^+	CO_3^{2-}	HCO_3^-	NH_3
initial (M)	0.16	0.080	0	0.16
change (M)	$-x$	$-x$	$+x$	$+x$
equil (M)	$0.16 - x$	$0.080 - x$	x	$0.16 + x$

$$K = \frac{[HCO_3^-][NH_3]}{[NH_4^+][CO_3^{2-}]} = \frac{1.8 \times 10^{-4}}{1.8 \times 10^{-5}} = 10 = \frac{x(0.16 + x)}{(0.16 - x)(0.080 - x)}$$

$9x^2 - 2.56x + 0.128 = 0$

Use the quadratic formula to solve for x.

$$x = \frac{-(-2.56) \pm \sqrt{(-2.56)^2 - (4)(9)(0.128)}}{2(9)} = \frac{2.56 \pm 1.395}{18}$$

$x = 0.220$ and 0.0647

Of the two solutions for x, only 0.0647 has physical meaning because 0.220 leads to negative concentrations.

$[NH_4^+] = 0.16 - x = 0.16 - 0.0647 = 0.0953\ M = 0.095\ M$

$[NH_3] = 0.16 + x = 0.16 + 0.0647 = 0.225\ M = 0.22\ M$

$[CO_3^{2-}] = 0.080 - x = 0.080 - 0.0647 = 0.0153\ M = 0.015\ M$

$[HCO_3^-] = x = 0.0647\ M = 0.065\ M$

The solution is a buffer containing two different sets of conjugate acid-base pairs. Either pair can be used to calculate the pH.

For NH_4^+, $K_a = 5.6 \times 10^{-10}$ and $pK_a = 9.25$

$$pH = pK_a + \log \frac{[NH_3]}{[NH_4^+]} = 9.25 + \log \frac{(0.225)}{(0.0953)} = 9.62$$

$[H_3O^+] = 10^{-pH} = 10^{-9.62} = 2.4 \times 10^{-10}\ M$

$$[OH^-] = \frac{1.0 \times 10^{-14}}{2.4 \times 10^{-10}} = 4.2 \times 10^{-5}\ M$$

$$[H_2CO_3] = \frac{[HCO_3^-][H_3O^+]}{K_a} = \frac{(0.647)(2.4 \times 10^{-10})}{(4.3 \times 10^{-7})} = 3.6 \times 10^{-4}\ M$$

(c) For MCO_3, $IP = [M^{2+}][CO_3^{2-}] = (0.010)(0.0153) = 1.5 \times 10^{-4}$

$K_{sp}(CaCO_3) = 5.0 \times 10^{-9}$, $10^3 K_{sp} = 5.0 \times 10^{-6}$

$K_{sp}(BaCO_3) = 2.6 \times 10^{-9}$, $10^3 K_{sp} = 2.6 \times 10^{-6}$

$K_{sp}(MgCO_3) = 6.8 \times 10^{-6}$, $10^3 K_{sp} = 6.8 \times 10^{-3}$

$IP > 10^3 K_{sp}$ for $CaCO_3$ and $BaCO_3$, but $IP < 10^3 K_{sp}$ for $MgCO_3$ so the $[CO_3^{2-}]$ is large enough to give observable precipitation of $CaCO_3$ and $BaCO_3$, but not $MgCO_3$.

(d) For $M(OH)_2$, $IP = [M^{2+}][OH^-]^2 = (0.010)(4.17 \times 10^{-5})^2 = 1.7 \times 10^{-11}$

$K_{sp}(Ca(OH)_2) = 4.7 \times 10^{-6}$, $10^3 K_{sp} = 4.7 \times 10^{-3}$

$K_{sp}(Ba(OH)_2) = 5.0 \times 10^{-3}$, $10^3 K_{sp} = 5.0$

$K_{sp}(Mg(OH)_2) = 5.6 \times 10^{-12}$, $10^3 K_{sp} = 5.6 \times 10^{-9}$

$IP < 10^3 K_{sp}$ for all three $M(OH)_2$. None precipitate.

(e)

	$CO_3^{2-}(aq)$	+	$H_2O(l)$	$\rightleftharpoons$	$HCO_3^-(aq)$	+	$OH^-(aq)$
initial (M)	0.08				0		~0
change (M)	−x				+x		+x
equil (M)	0.08 − x				x		x

$$K_b = \frac{[HCO_3^-][OH^-]}{[CO_3^{2-}]} = 1.8 \times 10^{-4} = \frac{x^2}{(0.08 - x)}$$

$x^2 + 1.8 \times 10^{-4}x - 1.44 \times 10^{-5} = 0$

Use the quadratic formula to solve for x.

$$x = \frac{-(1.8 \times 10^{-4}) \pm \sqrt{(1.8 \times 10^{-4})^2 - (4)(1)(-1.44 \times 10^{-5})}}{2(1)} = \frac{-(1.8 \times 10^{-4}) \pm 7.59 \times 10^{-3}}{2}$$

$x = 0.0037$ and -0.0039

Of the two solutions for x, only 0.0037 has physical meaning because −0.0039 leads to negative concentrations.

$[OH^-] = x = 3.7 \times 10^{-3}$ M

For MCO_3, $IP = [M^{2+}][CO_3^{2-}] = (0.010)(0.08) = 8.0 \times 10^{-4}$

For $M(OH)_2$, $IP = [M^{2+}][OH^-]^2 = (0.010)(3.7 \times 10^{-3})^2 = 1.4 \times 10^{-7}$

Comparing IP's here and $10^3 K_{sp}$'s in (c) and (d) above, Ca^{2+} and Ba^{2+} cannot be separated from Mg^{2+} using 0.08 M Na_2CO_3. Na_2CO_3 is more basic than $(NH_4)_2CO_3$ and $Mg(OH)_2$ would precipitate along with $CaCO_3$ and $BaCO_3$.

16.148 $Pb(CH_3CO_2)_2$, 325.29 amu; PbS, 239.27 amu

(a) mass PbS = $(2 \text{ mL})(1 \text{ g/mL})(0.003) \times \dfrac{1 \text{ mol } Pb(CH_3CO_2)_2}{325.29 \text{ g } Pb(CH_3CO_2)_2} \times$

$\dfrac{1 \text{ mol PbS}}{1 \text{ mol } Pb(CH_3CO_2)_2} \times \dfrac{239.27 \text{ g PbS}}{1 \text{ mol PbS}} \times (30/100) = 0.0013$ g

= 1.3 mg PbS per dye application

(b) $[H_3O^+] = 10^{-pH} = 10^{-5.50} = 3.16 \times 10^{-6}$ M

	$PbS(s)$	+	$2 H_3O^+(aq)$	$\rightleftharpoons$	$Pb^{2+}(aq)$	+	$H_2S(aq)$	+	$2 H_2O(l)$
initial (M)			3.16×10^{-6}		0		0		
change (M)			−2x		+x		+x		
equil (M)			$3.16 \times 10^{-6} - 2x$		x		x		

$$K_{spa} = \frac{[Pb^{2+}][H_2S]}{[H_3O^+]^2} = \frac{x^2}{(3.16 \times 10^{-6} - 2x)^2} \approx \frac{x^2}{(3.16 \times 10^{-6})^2} = 3 \times 10^{-7}$$

$x^2 = (3.16 \times 10^{-6})^2(3 \times 10^{-7}) = 3.0 \times 10^{-18}$

$x = 1.7 \times 10^{-9}$ M = [Pb^{2+}] for a saturated solution.

mass of PbS dissolved per washing =

$$(3 \text{ gal})(3.7854 \text{ L}/1 \text{ gal})(1.7 \times 10^{-9} \text{ mol/L}) \times \frac{239.27 \text{ g PbS}}{1 \text{ mol PbS}} = 4.7 \times 10^{-6} \text{ g PbS/washing}$$

Number of washings required to remove 50% of the PbS from one application =

$$\frac{(0.0013 \text{ g PbS})(50/100)}{(4.7 \times 10^{-6} \text{ g PbS/washing})} = 1.4 \times 10^2 \text{ washings}$$

(c) The number of washings does not look reasonable. It seems too high considering that frequent dye application is recommended. If the PbS is located mainly on the surface of the hair, as is believed to be the case, solid particles of PbS can be lost by abrasion during shampooing.

Thermodynamics: Entropy, Free Energy, and Equilibrium

17.1   (a) spontaneous;   (b), (c), and (d) nonspontaneous

17.2   (a) $H_2O(g) \rightarrow H_2O(l)$
A liquid is more ordered than a gas. Therefore, ΔS is negative.
(b) $I_2(g) \rightarrow 2\ I(g)$
ΔS is positive because the reaction increases the number of gaseous particles from 1 mol to 2 mol.
(c) $CaCO_3(s) \rightarrow CaO(s) + CO_2(g)$
ΔS is positive because the reaction increases the number of gaseous molecules.
(d) $Ag^+(aq) + Br^-(aq) \rightarrow AgBr(s)$
A solid is more ordered than +1 and −1 charged ions in an aqueous solution. Therefore, ΔS is negative.

17.3   (a) $A_2 + AB_3 \rightarrow 3\ AB$
(b) ΔS is positive because the reaction increases the number of gaseous molecules.

17.4   (a) disordered N_2O
(b) silica glass (amorphous solid, more disorder)
(c) 1 mole N_2 at STP (larger volume, more disorder)
(d) 1 mole N_2 at 273 K and 0.25 atm (larger volume, more disorder)

17.5 $CaCO_3(s) \rightarrow CaO(s) + CO_2(g)$
$\Delta S° = [S°(CaO) + S°(CO_2)] - S°(CaCO_3)$
$\Delta S° = [(1\ mol)(39.7\ J/(K\cdot mol)) + (1\ mol)(213.6\ J/(K \cdot mol))]$
$\quad\quad\quad\quad\quad\quad\quad\quad - (1\ mol)(92.9\ J/(K \cdot mol)) = +160.4\ J/K$

17.6   From Problem 17.5, $\Delta S_{sys} = \Delta S° = 160.4\ J/K$
$CaCO_3(s) \rightarrow CaO(s) + CO_2(g)$
$\Delta H° = [\Delta H°_f(CaO) + \Delta H°_f(CO_2)] - \Delta H°_f(CaCO_3)$
$\Delta H° = [(1\ mol)(-635.1\ kJ/mol) + (1\ mol)(-393.5\ kJ/mol)]$
$\quad\quad\quad\quad\quad\quad\quad\quad - (1\ mol)(-1206.9\ kJ/mol) = +178.3\ kJ$

$\Delta S_{surr} = \dfrac{-\Delta H°}{T} = \dfrac{-178,300\ J}{298\ K} = -598\ J/K$

$\Delta S_{total} = \Delta S_{sys} + \Delta S_{surr} = 160.4\ J/K + (-598\ J/K) = -438\ J/K$
Because ΔS_{total} is negative, the reaction is not spontaneous under standard-state conditions at 25°C.

17.7   (a) $\Delta G = \Delta H - T\Delta S = 57.1\ kJ - (298\ K)(0.1758\ kJ/K) = +4.7\ kJ$
Because $\Delta G > 0$, the reaction is nonspontaneous at 25°C (298 K)

(b) Set $\Delta G = 0$ and solve for T.

$$0 = \Delta H - T\Delta S; \qquad T = \frac{\Delta H}{\Delta S} = \frac{57.1 \text{ kJ}}{0.1758 \text{ kJ/K}} = 325 \text{ K} = 52°C$$

17.8 (a) $\Delta G = \Delta H - T\Delta S = 58.5 \text{ kJ/mol} - (598 \text{ K})[0.0929 \text{ kJ/(K} \cdot \text{mol)}] = +2.9 \text{ kJ/mol}$
Because $\Delta G > 0$, Hg does not boil at 325°C and 1 atm.
(b) The boiling point (phase change) is associated with an equilibrium. Set $\Delta G = 0$ and solve for T, the boiling point.

$$0 = \Delta H_{vap} - T\Delta S_{vap}; \qquad T_{bp} = \frac{\Delta H_{vap}}{\Delta S_{vap}} = \frac{58.5 \text{ kJ/mol}}{0.0929 \text{ kJ/(K} \cdot \text{mol)}} = 630 \text{ K} = 357°C$$

17.9 $\Delta H < 0$ (reaction involves bond making -- exothermic)
$\Delta S < 0$ (the reaction becomes more ordered in going from reactants (2 atoms) to products (1 molecule)
$\Delta G < 0$ (the reaction is spontaneous)

17.10 From Problems 17.5 and 17.6: $\Delta H° = 178.3 \text{ kJ}$ and $\Delta S° = 160.4 \text{ J/K} = 0.1604 \text{ kJ/K}$
(a) $\Delta G° = \Delta H° - T\Delta S° = 178.3 \text{ kJ} - (298 \text{ K})(0.1604 \text{ kJ/K}) = +130.5 \text{ kJ}$
(b) Because $\Delta G > 0$, the reaction is nonspontaneous at 25°C (298 K).
(c) Set $\Delta G = 0$ and solve for T, the temperature above which the reaction becomes spontaneous.

$$0 = \Delta H - T\Delta S; \qquad T = \frac{\Delta H}{\Delta S} = \frac{178.3 \text{ kJ}}{0.1604 \text{ kJ/K}} = 1112 \text{ K} = 839°C$$

17.11 $2 \text{ AB}_2 \rightarrow \text{A}_2 + 2 \text{ B}_2$
(a) $\Delta S°$ is positive because the reaction increases the number of molecules.
(b) $\Delta H°$ is positive because the reaction is endothermic.
$\Delta G° = \Delta H° - T\Delta S°$
For the reaction to be spontaneous, $\Delta G°$ must be negative. This will only occur at high temperature where $T\Delta S°$ is greater than $\Delta H°$.

17.12 (a) $\text{CaC}_2(s) + 2 \text{ H}_2\text{O}(l) \rightarrow \text{C}_2\text{H}_2(g) + \text{Ca(OH)}_2(s)$
$\Delta G° = [\Delta G°_f(\text{C}_2\text{H}_2) + \Delta G°_f(\text{Ca(OH)}_2)] - [\Delta G°_f(\text{CaC}_2) + 2 \Delta G°_f(\text{H}_2\text{O})]$
$\Delta G° = [(1 \text{ mol})(209.2 \text{ kJ/mol}) + (1 \text{ mol})(-898.6 \text{ kJ/mol})]$
$\qquad - [(1 \text{ mol})(-64.8 \text{ kJ/mol}) + (2 \text{ mol})(-237.2 \text{ kJ/mol})] = -150.2 \text{ kJ}$
This reaction can be used for the synthesis of C_2H_2 because $\Delta G < 0$.
(b) It is not possible to synthesize acetylene from solid graphite and gaseous H_2 at 25°C and 1 atm because $\Delta G°_f(\text{C}_2\text{H}_2) > 0$.

17.13 $\text{C}(s) + 2 \text{ H}_2(g) \rightarrow \text{C}_2\text{H}_4(g)$

$$Q_p = \frac{P_{\text{C}_2\text{H}_4}}{(P_{\text{H}_2})^2} = \frac{(0.10)}{(100)^2} = 1.0 \times 10^{-5}$$

$\Delta G = \Delta G° + RT \ln Q_p$
$\Delta G = 68.1$ kJ/mol $+ [8.314 \times 10^{-3}$ kJ/(K $\cdot$ mol)$](298$ K$)\ln(1.0 \times 10^{-5}) = +39.6$ kJ/mol
Because $\Delta G > 0$, the reaction is spontaneous in the reverse direction.

17.14 $\Delta G = \Delta G° + RT \ln Q$ and $\Delta G° = 15$ kJ

For A$_2$(g) $+$ B$_2$(g) $\rightleftharpoons$ 2 AB(g), $Q_p = \dfrac{(P_{AB})^2}{(P_{A_2})(P_{B_2})}$

Let the number of molecules be proportional to the partial pressure.
(1) $Q_p = 1.5$ (2) $Q_p = 0.0667$ (3) $Q_p = 18$
(a) Reaction (3) has the largest ΔG because Q_p is the largest. Reaction (1) has the smallest ΔG because Q_p is the smallest.
(b) $\Delta G = \Delta G° = 15$ kJ because $Q_p = 1$ and $\ln (1) = 0$.

17.15   From Problem 17.10, $\Delta G° = +130.5$ kJ
$\Delta G° = -RT \ln K_p$

$\ln K_p = \dfrac{-\Delta G°}{RT} = \dfrac{-130.5 \text{ kJ/mol}}{[8.314 \times 10^{-3} \text{ kJ/(K} \cdot \text{mol)}](298 \text{ K})} = -52.7$

$K_p = e^{-52.7} = 1 \times 10^{-23}$

17.16 $H_2O(l) \rightleftharpoons H_2O(g)$
$K_p = P_{H_2O}$; K_p is equal to the vapor pressure for H_2O.

$\Delta G° = \Delta G°_f(H_2O(g)) - \Delta G°_f(H_2O(l))$
$\Delta G° = (1 \text{ mol})(-228.6 \text{ kJ/mol}) - (1 \text{ mol})(-237.2 \text{ kJ/mol}) = +8.6$ kJ
$\Delta G° = - RT \ln K_p$

$\ln K_p = \dfrac{-\Delta G°}{RT} = \dfrac{-8.6 \text{ kJ/mol}}{[8.314 \times 10^{-3} \text{ kJ/(K} \cdot \text{mol)}](298 \text{ K})} = -3.5$

$K_p = P_{H_2O} = e^{-3.5} = 0.03$ atm

17.17 $\Delta G° = - RT \ln K = -[8.314 \times 10^{-3}$ kJ/(K $\cdot$ mol)$](298$ K$) \ln (1.0 \times 10^{-14}) = 80$ kJ/mol

17.18   Photosynthetic cells in plants use the sun's energy to make glucose, which is then used by animals as their primary source of energy. The energy an animal obtains from glucose is then used to build and organize complex molecules, resulting in a decrease in entropy for the animal. At the same time, however, the entropy of the surroundings increases as the animal releases small, simple waste products such as CO_2 and H_2O. Furthermore, heat is released by the animal, further increasing the entropy of the surroundings. Thus, an organism pays for its decrease in entropy by increasing the entropy of the rest of the universe.

17.19   You would expect to see violations of the second law if you watched a movie run backwards. Consider an action-adventure movie with a lot of explosions. An explosion is a spontaneous process that increases the entropy of the universe. You would see an explosion go backwards if you run the the movie backwards but this is impossible because it would decrease the entropy of the universe.

Understanding Key Concepts

17.20 (a)

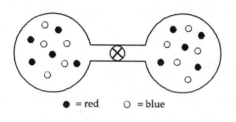

● = red   ○ = blue

(b) $\Delta H = 0$ (no heat is gained or lost in the mixing of ideal gases)
$\Delta S > 0$ (the mixture of the two gases is more disordered)
$\Delta G < 0$ (the mixing of the two gases is spontaneous)
(c) For an isolated system, $\Delta S_{surr} = 0$ and $\Delta S_{sys} = \Delta S_{Total} > 0$ for the spontaneous process.
(d) $\Delta G > 0$ and the process is nonspontaneous.

17.22 $\Delta H < 0$ (heat is lost during condensation)
$\Delta S < 0$ (liquid is more ordered than vapor)
$\Delta G < 0$ (the reaction is spontaneous)

17.24 (a) $2 A_2 + B_2 \rightarrow 2 A_2B$
(b) $\Delta H < 0$ (because ΔS is negative, ΔH must also be negative in order for ΔG to be negative)
$\Delta S < 0$ (the reaction becomes more ordered in going from reactants (3 molecules) to products (2 molecules))
$\Delta G < 0$ (the reaction is spontaneous)

17.26 (a) $\Delta H° > 0$ (reaction involves bond breaking - endothermic)
$\Delta S° > 0$ (2 A's are more disordered than A_2)
(b) $\Delta S°$ is for the complete conversion of 1 mole of A_2 in its standard state to 2 moles of A in its standard state.
(c) There is not enough information to say anything about the sign of $\Delta G°$. $\Delta G°$ decreases (becomes less positive or more negative) as the temperature increases.
(d) K_p increases as the temperature increases. As the temperature increases there will be more A and less A_2.
(e) $\Delta G = 0$ at equilibrium.

17.28 $\Delta G° = -RT \ln K$ where $K = \dfrac{[X]}{[A]}$ or $\dfrac{[Y]}{[A]}$ or $\dfrac{[Z]}{[A]}$
Let the number of molecules be proportional to the concentration.
(1) $K = 1$, $\ln K = 0$, and $\Delta G° = 0$.
(2) $K > 1$, $\ln K$ is positive, and $\Delta G°$ is negative.
(3) $K < 1$, $\ln K$ is negative, and $\Delta G°$ is positive.

Additional Problems
Spontaneous Processes

17.30 A spontaneous process is one that proceeds on its own without any external influence.
For example: $H_2O(s) \rightarrow H_2O(l)$ at 25°C
A nonspontaneous process takes place only in the presence of some continuous external influence.
For example: $2\,NaCl(s) \rightarrow 2\,Na(s) + Cl_2(g)$

17.32 (a) and (d) nonspontaneous; (b) and (c) spontaneous

17.34 (b) and (d) spontaneous (because of the large positive K_p's)

Entropy

17.36 Molecular randomness or disorder is called entropy. For the following reaction, the entropy (disorder) increases: $H_2O(s) \rightarrow H_2O(l)$ at 25°C.

17.38 (a) + (solid → gas)   (b) – (liquid → solid)
(c) – (aqueous ions → solid)   (d) + (CO_2(aq) → CO_2(g))

17.40 (a) – (liquid → solid)
(b) – (decrease in number of O_2 molecules)
(c) + (gas is more disordered in larger volume)
(d) – (aqueous ions → solid)

17.42 $S = k \ln W$, $k = 1.38 \times 10^{-23}$ J/K
(a) $S = (1.38 \times 10^{-23}\ \text{J/K}) \ln (4^{12}) = 2.30 \times 10^{-22}$ J/K
(b) $S = (1.38 \times 10^{-23}\ \text{J/K}) \ln (4^{120}) = 2.30 \times 10^{-21}$ J/K
(c) $S = (1.38 \times 10^{-23}\ \text{J/K}) \ln (4^{6.02 \times 10^{23}}) = 11.5$ J/K
If all C–D bonds point in the same direction, $S = 0$.

17.44 (a) H_2 at 25°C in 50 L (larger volume)
(b) O_2 at 25°C, 1 atm (larger volume)
(c) H_2 at 100°C, 1 atm (larger volume and higher T)
(d) CO_2 at 100°C, 0.1 atm (larger volume and higher T)

Standard Molar Entropies and Standard Entropies of Reaction

17.46 The standard molar entropy of a substance is the entropy of 1 mol of the pure substance at 1 atm pressure and 25°C.
$\Delta S° = S°$(products) – $S°$(reactants)

17.48 (a) $C_2H_6(g)$; more atoms/molecule
(b) $CO_2(g)$; more atoms/molecule

(c) $I_2(g)$; gas is more disordered than the solid

(d) $CH_3OH(g)$; gas is more disordered than the liquid.

17.50 (a) $2 H_2O_2(l) \rightarrow 2 H_2O(l) + O_2(g)$
$\Delta S° = [2 S°(H_2O(l)) + S°(O_2)] - 2 S°(H_2O_2)$
$\Delta S° = [(2 \text{ mol})(69.9 \text{ J/(K} \cdot \text{mol})) + (1 \text{ mol})(205.0 \text{ J/(K} \cdot \text{mol}))]$
$\qquad\qquad - (2 \text{ mol})(110 \text{ J/(K} \cdot \text{mol})) = +125 \text{ J/K}$ (+, because moles of gas increase)

(b) $2 Na(s) + Cl_2(g) \rightarrow 2 NaCl(s)$
$\Delta S° = 2 S°(NaCl) - [2 S°(Na) + S°(Cl_2)]$
$\Delta S° = (2 \text{ mol})(72.1 \text{ J/(K} \cdot \text{mol})) - [(2 \text{ mol})(51.2 \text{ J/(K} \cdot \text{mol})) + (1 \text{ mol})(223.0 \text{ J/(K} \cdot \text{mol}))]$
$\Delta S° = -181.2 \text{ J/K}$ (–, because moles of gas decrease)

(c) $2 O_3(g) \rightarrow 3 O_2(g)$
$\Delta S° = 3 S°(O_2) - 2 S°(O_3)$
$\Delta S° = (3 \text{ mol})(205.0 \text{ J/(K} \cdot \text{mol})) - (2 \text{ mol})(238.8 \text{ J/(K} \cdot \text{mol}))$
$\Delta S° = +137.4 \text{ J/K}$ (+, because moles of gas increase)

(d) $4 Al(s) + 3 O_2(g) \rightarrow 2 Al_2O_3(s)$
$\Delta S° = 2 S°(Al_2O_3) - [4 S°(Al) + 3 S°(O_2)]$
$\Delta S° = (2 \text{ mol})(50.9 \text{ J/(K} \cdot \text{mol})) - [(4 \text{ mol})(28.3 \text{ J/(K} \cdot \text{mol})) + (3 \text{ mol})(205.0 \text{ J/(K} \cdot \text{mol}))]$
$\Delta S° = -626.4 \text{ J/K}$ (–, because moles of gas decrease)

Entropy and the Second Law of Thermodynamics

17.52 In any spontaneous process, the total entropy of a system and its surroundings always increases.

17.54 $\Delta S_{surr} = \dfrac{-\Delta H}{T}$; the temperature (T) is always positive.

(a) For an exothermic reaction, ΔH is negative and ΔS_{surr} is positive.
(b) For an endothermic reaction, ΔH is positive and ΔS_{surr} is negative.

17.56 $N_2(g) + 2 O_2(g) \rightarrow N_2O_4(g)$
$\Delta H° = \Delta H°_f(N_2O_4) = 9.16 \text{ kJ}$
$\Delta S_{sys} = \Delta S° = S°(N_2O_4) - [S°(N_2) + 2 S°(O_2)]$
$\Delta S_{sys} = (1 \text{ mol})(304.2 \text{ J/(K} \cdot \text{mol}))$
$\qquad\qquad - [(1 \text{ mol})(191.5 \text{ J/(K} \cdot \text{mol})) + (2 \text{ mol})(205.0 \text{ J/(K} \cdot \text{mol}))] = -297.3 \text{ J/K}$
$\Delta S_{surr} = \dfrac{-\Delta H°}{T} = \dfrac{-9.16 \text{ kJ}}{298 \text{ K}} = -0.0307 \text{ kJ/K} = -30.7 \text{ J/K}$
$\Delta S_{total} = \Delta S_{sys} + \Delta S_{surr} = -297.3 \text{ J/K} + (-30.7 \text{ J/K}) = -328.0 \text{ J/K}$
Because $\Delta S_{total} < 0$, the reaction is nonspontaneous.

17.58 (a) $\Delta S_{surr} = \dfrac{-\Delta H_{vap}}{T} = \dfrac{-30,700 \text{ J/mol}}{343 \text{ K}} = -89.5 \text{ J/(K} \cdot \text{mol})$

$\Delta S_{total} = \Delta S_{vap} + \Delta S_{surr} = 87.0 \text{ J/(K} \cdot \text{mol}) + (-89.5 \text{ J/(K} \cdot \text{mol})) = -2.5 \text{ J/(K} \cdot \text{mol})$

(b) $\Delta S_{surr} = \dfrac{-\Delta H_{vap}}{T} = \dfrac{-30,700 \text{ J/mol}}{353 \text{ K}} = -87.0 \text{ J/(K} \cdot \text{mol)}$

$\Delta S_{total} = \Delta S_{vap} + \Delta S_{surr} = 87.0 \text{ J/(K} \cdot \text{mol)} + (-87.0 \text{ J/(K} \cdot \text{mol)}) = 0$

(c) $\Delta S_{surr} = \dfrac{-\Delta H_{vap}}{T} = \dfrac{-30,700 \text{ J/mol}}{363 \text{ K}} = -84.6 \text{ J/(K} \cdot \text{mol)}$

$\Delta S_{total} = \Delta S_{vap} + \Delta S_{surr} = 87.0 \text{ J/(K} \cdot \text{mol)} + (-84.6 \text{ J/(K} \cdot \text{mol)}) = +2.4 \text{ J/(K} \cdot \text{mol)}$

Benzene does not boil at 70°C (343 K) because ΔS_{total} is negative.
The normal boiling point for benzene is 80°C (353 K), where $\Delta S_{total} = 0$.

Free Energy

17.60

ΔH	ΔS	$\Delta G = \Delta H - T\Delta S$	Reaction Spontaneity
–	+	–	Spontaneous at all temperatures
–	–	– or +	Spontaneous at low temperatures where $\lvert \Delta H \rvert > \lvert T\Delta S \rvert$ Nonspontaneous at high temperatures where $\lvert \Delta H \rvert < \lvert T\Delta S \rvert$
+	–	+	Nonspontaneous at all temperatures
+	+	– or +	Spontaneous at high temperatures where $T\Delta S > \Delta H$ Nonspontaneous at low temperature where $T\Delta S < \Delta H$

17.62 (a) 0°C (temperature is below mp); $\Delta H > 0$, $\Delta S > 0$, $\Delta G > 0$
(b) 15°C (temperature is above mp); $\Delta H > 0$, $\Delta S > 0$, $\Delta G < 0$

17.64 $\Delta H_{vap} = 30.7 \text{ kJ/mol}$
$\Delta S_{vap} = 87.0 \text{ J/(K} \cdot \text{mol)} = 87.0 \times 10^{-3} \text{ kJ/(K} \cdot \text{mol)}$
$\Delta G_{vap} = \Delta H_{vap} - T\Delta S_{vap}$
(a) $\Delta G_{vap} = 30.7 \text{ kJ/mol} - (343 \text{ K})(87.0 \times 10^{-3} \text{ kJ/(K} \cdot \text{mol)}) = +0.9 \text{ kJ/mol}$
At 70°C (343 K), benzene does not boil because ΔG_{vap} is positive.
(b) $\Delta G_{vap} = 30.7 \text{ kJ/mol} - (353 \text{ K})(87.0 \times 10^{-3} \text{ kJ/(K} \cdot \text{mol)}) = 0$
80°C (353 K) is the boiling point for benzene because $\Delta G_{vap} = 0$
(c) $\Delta G_{vap} = 30.7 \text{ kJ/mol} - (363 \text{ K})(87.0 \times 10^{-3} \text{ kJ/(K} \cdot \text{mol)}) = -0.9 \text{ kJ/mol}$
At 90°C (363 K), benzene boils because ΔG_{vap} is negative.

17.66 At the melting point (phase change), $\Delta G_{fusion} = 0$
$\Delta G_{fusion} = \Delta H_{fusion} - T\Delta S_{fusion}$

$0 = \Delta H_{fusion} - T\Delta S_{fusion}$; $T = \dfrac{\Delta H_{fusion}}{\Delta S_{fusion}} = \dfrac{17.3 \text{ kJ/mol}}{43.8 \times 10^{-3} \text{ kJ/(K} \cdot \text{mol)}} = 395 \text{ K} = 122°C$

Standard Free-Energy Changes and Standard Free Energies of Formation

17.68   (a) ΔG° is the change in free energy that occurs when reactants in their standard states are converted to products in their standard states.
(b) ΔG°_f is the free-energy change for formation of one mole of a substance in its standard state from the most stable form of the constituent elements in their standard states.

17.70   (a) $N_2(g) + 2 O_2(g) \rightarrow 2 NO_2(g)$
$\Delta H^\circ = 2 \Delta H^\circ_f(NO_2) = (2 \text{ mol})(33.2 \text{ kJ/mol}) = 66.4 \text{ kJ}$
$\Delta S^\circ = 2 S^\circ(NO_2) - [S^\circ(N_2) + 2 S^\circ(O_2)]$
$\Delta S^\circ = (2 \text{ mol})(240.0 \text{ J/(K} \cdot \text{mol)}) - [(1 \text{ mol})(191.5 \text{ J/(K} \cdot \text{mol)}) + (2 \text{ mol})(205.0 \text{ J/(K} \cdot \text{mol)})]$
$\Delta S^\circ = -121.5 \text{ J/K} = -121.5 \times 10^{-3} \text{ kJ/K}$
$\Delta G^\circ = \Delta H^\circ - T\Delta S^\circ = 66.4 \text{ kJ} - (298 \text{ K})(-121.5 \times 10^{-3} \text{ kJ/K}) = +102.6 \text{ kJ}$
Because ΔG° is positive, the reaction is nonspontaneous under standard-state conditions at 25°C.
(b) $2 KClO_3(s) \rightarrow 2 KCl(s) + 3 O_2(g)$
$\Delta H^\circ = 2 \Delta H^\circ_f(KCl) - 2 \Delta H^\circ_f(KClO_3)$
$\Delta H^\circ = (2 \text{ mol})(-436.7 \text{ kJ/mol}) - (2 \text{ mol})(-397.7 \text{ kJ/mol}) = -78.0 \text{ kJ}$
$\Delta S^\circ = [2 S^\circ(KCl) + 3 S^\circ(O_2)] - 2 S^\circ(KClO_3)$
$\Delta S^\circ = [(2 \text{ mol})(82.6 \text{ J/(K} \cdot \text{mol)}) + (3 \text{ mol})(205.0 \text{ J/(K} \cdot \text{mol)})]$
$\quad\quad - (2 \text{ mol})(143 \text{ J/(K} \cdot \text{mol)})$
$\Delta S^\circ = 494.2 \text{ J/(K} \cdot \text{mol)} = 494.2 \times 10^{-3} \text{ kJ/(K} \cdot \text{mol)}$
$\Delta G^\circ = \Delta H^\circ - T\Delta S^\circ = -78.0 \text{ kJ} - (298 \text{ K})(494.2 \times 10^{-3} \text{ kJ/(K} \cdot \text{mol)}) = -225.3 \text{ kJ}$
Because ΔG° is negative, the reaction is spontaneous under standard-state conditions at 25°C.
(c) $CH_3CH_2OH(l) + O_2(g) \rightarrow CH_3CO_2H(l) + H_2O(l)$
$\Delta H^\circ = [\Delta H^\circ_f(CH_3CO_2H) + \Delta H^\circ_f(H_2O)] - \Delta H^\circ_f(CH_3CH_2OH)$
$\Delta H^\circ = [(1 \text{ mol})(-484.5 \text{ kJ/mol}) + (1 \text{ mol})(-285.8 \text{ kJ/mol})]$
$\quad\quad - (1 \text{ mol})(-277.7 \text{ kJ/mol}) = -492.6 \text{ kJ}$
$\Delta S^\circ = [S^\circ(CH_3CO_2H) + S^\circ(H_2O)] - [S^\circ(CH_3CH_2OH) + S^\circ(O_2)]$
$\Delta S^\circ = [(1 \text{ mol})(160 \text{ J/(K} \cdot \text{mol)}) + (1 \text{ mol})(69.9 \text{ J/(K} \cdot \text{mol)})]$
$\quad\quad - [(1 \text{ mol})(161 \text{ J/(K} \cdot \text{mol)}) + (1 \text{ mol})(205.0 \text{ J/(K} \cdot \text{mol)})]$
$\Delta S^\circ = -136.1 \text{ J/(K} \cdot \text{mol)} = -136.1 \times 10^{-3} \text{ kJ/(K} \cdot \text{mol)}$
$\Delta G^\circ = \Delta H^\circ - T\Delta S^\circ = -492.6 \text{ kJ} - (298 \text{ K})(-136.1 \times 10^{-3} \text{ kJ/(K} \cdot \text{mol)}) = -452.0 \text{ kJ}$
Because ΔG° is negative, the reaction is spontaneous under standard-state conditions at 25°C.

17.72   (a) $N_2(g) + 2 O_2(g) \rightarrow 2 NO_2(g)$
$\Delta G^\circ = 2 \Delta G^\circ_f(NO_2) = (2 \text{ mol})(51.3 \text{ kJ/mol}) = +102.6 \text{ kJ}$
(b) $2 KClO_3(s) \rightarrow 2 KCl(s) + 3 O_2(g)$
$\Delta G^\circ = 2 \Delta G^\circ_f(KCl) - 2 \Delta G^\circ_f(KClO_3)$
$\Delta G^\circ = (2 \text{ mol})(-409.2 \text{ kJ/mol}) - (2 \text{ mol})(-296.3 \text{ kJ/mol}) = -225.8 \text{ kJ}$
(c) $CH_3CH_2OH(l) + O_2(g) \rightarrow CH_3CO_2H(l) + H_2O(l)$
$\Delta G^\circ = [\Delta G^\circ_f(CH_3CO_2H) + \Delta G^\circ_f(H_2O)] - \Delta G^\circ_f(CH_3CH_2OH)$
$\Delta G^\circ = [(1 \text{ mol})(-390 \text{ kJ/mol}) + (1 \text{ mol})(-237.2 \text{ kJ/mol})] - (1 \text{ mol})(-174.9 \text{ kJ/mol}) = -452 \text{ kJ}$

17.74 A compound is thermodynamically stable with respect to its constituent elements at 25°C
if ΔG°_f is negative.

	ΔG°_f (kJ/mol)	Stable
(a) $BaCO_3(s)$	–1138	yes
(b) $HBr(g)$	–53.4	yes
(c) $N_2O(g)$	+104.2	no
(d) $C_2H_4(g)$	+68.1	no

17.76 $CH_2=CH_2(g) + H_2O(l) \rightarrow CH_3CH_2OH(l)$

$\Delta H^\circ = \Delta H^\circ_f(CH_3CH_2OH) - [\Delta H^\circ_f(CH_2=CH_2) + \Delta H^\circ_f(H_2O)]$

$\Delta H^\circ = (1\ mol)(-277.7\ kJ/mol) - [(1\ mol)(52.3\ kJ/mol) + (1\ mol)(-285.8\ kJ/mol)]$

$\Delta H^\circ = -44.2\ kJ$

$\Delta S^\circ = S^\circ(CH_3CH_2OH) - [S^\circ(CH_2=CH_2) + S^\circ(H_2O)]$

$\Delta S^\circ = (1\ mol)(161\ J/(K \cdot mol)) - [(1\ mol)(219.5\ J/(K \cdot mol))$
$\qquad\qquad + (1\ mol)(69.9\ J/(K \cdot mol))]$

$\Delta S^\circ = -128\ J/(K \cdot mol) = -128 \times 10^{-3}\ kJ/(K \cdot mol)$

$\Delta G^\circ = \Delta H^\circ - T\Delta S^\circ = -44.2\ kJ - (298\ K)(-128 \times 10^{-3}\ kJ/K) = -6.1\ kJ$

Because ΔG° is negative, the reaction is spontaneous under standard-state conditions at 25°C.
The reaction becomes nonspontaneous at high temperatures because ΔS° is negative.
To find the crossover temperature, set $\Delta G = 0$ and solve for T.

$$T = \frac{\Delta H^\circ}{\Delta S^\circ} = \frac{-44,200\ J}{-128\ J/K} = 345\ K = 72°C$$

The reaction becomes nonspontaneous at 72°C.

17.78 $3\ C_2H_2(g) \rightarrow C_6H_6(l)$

$\Delta G^\circ = \Delta G^\circ_f(C_6H_6) - 3\ \Delta G^\circ_f(C_2H_2)$

$\Delta G^\circ = (1\ mol)(124.5\ kJ/mol) - (3\ mol)(209.2\ kJ/mol) = -503.1\ kJ$

Because ΔG° is negative, the reaction is possible. Look for a catalyst.

Because ΔG°_f for benzene is positive (+124.5 kJ/mol), the synthesis of benzene from
graphite and gaseous H_2 at 25°C and 1 atm pressure is not possible.

Free Energy, Composition, and Chemical Equilibrium

17.80 $\Delta G = \Delta G^\circ + RT \ln Q$

17.82 $\Delta G = \Delta G^\circ + RT \ln \left[\dfrac{(P_{SO_3})^2}{(P_{SO_2})^2(P_{O_2})} \right]$

(a) $\Delta G = (-141.8\ kJ/mol) + [8.314 \times 10^{-3}\ kJ/(K \cdot mol)](298\ K)\ln\left[\dfrac{(1.0)^2}{(100)^2(100)}\right] = -176.0\ kJ/m$

(b) $\Delta G = (-141.8\ kJ/mol) + [8.314 \times 10^{-3}\ kJ/(K \cdot mol)](298\ K)\ln\left[\dfrac{(10)^2}{(2.0)^2(1.0)}\right] = -133.8\ kJ/mo$

(c) $Q = 1$, $\ln Q = 0$, $\Delta G = \Delta G^\circ = -141.8\ kJ/mol$

17.84 $\Delta G° = -RT \ln K$
   (a) If $K > 1$, $\Delta G°$ is negative.   (b) If $K = 1$, $\Delta G° = 0$.   (c) If $K < 1$, $\Delta G°$ is positive.

17.86 $\Delta G° = -RT \ln K_p = -141.8$ kJ

$$\ln K_p = \frac{-\Delta G°}{RT} = \frac{-(-141.8 \text{ kJ/mol})}{[8.314 \times 10^{-3} \text{ kJ/(K} \cdot \text{mol)}](298 \text{ K})} = 57.23$$

$K_p = e^{57.23} = 7.1 \times 10^{24}$

17.88 $C_2H_5OH(l) \rightleftharpoons C_2H_5OH(g)$
   $\Delta G° = \Delta G°_f(C_2H_5OH(g)) - \Delta G°_f(C_2H_5OH(l))$
   $\Delta G° = (1 \text{ mol})(-168.6 \text{ kJ/mol}) - (1 \text{ mol})(-174.9 \text{ kJ/mol}) = +6.3$ kJ
   $\Delta G° = -RT \ln K$

$$\ln K = \frac{-\Delta G°}{RT} = \frac{-(6.3 \text{ kJ/mol})}{[8.314 \times 10^{-3} \text{ kJ/(K} \cdot \text{mol)}](298 \text{ K})} = -2.54$$

$K = e^{-2.54} = 0.079;$ $K = K_p = P_{C_2H_5OH} = 0.079$ atm

17.90 $2 \text{ CH}_2\text{=CH}_2(g) + O_2(g) \rightarrow 2 \text{ C}_2\text{H}_4\text{O}(g)$
   $\Delta G° = 2 \Delta G°_f(C_2H_4O) - 2 \Delta G°_f(CH_2=CH_2)$
   $\Delta G° = (2 \text{ mol})(-13.1 \text{ kJ/mol}) - (2 \text{ mol})(68.1 \text{ kJ/mol}) = -162.4$ kJ
   $\Delta G° = -RT \ln K$

$$\ln K = \frac{-\Delta G°}{RT} = \frac{-(-162.4 \text{ kJ/mol})}{[8.314 \times 10^{-3} \text{ kJ/(K} \cdot \text{mol)}](298 \text{ K})} = 65.55$$

$K = K_p = e^{65.55} = 2.9 \times 10^{28}$

General Problems

17.92   The kinetic parameters [(a), (b), and (h)] are affected by a catalyst. The thermodynamic and equilibrium parameters [(c), (d), (e), (f), and (g)] are not affected by a catalyst.

17.94   (a) Spontaneous does not mean fast, just possible.
   (b) For a spontaneous reaction $\Delta S_{total} > 0$. ΔS_{sys} can be positive or negative.
   (c) An endothermic reaction can be spontaneous if $\Delta S_{sys} > 0$.
   (d) This statement is true because the sign of ΔG changes when the direction of a reaction is reversed.

17.96
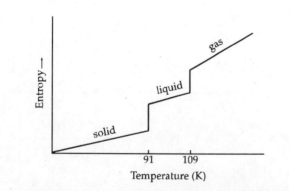

17.98   At the normal boiling point, $\Delta G = 0$.

$$\Delta G_{vap} = \Delta H_{vap} - T\Delta S_{vap}; \qquad T = \frac{\Delta H_{vap}}{\Delta S_{vap}} = \frac{38,600 \text{ J}}{110 \text{ J/K}} = 351 \text{ K} = 78°C$$

17.100 $\Delta G = \Delta H - T\Delta S$

(a) ΔH must be positive (endothermic) and greater than $T\Delta S$ in order for ΔG to be positive (nonspontaneous reaction).

(b) Set $\Delta G = 0$ and solve for ΔH.

$\Delta G = 0 = \Delta H - T\Delta S = \Delta H - (323 \text{ K})(104 \text{ J/K}) = \Delta H - (33592 \text{ J}) = \Delta H - (33.6 \text{ kJ})$

$\Delta H = 33.6 \text{ kJ}$

ΔH must be greater than 33.6 kJ.

17.102   (a) $2 \text{ Mg(s)} + O_2(g) \rightarrow 2 \text{ MgO(s)}$

$\Delta H° = 2 \Delta H°_f(MgO) = (2 \text{ mol})(-601.7 \text{ kJ/mol}) = -1203.4 \text{ kJ}$

$\Delta S° = 2 \text{ S°(MgO)} - [2 \text{ S°(Mg)} + \text{S°}(O_2)]$

$\Delta S° = (2 \text{ mol})(26.9 \text{ J/(K} \cdot \text{mol)})$

$\qquad\qquad - [(2 \text{ mol})(32.7 \text{ J/(K} \cdot \text{mol)}) + (1 \text{ mol})(205.0 \text{ J/(K} \cdot \text{mol)})]$

$\Delta S° = -216.6 \text{ J/K} = -216.6 \times 10^{-3} \text{ kJ/K}$

$\Delta G° = \Delta H° - T\Delta S° = -1203.4 \text{ kJ} - (298 \text{ K})(-216.6 \times 10^{-3} \text{ kJ/K}) = -1138.8 \text{ kJ}$

Because $\Delta G°$ is negative, the reaction is spontaneous at 25°C. $\Delta G°$ becomes less negative as the temperature is raised.

(b) $\text{MgCO}_3(s) \rightarrow \text{MgO(s)} + CO_2(g)$

$\Delta H° = [\Delta H°_f(MgO) + \Delta H°_f(CO_2)] - \Delta H°_f(MgCO_3)$

$\Delta H° = [(1 \text{ mol})(-601.1 \text{ kJ/mol}) + (1 \text{ mol})(-393.5 \text{ kJ/mol})]$

$\qquad\qquad - (1 \text{ mol})(-1096 \text{ kJ/mol}) = +101 \text{ kJ}$

$\Delta S° = [\text{S°(MgO)} + \text{S°}(CO_2)] - \text{S°}(MgCO_3)$

$\Delta S° = [(1 \text{ mol})(26.9 \text{ J/(K} \cdot \text{mol)}) + (1 \text{ mol})(213.6 \text{ J/(K} \cdot \text{mol)})]$

$\qquad\qquad - (1 \text{ mol})(65.7 \text{ J/(K} \cdot \text{mol)})$

$\Delta S° = 174.8 \text{ J/K} = 174.8 \times 10^{-3} \text{ kJ/K}$

$\Delta G° = \Delta H° - T\Delta S° = 101 \text{ kJ} - (298 \text{ K})(174.8 \times 10^{-3} \text{ kJ/K}) = +49 \text{ kJ}$

Because $\Delta G°$ is positive, the reaction is not spontaneous at 25°C. $\Delta G°$ becomes less positive as the temperature is raised.

(c) $Fe_2O_3(s) + 2 \text{ Al(s)} \rightarrow Al_2O_3(s) + 2 \text{ Fe(s)}$

$\Delta H° = \Delta H°_f(Al_2O_3) - \Delta H°_f(Fe_2O_3)$

$\Delta H° = (1 \text{ mol})(-1676 \text{ kJ/mol}) - (1 \text{ mol})(-824.2 \text{ kJ/mol}) = -852 \text{ kJ}$

$\Delta S° = [\text{S°}(Al_2O_3) + 2 \text{ S°(Fe)}] - [\text{S°}(Fe_2O_3) + 2 \text{ S°(Al)}]$

$\Delta S° = [(1 \text{ mol})(50.9 \text{ J/(K} \cdot \text{mol)}) + (2 \text{ mol})(27.3 \text{ J/(K} \cdot \text{mol)})]$

$\qquad - [(1 \text{ mol})(87.4 \text{ J/(K} \cdot \text{mol)}) + (2 \text{ mol})(28.3 \text{ J/(K} \cdot \text{mol)})]$

$\Delta S° = -38.5 \text{ J/K} = -38.5 \times 10^{-3} \text{ kJ/K}$

$\Delta G° = \Delta H° - T\Delta S° = -852 \text{ kJ} - (298 \text{ K})(-38.5 \times 10^{-3} \text{ kJ/K}) = -840 \text{ kJ}$

Because $\Delta G°$ is negative, the reaction is spontaneous at 25°C. $\Delta G°$ becomes less negative as the temperature is raised.

(d) $2 \text{ NaHCO}_3(s) \rightarrow Na_2CO_3(s) + CO_2(g) + H_2O(g)$

$\Delta H° = [\Delta H°_f(Na_2CO_3) + \Delta H°_f(CO_2) + \Delta H°_f(H_2O)] - 2 \Delta H°_f(NaHCO_3)$

$\Delta H^\circ = [(1 \text{ mol})(-1130.7 \text{ kJ/mol}) + (1 \text{ mol})(-393.5 \text{ kJ/mol})$
$\qquad + (1 \text{ mol})(-241.8 \text{ kJ/mol})] - (2 \text{ mol})(-950.8 \text{ kJ/mol}) = +135.6 \text{ kJ}$
$\Delta S^\circ = [S^\circ(Na_2CO_3) + S^\circ(CO_2) + S^\circ(H_2O)] - 2\, S^\circ(NaHCO_3)$
$\Delta S^\circ = [(1 \text{ mol})(135.0 \text{ J/(K} \cdot \text{mol)}) + (1 \text{ mol})(213.6 \text{ J/(K} \cdot \text{mol)})$
$\qquad + (1 \text{ mol})(188.7 \text{ J/(K} \cdot \text{mol)})] - (2 \text{ mol})(102 \text{ J/(K} \cdot \text{mol)})$
$\Delta S^\circ = +333 \text{ J/K} = +333 \times 10^{-3} \text{ kJ/K}$
$\Delta G^\circ = \Delta H^\circ - T\Delta S^\circ = +135.6 \text{ kJ} - (298 \text{ K})(+333 \times 10^{-3} \text{ kJ/K}) = +36.4 \text{ kJ}$
Because ΔG° is positive, the reaction is not spontaneous at 25°C. ΔG° becomes less positive as the temperature is raised.

17.104 (a) $6 \text{ C(s)} + 3 \text{ H}_2(g) \rightarrow C_6H_6(l)$
$\Delta S^\circ_f = S^\circ(C_6H_6) - [6\, S^\circ(C) + 3\, S^\circ(H_2)]$
$\Delta S^\circ_f = (1 \text{ mol})(172.8 \text{ J/(K} \cdot \text{mol)})$
$\qquad - [(6 \text{ mol})(5.7 \text{ J/(K} \cdot \text{mol)}) + (3 \text{ mol})(130.6 \text{ J/(K} \cdot \text{mol)})]$
$\Delta S^\circ_f = -253 \text{ J/K} = -253 \text{ J/(K} \cdot \text{mol)}$
$\Delta G^\circ_f = \Delta H^\circ_f - T\Delta S^\circ_f$
$\Delta S^\circ_f = \dfrac{\Delta H^\circ_f - \Delta G^\circ_f}{T} = \dfrac{49.0 \text{ kJ/mol} - 124.5 \text{ kJ/mol}}{298 \text{ K}} = -0.253 \text{ kJ/(K} \cdot \text{mol)}$
$\Delta S^\circ_f = -253 \text{ J/(K} \cdot \text{mol)}$
Both calculations lead to the same value of ΔS°_f.
(b) $Ca(s) + S(s) + 2 O_2(g) \rightarrow CaSO_4(s)$
$\Delta S^\circ_f = S^\circ(CaSO_4) - [S^\circ(Ca) + S^\circ(S) + 2\, S^\circ(O_2)]$
$\Delta S^\circ_f = (1 \text{ mol})(107 \text{ J/(K} \cdot \text{mol)})$
$\qquad - [(1 \text{ mol})(41.4 \text{ J/(K} \cdot \text{mol)}) + (1 \text{ mol})(31.8 \text{ J/(K} \cdot \text{mol)}) + (2 \text{ mol})(205.0 \text{ J/(K} \cdot \text{mol)})]$
$\Delta S^\circ_f = -376 \text{ J/K} = -376 \text{ J/(K} \cdot \text{mol)}$
$\Delta G^\circ_f = \Delta H^\circ_f - T\Delta S^\circ_f$
$\Delta S^\circ_f = \dfrac{\Delta H^\circ_f - \Delta G^\circ_f}{T} = \dfrac{-1434.1 \text{ kJ/mol} - (-1321.9 \text{ kJ/mol})}{298 \text{ K}} = -0.376 \text{ kJ/(K} \cdot \text{mol)}$
$\Delta S^\circ_f = -376 \text{ J/(K} \cdot \text{mol)}$
Both calculations lead to the same value of ΔS°_f.
(c) $2 \text{ C(s)} + 3 \text{ H}_2(g) + 1/2 \text{ O}_2(g) \rightarrow C_2H_5OH(l)$
$\Delta S^\circ_f = S^\circ(C_2H_5OH) - [S^\circ(C) + S^\circ(H_2) + 1/2\, S^\circ(O_2)]$
$\Delta S^\circ_f = (1 \text{ mol})(161 \text{ J/(K} \cdot \text{mol)})$
$\qquad - [(2 \text{ mol})(5.7 \text{ J/(K} \cdot \text{mol)}) + (3 \text{ mol})(130.6 \text{ J/(K} \cdot \text{mol)}) + (0.5 \text{ mol})(205.0 \text{ J/(K} \cdot \text{mol)})]$
$\Delta S^\circ_f = -345 \text{ J/K} = -345 \text{ J/(K} \cdot \text{mol)}$
$\Delta G^\circ_f = \Delta H^\circ_f - T\Delta S^\circ_f$
$\Delta S^\circ_f = \dfrac{\Delta H^\circ_f - \Delta G^\circ_f}{T} = \dfrac{-277.7 \text{ kJ/mol} - (-174.9 \text{ kJ/mol})}{298 \text{ K}} = -0.345 \text{ kJ/(K} \cdot \text{mol)}$
$\Delta S^\circ_f = -345 \text{ J/(K} \cdot \text{mol)}$
Both calculations lead to the same value of ΔS°_f.

17.106 $\Delta G^\circ = -RT \ln K_b$
At 20 °C: $\Delta G^\circ = -[8.314 \times 10^{-3} \text{ kJ/(K} \cdot \text{mol)}](293 \text{ K}) \ln(1.710 \times 10^{-5}) = +26.74 \text{ kJ/mol}$
At 50 °C: $\Delta G^\circ = -[8.314 \times 10^{-3} \text{ kJ/(K} \cdot \text{mol)}](323 \text{ K}) \ln(1.892 \times 10^{-5}) = +29.20 \text{ kJ/mol}$
$\Delta G^\circ = \Delta H^\circ - T\Delta S^\circ$

$26.74 = \Delta H^\circ - 293\Delta S^\circ$

$29.20 = \Delta H^\circ - 323\Delta S^\circ$ Solve these two equations simultaneously for ΔH° and ΔS°.

$26.74 + 293\Delta S^\circ = \Delta H^\circ$

$29.20 + 323\Delta S^\circ = \Delta H^\circ$ Set these two equations equal to each other.

$26.74 + 293\Delta S^\circ = 29.20 + 323\Delta S^\circ$

$26.74 - 29.20 = 323\Delta S^\circ - 293\,\Delta S^\circ$

$-2.46 = 30\Delta S^\circ$

$\Delta S^\circ = -2.46/30 = -0.0820 = -0.0820\text{ kJ/K} = -82.0\text{ J/K}$

$26.74 + 293\Delta S^\circ = 26.74 + 293(-0.0820) = \Delta H^\circ = +2.71\text{ kJ}$

17.108 (a) $\Delta H^\circ = [\Delta H^\circ_f(Ag^+(aq)) + \Delta H^\circ_f(Br^-(aq))] - \Delta H^\circ_f(AgBr(s))$

$\Delta H^\circ = [(1\text{ mol})(105.6\text{ kJ/mol}) + (1\text{ mol})(-121.5\text{ kJ/mol})]$
$- (1\text{ mol})(-100.4\text{ kJ/mol}) = +84.5\text{ kJ}$

$\Delta S^\circ = [S^\circ(Ag^+(aq)) + S^\circ(Br^-(aq))] - S^\circ(AgBr(s))$

$\Delta S^\circ = [(1\text{ mol})(72.7\text{ J/(K} \cdot \text{mol})) + (1\text{ mol})(82.4\text{ J/(K} \cdot \text{mol}))]$
$- (1\text{ mol})(107\text{ J/(K} \cdot \text{mol})) = +48.1\text{ J}$

$\Delta G^\circ = \Delta H^\circ - T\Delta S^\circ = 84.5\text{ kJ} - (298\text{ K})(48.1 \times 10^{-3}\text{ kJ/K}) = +70.2\text{ kJ}$

(b) $\Delta G^\circ = -RT\ln K_{sp}$

$\ln K_{sp} = \dfrac{-\Delta G^\circ}{RT} = \dfrac{-70.2\text{ kJ/mol}}{[8.314 \times 10^{-3}\text{ kJ/(K} \cdot \text{mol})](298\text{ K})} = -28.3$

$K_{sp} = e^{-28.3} = 5 \times 10^{-13}$

(c) $Q = [Ag^+][Br^-] = (1.00 \times 10^{-5})(1.00 \times 10^{-5}) = 1.00 \times 10^{-10}$

$\Delta G = \Delta G^\circ + RT\ln Q$

$\Delta G = 70.2\text{ kJ/mol} + [8.314 \times 10^{-3}\text{ kJ/(K} \cdot \text{mol})](298\text{ K})\ln(1.00 \times 10^{-10}) = 13.2\text{ kJ/mol}$

A positive value of ΔG means that the forward reaction is nonspontaneous under these conditions. The reverse reaction is therefore spontaneous, which is consistent with the fact that $Q > K_{sp}$.

17.110 $Br_2(l) \;\rightleftharpoons\; Br_2(g)$

$\Delta S^\circ = S^\circ(Br_2(g)) - S^\circ(Br_2(l))$

$\Delta S^\circ = (1\text{ mol})(245.4\text{ J/(K} \cdot \text{mol})) - (1\text{ mol})(152.2\text{ J/(K} \cdot \text{mol})) = 93.2\text{ J/K} = 93.2 \times 10^{-3}\text{ kJ/K}$

$\Delta G = \Delta H^\circ - T\Delta S^\circ$

At the boiling point, $\Delta G = 0$.

$0 = \Delta H^\circ - T_{bp}\Delta S^\circ$

$T_{bp} = \dfrac{\Delta H^\circ}{\Delta S^\circ}$

$\Delta H^\circ = T_{bp}\,\Delta S^\circ = (332\text{ K})(93.2 \times 10^{-3}\text{ kJ/K}) = 30.9\text{ kJ}$

$K_p = P_{Br_2} = \left(227\text{ mm Hg} \times \dfrac{1\text{ atm}}{760\text{ mm Hg}}\right) = 0.299\text{ atm}$

$\Delta G^\circ = -RT\ln K_p$ and $\Delta G^\circ = \Delta H^\circ - T\Delta S^\circ$ (set equations equal to each other)

$\Delta H^\circ - T\Delta S^\circ = -RT\ln K_p$ (rearrange)

$$\ln K_p = \frac{-\Delta H^\circ}{R} \frac{1}{T} + \frac{\Delta S^\circ}{R} \quad \text{(solve for T)}$$

$$T = \frac{\left(\dfrac{-\Delta H^\circ}{R}\right)}{\left(\ln K_p - \dfrac{\Delta S^\circ}{R}\right)} = \frac{\left(\dfrac{-30.9 \text{ kJ/mol}}{8.314 \times 10^{-3} \text{ kJ/(K} \cdot \text{mol)}}\right)}{\left(\ln(0.299) - \dfrac{93.2 \times 10^{-3} \text{ kJ/(K} \cdot \text{mol)}}{8.314 \times 10^{-3} \text{ kJ/(K} \cdot \text{mol)}}\right)} = 299 \text{ K} = 26^\circ\text{C}$$

$Br_2(l)$ has a vapor pressure of 227 mm Hg at 26°C.

17.112 $\Delta H^\circ = [2 \ \Delta H^\circ_f(Cl^-(aq))] - [2 \ \Delta H^\circ_f(Br^-(aq))]$
$\Delta H^\circ = [(2 \text{ mol})(-167.2 \text{ kJ/mol})] - [(2 \text{ mol})(-121.5 \text{ kJ/mol})] = -91.4 \text{ kJ}$
$\Delta S^\circ = [S^\circ(Br_2(l)) + 2 \ S^\circ(Cl^-(aq))] - [2 \ S^\circ(Br^-(aq)) + S^\circ(Cl_2(g))]$
$\Delta S^\circ = [(1 \text{ mol})(152.2 \text{ J/(K} \cdot \text{mol)}) + (2 \text{ mol})(56.5 \text{ J/(K} \cdot \text{mol)})]$
$\qquad\qquad - [(2 \text{ mol})(82.4 \text{ J/(K} \cdot \text{mol)}) + (1 \text{ mol})(223.0 \text{ J/(K} \cdot \text{mol)})] = -122.6 \text{ J/K}$
$80^\circ\text{C} = 80 + 273 = 353 \text{ K}$
$\Delta G^\circ = \Delta H^\circ - T\Delta S^\circ = -91.4 \text{ kJ} - (353 \text{ K})(-122.6 \times 10^{-3} \text{ kJ/K}) = -48.1 \text{ kJ}$
$\Delta G^\circ = -RT \ln K$

$$\ln K = \frac{-\Delta G^\circ}{RT} = \frac{-(-48.1 \text{ kJmol})}{[8.314 \times 10^{-3} \text{ kJ/(K} \cdot \text{mol)}](353 \text{ K})} = 16.4$$

$K = e^{16.4} = 1.3 \times 10^7$

17.114 $35^\circ\text{C} = 35 + 273 = 308 \text{ K}$
$\Delta G^\circ = \Delta H^\circ - T\Delta S^\circ = -352 \text{ kJ} - (308 \text{ K})(-899 \times 10^{-3} \text{ kJ/K}) = -75.1 \text{ kJ}$
$\Delta G^\circ = -RT \ln K_p$

$$\ln K_p = \frac{-\Delta G^\circ}{RT} = \frac{-(-75.1 \text{ kJ/mol})}{[8.314 \times 10^{-3} \text{ kJ/(K} \cdot \text{mol)}](308 \text{ K})} = 29.33$$

$K_p = e^{29.33} = 5.5 \times 10^{12}$

$$K_p = \frac{1}{(P_{H_2O})^6} = 5.5 \times 10^{12}$$

$$P_{H_2O} = \sqrt[6]{\frac{1}{5.5 \times 10^{12}}} = 0.0075 \text{ atm}$$

$$P_{H_2O} = 0.0075 \text{ atm} \times \frac{760 \text{ mm Hg}}{1 \text{ atm}} = 5.7 \text{ mm Hg}$$

17.116 $N_2O_4(g) \rightleftarrows 2 \ NO_2(g)$
$\Delta H^\circ = 2 \ \Delta H^\circ_f(NO_2) - \Delta H^\circ_f(N_2O_4) = (2 \text{ mol})(33.2 \text{ kJ}) - (1 \text{ mol})(9.16 \text{ kJ}) = 57.2 \text{ kJ}$
$\Delta S^\circ = 2 \ S^\circ(NO_2) - S^\circ(N_2O_4) = (2 \text{ mol})(240.0 \text{ J/(K} \cdot \text{mol)}) - (1 \text{ mol})(304.2 \text{ J/(K} \cdot \text{mol)})$
$\Delta S^\circ = 175.8 \text{ J/K} = 175.8 \times 10^{-3} \text{ kJ/K}$
$\Delta G^\circ = \Delta H^\circ - T\Delta S^\circ$ and $\Delta G^\circ = -RT \ln K_p$; Set these two equations equal to each other and solve for T.
$\Delta H^\circ - T\Delta S^\circ = -RT \ln K_p$
$\Delta H^\circ = T\Delta S^\circ - RT \ln K_p = T(\Delta S^\circ - R \ln K_p)$

$$T = \frac{\Delta H^\circ}{\Delta S^\circ - R \ln K_p}$$

(a) $P_{N_2O_4} + P_{NO_2} = 1.00$ atm and $P_{NO_2} = 2 P_{N_2O_4}$

$P_{N_2O_4} + 2 P_{N_2O_4} = 3 P_{N_2O_4} = 1.00$ atm

$P_{N_2O_4} = 1.00$ atm/3 $= 0.333$ atm

$P_{NO_2} = 1.00$ atm $- P_{N_2O_4} = 1.00 - 0.333 = 0.667$ atm

$$K_p = \frac{(P_{NO_2})^2}{P_{N_2O_4}} = \frac{(0.667)^2}{(0.333)} = 1.34$$

$$T = \frac{\Delta H^\circ}{\Delta S^\circ - R \ln K_p}$$

$$T = \frac{57.2 \text{ kJ/mol}}{[175.8 \times 10^{-3} \text{ kJ/(K} \cdot \text{mol)}] - [8.314 \times 10^{-3} \text{ kJ/(K} \cdot \text{mol)}] \ln(1.34)} = 330 \text{ K}$$

$T = 330$ K $= 330 - 273 = 57°C$

(b) $P_{N_2O_4} + P_{NO_2} = 1.00$ atm and $P_{NO_2} = P_{N_2O_4}$ so $P_{NO_2} = P_{N_2O_4} = 0.50$ atm

$$K_p = \frac{(P_{NO_2})^2}{P_{N_2O_4}} = \frac{(0.500)^2}{(0.500)} = 0.500$$

$$T = \frac{\Delta H^\circ}{\Delta S^\circ - R \ln K_p}$$

$$T = \frac{57.2 \text{ kJ/mol}}{[175.8 \times 10^{-3} \text{ kJ/(K} \cdot \text{mol)}] - [8.314 \times 10^{-3} \text{ kJ/(K} \cdot \text{mol)}] \ln(0.500)} = 315 \text{ K}$$

$T = 315$ K $= 315 - 273 = 42°C$

Multi-Concept Problems

17.118 $N_2(g) + 3 H_2(g) \rightleftharpoons 2 NH_3(g)$

$\Delta H^\circ = 2 \Delta H^\circ_f(NH_3) - [\Delta H^\circ_f(N_2) + 3 \Delta H^\circ_f(H_2)] = (2 \text{ mol})(-46.1 \text{ kJ}) - [0] = -92.2 \text{ kJ}$

$\Delta S^\circ = 2 S^\circ(NH_3) - [S^\circ(N_2) + 3 S^\circ(H_2)]$

$\Delta S^\circ = (2 \text{ mol})(192.3 \text{ J/(K} \cdot \text{mol)})$

$\quad\quad\quad - [(1 \text{ mol})(191.5 \text{ J/(K} \cdot \text{mol)}) + (3 \text{ mol})(130.6 \text{ J/(K} \cdot \text{mol)})] = -198.7 \text{ J/K}$

$\Delta G^\circ = \Delta H^\circ - T\Delta S^\circ = -92.2 \text{ kJ} - (673 \text{ K})(-198.7 \times 10^{-3} \text{ kJ/K}) = 41.5 \text{ kJ}$

$\Delta G^\circ = -RT \ln K_p$

$$\ln K_p = \frac{-\Delta G^\circ}{RT} = \frac{-41.5 \text{ kJ/mol}}{[8.314 \times 10^{-3} \text{ kJ/(K} \cdot \text{mol)}](673 \text{ K})} = -7.42$$

$K_p = e^{-7.42} = 6.0 \times 10^{-4}$

Because $K_p = K_c(RT)^{\Delta n}$, $K_c = K_p(RT)^{-\Delta n}$

$K_c = K_p(RT)^2 = (6.0 \times 10^{-4})[(0.082\ 06)(673)]^2 = 1.83$

N_2, 28.01 amu; H_2, 2.016 amu

Initial concentrations:

$$[N_2] = \frac{(14.0 \text{ g})\left(\dfrac{1 \text{ mol}}{28.01 \text{ g}}\right)}{5.00 \text{ L}} = 0.100 \text{ M} \quad \text{and} \quad [H_2] = \frac{(3.024 \text{ g})\left(\dfrac{1 \text{ mol}}{2.016 \text{ g}}\right)}{5.00 \text{ L}} = 0.300 \text{ M}$$

	$N_2(g)$	+	$3 H_2(g)$	⇌	$2 NH_3(g)$
initial (M)	0.100		0.300		0
change (M)	–x		–3x		+2x
equil (M)	0.100 – x		0.300 – 3x		2x

$$K_c = \frac{[NH_3]^2}{[N_2][H_2]^3} = \frac{(2x)^2}{(0.100 - x)(0.300 - 3x)^3} = \frac{4x^2}{27(0.100 - x)^4} = 1.83$$

$$\left(\frac{x}{(0.100 - x)^2}\right)^2 = \frac{(27)(1.83)}{4} = 12.35; \qquad \frac{x}{(0.100 - x)^2} = \sqrt{12.35} = 3.515$$

$3.515x^2 - 1.703x + 0.03515 = 0$

Use the quadratic formula to solve for x.

$$x = \frac{-(-1.703) \pm \sqrt{(-1.703)^2 - (4)(3.515)(0.03515)}}{2(3.515)} = \frac{1.703 \pm 1.551}{7.030}$$

x = 0.463 and 0.0216

Of the two solutions for x, only 0.0216 has physical meaning because 0.463 would lead to negative concentrations of N_2 and H_2.

$[N_2] = 0.100 - x = 0.100 - 0.0216 = 0.078$ M

$[H_2] = 0.300 - 3x = 0.300 - 3(0.0216) = 0.235$ M

$[NH_3] = 2x = 2(0.0216) = 0.043$ M

17.120 $Pb(s) + PbO_2(s) + 2 H^+(aq) + 2 HSO_4^-(aq) \rightarrow 2 PbSO_4(s) + 2 H_2O(l)$

(a) $\Delta G° = [2 \Delta G°_f(PbSO_4) + 2 \Delta G°_f(H_2O)] - [\Delta G°_f(PbO_2) + 2 \Delta G°_f(HSO_4^-)]$

$\Delta G° = (2 \text{ mol})(-813.2 \text{ kJ/mol}) + (2 \text{ mol})(-237.2 \text{ kJ/mol})]$

$\qquad - [(1 \text{ mol})(-217.4 \text{ kJ/mol}) + (2 \text{ mol})(-756.0 \text{ kJ/mol})] = -371.4$ kJ

(b) $°C = 5/9(°F - 32) = 5/9(10 - 32) = -12.2°C; \quad -12.2°C = 261$ K

$\Delta H° = [2 \Delta H°_f(PbSO_4) + 2 \Delta H°_f(H_2O)] - [\Delta H°_f(PbO_2) + 2 \Delta H°_f(HSO_4^-)]$

$\Delta H° = [(2 \text{ mol})(-919.9 \text{ kJ/mol}) + (2 \text{ mol})(-285.8 \text{ kJ/mol})]$

$\qquad - [(1 \text{ mol})(-277 \text{ kJ/mol}) + (2 \text{ mol})(-887.3 \text{ kJ/mol})] = -359.8$ kJ

$\Delta S° = [2 S°(PbSO_4) + 2 S°(H_2O)] - [S°(Pb) + S°(PbO_2) + 2 S°(H^+) + 2 S°(HSO_4^-)]$

$\Delta S° = [(2 \text{ mol})(148.6 \text{ J/(K} \cdot \text{mol)}) + (2 \text{ mol})(69.9 \text{ J/(K} \cdot \text{mol)})]$

$\qquad - [(1 \text{ mol})(64.8 \text{ J/(K} \cdot \text{mol)}) + (1 \text{ mol})(68.6 \text{ J/(K} \cdot \text{mol)})$

$\qquad + (2 \text{ mol})(132 \text{ J/(K} \cdot \text{mol)})] = 39.6 \text{ J/K} = 39.6 \times 10^{-3}$ kJ/K

$\Delta G° = \Delta H° - T\Delta S° = -359.8 \text{ kJ} - (261 \text{ K})(39.6 \times 10^{-3} \text{ kJ/K}) = -370.1$ kJ at 261 K

	$HSO_4^-(aq)$	+	$H_2O(l)$	⇌	$H_3O^+(aq)$	+	$SO_4^{2-}(aq)$
initial (M)	0.100				0.100		0
change (M)	–x				+x		+x
equil (M)	0.100 – x				0.100 + x		x

$$K_{a2} = \frac{[H_3O^+][SO_4^{2-}]}{[HSO_4^-]} = 1.2 \times 10^{-2} = \frac{(0.100 + x)x}{0.100 - x}$$

$x^2 + 0.112x - (1.2 \times 10^{-3}) = 0$

Use the quadratic formula to solve for x.

$$x = \frac{-(0.112) \pm \sqrt{(0.112)^2 - (4)(1)(-1.2 \times 10^{-3})}}{2(1)} = \frac{-0.112 \pm 0.132}{2}$$

x = −0.122 and 0.010

Of the two solutions for x, only 0.010 has physical meaning because −0.122 would lead to a negative concentration of H_3O^+.

$[H^+] = 0.100 + x = 0.100 + 0.010 = 0.110$ M

$[HSO_4^-] = 0.100 - x = 0.100 - 0.010 = 0.090$ M

$$\Delta G = \Delta G^\circ + RT \ln \frac{1}{[H^+]^2[HSO_4^-]^2}$$

$$\Delta G = (-370.1 \text{ kJ/mol}) + [8.314 \times 10^{-3} \text{ kJ/(K} \cdot \text{mol})](261 \text{ K}) \ln \frac{1}{(0.110)^2(0.090)^2}$$

$\Delta G = -350.1$ kJ/mol

17.122   PV = nRT

$$n_{NH_3} = \frac{PV}{RT} = \frac{\left(744 \text{ mm Hg} \times \dfrac{1.00 \text{ atm}}{760 \text{ mm Hg}}\right)(1.00 \text{ L})}{\left(0.082\ 06 \dfrac{L \cdot atm}{K \cdot mol}\right)(298.1 \text{ K})} = 0.0400 \text{ mol NH}_3$$

500.0 mL = 0.5000 L

$[NH_3] = 0.0400$ mol/0.5000 L = 0.0800 M

$NH_3(aq) + H_2O(l) \rightleftharpoons NH_4^+(aq) + OH^-(aq)$

$\Delta H^\circ = [\Delta H^\circ_f(NH_4^+) + \Delta H^\circ_f(OH^-)] - [\Delta H^\circ_f(NH_3) + \Delta H^\circ_f(H_2O)]$

$\Delta H^\circ = [(1 \text{ mol})(-132.5 \text{ kJ/mol}) + (1 \text{ mol})(-230.0 \text{ kJ/mol})]$

$\qquad - [(1 \text{ mol})(-80.3 \text{ kJ/mol}) + (1 \text{ mol})(-285.8 \text{ kJ/mol})] = +3.6 \text{ kJ}$

$\Delta S^\circ = [S^\circ(NH_4^+) + S^\circ(OH^-)] - [S^\circ(NH_3) + S^\circ(H_2O)]$

$\Delta S^\circ = [(1 \text{ mol})(113 \text{ J/(K} \cdot \text{mol})) + (1 \text{ mol})(-10.8 \text{ J/(K} \cdot \text{mol}))]$

$\qquad - [(1 \text{ mol})(111 \text{ J/(K} \cdot \text{mol})) + (1 \text{ mol})(69.9 \text{ J/(K} \cdot \text{mol}))] = -78.7 \text{ J/K}$

T = 2.0°C = 2.0 + 273.1 = 275.1 K

$\Delta G^\circ = \Delta H^\circ - T\Delta S^\circ = 3.6 \text{ kJ} - (275.1 \text{ K})(-78.7 \times 10^{-3} \text{ kJ/K}) = 25.3 \text{ kJ}$

$\Delta G^\circ = -RT \ln K_b$

$$\ln K_b = \frac{-\Delta G^\circ}{RT} = \frac{-25.3 \text{ kJ/mol}}{[8.314 \times 10^{-3} \text{ kJ/(K} \cdot \text{mol})](275.1 \text{ K})} = -11.06$$

$K_b = e^{-11.06} = 1.6 \times 10^{-5}$

	$NH_3(aq)$ +	$H_2O(l)$ $\rightleftharpoons$	$NH_4^+(aq)$ +	$OH^-(aq)$
initial (M)	0.0800		0	~0
change (M)	−x		+x	+x
equil (M)	0.0800 − x		x	x

at 2°C, $K_b = \dfrac{[NH_4^+][OH^-]}{[NH_3]} = 1.6 \times 10^{-5} = \dfrac{x^2}{0.0800 - x} \approx \dfrac{x^2}{0.0800}$

$x^2 = (1.6 \times 10^{-5})(0.0800)$

$x = [OH^-] = \sqrt{(1.6 \times 10^{-5})(0.0800)} = 1.13 \times 10^{-3} \text{ M}$

$[H_3O^+] = \dfrac{1.0 \times 10^{-14}}{1.13 \times 10^{-3}} = 8.85 \times 10^{-12} \text{ M}$

$pH = -\log[H_3O^+] = -\log(8.85 \times 10^{-12}) = 11.05$

Electrochemistry

18.1 $2 Ag^+(aq) + Ni(s) \rightarrow 2 Ag(s) + Ni^{2+}(aq)$

There is a Ni anode in an aqueous solution of Ni^{2+}, and a Ag cathode in an aqueous solution of Ag^+. A salt bridge connects the anode and cathode compartment. The electrodes are connected through an external circuit.

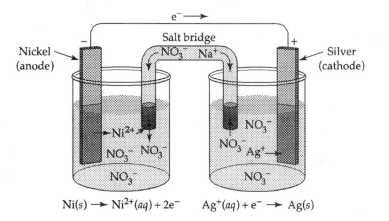

18.2 $Fe(s)|Fe^{2+}(aq)\|Sn^{2+}(aq)|Sn(s)$

18.3 $Pb(s) + Br_2(l) \rightarrow Pb^{2+}(aq) + 2 Br^-(aq)$
There is a Pb anode in an aqueous solution of Pb^{2+}. The cathode is a Pt wire that dips into a pool of liquid Br_2 and an aqueous solution that is saturated with Br_2. A salt bridge connects the anode and cathode compartment. The electrodes are connected through an external circuit.

18.4   (a) and (b)

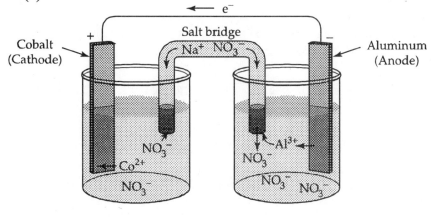

(c) $2 Al(s) + 3 Co^{2+}(aq) \rightarrow 2 Al^{3+}(aq) + 3 Co(s)$
(d) $Al(s)|Al^{3+}(aq)\|Co^{2+}(aq)|Co(s)$

18.5 $Al(s) + Cr^{3+}(aq) \rightarrow Al^{3+}(aq) + Cr(s)$

$$\Delta G° = -nFE° = -(3 \text{ mol } e^-)\left(\frac{96,500 \text{ C}}{1 \text{ mol } e^-}\right)(0.92 \text{ V})\left(\frac{1 \text{ J}}{1 \text{ C} \cdot \text{V}}\right) = -266,340 \text{ J} = -270 \text{ kJ}$$

18.6   oxidation: $Al(s) \rightarrow Al^{3+}(aq) + 3 e^-$ $E° = 1.66$ V
          reduction: $\underline{Cr^{3+}(aq) + 3 e^- \rightarrow Cr(s)}$ $\underline{E° = ?}$
          overall $Al(s) + Cr^{3+}(aq) \rightarrow Al^{3+}(aq) + Cr(s)$ $E° = 0.92$ V
          The standard reduction potential for the Cr^{3+}/Cr half cell is:
          $E° = 0.92 - 1.66 = -0.74$ V

18.7   (a) $Cl_2(g) + 2 e^- \rightarrow 2 Cl^-(aq)$ $E° = 1.36$ V
              $Ag^+(aq) + e^- \rightarrow Ag(s)$ $E° = 0.80$ V
          Cl_2 has the greater tendency to be reduced (larger $E°$). The species that has the greater tendency to be reduced is the stronger oxidizing agent. Cl_2 is the stronger oxidizing agent.

          (b) $Fe^{2+}(aq) + 2 e^- \rightarrow Fe(s)$ $E° = -0.45$ V
              $Mg^{2+}(aq) + 2 e^- \rightarrow Mg(s)$ $E° = -2.37$ V
          The second half-reaction has the lesser tendency to occur in the forward direction (more negative $E°$) and the greater tendency to occur in the reverse direction. Therefore, Mg is the stronger reducing agent.

18.8   (a) $2 Fe^{3+}(aq) + 2 I^-(aq) \rightarrow 2 Fe^{2+}(aq) + I_2(s)$
          reduction: $Fe^{3+}(aq) + e^- \rightarrow Fe^{2+}(aq)$ $E° = 0.77$ V
          oxidation: $2 I^-(aq) \rightarrow I_2(s) + 2 e^-$ $\underline{E° = -0.54 \text{ V}}$
                                                  overall $E° = 0.23$ V
          Because $E°$ for the overall reaction is positive, this reaction can occur under standard-state conditions.

          (b) $3 Ni(s) + 2 Al^{3+}(aq) \rightarrow 3 Ni^{2+}(aq) + 2 Al(s)$
          oxidation: $Ni(s) \rightarrow Ni^{2+}(aq) + 2 e^-$ $E° = 0.26$ V
          reduction: $Al^{3+}(aq) + 3 e^- \rightarrow Al(s)$ $\underline{E° = -1.66 \text{ V}}$
                                                  overall $E° = -1.40$ V
          Because $E°$ for the overall reaction is negative, this reaction cannot occur under standard-state conditions. This reaction can occur in the reverse direction.

18.9   (a) D is the strongest reducing agent. D^+ has the most negative standard reduction potential. A^{3+} is the strongest oxidizing agent. It has the most positive standard reduction potential.
          (b) An oxidizing agent can oxidize any reducing agent that is below it in the table. B^{2+} can oxidize C and D.
          A reducing agent can reduce any oxidizing agent that is above it in the table. C can reduce A^{3+} and B^{2+}.

(c) Use the two half-reactions that have the most positive and the most negative standard reduction potentials, respectively.

$$A^{3+} \; + \; 2\,e^- \; \rightarrow \; A^+ \qquad\qquad 1.47\ V$$
$$\underline{2 \times (D \; \rightarrow \; D^+ \; + \; e^-)} \qquad\qquad \underline{1.38\ V}$$
$$A^{3+} \; + \; 2\,D \; \rightarrow \; A^+ \; + \; 2\,D^+ \qquad\qquad 2.85\ V$$

18.10 $Cu(s) \; + \; 2\,Fe^{3+}(aq) \; \rightarrow \; Cu^{2+}(aq) \; + \; 2\,Fe^{2+}(aq)$

$$E^\circ = E^\circ_{Cu \rightarrow Cu^{2+}} + E^\circ_{Fe^{3+} \rightarrow Fe^{2+}} = -0.34\ V + 0.77\ V = 0.43\ V; \qquad n = 2\ mol\ e^-$$

$$E = E^\circ - \frac{0.0592\ V}{n} \log \frac{[Cu^{2+}][Fe^{2+}]^2}{[Fe^{3+}]^2} = 0.43\ V - \frac{(0.0592\ V)}{2} \log \frac{(0.25)(0.20)^2}{(1.0 \times 10^{-4})^2} = 0.25\ V$$

18.11 $5\,[Cu(s) \; \rightarrow \; Cu^{2+}(aq) \; + \; 2\,e^-] \qquad$ oxidation half reaction
$2\,[5\,e^- \; + \; 8\,H^+(aq) \; + \; MnO_4^-(aq) \; \rightarrow \; Mn^{2+}(aq) \; + \; 4\,H_2O(l)] \qquad$ reduction half reaction

$$5\,Cu(s) \; + \; 16\,H^+(aq) \; + \; 2\,MnO_4^-(aq) \; \rightarrow \; 5\,Cu^{2+}(aq) \; + \; 2\,Mn^{2+}(aq) \; + \; 8\,H_2O(l)$$

$$\Delta E = -\frac{0.0592\ V}{n} \log \frac{[Cu^{2+}]^5[Mn^{2+}]^2}{[MnO_4^-]^2[H^+]^{16}}$$

(a) The anode compartment contains Cu^{2+}.

$$\Delta E = -\frac{0.0592\ V}{10} \log \frac{(0.01)^5(1)^2}{(1)^2(1)^{16}} = +0.059\ V$$

(b) The cathode compartment contains Mn^{2+}, MnO_4^-, and H^+.

$$\Delta E = -\frac{0.0592\ V}{10} \log \frac{(1)^5(0.01)^2}{(0.01)^2(0.01)^{16}} = -0.19\ V$$

18.12 $H_2(g) \; + \; Pb^{2+}(aq) \; \rightarrow \; 2\,H^+(aq) \; + \; Pb(s)$

$$E^\circ = E^\circ_{H_2 \rightarrow H^+} + E^\circ_{Pb^{2+} \rightarrow Pb} = 0\ V + (-0.13\ V) = -0.13\ V; \qquad n = 2\ mol\ e^-$$

$$E = E^\circ - \frac{0.0592\ V}{n} \log \frac{[H_3O^+]^2}{[Pb^{2+}](P_{H_2})}$$

$$0.28\ V = -0.13\ V - \frac{(0.0592\ V)}{2} \log \frac{[H_3O^+]^2}{(1)(1)} = -0.13\ V - (0.0592\ V) \log [H_3O^+]$$

$pH = -\log[H_3O^+]$ therefore $0.28\ V = -0.13\ V + (0.0592\ V)\ pH$

$$pH = \frac{(0.28\ V + 0.13\ V)}{0.0592\ V} = 6.9$$

18.13 $4\,Fe^{2+}(aq) \; + \; O_2(g) \; + \; 4\,H^+(aq) \; \rightarrow \; 4\,Fe^{3+}(aq) \; + \; 2\,H_2O(l)$

$$E^\circ = E^\circ_{Fe^{2+} \rightarrow Fe^{3+}} + E^\circ_{O_2 \rightarrow H_2O} = -0.77\ V + 1.23\ V = 0.46\ V; \qquad n = 4\ mol\ e^-$$

$$E^\circ = \frac{0.0592\ V}{n} \log K$$

$$\log K = \frac{nE^\circ}{0.0592\text{ V}} = \frac{(4)(0.46\text{ V})}{0.0592\text{ V}} = 31; \quad K = 10^{31} \text{ at } 25^\circ C$$

18.14 $E^\circ = \dfrac{0.0592\text{ V}}{n} \log K = \dfrac{0.0592\text{ V}}{2} \log(1.8 \times 10^{-5}) = -0.140\text{ V}$

18.15   (a) $Zn(s) + 2\,MnO_2(s) + 2\,NH_4^+(aq) \rightarrow Zn^{2+}(aq) + Mn_2O_3(s) + 2\,NH_3(aq) + H_2O(l)$
   (b) $Zn(s) + 2\,MnO_2(s) \rightarrow ZnO(s) + Mn_2O_3(s)$
   (c) $Zn(s) + HgO(s) \rightarrow ZnO(s) + Hg(l)$
   (d) $Cd(s) + 2\,NiO(OH)(s) + 2\,H_2O(l) \rightarrow Cd(OH)_2(s) + 2\,Ni(OH)_2(s)$

18.16   (a) $[Mg(s) \rightarrow Mg^{2+}(aq) + 2\,e^-] \times 2$
      $\underline{O_2(g) + 4\,H^+(aq) + 4\,e^- \rightarrow 2\,H_2O(l)}$
      $2\,Mg(s) + O_2(g) + 4\,H^+(aq) \rightarrow 2\,Mg^{2+}(aq) + 2\,H_2O(l)$
   (b) $[Fe(s) \rightarrow Fe^{2+}(aq) + 2\,e^-] \times 4$
      $[O_2(g) + 4\,H^+(aq) + 4\,e^- \rightarrow 2\,H_2O(l)] \times 2$
      $4\,Fe^{2+}(aq) + O_2(g) + 4\,H^+(aq) \rightarrow 4\,Fe^{3+}(aq) + 2\,H_2O(l)$
      $\underline{[2\,Fe^{3+}(aq) + 4\,H_2O(l) \rightarrow Fe_2O_3\cdot H_2O(s) + 6\,H^+(aq)] \times 2}$
      $4\,Fe(s) + 3\,O_2(g) + 2\,H_2O(l) \rightarrow 2\,Fe_2O_3\cdot H_2O(s)$

18.17   (a)

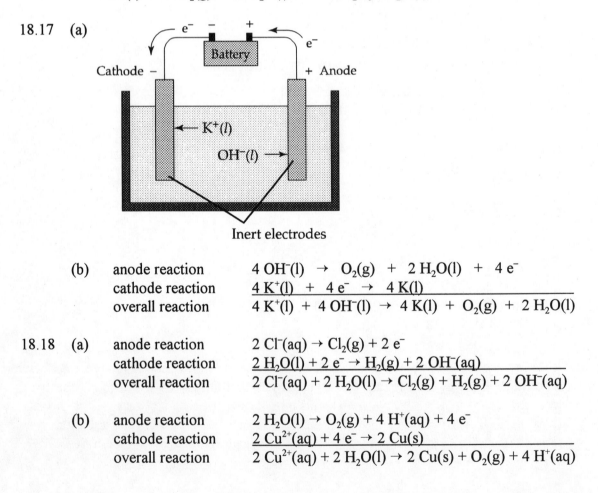

Inert electrodes

   (b)   anode reaction $4\,OH^-(l) \rightarrow O_2(g) + 2\,H_2O(l) + 4\,e^-$
      cathode reaction $\underline{4\,K^+(l) + 4\,e^- \rightarrow 4\,K(l)}$
      overall reaction $4\,K^+(l) + 4\,OH^-(l) \rightarrow 4\,K(l) + O_2(g) + 2\,H_2O(l)$

18.18   (a)   anode reaction $2\,Cl^-(aq) \rightarrow Cl_2(g) + 2\,e^-$
      cathode reaction $\underline{2\,H_2O(l) + 2\,e^- \rightarrow H_2(g) + 2\,OH^-(aq)}$
      overall reaction $2\,Cl^-(aq) + 2\,H_2O(l) \rightarrow Cl_2(g) + H_2(g) + 2\,OH^-(aq)$

   (b)   anode reaction $2\,H_2O(l) \rightarrow O_2(g) + 4\,H^+(aq) + 4\,e^-$
      cathode reaction $\underline{2\,Cu^{2+}(aq) + 4\,e^- \rightarrow 2\,Cu(s)}$
      overall reaction $2\,Cu^{2+}(aq) + 2\,H_2O(l) \rightarrow 2\,Cu(s) + O_2(g) + 4\,H^+(aq)$

18.19

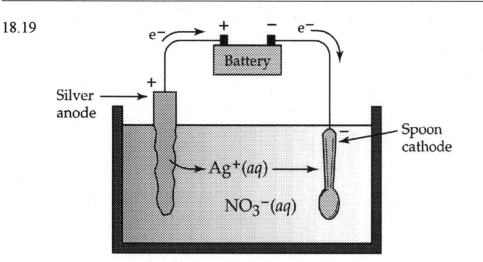

anode reaction $\qquad$ $Ag(s) \rightarrow Ag^+(aq) + e^-$
cathode reaction $\qquad$ $Ag^+(aq) + e^- \rightarrow Ag(s)$
The overall reaction is transfer of silver metal from the silver anode to the spoon.

18.20 $\quad$ Charge $= \left(1.00 \times 10^5 \, \dfrac{C}{s}\right)(8.00 \text{ h})\left(\dfrac{60 \text{ min}}{h}\right)\left(\dfrac{60 \text{ s}}{\text{min}}\right) = 2.88 \times 10^9 \text{ C}$

Moles of $e^- = (2.88 \times 10^9 \text{ C})\left(\dfrac{1 \text{ mol } e^-}{96,500 \text{ C}}\right) = 2.98 \times 10^4 \text{ mol } e^-$

cathode reaction: $Al^{3+} + 3 e^- \rightarrow Al$

mass Al $= (2.98 \times 10^4 \text{ mol } e^-) \times \dfrac{1 \text{ mol Al}}{3 \text{ mol } e^-} \times \dfrac{26.98 \text{ g Al}}{1 \text{ mol Al}} \times \dfrac{1 \text{ kg}}{1000 \text{ g}} = 268 \text{ kg Al}$

18.21 $\quad$ 3.00 g Ag $\times \dfrac{1 \text{ mol Ag}}{107.9 \text{ g Ag}} = 0.0278 \text{ mol Ag}$

cathode reaction: $Ag^+(aq) + e^- \rightarrow Ag(s)$

Charge $= (0.0278 \text{ mol Ag})\left(\dfrac{1 \text{ mol } e^-}{1 \text{ mol Ag}}\right)\left(\dfrac{96,500 \text{ C}}{1 \text{ mol } e^-}\right) = 2682.7 \text{ C}$

Time $= \dfrac{C}{A} = \left(\dfrac{2682.7 \text{ C}}{0.100 \text{ C/s}} \times \dfrac{1 \text{ h}}{3600 \text{ s}}\right) = 7.45 \text{ h}$

18.22 $\quad$ When a beam of white light strikes the anodized surface, part of the light is reflected from the outer TiO_2, while part penetrates through the semitransparent TiO_2 and is reflected from the inner metal. If the two reflections of a particular wavelength are out of phase, they interfere destructively and that wavelength is canceled from the reflected light. Because $n\lambda = 2d \times \sin \theta$, the canceled wavelength depends on the thickness of the TiO_2 layer.

18.23 $\text{volume} = \left(0.0100 \text{ mm} \times \dfrac{1 \text{ cm}}{10 \text{ mm}}\right)(10.0 \text{ cm})^2 = 0.100 \text{ cm}^3$

$\text{mol Al}_2\text{O}_3 = (0.100 \text{ cm}^3)(3.97 \text{ g/cm}^3)\dfrac{1 \text{ mol Al}_2\text{O}_3}{102.0 \text{ g Al}_2\text{O}_3} = 3.892 \times 10^{-3} \text{ mol Al}_2\text{O}_3$

$\text{mole e}^- = 3.892 \times 10^{-3} \text{ mol Al}_2\text{O}_3 \times \dfrac{6 \text{ mol e}^-}{1 \text{ mol Al}_2\text{O}_3} = 0.02335 \text{ mol e}^-$

$\text{coulombs} = 0.02335 \text{ mol e}^- \times \dfrac{96,500 \text{ C}}{1 \text{ mol e}^-} = 2253 \text{ C}$

$\text{time} = \dfrac{\text{C}}{\text{A}} = \dfrac{2253 \text{ C}}{0.600 \text{ C/s}} \times \dfrac{1 \text{ min}}{60 \text{ s}} = 62.6 \text{ min}$

Understanding Key Concepts

18.24 (a) - (d)

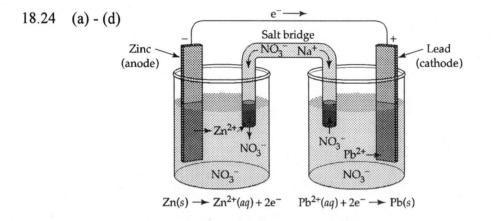

$\text{Zn}(s) \longrightarrow \text{Zn}^{2+}(aq) + 2e^- \qquad \text{Pb}^{2+}(aq) + 2e^- \longrightarrow \text{Pb}(s)$

(e) anode reaction $\quad \text{Zn}(s) \rightarrow \text{Zn}^{2+}(aq) + 2 e^-$
cathode reaction $\quad \underline{\text{Pb}^{2+}(aq) + 2 e^- \rightarrow \text{Pb}(s)}$
overall reaction $\quad \text{Zn}(s) + \text{Pb}^{2+}(aq) \rightarrow \text{Zn}^{2+}(aq) + \text{Pb}(s)$

18.26 (a) The three cell reactions are the same except for cation concentrations.

anode reaction	$\text{Cu}(s) \rightarrow \text{Cu}^{2+}(aq) + 2 e^-$	$E° = -0.34 \text{ V}$
cathode reaction	$\underline{2 \text{ Fe}^{3+}(aq) + 2 e^- \rightarrow 2 \text{ Fe}^{2+}(aq)}$	$\underline{E° = 0.77 \text{ V}}$
overall reaction	$\text{Cu}(s) + 2 \text{ Fe}^{3+}(aq) \rightarrow \text{Cu}^{2+}(aq) + 2 \text{ Fe}^{2+}(aq)$	$E° = 0.43 \text{ V}$

(b)

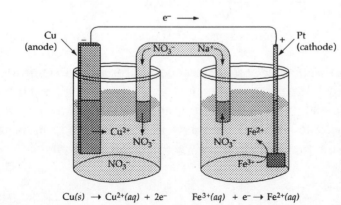

$\text{Cu}(s) \rightarrow \text{Cu}^{2+}(aq) + 2e^- \qquad \text{Fe}^{3+}(aq) + e^- \rightarrow \text{Fe}^{2+}(aq)$

(c) $E = E° - \dfrac{0.0592\,V}{n} \log \dfrac{[Cu^{2+}][Fe^{2+}]^2}{[Fe^{2+}]^2}$; $n = 2$ mol e^-

(1) $E = E° = 0.43$ V because all cation concentrations are 1 M.

(2) $E = E° - \dfrac{0.0592\,V}{2} \log \dfrac{(1)(5)^2}{(1)^2} = 0.39$ V

(3) $E = E° - \dfrac{0.0592\,V}{2} \log \dfrac{(0.1)(0.1)^2}{(0.1)^2} = 0.46$ V

Cell (3) has the largest potential, while cell (2) has the smallest as calculated from the Nernst equation.

18.28   (a) This is an electrolytic cell that has a battery connected between two inert electrodes.

(b)

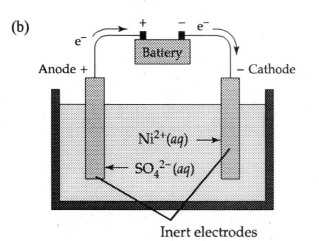

Inert electrodes

(c)   anode reaction $2\,H_2O(l) \;\rightarrow\; O_2(g) + 4\,H^+(aq) + 4\,e^-$
      cathode reaction $Ni^{2+}(aq) + 2\,e^- \;\rightarrow\; Ni(s)$
      overall reaction $2\,Ni^{2+}(aq) + 2\,H_2O(l) \rightarrow 2\,Ni(s) + O_2(g) + 4\,H^+(aq)$

18.30 $Zn(s) + Cu^{2+}(aq) \rightarrow Zn^{2+}(aq) + Cu(s)$; $E = E° - \dfrac{0.0592\ V}{2} \log \dfrac{[Zn^{2+}]}{[Cu^{2+}]}$

(a) E increases because increasing $[Cu^{2+}]$ decreases $\log \dfrac{[Zn^{2+}]}{[Cu^{2+}]}$.

(b) E will decrease because addition of H_2SO_4 increases the volume which decreases $[Cu^{2+}]$ and increases $\log \dfrac{[Zn^{2+}]}{[Cu^{2+}]}$.

(c) E decreases because increasing $[Zn^{2+}]$ increases $\log \dfrac{[Zn^{2+}]}{[Cu^{2+}]}$.

(d) Because there is no change in $[Zn^{2+}]$, there is no change in E.

Additional Problems
Galvanic Cells

18.32   The electrode where oxidation takes place is called the anode. For example, the lead
electrode in the lead storage battery.
The electrode where reduction takes place is called the cathode. For example, the PbO_2
electrode in the lead storage battery.

18.34   The cathode of a galvanic cell is considered to be the positive electrode because electrons
flow through the external circuit toward the positive electrode (the cathode).

18.36   (a) $Cd(s) + Sn^{2+}(aq) \rightarrow Cd^{2+}(aq) + Sn(s)$

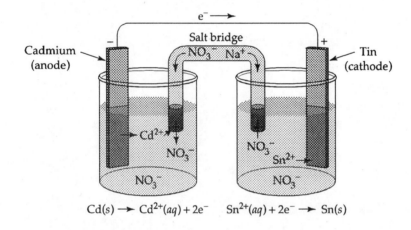

$$Cd(s) \rightarrow Cd^{2+}(aq) + 2e^- \qquad Sn^{2+}(aq) + 2e^- \rightarrow Sn(s)$$

(b) $2\,Al(s) + 3\,Cd^{2+}(aq) \rightarrow 2\,Al^{3+}(aq) + 3\,Cd(s)$

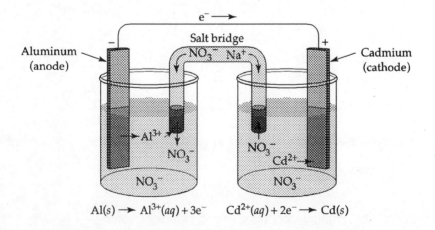

$$Al(s) \rightarrow Al^{3+}(aq) + 3e^- \qquad Cd^{2+}(aq) + 2e^- \rightarrow Cd(s)$$

322

(c) $6\ Fe^{2+}(aq) + Cr_2O_7^{2-}(aq) + 14\ H^+(aq) \rightarrow 6\ Fe^{3+}(aq) + 2\ Cr^{3+}(aq) + 7\ H_2O(l)$

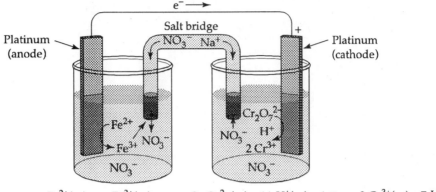

$$Fe^{2+}(aq) \rightarrow Fe^{3+}(aq) + e^- \qquad Cr_2O_7^{2-}(aq) + 14\ H^+(aq) + 6e^- \rightarrow 2\ Cr^{3+}(aq) + 7\ H_2O\ (l)$$

18.38   (a) $Cd(s)\,|\,Cd^{2+}(aq)\,\|\,Sn^{2+}(aq)\,|\,Sn(s)$
      (b) $Al(s)\,|\,Al^{3+}(aq)\,\|\,Cd^{2+}(aq)\,|\,Cd(s)$
      (c) $Pt(s)\,|\,Fe^{2+}(aq),\ Fe^{3+}(aq)\,\|\,Cr_2O_7^{2-}(aq),\ Cr^{3+}(aq)\,|\,Pt(s)$

18.40   (a)

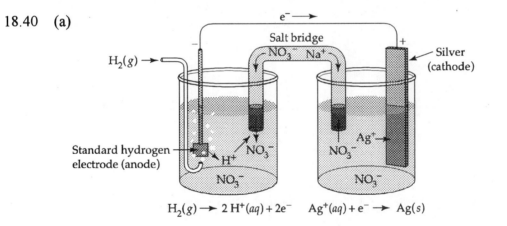

$$H_2(g) \rightarrow 2\ H^+(aq) + 2e^- \qquad Ag^+(aq) + e^- \rightarrow Ag(s)$$

      (b) anode reaction $H_2(g) \rightarrow 2\ H^+(aq) + 2\ e^-$
            cathode reaction $\underline{2\ Ag^+(aq) + 2\ e^- \rightarrow 2\ Ag(s)}$
            overall reaction $H_2(g) + 2\ Ag^+(aq) \rightarrow 2\ H^+(aq) + 2\ Ag(s)$

      (c) $Pt(s)\,|\,H_2(g)\,|\,H^+(aq)\,\|\,Ag^+(aq)\,|\,Ag(s)$

18.42   (a) anode reaction $Co(s) \rightarrow Co^{2+}(aq) + 2\ e^-$
           cathode reaction $\underline{Cu^{2+}(aq) + 2\ e^- \rightarrow Cu(s)}$
           overall reaction $Co(s) + Cu^{2+}(aq) \rightarrow Co^{2+}(aq) + Cu(s)$

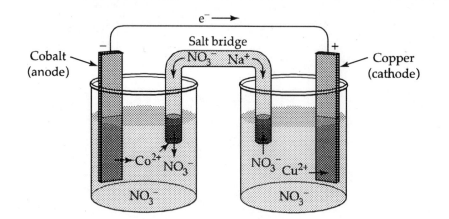

(b) anode reaction $2 \text{ Fe}(s) \rightarrow 2 \text{ Fe}^{2+}(aq) + 4 \text{ e}^-$
cathode reaction $O_2(g) + 4 \text{ H}^+(aq) + 4 \text{ e}^- \rightarrow 2 \text{ H}_2O(l)$
overall reaction $2 \text{ Fe}(s) + O_2(g) + 4 \text{ H}^+(aq) \rightarrow 2 \text{ Fe}^{2+}(aq) + 2 \text{ H}_2O(l)$

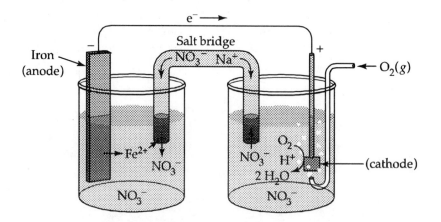

Cell Potentials and Free-Energy Changes; Standard Reduction potentials

18.44 The SI unit of electrical potential is the volt (V).
The SI unit of charge is the coulomb (C).
The SI unit of energy is the joule (J).
$1 \text{ J} = 1 \text{ C} \cdot 1 \text{ V}$

18.46 E is the standard cell potential (E°) when all reactants and products are in their standard states--solutes at 1 M concentrations, gases at a partial pressure of 1 atm, solids and liquids in pure form, all at 25°C.

18.48 $\text{Zn}(s) + \text{Ag}_2O(s) \rightarrow \text{ZnO}(s) + 2 \text{ Ag}(s);$ $n = 2 \text{ mol e}^-$

$$\Delta G = -nFE = -(2 \text{ mol e}^-)\left(\frac{96{,}500 \text{ C}}{1 \text{ mol e}^-}\right)(1.60 \text{ V})\left(\frac{1 \text{ J}}{1 \text{ C} \cdot \text{V}}\right) = -308{,}800 \text{ J} = -309 \text{ kJ}$$

18.50 $2 H_2(g) + O_2(g) \rightarrow 2 H_2O(l)$; $n = 4$ mol e^- and $1 V = 1 J/C$

$\Delta G° = 2 \Delta G°_f(H_2O(l)) = (2 \text{ mol})(-237.2 \text{ kJ/mol}) = -474.4 \text{ kJ}$

$\Delta G° = -nFE°$

$$E° = \frac{-\Delta G°}{nF} = \frac{-(-474,400 \text{ J})}{(4 \text{ mol } e^-)\left(\dfrac{96,500 \text{ C}}{1 \text{ mol } e^-}\right)} = +1.23 \text{ J/C} = +1.23 \text{ V}$$

18.52 oxidation: $Zn(s) \rightarrow Zn^{2+}(aq) + 2 e^-$ $E° = 0.76$ V
       reduction: $\underline{Eu^{3+}(aq) + e^- \rightarrow Eu^{2+}(aq)}$ $\underline{E° = ?}$
       overall $Zn(s) + 2 Eu^{3+}(aq) \rightarrow Zn^{2+}(aq) + 2 Eu^{2+}(aq)$ $E° = 0.40$ V
       The standard reduction potential for the Eu^{3+}/Eu^{2+} half cell is:
       $E° = 0.40 - 0.76 = -0.36$ V

18.54 $Sn^{4+}(aq) < Br_2(l) < MnO_4^-$

18.56 $Cr_2O_7^{2-}(aq)$ is highest in the table of standard reduction potentials, therefore it is the strongest oxidizing agent.
       $Fe^{2+}(aq)$ is lowest in the table of standard reduction potentials, therefore it is the weakest oxidizing agent.

18.58 (a) $Cd(s) + Sn^{2+}(aq) \rightarrow Cd^{2+}(aq) + Sn(s)$
       oxidation: $Cd(s) \rightarrow Cd^{2+}(aq) + 2 e^-$ $E° = 0.40$ V
       reduction: $Sn^{2+}(aq) + 2 e^- \rightarrow Sn(s)$ $\underline{E° = -0.14 \text{ V}}$
                                                           overall $E° = 0.26$ V

       $n = 2$ mol e^-

$$\Delta G° = -nFE° = -(2 \text{ mol } e^-)\left(\frac{96,500 \text{ C}}{1 \text{ mol } e^-}\right)(0.26 \text{ V})\left(\frac{1 \text{ J}}{1 \text{ C} \cdot \text{V}}\right) = -50,180 \text{ J} = -50 \text{ kJ}$$

       (b) $2 Al(s) + 3 Cd^{2+}(aq) \rightarrow 2 Al^{3+}(aq) + 3 Cd(s)$
       oxidation: $2 Al(s) \rightarrow 2 Al^{3+}(aq) + 6 e^-$ $E° = 1.66$ V
       reduction: $3 Cd^{2+}(aq) + 6 e^- \rightarrow 3 Cd(s)$ $\underline{E° = -0.40 \text{ V}}$
                                                           overall $E° = 1.26$ V

       $n = 6$ mol e^-

$$\Delta G° = -nFE° = -(6 \text{ mol } e^-)\left(\frac{96,500 \text{ C}}{1 \text{ mol } e^-}\right)(1.26 \text{ V})\left(\frac{1 \text{ J}}{1 \text{ C} \cdot \text{V}}\right) = -729,540 \text{ J} = -730 \text{ kJ}$$

       (c) $6 Fe^{2+}(aq) + Cr_2O_7^{2-}(aq) + 14 H^+(aq) \rightarrow 6 Fe^{3+}(aq) + 2 Cr^{3+}(aq) + 7 H_2O(l)$
       oxidation: $6 Fe^{2+}(aq) \rightarrow 6 Fe^{3+}(aq) + 6 e^-$ $E° = -0.77$ V
       reduction: $Cr_2O_7^{2-}(aq) + 14 H^+(aq) + 6 e^- + \rightarrow 2 Cr^{3+}(aq) + 7 H_2O(l)$ $\underline{E° = 1.33 \text{ V}}$
                                                           overall $E° = 0.56$ V

$$n = 6 \text{ mol e}^-$$

$$\Delta G^\circ = -nFE^\circ = -(6 \text{ mol e}^-)\left(\frac{96,500 \text{ C}}{1 \text{ mol e}^-}\right)(0.56 \text{ V})\left(\frac{1 \text{ J}}{1 \text{ C} \cdot \text{V}}\right) = -324,240 \text{ J} = -324 \text{ kJ}$$

18.60 (a) $2 \text{ Fe}^{2+}(aq) + \text{Pb}^{2+}(aq) \rightarrow 2 \text{ Fe}^{3+}(aq) + \text{Pb}(s)$

oxidation: $2 \text{ Fe}^{2+}(aq) \rightarrow 2 \text{ Fe}^{3+}(aq) + 2 \text{ e}^-$ $E^\circ = -0.77 \text{ V}$

reduction: $\text{Pb}^{2+}(aq) + 2 \text{ e}^- \rightarrow \text{Pb}(s)$ $\underline{E^\circ = -0.13 \text{ V}}$

overall $E^\circ = -0.90 \text{ V}$

Because E° is negative, this reaction is nonspontaneous.

(b) $\text{Mg}(s) + \text{Ni}^{2+}(aq) \rightarrow \text{Mg}^{2+}(aq) + \text{Ni}(s)$

oxidation: $\text{Mg}(s) \rightarrow \text{Mg}^{2+}(aq) + 2 \text{ e}^-$ $E^\circ = 2.37 \text{ V}$

reduction: $\text{Ni}^{2+}(aq) + 2 \text{ e}^- \rightarrow \text{Ni}(s)$ $\underline{E^\circ = -0.26 \text{ V}}$

overall $E^\circ = 2.11 \text{ V}$

Because E° is positive, this reaction is spontaneous.

18.62 (a) oxidation: $\text{Sn}^{2+}(aq) \rightarrow \text{Sn}^{4+}(aq) + 2 \text{ e}^-$ $E^\circ = -0.15 \text{ V}$

reduction: $\text{Br}_2(l) + 2 \text{ e}^- \rightarrow 2 \text{ Br}^-(aq)$ $\underline{E^\circ = 1.09 \text{ V}}$

overall $E^\circ = +0.94 \text{ V}$

Because the overall E° is positive, $\text{Sn}^{2+}(aq)$ can be oxidized by $\text{Br}_2(l)$.

(b) oxidation: $\text{Sn}^{2+}(aq) \rightarrow \text{Sn}^{4+}(aq) + 2 \text{ e}^-$ $E^\circ = -0.15 \text{ V}$

reduction: $\text{Ni}^{2+}(aq) + 2 \text{ e}^- \rightarrow \text{Ni}(s)$ $\underline{E^\circ = -0.26 \text{ V}}$

overall $E^\circ = -0.41 \text{ V}$

Because the overall E° is negative, $\text{Ni}^{2+}(aq)$ cannot be reduced by $\text{Sn}^{2+}(aq)$.

(c) oxidation: $2 \text{ Ag}(s) \rightarrow 2 \text{ Ag}^+(aq) + 2 \text{ e}^-$ $E^\circ = -0.80 \text{ V}$

reduction: $\text{Pb}^{2+}(aq) + 2 \text{ e}^- \rightarrow \text{Pb}(s)$ $\underline{E^\circ = -0.13 \text{ V}}$

overall $E^\circ = -0.93 \text{ V}$

Because the overall E° is negative, $\text{Ag}(s)$ cannot be oxidized by $\text{Pb}^{2+}(aq)$.

(d) oxidation: $\text{H}_2\text{SO}_3(aq) + \text{H}_2\text{O}(l) \rightarrow \text{SO}_4^{2-}(aq) + 4 \text{ H}^+(aq) + 2 \text{ e}^-$ $E^\circ = -0.17 \text{ V}$

reduction: $\text{I}_2(s) + 2 \text{ e}^- \rightarrow 2 \text{ I}^-(aq)$ $\underline{E^\circ = 0.54 \text{ V}}$

overall $E^\circ = +0.37 \text{ V}$

Because the overall E° positive, $\text{I}_2(s)$ can be reduced by H_2SO_3.

The Nernst Equation

18.64 $2 \text{ Ag}^+(aq) + \text{Sn}(s) \rightarrow 2 \text{ Ag}(s) + \text{Sn}^{2+}(aq)$

oxidation: $\text{Sn}(s) \rightarrow \text{Sn}^{2+}(aq) + 2 \text{ e}^-$ $E^\circ = 0.14 \text{ V}$

reduction: $2 \text{ Ag}^+(aq) + 2 \text{ e}^- \rightarrow 2 \text{ Ag}(s)$ $\underline{E^\circ = 0.80 \text{ V}}$

overall $E^\circ = 0.94 \text{ V}$

$$E = E^\circ - \frac{0.0592 \text{ V}}{n} \log \frac{[\text{Sn}^{2+}]}{[\text{Ag}^+]^2} = 0.94 \text{ V} - \frac{(0.0592 \text{ V})}{2} \log \frac{(0.020)}{(0.010)^2} = 0.87 \text{ V}$$

18.66 $Pb(s) + Cu^{2+}(aq) \rightarrow Pb^{2+}(aq) + Cu(s)$

       oxidation: $Pb(s) \rightarrow Pb^{2+}(aq) + 2\,e^-$ $E° = 0.13\ V$

       reduction: $Cu^{2+}(aq) + 2\,e^- \rightarrow Cu(s)$ $\underline{E° = 0.34\ V}$

                                                 overall $E° = 0.47\ V$

$$E = E° - \frac{0.0592\ V}{n} \log \frac{[Pb^{2+}]}{[Cu^{2+}]} = 0.47\ V - \frac{(0.0592\ V)}{2} \log \frac{1.0}{(1.0 \times 10^{-4})} = 0.35\ V$$

$$\text{When } E = 0, \quad 0 = E° - \frac{0.0592\ V}{n} \log \frac{[Pb^{2+}]}{[Cu^{2+}]} = 0.47\ V - \frac{(0.0592\ V)}{2} \log \frac{1.0}{[Cu^{2+}]}$$

$$0 = 0.47\ V + \frac{(0.0592\ V)}{2} \log [Cu^{2+}]$$

$$\log [Cu^{2+}] = (-0.47\ V)\left(\frac{2}{0.0592\ V}\right) = -15.88; \qquad [Cu^{2+}] = 10^{-15.88} = 1 \times 10^{-16}\ M$$

18.68   (a) $\ E = E° - \dfrac{0.0592\ V}{n} \log [I^-]^2 = 0.54\ V - \dfrac{(0.0592\ V)}{2} \log (0.020)^2 = 0.64\ V$

       (b) $\ E = E° - \dfrac{0.0592\ V}{n} \log \dfrac{[Fe^{2+}]}{[Fe^{3+}]} = 0.77\ V - \dfrac{(0.0592\ V)}{1} \log \left(\dfrac{0.10}{0.10}\right) = 0.77\ V$

       (c) $\ E = E° - \dfrac{0.0592\ V}{n} \log \dfrac{[Sn^{4+}]}{[Sn^{2+}]} = -0.15\ V - \dfrac{(0.0592\ V)}{2} \log \left(\dfrac{0.40}{0.0010}\right) = -0.23\ V$

       (d) $\ E = E° - \dfrac{0.0592\ V}{n} \log \dfrac{[Cr_2O_7^{2-}][H^+]^{14}}{[Cr^{3+}]^2} = -1.33\ V - \dfrac{(0.0592\ V)}{6} \log \left(\dfrac{(1.0)(0.010)^{14}}{1.0}\right)$

$$E = -1.33\ V - \frac{(0.0592\ V)}{6} (14) \log (0.010) = -1.05\ V$$

18.70 $H_2(g) + Ni^{2+}(aq) \rightarrow 2\,H^+(aq) + Ni(s)$

$$E° = E°_{H_2 \rightarrow H^+} + E°_{Ni^{2+} \rightarrow Ni} = 0\ V + (-0.26\ V) = -0.26\ V$$

$$E = E° - \frac{0.0592\ V}{n} \log \frac{[H_3O^+]^2}{[Ni^{2+}](P_{H_2})}$$

$$0.27\ V = -0.26\ V - \frac{(0.0592\ V)}{2} \log \frac{[H_3O^+]^2}{(1)(1)}$$

$$0.27\ V = -0.26\ V - (0.0592\ V) \log [H_3O^+]$$

$pH = -\log [H_3O^+]$ therefore $0.27\ V = -0.26\ V + (0.0592\ V)\ pH$

$$pH = \frac{(0.27\ V + 0.26\ V)}{0.0592\ V} = 9.0$$

Standard Cell Potentials and Equilibrium Constants

18.72 $\Delta G° = -nFE°$

Because n and F are always positive, $\Delta G°$ is negative when $E°$ is positive because of the negative sign in the equation.

$$E° = \frac{0.0592\ V}{n}\ \log K; \quad \log K = \frac{nE°}{0.0592\ V}; \quad K = 10^{\frac{nE°}{0.0592}}$$

If $E°$ is positive, the exponent is positive (because n is positive), and K is greater than 1.

18.74 $Ni(s) + 2\ Ag^+(aq) \rightarrow Ni^{2+}(aq) + 2\ Ag(s)$
oxidation: $Ni(s) \rightarrow Ni^{2+}(aq) + 2\ e^-$ $\qquad\qquad E° = 0.26\ V$
reduction: $2\ Ag^+(aq) + 2\ e^- \rightarrow 2\ Ag(s)$ $\qquad\quad \underline{E° = 0.80\ V}$
$\qquad\qquad\qquad\qquad\qquad\qquad\qquad$ overall $E° = 1.06\ V$

$$E° = \frac{0.0592\ V}{n}\ \log K;\ \log K = \frac{nE°}{0.0592\ V} = \frac{(2)(1.06\ V)}{0.0592\ V} = 35.8;\ \ K = 10^{35.8} = 6 \times 10^{35}$$

18.76 $E°$ and n are from Problem 18.58.

$$E° = \frac{0.0592\ V}{n}\ \log K; \quad \log K = \frac{nE°}{0.0592\ V}$$

(a) $Cd(s) + Sn^{2+}(aq) \rightarrow Cd^{2+}(aq) + Sn(s);$ $\qquad E° = 0.26\ V$ and $n = 2$ mol e^-

$$\log K = \frac{(2)(0.26\ V)}{0.0592\ V} = 8.8; \qquad K = 10^{8.8} = 6 \times 10^8$$

(b) $2\ Al(s) + 3\ Cd^{2+}(aq) \rightarrow 2\ Al^{3+}(aq) + 3\ Cd(s);$ $\qquad E° = 1.26\ V$ and $n = 6$ mol e^-

$$\log K = \frac{(6)(1.26\ V)}{0.0592\ V} = 128; \qquad K = 10^{128}$$

(c) $6\ Fe^{2+}(aq) + Cr_2O_7^{2-}(aq) + 14\ H^+(aq) \rightarrow 6\ Fe^{3+}(aq) + 2\ Cr^{3+}(aq) + 7\ H_2O(l)$
$E° = 0.56\ V$ and $n = 6$ mol e^-

$$\log K = \frac{(6)(0.56\ V)}{0.0592\ V} = 57; \qquad K = 10^{57}$$

18.78 $Hg_2^{2+}(aq) \rightarrow Hg(l) + Hg^{2+}(aq)$
oxidation: $\frac{1}{2}[Hg_2^{2+}(aq) \rightarrow 2\ Hg^{2+}(aq) + 2\ e^-]$ $\qquad\quad E° = -0.92\ V$
reduction: $\frac{1}{2}[Hg_2^{2+}(aq) + 2\ e^- \rightarrow 2\ Hg(l)]$ $\qquad\quad \underline{E° = \ \ 0.80\ V}$
$\qquad\qquad\qquad\qquad\qquad\qquad\qquad$ overall $E° = -0.12\ V$

$$E° = \frac{0.0592\ V}{n}\ \log K$$

$$\log K = \frac{nE°}{0.0592\ V} = \frac{(1)(-0.12\ V)}{0.0592\ V} = -2.027; \qquad K = 10^{-2.027} = 9 \times 10^{-3}$$

Batteries; Corrosion

18.80   Rust is a hydrated form of iron(III) oxide ($Fe_2O_3 \cdot H_2O$). Rust forms from the oxidation of Fe in the presence of O_2 and H_2O. Rust can be prevented by coating Fe with Zn (galvanizing).

18.82   Cathodic protection is the attachment of a more easily oxidized metal to the metal you want to protect. This forces the metal you want to protect to be the cathode, hence the name, cathodic protection.
Zn and Al can offer cathodic protection to Fe (Ni and Sn cannot).

18.84   (a)

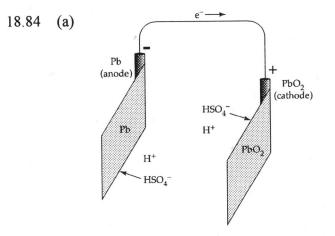

(b)   Anode: $Pb(s) + HSO_4^-(aq) \rightarrow PbSO_4(s) + H^+(aq) + 2\ e^-$ $E° = 0.296$ V
Cathode: $\underline{PbO_2(s) + 3\ H^+(aq) + HSO_4^-(aq) + 2\ e^- \rightarrow PbSO_4(s) + 2\ H_2O(l)}$ $\underline{E° = 1.628\ V}$
Overall $Pb(s) + PbO_2(s) + 2\ H^+(aq) + 2\ HSO_4^-(aq) \rightarrow 2\ PbSO_4(s) + 2\ H_2O(l)$ $E° = 1.924$ V

(c) $E° = \dfrac{0.0592\ V}{n} \log K$; $\log K = \dfrac{nE°}{0.0592\ V} = \dfrac{(2)(1.924\ V)}{0.0592\ V} = 65.0$; $K = 1 \times 10^{65}$

(d) When the cell reaction reaches equilibrium the cell voltage $= 0$.

18.86 $Zn(s) + HgO(s) \rightarrow ZnO(s) + Hg(l)$;   Zn, 65.39 amu; HgO, 216.59 amu

mass HgO $= 2.00$ g Zn x $\dfrac{1\ mol\ Zn}{65.39\ g\ Zn}$ x $\dfrac{1\ mol\ HgO}{1\ mol\ Zn}$ x $\dfrac{216.59\ g\ HgO}{1\ mol\ HgO} = 6.62$ g HgO

Electrolysis

18.88   (a)

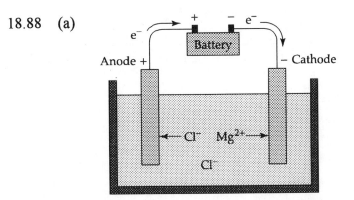

(b) anode: $2 \, Cl^-(l) \rightarrow Cl_2(g) + 2 \, e^-$
   cathode: $\underline{Mg^{2+}(l) + 2 \, e^- \rightarrow Mg(l)}$
   overall: $Mg^{2+}(l) + 2 \, Cl^-(l) \rightarrow Mg(l) + Cl_2(g)$

18.90   possible anode reactions:
   $2 \, Cl^-(aq) \rightarrow Cl_2(g) + 2 \, e^-$
   $2 \, H_2O(l) \rightarrow O_2(g) + 4 \, H^+(aq) + 4 \, e^-$
   possible cathode reactions:
   $2 \, H_2O(l) + 2 \, e^- \rightarrow H_2(g) + 2 \, OH^-(aq)$
   $Mg^{2+}(aq) + 2 \, e^- \rightarrow Mg(s)$
   actual reactions:
   anode: $2 \, Cl^-(aq) \rightarrow Cl_2(g) + 2 \, e^-$
   cathode: $2 \, H_2O(l) + 2 \, e^- \rightarrow H_2(g) + 2 \, OH^-(aq)$
   This anode reaction takes place instead of $2 \, H_2O(l) \rightarrow O_2(g) + 4 \, H^+(aq) + 4 \, e^-$ because of a high overvoltage for formation of gaseous O_2.
   This cathode reaction takes place instead of $Mg^{2+}(aq) + 2 \, e^- \rightarrow Mg(s)$ because H_2O is easier to reduce than Mg^{2+}.

18.92   (a) NaBr
   anode: $2 \, Br^-(aq) \rightarrow Br_2(l) + 2 \, e^-$
   cathode: $\underline{2 \, H_2O(l) + 2 \, e^- \rightarrow H_2(g) + 2 \, OH^-(aq)}$
   overall: $2 \, H_2O(l) + 2 \, Br^-(aq) \rightarrow Br_2(l) + H_2(g) + 2 \, OH^-(aq)$
   (b) $CuCl_2$
   anode: $2 \, Cl^-(aq) \rightarrow Cl_2(g) + 2 \, e^-$
   cathode: $\underline{Cu^{2+}(aq) + 2 \, e^- \rightarrow Cu(s)}$
   overall: $Cu^{2+}(aq) + 2 \, Cl^-(aq) \rightarrow Cu(s) + Cl_2(g)$
   (c) LiOH
   anode: $4 \, OH^-(aq) \rightarrow O_2(g) + 2 \, H_2O(l) + 4 \, e^-$
   cathode: $\underline{4 \, H_2O(l) + 4 \, e^- \rightarrow 2 \, H_2(g) + 4 \, OH^-(aq)}$
   overall: $2 \, H_2O(l) \rightarrow O_2(g) + 2 \, H_2(g)$

18.94 $Ag^+(aq) + e^- \rightarrow Ag(s); \qquad 1 \, A = 1 \, C/s$

$$\text{mass Ag} = 2.40 \, \frac{C}{s} \times 20.0 \, \text{min} \times \frac{60 \, s}{1 \, \text{min}} \times \frac{1 \, \text{mol e}^-}{96{,}500 \, C} \times \frac{1 \, \text{mol Ag}}{1 \, \text{mol e}^-} \times \frac{107.87 \, \text{g Ag}}{1 \, \text{mol Ag}} = 3.22 \, g$$

18.96 $2 \, Na^+(l) + 2 \, Cl^-(l) \rightarrow 2 \, Na(l) + Cl_2(g)$
   $Na^+(l) + e^- \rightarrow Na(l); \qquad 1 \, A = 1 \, C/s; \qquad 1.00 \times 10^3 \, kg = 1.00 \times 10^6 \, g$

$$\text{Charge} = 1.00 \times 10^6 \, \text{g Na} \times \frac{1 \, \text{mol Na}}{22.99 \, \text{g Na}} \times \frac{1 \, \text{mol e}^-}{1 \, \text{mol Na}} \times \frac{96{,}500 \, C}{1 \, \text{mol e}^-} = 4.20 \times 10^9 \, C$$

$$\text{Time} = \frac{4.20 \times 10^9 \, C}{30{,}000 \, C/s} \times \frac{1 \, h}{3600 \, s} = 38.9 \, h$$

$$1.00 \times 10^6 \, \text{g Na} \times \frac{1 \, \text{mol Na}}{22.99 \, \text{g Na}} \times \frac{1 \, \text{mol Cl}_2}{2 \, \text{mol Na}} = 21{,}748.6 \, \text{mol Cl}_2$$

$PV = nRT$

$$V = \frac{nRT}{P} = \frac{(21{,}748.6 \text{ mol})\left(0.08206\,\frac{L \cdot atm}{K \cdot mol}\right)(273.15 \text{ K})}{1.00 \text{ atm}} = 4.87 \times 10^5 \text{ L Cl}_2$$

18.98 $PbSO_4(s) + H^+(aq) + 2\,e^- \rightarrow Pb(s) + HSO_4^-(aq)$

mass $PbSO_4 = 10.0$ C/s $\times$ 1.50 h $\times \dfrac{3600 \text{ s}}{1 \text{ h}} \times \dfrac{1 \text{ mol } e^-}{96{,}500 \text{ C}} \times \dfrac{1 \text{ mol } PbSO_4}{2 \text{ mol } e^-} \times \dfrac{303.3 \text{ g } PbSO_4}{1 \text{ mol } PbSO_4}$

mass $PbSO_4 = 84.9$ g $PbSO_4$

General Problems

18.100 (a) $2\,MnO_4^-(aq) + 16\,H^+(aq) + 5\,Sn^{2+}(aq) \rightarrow 2\,Mn^{2+}(aq) + 5\,Sn^{4+}(aq) + 8\,H_2O(l)$
(b) MnO_4^- is the oxidizing agent; Sn^{2+} is the reducing agent.
(c) $E° = 1.51$ V $+ (-0.15$ V$) = 1.36$ V

18.102 (a) Ag^+ is the strongest oxidizing agent because Ag^+ has the most positive standard reduction potential.
Pb is the strongest reducing agent because Pb^{2+} has the most negative standard reduction potential.

(b)

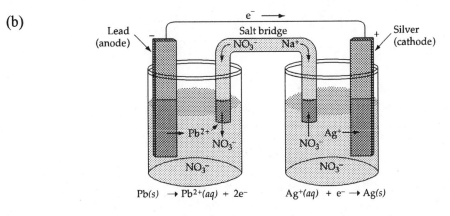

(c) $Pb(s) + 2\,Ag^+(aq) \rightarrow Pb^{2+}(aq) + 2\,Ag(s);$ $n = 2$ mol e^-
$E° = E°_{ox} + E°_{red} = 0.13$ V $+ 0.80$ V $= 0.93$ V

$$\Delta G° = -nFE° = -(2 \text{ mol } e^-)\left(\frac{96{,}500 \text{ C}}{1 \text{ mol } e^-}\right)(0.93 \text{ V})\left(\frac{1 \text{ J}}{1 \text{ C} \cdot \text{V}}\right) = -179{,}490 \text{ J} = -180 \text{ kJ}$$

$$E° = \frac{0.0592 \text{ V}}{n}\log K; \quad \log K = \frac{nE°}{0.0592 \text{ V}} = \frac{(2)(0.93 \text{ V})}{0.0592 \text{ V}} = 31; \quad K = 10^{31}$$

$$\text{(d) } E = E° - \frac{0.0592 \text{ V}}{n}\log\frac{[Pb^{2+}]}{[Ag^+]^2} = 0.93 \text{ V} - \frac{0.0592 \text{ V}}{2}\log\left(\frac{0.01}{(0.01)^2}\right) = 0.87 \text{ V}$$

18.104 (a)

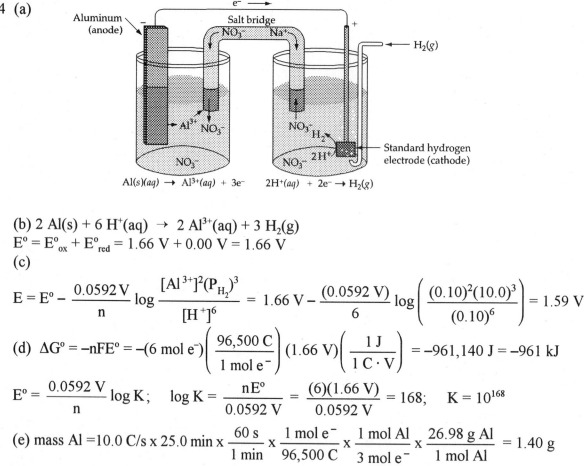

$$Al(s)(aq) \rightarrow Al^{3+}(aq) + 3e^- \qquad 2H^+(aq) + 2e^- \rightarrow H_2(g)$$

(b) $2\,Al(s) + 6\,H^+(aq) \rightarrow 2\,Al^{3+}(aq) + 3\,H_2(g)$

$E^\circ = E^\circ_{ox} + E^\circ_{red} = 1.66\,V + 0.00\,V = 1.66\,V$

(c)

$$E = E^\circ - \frac{0.0592\,V}{n}\log\frac{[Al^{3+}]^2(P_{H_2})^3}{[H^+]^6} = 1.66\,V - \frac{(0.0592\,V)}{6}\log\left(\frac{(0.10)^2(10.0)^3}{(0.10)^6}\right) = 1.59\,V$$

(d) $\Delta G^\circ = -nFE^\circ = -(6\text{ mol e}^-)\left(\dfrac{96{,}500\text{ C}}{1\text{ mol e}^-}\right)(1.66\,V)\left(\dfrac{1\text{ J}}{1\text{ C}\cdot V}\right) = -961{,}140\text{ J} = -961\text{ kJ}$

$E^\circ = \dfrac{0.0592\text{ V}}{n}\log K; \quad \log K = \dfrac{nE^\circ}{0.0592\text{ V}} = \dfrac{(6)(1.66\text{ V})}{0.0592\text{ V}} = 168; \quad K = 10^{168}$

(e) mass Al $=10.0$ C/s x 25.0 min x $\dfrac{60\text{ s}}{1\text{ min}}$ x $\dfrac{1\text{ mol e}^-}{96{,}500\text{ C}}$ x $\dfrac{1\text{ mol Al}}{3\text{ mol e}^-}$ x $\dfrac{26.98\text{ g Al}}{1\text{ mol Al}} = 1.40$ g

18.106 $2\,Cl^-(aq) \rightarrow Cl_2(g) + 2\,e^-$

13 million tons $= 13 \times 10^6$ tons; $\qquad Cl_2$, 70.91 amu

13×10^6 tons x $\dfrac{907{,}200\text{ g}}{1\text{ ton}}$ x $\dfrac{1\text{ mol }Cl_2}{70.91\text{ g }Cl_2} = 1.66 \times 10^{11}$ mol Cl_2

Charge $= 1.66 \times 10^{11}$ mol Cl_2 x $\dfrac{2\text{ mol e}^-}{1\text{ mol }Cl_2}$ x $\dfrac{96{,}500\text{ C}}{1\text{ mol e}^-} = 3.20 \times 10^{16}$ C

1 J = 1 C x 1 V; $\qquad$ Energy $= (3.20 \times 10^{16}\text{ C})(4.5\text{ V}) = 1.44 \times 10^{17}$ J

kWh $= (1.44 \times 10^{17}\text{ J})\left(\dfrac{1\text{ kWh}}{3.6 \times 10^6\text{ J}}\right) = 4.0 \times 10^{10}$ kWh

18.108 (a) oxidizing agents: PbO_2, H^+, $Cr_2O_7^{2-}$; reducing agents: Al, Fe, Ag

(b) PbO_2 is the strongest oxidizing agent. H^+ is the weakest oxidizing agent.

(c) Al is the strongest reducing agent. Ag is the weakest reducing agent.

(d) oxidized by Cu^{2+}: Fe and Al; reduced by H_2O_2: PbO_2 and $Cr_2O_7^{2-}$

18.110 (a) $3 CH_3CH_2OH(aq) + 2 Cr_2O_7{}^{2-}(aq) + 16 H^+(aq) \rightarrow$
$$3 CH_3CO_2H(aq) + 4 Cr^{3+}(aq) + 11 H_2O(l)$$

oxidation:
$3 CH_3CH_2OH(aq) + 3 H_2O(l) \rightarrow 3 CH_3CO_2H(aq) + 12 H^+(aq) + 12 e^-$ $\quad E° = -0.058$ V
reduction:
$2 Cr_2O_7{}^{2-}(aq) + 28 H^+(aq) + 12 e^- + \rightarrow 4 Cr^{3+}(aq) + 14 H_2O(l)$ $\quad \underline{E° = \quad 1.33 \text{ V}}$
overall $E° = \quad 1.27$ V

(b) $E = E° - \dfrac{0.0592 \text{ V}}{n} \log \dfrac{[CH_3CO_2H]^3[Cr^{3+}]^4}{[CH_3CH_2OH]^3[Cr_2O_7{}^{2-}]^2[H^+]^{16}}$

pH = 4.00, $[H^+] = 0.000 \ 10$ M

$E = 1.27 \text{ V} - \dfrac{(0.0592 \text{ V})}{12} \log \left(\dfrac{(1.0)^3(1.0)^4}{(1.0)^3(1.0)^2(0.000 \ 10)^{16}} \right)$

$E = 1.27 \text{ V} - \dfrac{(0.0592 \text{ V})}{12} \log \dfrac{1}{(0.000 \ 10)^{16}} = 0.95$ V

18.112 anode: $\quad Ag(s) + Cl^-(aq) \rightarrow AgCl(s) + e^-$
cathode: $\quad \underline{Ag^+(aq) + e^- \rightarrow Ag(s)}$
overall: $\quad Ag^+(aq) + Cl^-(aq) \rightarrow AgCl(s)$ $\quad E° = 0.578$ V

For $AgCl(s) \rightleftharpoons Ag^+(aq) + Cl^-(aq)$ $\quad E° = -0.578$ V

$E° = \dfrac{0.0592 \text{ V}}{n} \log K; \quad \log K = \dfrac{nE°}{0.0592 \text{ V}} = \dfrac{(1)(-0.578 \text{ V})}{0.0592 \text{ V}} = -9.76$

$K = K_{sp} = 10^{-9.76} = 1.7 \times 10^{-10}$

18.114 $4 Fe^{2+}(aq) + O_2(g) + 4 H^+(aq) \rightarrow 4 Fe^{3+}(aq) + 2 H_2O(l)$

oxidation $\quad 4 Fe^{2+}(aq) \rightarrow 4 Fe^{3+}(aq) + 4 e^-$ $\quad E° = -0.77$ V
reduction $\quad O_2(g) + 4 H^+(aq) + 4 e^- \rightarrow 2 H_2O(l)$ $\quad \underline{E° = 1.23 \text{ V}}$
overall $\quad E° = 0.46$ V

$P_{O_2} = 160 \text{ mm Hg} \times \dfrac{1.00 \text{ atm}}{760 \text{ mm Hg}} = 0.211$ atm

$E = E° - \dfrac{0.0592 \text{ V}}{n} \log \dfrac{[Fe^{3+}]^4}{[Fe^{2+}]^4[H^+]^4(P_{O_2})}$

$E = 0.46 \text{ V} - \dfrac{0.0592 \text{ V}}{4} \log \dfrac{(1 \times 10^{-7})^4}{(1 \times 10^{-7})^4(1 \times 10^{-7})^4(0.211)}$

$E = 0.46 \text{ V} - 0.42 \text{ V} = 0.04$ V
Because E is positive, the reaction is spontaneous.

Chapter 18 – Electrochemistry

18.116 First calculate $E°$ for the galvanic cell in order to determine $E°_1$.

anode: $5 [2\ Hg(l) + 2\ Br^-(aq) \rightarrow Hg_2Br_2(s) + 2\ e^-]$ $E°_1 = ?$

cathode: $2 [MnO_4^-(aq) + 8\ H^+(aq) + 5\ e^- \rightarrow Mn^{2+}(aq) + 4\ H_2O(l)]$ $E°_2 = 1.51\ V$

overall: $2\ MnO_4^-(aq) + 10\ Hg(l) + 10\ Br^-(aq) + 16\ H^+(aq) \rightarrow$
$2\ Mn^{2+}(aq) + 5\ Hg_2Br_2(s) + 8\ H_2O(l)$

$n = 10\ mol\ e^-$

$$E = E° - \frac{0.0592\ V}{n} \log \frac{[Mn^{2+}]^2}{[Br^-]^{10}[MnO_4^-]^2[H^+]^{16}}$$

$$1.214\ V = E° - \frac{(0.0592\ V)}{10} \log \left(\frac{(0.10)^2}{(0.10)^{10}(0.10)^2(0.10)^{16}} \right)$$

$$1.214\ V = E° - \frac{(0.0592\ V)}{10} \log \frac{1}{(0.10)^{26}} = E° - 0.154\ V$$

$E° = 1.214 + 0.154 = 1.368\ V$

$E°_1 + E°_2 = 1.368\ V;$ $E°_1 + 1.51\ V = 1.368\ V;$ $E°_1 = 1.368\ V - 1.51\ V = -0.142\ V$

oxidation: $2\ Hg(l) \rightarrow Hg_2^{2+}(aq) + 2\ e^-$ $E° = -0.80\ V$ (Appendix D)

reduction: $Hg_2Br_2(s) + 2\ e^- \rightarrow 2\ Hg(l) + 2\ Br^-(aq)$ $E° = -0.142\ V$ (from $E°_1$)

overall: $Hg_2Br_2(s) \rightarrow Hg_2^{2+}(aq) + 2\ Br^-(aq)$ $E° = -0.658\ V$

$$E° = \frac{0.0592\ V}{n} \log K;\quad \log K = \frac{nE°}{0.0592\ V} = \frac{(2)(-0.658\ V)}{0.0592\ V} = -22.2$$

$K = K_{sp} = 10^{-22.2} = 6 \times 10^{-23}$

18.118 (a) anode: $4[Al(s) \rightarrow Al^{3+}(aq) + 3\ e^-]$ $E° = 1.66\ V$

cathode: $3[O_2(g) + 4\ H^+(aq) + 4\ e^- \rightarrow 2\ H_2O(l)]$ $E° = 1.23\ V$

overall: $4\ Al(s) + 3\ O_2(g) + 12\ H^+(aq) \rightarrow 4\ Al^{3+}(aq) + 6\ H_2O(l)$ $E° = 2.89\ V$

(b) & (c) $E = E° - \dfrac{2.303\ RT}{nF} \log \dfrac{[Al^{3+}]^4}{(P_{O_2})^3[H^+]^{12}}$

$$E = 2.89\ V - \frac{(2.303)\left(8.314 \times 10^{-3}\ \frac{kJ}{K \cdot mol}\right)(310\ K)}{(12\ mol\ e^-)(96,500\ C/mol\ e^-)} \log \left(\frac{(1.0 \times 10^{-9})^4}{(0.20)^3(1.0 \times 10^{-7})^{12}} \right)$$

$E = 2.89\ V - 0.257\ V = 2.63\ V$

Multi-Concept Problems

18.120 (a) $4\ CH_2=CHCN + 2\ H_2O \rightarrow 2\ NC(CH_2)_4CN + O_2$

(b) $mol\ e^- = 3000\ C/s \times 10.0\ h \times \dfrac{3600\ s}{1\ h} \times \dfrac{1\ mol\ e^-}{96,500\ C} = 1119.2\ mol\ e^-$

mass adiponitrile =

$1119.2\ mol\ e^- \times \dfrac{1\ mol\ adiponitrile}{2\ mol\ e^-} \times \dfrac{108.14\ g\ adiponitrile}{1\ mol\ adiponitrile} \times \dfrac{1.0\ kg}{1000\ g} = 60.5\ kg$

(c) $1119.2\ mol\ e^- \times \dfrac{1\ mol\ O_2}{4\ mol\ e^-} = 279.8\ mol\ O_2$

334

$$PV = nRT$$

$$V = \frac{nRT}{P} = \frac{(279.8 \text{ mol})\left(0.08206 \dfrac{L \cdot atm}{K \cdot mol}\right)(298 \text{ K})}{\left(740 \text{ mm Hg} \times \dfrac{1 \text{ atm}}{760 \text{ mm Hg}}\right)} = 7030 \text{ L } O_2$$

18.122 (a) $Cr_2O_7^{2-}(aq) + 6 Fe^{2+}(aq) + 14 H^+(aq) \rightarrow 2 Cr^{3+}(aq) + 6 Fe^{3+}(aq) + 7 H_2O(l)$
(b) The two half reactions are:

oxidation: $\quad Fe^{2+}(aq) \rightarrow Fe^{3+}(aq) + e^- \qquad\qquad\qquad\qquad E° = -0.77 \text{ V}$
reduction: $\quad Cr_2O_7^{2-}(aq) + 14 H^+(aq) + 6 e^- \rightarrow 2 Cr^{3+}(aq) + 7 H_2O(l) \quad E° = \;\; 1.33 \text{ V}$

At the equivalence point the potential is given by either of the following expressions:

(1) $E = 1.33 \text{ V} - \dfrac{0.0592 \text{ V}}{6} \log \dfrac{[Cr^{3+}]^2}{[Cr_2O_7^{2-}][H^+]^{14}}$

(2) $E = 0.77 \text{ V} - \dfrac{0.0592 \text{ V}}{1} \log \dfrac{[Fe^{2+}]}{[Fe^{3+}]}$

where E is the same in both because equilibrium is reached and the solution can have only one potential. Multiplying (1) by 6, adding it to (2), and using some stoichiometric relationships at the equivalence point will simplify the log term.

$$7E = [(6 \times 1.33 \text{ V}) + 0.77 \text{ V}] - (0.0592 \text{ V}) \log \frac{[Fe^{2+}][Cr^{3+}]^2}{[Fe^{3+}][Cr_2O_7^{2-}][H^+]^{14}}$$

At the equivalence point, $[Fe^{2+}] = 6[Cr_2O_7^{2-}]$ and $[Fe^{3+}] = 3[Cr^{3+}]$. Substitute these equalities into the previous equation.

$$7E = [(6 \times 1.33 \text{ V}) + 0.77 \text{ V}] - (0.0592 \text{ V}) \log \frac{6[Cr_2O_7^{2-}][Cr^{3+}]^2}{3[Cr^{3+}][Cr_2O_7^{2-}][H^+]^{14}}$$

Cancel identical terms.

$$7E = [(6 \times 1.33 \text{ V}) + 0.77 \text{ V}] - (0.0592 \text{ V}) \log \frac{6[Cr^{3+}]}{3[H^+]^{14}}$$

mol $Fe^{2+} = (0.120 \text{ L})(0.100 \text{ mol/L}) = 0.0120 \text{ mol } Fe^{2+}$

$$\text{mol } Cr_2O_7^{2-} = 0.0120 \text{ mol } Fe^{2+} \times \frac{1 \text{ mol } Cr_2O_7^{2-}}{6 \text{ mol } Fe^{2+}} = 0.00200 \text{ mol } Cr_2O_7^{2-}$$

$$\text{volume } Cr_2O_7^{2-} = 0.00200 \text{ mol} \times \frac{1 \text{ L}}{0.120 \text{ mol}} = 0.0167 \text{ L}$$

At the equivalence point assume mol Fe^{3+} = initial mol Fe^{2+} = 0.0120 mol
Total volume at the equivalence point is 0.120 L + 0.0167 L = 0.1367 L

$$[Fe^{3+}] = \frac{0.0120 \text{ mol}}{0.1367 \text{ L}} = 0.0878 \text{ M}; \quad [Cr^{3+}] = [Fe^{3+}]/3 = (0.0878 \text{ M})/3 = 0.0293 \text{ M}$$

$[H^+] = 10^{-pH} = 10^{-2.00} = 0.010 \text{ M}$

$$7E = [(6 \times 1.33 \text{ V}) + 0.77 \text{ V}] - (0.0592 \text{ V})\log \frac{6(0.0293)}{3(0.010)^{14}} = 8.75 - 1.585 = 7.165 \text{ V}$$

$$E = \frac{7.165 \text{ V}}{7} = 1.02 \text{ V at the equivalence point.}$$

18.124 (a) $Zn(s) + 2 Ag^+(aq) + H_2O(l) \rightarrow ZnO(s) + 2 Ag(s) + 2 H^+(aq)$

$\Delta H^\circ_{rxn} = \Delta H^\circ_f(ZnO) - [2 \Delta H^\circ_f(Ag^+) + \Delta H^\circ_f(H_2O)]$

$\Delta H^\circ_{rxn} = [(1 \text{ mol})(-348.3 \text{ kJ/mol})] - [(2 \text{ mol})(105.6 \text{ kJ/mol}) + (1 \text{ mol})(-285.8 \text{ kJ/mol})]$

$\Delta H^\circ_{rxn} = -273.7 \text{ kJ}$

$\Delta S^\circ = [S^\circ(ZnO) + 2 S^\circ(Ag)] - [S^\circ(Zn) + 2 S^\circ(Ag^+) + S^\circ(H_2O)]$

$\Delta S^\circ = [(1 \text{ mol})(43.6 \text{ J/(K}\cdot\text{mol)}) + (2 \text{ mol})(42.6 \text{ J/(K}\cdot\text{mol)})]$

$\qquad - [(1 \text{ mol})(41.6 \text{ J/(K}\cdot\text{mol)}) + (2 \text{ mol})(72.7 \text{ J/(K}\cdot\text{mol)}) + (1 \text{ mol})(69.9 \text{ J/(K}\cdot\text{mol)})]$

$\Delta S^\circ = -128.1 \text{ J/K}$

$\Delta G^\circ = \Delta H^\circ - T\Delta S^\circ = -273.7 \text{ kJ} - (298 \text{ K})(-128.1 \times 10^{-3} \text{ kJ/K}) = -235.5 \text{ kJ}$

(b) 1 V = 1 J/C

$$\Delta G^\circ = -nFE^\circ \quad E^\circ = \frac{-\Delta G^\circ}{nF} = \frac{-(-235.5 \times 10^3 \text{ J})}{(2 \text{ mol e}^-)\left(\dfrac{96{,}500 \text{ C}}{1 \text{ mol e}^-}\right)} = 1.220 \text{ J/C} = 1.220 \text{ V}$$

$$E^\circ = \frac{0.0592 \text{ V}}{n} \log K; \quad \log K = \frac{nE^\circ}{0.0592 \text{ V}} = \frac{(2)(1.220 \text{ V})}{0.0592 \text{ V}} = 41.22$$

$K = 10^{41.22} = 2 \times 10^{41}$

(c) $E = E^\circ - \dfrac{0.0592 \text{ V}}{n} \log \dfrac{[H^+]^2}{[Ag^+]^2}$

The addition of NH_3 to the cathode compartment would result in the formation of the $Ag(NH_3)_2^+$ complex ion which results in a decrease in Ag^+ concentration. The log term in the Nernst equation becomes larger and the cell voltage decreases.

On mixing equal volumes of two solutions, the concentrations of both solutions are cut in half.

	$Ag^+(aq)$	+	$2 NH_3(aq)$	$\rightleftharpoons$	$Ag(NH_3)_2^+(aq)$
before reaction (M)	0.0500		2.00		0
assume 100% reaction	−0.0500		−2(0.0500)		+0.0500
after reaction (M)	0		1.90		0.0500
assume small back rxn	+x		+2x		−x
equil (M)	x		1.90 + 2x		0.0500 − x

$$K_f = 1.7 \times 10^7 = \frac{[Ag(NH_3)_2^+]}{[Ag^+][NH_3]^2} = \frac{(0.0500 - x)}{(x)(1.90 + 2x)^2} \approx \frac{0.0500}{(x)(1.90)^2}$$

Solve for x. $x = [Ag^+] = 8.15 \times 10^{-10} \text{ M}$

$$E = E^\circ - \frac{0.0592 \text{ V}}{n} \log \frac{[H^+]^2}{[Ag^+]^2} = 1.220 \text{ V} - \frac{0.0592 \text{ V}}{2} \log \frac{(1.00 \text{ M})^2}{(8.15 \times 10^{-10} \text{ M})^2} = 0.682 \text{ V}$$

(d) Calculate new initial concentrations because of dilution to 110.0 mL.

$$M_i \times V_i = M_f \times V_f; \quad M_f = [Cl^-] = \frac{M_i \times V_i}{V_f} = \frac{0.200 \text{ M} \times 10.0 \text{ mL}}{110.0 \text{ mL}} = 0.0182 \text{ M}$$

$$M_i \times V_i = M_f \times V_f; \quad M_f = [Ag^+] = \frac{M_i \times V_i}{V_f} = \frac{0.0500 \text{ M} \times 100.0 \text{ mL}}{110.0 \text{ mL}} = 0.0455 \text{ M}$$

$$M_i \times V_i = M_f \times V_f; \quad M_f = [NH_3] = \frac{M_i \times V_i}{V_f} = \frac{2.00 \text{ M} \times 100.0 \text{ mL}}{110.0 \text{ mL}} = 1.82 \text{ M}$$

Now calculate the $[Ag^+]$ as a result of the following equilibrium:

	$Ag^+(aq)$	+	$2 NH_3(aq)$	⇌	$Ag(NH_3)_2^+(aq)$
before reaction (M)	0.0455		1.82		0
assume 100% reaction	−0.0455		−2(0.0455)		+0.0455
after reaction (M)	0		1.73		0.0455
assume small back rxn	+x		+2x		−x
equil (M)	x		1.73 + 2x		0.0455 − x

$$K_f = 1.7 \times 10^7 = \frac{[Ag(NH_3)_2^+]}{[Ag^+][NH_3]^2} = \frac{(0.0455 - x)}{(x)(1.73 + 2x)^2} \approx \frac{0.0455}{(x)(1.73)^2}$$

Solve for x. $x = [Ag^+] = 8.94 \times 10^{-10} \text{ M}$

For AgCl, $K_{sp} = 1.8 \times 10^{-10}$

$IP = [Ag^+][Cl^-] = (8.94 \times 10^{-10} \text{ M})(0.0182 \text{ M}) = 1.6 \times 10^{-11}$

$IP < K_{sp}$, AgCl will not precipitate.

Now calculate new initial concentrations because of dilution to 120.0 mL.

$$M_i \times V_i = M_f \times V_f; \quad M_f = [Br^-] = \frac{M_i \times V_i}{V_f} = \frac{0.200 \text{ M} \times 10.0 \text{ mL}}{120.0 \text{ mL}} = 0.0167 \text{ M}$$

$$M_i \times V_i = M_f \times V_f; \quad M_f = [Ag^+] = \frac{M_i \times V_i}{V_f} = \frac{0.0500 \text{ M} \times 100.0 \text{ mL}}{120.0 \text{ mL}} = 0.0417 \text{ M}$$

$$M_i \times V_i = M_f \times V_f; \quad M_f = [NH_3] = \frac{M_i \times V_i}{V_f} = \frac{2.00 \text{ M} \times 100.0 \text{ mL}}{120.0 \text{ mL}} = 1.67 \text{ M}$$

Now calculate the $[Ag^+]$ as a result of the following equilibrium:

	$Ag^+(aq)$	+	$2 NH_3(aq)$	⇌	$Ag(NH_3)_2^+(aq)$
before reaction (M)	0.0417		1.67		0
assume 100% reaction	−0.0417		−2(0.0417)		+0.0417
after reaction (M)	0		1.59		0.0417
assume small back rxn	+x		+2x		−x
equil (M)	x		1.59 + 2x		0.0417 − x

$$K_f = 1.7 \times 10^7 = \frac{[Ag(NH_3)_2^+]}{[Ag^+][NH_3]^2} = \frac{(0.0417 - x)}{(x)(1.59 + 2x)^2} \approx \frac{0.0417}{(x)(1.59)^2}$$

Solve for x. $x = [Ag^+] = 9.70 \times 10^{-10} \text{ M}$

For AgBr, $K_{sp} = 5.4 \times 10^{-13}$

$IP = [Ag^+][Br^-] = (9.70 \times 10^{-10} \text{ M})(0.0167 \text{ M}) = 1.6 \times 10^{-11}$

$IP > K_{sp}$, AgBr will precipitate.

18.126 (a) Oxidation half reaction: $2 [C_4H_{10}(g) + 13 O^{2-}(s) \rightarrow 4 CO_2(g) + 5 H_2O(l) + 26 e^-]$
Reduction half reaction: $\underline{13 [O_2(g) + 4 e^- \rightarrow 2 O^{2-}(s)]}$
Cell reaction: $\qquad 2 C_4H_{10}(g) + 13 O_2(g) \rightarrow 8 CO_2(g) + 10 H_2O(l)$

(b) $\Delta H° = [8 \Delta H°_f(CO_2)) + 10 \Delta H°_f(H_2O)] - [2 \Delta H°_f(C_4H_{10})]$
$\Delta H° = [(8 \text{ mol})(-393.5 \text{ kJ/mol}) + (10 \text{ mol})(-285.8 \text{ kJ/mol})]$
$$- [(2 \text{ mol})(-126 \text{ kJ/mol})] = -5754 \text{ kJ}$$

$\Delta S° = [8 S°(CO_2) + 10 S°(H_2O)] - [2 S°(C_4H_{10}) + 13 S°(O_2)]$
$\Delta S° = [(8 \text{ mol})(213.6 \text{ J/(K} \cdot \text{mol)}) + (10 \text{ mol})(69.9 \text{ J/(K} \cdot \text{mol)})]$
$$- [(2 \text{ mol})(310 \text{ J/(K} \cdot \text{mol)}) + (13 \text{ mol})(205 \text{ J/(K} \cdot \text{mol)})] = -877.2 \text{ J/K}$$

$\Delta G° = \Delta H° - T\Delta S° = -5754 \text{ kJ} - (298 \text{ K})(-877.2 \times 10^{-3} \text{ kJ/K}) = -5493 \text{ kJ}$
$1 \text{ V} = 1 \text{ J/C}$

$$\Delta G° = -nFE°; \quad E° = -\frac{\Delta G°}{nF} = -\frac{-5493 \times 10^3 \text{ J}}{(52)(96,500 \text{ C})} = 1.09 \text{ J/C} = 1.09 \text{ V}$$

$\Delta G° = -RT \ln K$

$$\ln K = \frac{-\Delta G°}{RT} = \frac{-(-5493 \text{ kJ})}{(8.314 \times 10^{-3} \text{ kJ/K})(298 \text{ K})} = 2217$$

$K = e^{2217} = 7 \times 10^{962}$

On raising the temperature, both K and E° will decrease because the reaction is exothermic ($\Delta H° < 0$).

(c) C_4H_{10}, 58.12 amu; 10.5 A = 10.5 C/s

$$\text{mass } C_4H_{10} = 10.5 \text{ C/s} \times 8 \text{ hr} \times \frac{60 \text{ min}}{1 \text{ hr}} \times \frac{60 \text{ s}}{1 \text{ min}} \times \frac{1 \text{ mol e}^-}{96,500 \text{ C}} \times \frac{2 \text{ mol } C_4H_{10}}{52 \text{ mol e}^-} \times$$

$$\frac{58.12 \text{ g } C_4H_{10}}{1 \text{ mol } C_4H_{10}} = 7.00 \text{ g } C_4H_{10}$$

$$n = 7.00 \text{ g } C_4H_{10} \times \frac{1 \text{ mol } C_4H_{10}}{58.12 \text{ g } C_4H_{10}} = 0.120 \text{ mol } C_4H_{10}$$

$20°C = 20 + 273 = 293 \text{ K}$
$PV = nRT$

$$V = \frac{nRT}{P} = \frac{(0.120 \text{ mol})\left(0.082\ 06 \dfrac{\text{L} \cdot \text{atm}}{\text{K} \cdot \text{mol}}\right)(293 \text{ K})}{\left(815 \text{ mm Hg} \times \dfrac{1.00 \text{ atm}}{760 \text{ mm Hg}}\right)} = 2.69 \text{ L}$$

18.128 (a) $4 [Au(s) + 2 CN^-(aq) \rightarrow Au(CN)_2^-(aq) + e^-]$ oxidation half reaction

$O_2(g) \rightarrow 2 H_2O(l)$
$O_2(g) + 4 H^+(aq) \rightarrow 2 H_2O(l)$
$4 e^- + O_2(g) + 4 H^+(aq) \rightarrow 2 H_2O(l)$ reduction half reaction

Combine the two half reactions.

$4 Au(s) + 8 CN^-(aq) + O_2(g) + 4 H^+(aq) \rightarrow 4 Au(CN)_2^-(aq) + 2 H_2O(l)$

$4 Au(s) + 8 CN^-(aq) + O_2(g) + 4 H^+(aq) + 4 OH^-(aq)$
$$\rightarrow 4 Au(CN)_2^-(aq) + 2 H_2O(l) + 4 OH^-(aq)$$

$4 Au(s) + 8 CN^-(aq) + O_2(g) + 4 H_2O(l)$
$$\rightarrow 4 Au(CN)_2^-(aq) + 2 H_2O(l) + 4 OH^-(aq)$$

$4 Au(s) + 8 CN^-(aq) + O_2(g) + 2 H_2O(l) \rightarrow 4 Au(CN)_2^-(aq) + 4 OH^-(aq)$

(b) Add the following five reactions together. $\Delta G°$ is calculated below each reaction.

$4 [Au^+(aq) + 2 CN^-(aq) \rightarrow Au(CN)_2^-(aq)]$ $\qquad K = (K_f)^4$

$\Delta G° = -RT \ln K = -(8.314 \times 10^{-3} \text{ kJ/K})(298 \text{ K}) \ln (6.2 \times 10^{38})^4 = -885.2$ kJ

$O_2(g) + 4 H^+(aq) + 4 e^- \rightarrow 2 H_2O(l)$ $\qquad E° = 1.229$ V

$\Delta G° = -nFE° = -(4 \text{ mol } e^-)\left(\dfrac{96,500 \text{ C}}{1 \text{ mol } e^-}\right)(1.229 \text{ V})\left(\dfrac{1 \text{ J}}{1 \text{ C} \cdot \text{V}}\right) = -474,394 \text{ J} = -474.4$ kJ

$4 [H_2O(l) \rightleftharpoons H^+(aq) + OH^-(aq)]$ $\qquad K = (K_w)^4$

$\Delta G° = -RT \ln K = -(8.314 \times 10^{-3} \text{ kJ/K})(298 \text{ K}) \ln (1.0 \times 10^{-14})^4 = +319.5$ kJ

$4 [Au(s) \rightarrow Au^{3+}(aq) + 3 e^-]$ $\qquad E° = -1.498$ V

$\Delta G° = -nFE° = -(12 \text{ mol } e^-)\left(\dfrac{96,500 \text{ C}}{1 \text{ mol } e^-}\right)(-1.498 \text{ V})\left(\dfrac{1 \text{ J}}{1 \text{ C} \cdot \text{V}}\right) = +1,734,684 \text{ J} = +1,734.7$ kJ

$4 [Au^{3+}(aq) + 2 e^- \rightarrow Au^+(aq)]$ $\qquad E° = 1.401$ V

$\Delta G° = -nFE° = -(8 \text{ mol } e^-)\left(\dfrac{96,500 \text{ C}}{1 \text{ mol } e^-}\right)(1.401 \text{ V})\left(\dfrac{1 \text{ J}}{1 \text{ C} \cdot \text{V}}\right) = -1,081,572 \text{ J} = -1,081.6$ kJ

Overall reaction:

$4 Au(s) + 8 CN^-(aq) + O_2(g) + 2 H_2O(l) \rightarrow 4 Au(CN)_2^-(aq) + 4 OH^-(aq)$

$\Delta G° = -885.2 \text{ kJ} - 474.4 \text{ kJ} + 319.5 \text{ kJ} + 1,734.7 \text{ kJ} - 1,081.6 \text{ kJ} = -387.0$ kJ

The Main-Group Elements

19.1 (a) B is above Al in group 3A, and therefore B is more nonmetallic than Al.
(b) Ge and Br are in the same row of the periodic table, but Br (group 7A) is to the right of Ge (group 4A). Therefore, Br is more nonmetallic.
(c) Se (group 6A) is more nonmetallic than In because it is above and to the right of In (group 3A).
(d) Cl (group 7A) is more nonmetallic than Te because it is above and to the right of Te (group 6A).

19.2 (a) HNO_3 H_3PO_4

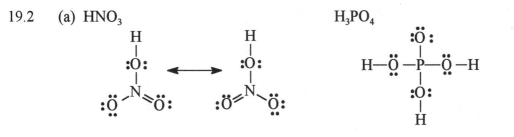

Nitrogen can form very strong pπ - pπ bonds. Phosphorus forms weaker pπ - pπ bonds, so it tends to form more single bonds.
(b) Sulfur can use empty 3d orbitals to form octahedral sp^3d^2 hybrid orbitals to bond with six fluorines in SF_6. With just 2s and 2p valence orbitals, oxygen cannot expand its valence orbitals beyond four. OF_2 has two single bonds and two lone pairs of electrons.

19.3 Carbon forms strong π bonds with oxygen. Silicon does not form strong π bonds with oxygen, and what results are chains of alternating silicon and oxygen singly bonded to each other.

19.4 An ethanelike structure is unlikely for diborane because it would require 14 valence electrons and diborane only has 12. The result is two three-center, two-electron bonds between the borons and the bridging hydrogen atoms.

19.5   H—C≡N: The carbon is sp hybridized.

19.6 $Hb–O_2 + CO \rightleftarrows Hb–CO + O_2$
Mild cases of carbon monoxide poisoning can be treated with O_2. Le Châtelier's principle says that adding a product (O_2) will cause the reaction to proceed in the reverse direction, back to $Hb–O_2$.

19.7 (a) $Si_8O_{24}^{16-}$ (b) $Si_2O_5^{2-}$

19.8 $\text{Formal charge} = \begin{pmatrix} \text{Number of} \\ \text{valence electrons} \\ \text{in free atom} \end{pmatrix} - \frac{1}{2}\begin{pmatrix} \text{Number of} \\ \text{bonding} \\ \text{electrons} \end{pmatrix} - \begin{pmatrix} \text{Number of} \\ \text{nonbonding} \\ \text{electrons} \end{pmatrix}$

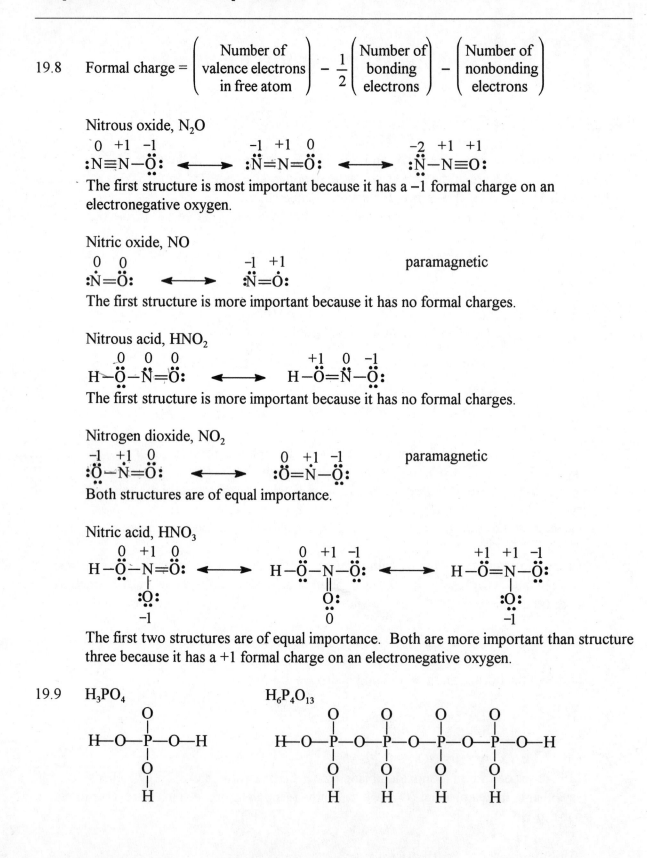

Nitrous oxide, N_2O

The first structure is most important because it has a −1 formal charge on an electronegative oxygen.

Nitric oxide, NO

paramagnetic

The first structure is more important because it has no formal charges.

Nitrous acid, HNO_2

The first structure is more important because it has no formal charges.

Nitrogen dioxide, NO_2

paramagnetic

Both structures are of equal importance.

Nitric acid, HNO_3

The first two structures are of equal importance. Both are more important than structure three because it has a +1 formal charge on an electronegative oxygen.

19.9 H_3PO_4 $H_6P_4O_{13}$

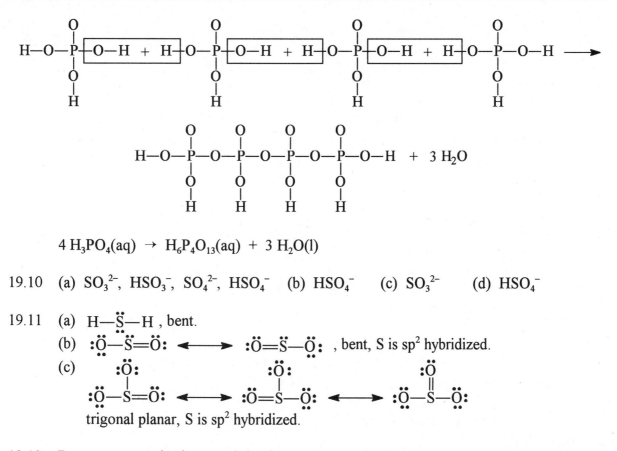

$$4 H_3PO_4(aq) \rightarrow H_6P_4O_{13}(aq) + 3 H_2O(l)$$

19.10 (a) SO_3^{2-}, HSO_3^-, SO_4^{2-}, HSO_4^- (b) HSO_4^- (c) SO_3^{2-} (d) HSO_4^-

19.11 (a) $H-\ddot{S}-H$, bent.

(b) $:\ddot{O}-\ddot{S}=\ddot{O}: \longleftrightarrow :\ddot{O}=\ddot{S}-\ddot{O}:$, bent, S is sp^2 hybridized.

(c)

$:\ddot{O}-\ddot{S}=\ddot{O}: \longleftrightarrow :\ddot{O}=\ddot{S}-\ddot{O}: \longleftrightarrow :\ddot{O}-\overset{\ddot{O}}{\underset{}{\overset{\|}{S}}}-\ddot{O}:$

trigonal planar, S is sp^2 hybridized.

19.12 Because copper is always conducting, a photocopier drum coated with copper would not hold any charge so no document image would adhere to the drum for copying.

Understanding Key Concepts

19.14

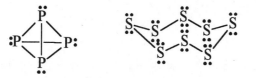

19.16 (a) N_2, O_2, F_2, P_4 (tetrahedral), S_8 (crown-shaped ring), Cl_2

(b) $:N\equiv N:$ $:\ddot{O}=\ddot{O}:$ $:\ddot{F}-\ddot{F}:$ $:\ddot{Cl}-\ddot{Cl}:$

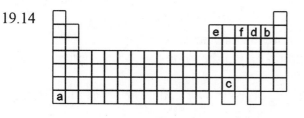

(c) The smaller N and O can form strong π bonds, whereas P and S cannot. In both F_2 and Cl_2, the atoms are joined by a single bond.

19.18   (a) H_2O, CH_4, HF, B_2H_6, NH_3

(b)   H—Ö—H   H   H—F̈:   H   H   H   H—N̈—H
                                  | \ . . / |
                       H—C—H   B   B   H
                                  | / . . \
                                  H   H   H   H

There is a problem in drawing an electron-dot structure for B_2H_6 because this molecule is electron deficient and has two three-center, two-electron bonds.

Additional Problems
General Properties and Periodic Trends

19.20   (a) Cl (group 7A) is to the right of S (group 6A) in the same row of the periodic table. Cl has the higher ionization energy.
(b) Si is above Ge in group 4A. Si has the higher ionization energy.
(c) O (group 6A) is above and to the right of In (group 3A) in the periodic table. O has the higher ionization energy.

19.22   (a) Al is below B in group 3A. Al has the larger atomic radius.
(b) P (group 5A) is to the left of S (group 6A) in the same row of the periodic table. P has the larger atomic radius.
(c) Pb (group 4A) is below and to the left of Br (group 7A) in the periodic table. Pb has the larger atomic radius.

19.24   (a) I (group 7A) is to the right of Te (group 6A) in the same row of the periodic table. I has the higher electronegativity.
(b) N is above P in group 5A. N has the higher electronegativity.
(c) F (group 7A) is above and to the right of In (group 3A) in the periodic table. F has the higher electronegativity.

19.26   (a) Sn is below Si in group 4A. Sn has more metallic character.
(b) Ge (group 4A) is to the left of Se (group 6A) in the same row of the periodic table. Ge has more metallic character.
(c) Bi (group 5A) is below and to the left of I (group 7A) in the periodic table. Bi has more metallic character.

19.28   In each case the more ionic compound is the one formed between a metal and nonmetal.
(a) CaH_2 (b) Ga_2O_3 (c) KCl   (d) $AlCl_3$

19.30   Molecular   (a) B_2H_6 (c) SO_3 (d) $GeCl_4$
Extended three-dimensional structure   (b) $KAlSi_3O_8$

19.32   Nonmetal and semimetal oxides are acidic. Metal oxides are basic or amphoteric.
(a) P_4O_{10} (b) B_2O_3 (c) SO_2
(d) N_2O_3 N_2O_3 is in the same group but above As_2O_3 in the periodic table.

19.34   (a) Sn   (b) Cl   (c) Sn   (d) Se   (e) B

19.36   Boron has only 2s and 2p valence orbitals and can form a maximum of four bonds. Aluminum has available 3d orbitals and can use octahedral sp^3d^2 hybrid orbitals to bond to six F^- ions.

19.38   In O_2 a π bond is formed by 2p orbitals on each O. S does not form strong π bonds with its 3p orbitals, which leads to the S_8 ring structure with single bonds.

The Group 3A Elements

19.40   +3 for B, Al, Ga and In;   +1 for Tl

19.42   Boron is a hard semiconductor with a high melting point. Boron forms only molecular compounds and does not form an aqueous B^{3+} ion. $B(OH)_3$ is an acid.

19.44 $2\ BBr_3(g) + 3\ H_2(g) \xrightarrow[1200°C]{Ta\ wire} 2\ B(s) + 6\ HBr(g)$

This reaction is used to produce crystalline boron.

19.46   (a) An electron deficient molecule is a molecule that doesn't have enough electrons to form a two center-two electron bond between each pair of bonded atoms. B_2H_6 is an electron deficient molecule.
(b) A three-center two-electron bond has three atoms bonded together using just two electrons. The B–H–B bridging bond in B_2H_6 is a three-center two-electron bond.

19.48   (a) Al   (b) Tl   (c) B   (d) B

The Group 4A Elements

19.50   (a) Pb   (b) C   (c) Si   (d) C

19.52   (a) $GeBr_4$, tetrahedral; Ge is sp^3 hybridized.
(b) CO_2, linear; C is sp hybridized.
(c) CO_3^{2-}, trigonal planar; C is sp^2 hybridized.
(d) $Sn(OH)_6^{2-}$, octahedral; Sn is sp^3d^2 hybridized.

19.54   Diamond is a very hard, high melting solid. It is an electrical insulator.
Diamond has a covalent network structure in which each C atom uses sp^3 hybrid orbitals to form a tetrahedral array of σ bonds. The interlocking, three-dimensional network of strong bonds makes diamond the hardest known substance with the highest melting point for an element. Because the valence electrons are localized in the σ bonds, diamond is an electrical insulator.

19.56   (a) carbon tetrachloride, CCl_4 (b) carbon monoxide, CO   (c) methane, CH_4

19.58   Some uses for CO_2 are:
(1) To provide the "bite" in soft drinks; $CO_2(aq) + H_2O(l) \rightleftharpoons H_2CO_3(aq)$
(2) CO_2 fire extinguishers; CO_2 is nonflammable and 1.5 times more dense than air.
(3) Refrigerant; dry ice, sublimes at –78°C.

19.60 $SiO_2(l) + 2\ C(s) \rightarrow Si(l) + 2\ CO(g)$
(sand)

Purification of silicon for semiconductor devices:
$Si(s) + 2\ Cl_2(g) \rightarrow SiCl_4(l)$; $SiCl_4$ is purified by distillation.

$$SiCl_4(g) + 2\ H_2(g) \xrightarrow{\text{heat}} Si(s) + 4\ HCl(g);\quad \text{Si is purified by zone refining.}$$

19.62   (a) $SiO_4{}^{4-}$ (b) $Si_4O_{13}{}^{10-}$

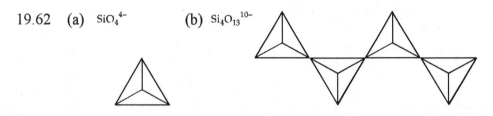

The charge on the anion is equal to the number of terminal O atoms.

19.64   (a) $Si_3O_{10}{}^{8-}$
(b) The charge on the anion is 8–. Because the Ca^{2+} to Cu^{2+} ratio is 1:1, there must be 2 Ca^{2+} and 2 Cu^{2+} ions in the formula for the mineral. There are also 2 waters. The formula of the mineral is: $Ca_2Cu_2Si_3O_{10} \cdot 2\ H_2O$

The Group 5A Elements

19.66   (a) P   (b) Sb and Bi   (c) N   (d) Bi

19.68   (a) N_2O, +1   (b) N_2H_4, –2   (c) Ca_3P_2, –3
(d) H_3PO_3, +3   (e) H_3AsO_4, +5

19.70 $:N \equiv N:$
N_2 is unreactive because of the large amount of energy necessary to break the $N \equiv N$ triple bond.

19.72   (a) $NO_2{}^-$, bent; N is sp^2 hybridized.
(b) PH_3, trigonal pyramidal; P is sp^3 hybridized.
(c) PF_5, trigonal bipyramidal; P is sp^3d hybridized.
(d) $PCl_4{}^+$, tetrahedral; P is sp^3 hybridized.

19.74   White phosphorus

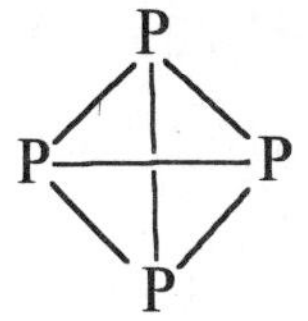

Red phosphorus

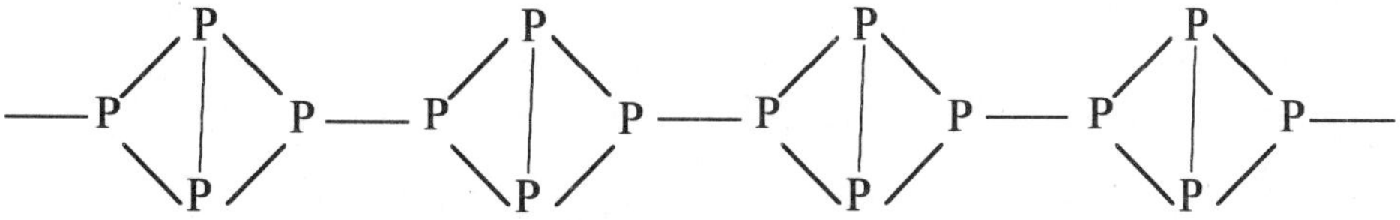

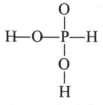

White phosphorus is reactive due to the considerable strain in the P_4 molecule.

19.76   (a) The structure for phosphorous acid is

$$\begin{array}{c} O \\ \| \\ H-O-P-H \\ | \\ O \\ | \\ H \end{array}$$

Only the two hydrogens bonded to oxygen are acidic.
(b) Nitrogen forms strong π bonds, and in N_2 the nitrogen atoms are triple bonded to each other. Phosphorus does not form strong $p\pi$ - $p\pi$ bonds, and so the P atoms are single bonded to each other in P_4

19.78   (a) $2\,NO(g)\ +\ O_2(g)\ \rightarrow\ 2\,NO_2(g)$
(b) $4\,HNO_3(aq)\ \rightarrow\ 4\,NO_2(aq)\ +\ O_2(g)\ +\ 2\,H_2O(l)$
(c) $3\,Ag(s)\ +\ 4\,H^+(aq)\ +\ NO_3^-(aq)\ \rightarrow\ 3\,Ag^+(aq)\ +\ NO(g)\ +\ 2\,H_2O(l)$
(d) $N_2H_4(aq)\ +\ 2\,I_2(aq)\ \rightarrow\ N_2(g)\ +\ 4\,H^+(aq)\ +\ 4\,I^-(aq)$

The Group 6A Elements

19.80   (a) O   (b) Te   (c) Po   (d) O

19.82   (a) rhombic sulfur -- yellow crystalline solid (mp 113°C) that contains crown-shaped S_8 rings.
(b) monoclinic sulfur -- an allotrope of sulfur in which the S_8 rings pack differently in the crystal.
(c) plastic sulfur -- when sulfur is cooled rapidly, the sulfur forms disordered, tangled chains, yielding an amorphous, rubbery material called plastic sulfur.
(d) Liquid sulfur between 160 and 195°C becomes dark reddish-brown and very viscous forming long polymer chains (S_n, n > 200,000).

19.84   (a) hydrogen sulfide, H_2S (b) sulfur dioxide, SO_2 (c) sulfur trioxide, SO_3
        lead(II) sulfide, PbS   sulfurous acid, H_2SO_3 sulfur hexafluoride, SF_6

19.86   Sulfuric acid (H_2SO_4) is manufactured by the contact process, a three-step reaction sequence in which (1) sulfur burns in air to give SO_2, (2) SO_2 is oxidized to SO_3 in the presence of a vanadium(V) oxide catalyst, and (3) SO_3 reacts with water to give H_2SO_4.
$$(1)\ S(s)\ +\ O_2(g)\ \rightarrow\ SO_2(g)$$

$$(2)\ 2\ SO_2(g)\ +\ O_2(g)\ \xrightarrow[V_2O_5\ catalyst]{heat}\ 2\ SO_3(g)$$

$$(3)\ SO_3(g)\ +\ H_2O\ (in\ conc\ H_2SO_4)\ \rightarrow\ H_2SO_4(l)$$

19.88   (a) $Zn(s) + 2\ H_3O^+(aq)\ \rightarrow\ Zn^{2+}(aq) + H_2(g) + 2\ H_2O(l)$
(b) $BaSO_3(s) + 2\ H_3O^+(aq) \rightarrow H_2SO_3(aq) + Ba^{2+}(aq) + 2\ H_2O(l)$
(c) $Cu(s) + 2\ H_2SO_4(l) \rightarrow Cu^{2+}(aq) + SO_4^{2-}(aq) + SO_2(g) + 2\ H_2O(l)$
(d) $H_2S(aq) + I_2(aq) \rightarrow S(s) + 2\ H^+(aq) + 2\ I^-(aq)$

19.90   (a) Acid strength increases as the number of O atoms increases.
(b) In comparison with S, O is much too electronegative to form compounds of O in the +4 oxidation state. Also, S uses sp^3d hybrid orbitals for bonding in SF_4, but O doesn't have valence d orbitals and so it can't form four bonds to F.
(c) Each S is sp^3 hybridized with two lone pairs of electrons. The bond angles are therefore 109.5°. A planar ring would require bond angles of 135°.

Halogen Oxoacids and Oxoacid Salts

19.92   (a) $HBrO_3$, +5   (b) HIO, +1   (c) $NaClO_2$, +3   (d) $NaIO_4$, +7

19.94   (a) iodic acid   (b) chlorous acid   (c) sodium hypobromite
(d) lithium perchlorate

19.96   (a) HIO_3 :Ö—İ—Ö—H   trigonal pyramidal

(b) ClO_2^- [:Ö—Ċl—Ö:]⁻   bent

(c) HOCl   H—Ö—Ċl: bent

(d) IO_6^{5-} 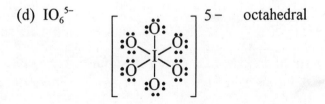 5–   octahedral

348

19.98   Oxygen atoms are highly electronegative. Increasing the number of oxygen atoms
increases the polarity of the O–H bond and increases the acid strength.

19.100 (a) $Br_2(l) + 2\ OH^-(aq)\ \rightarrow\ OBr^-(aq) + Br^-(aq) + H_2O(l)$
(b) $Cl_2(g) + H_2O(l)\ \rightarrow\ HOCl(aq) + H^+(aq) + Cl^-(aq)$
(c) $3\ Cl_2(g) + 6\ OH^-(aq)\ \rightarrow\ ClO_3^-(aq) + 5\ Cl^-(aq) + 3\ H_2O(l)$

General Problems

19.102 (a) $Na_2B_4O_7 \cdot 10\ H_2O$ (b) $Ca_3(PO_4)_2$
(c) elemental sulfur, FeS_2, PbS, HgS, $CaSO_4 \cdot 2\ H_2O$

19.104 (a) LiCl is an ionic compound. PCl_3 is a covalent molecular compound.
The ionic compound, LiCl, has the higher melting point.
(b) Carbon forms strong π bonds with oxygen, and CO_2 is a covalent molecular
compound with a low melting point. Silicon prefers to form single bonds with oxygen.
SiO_2 is three dimensional extended structure with alternating silicon and oxygen singly
bonded to each other. SiO_2 is a high melting solid.
(c) Nitrogen forms strong π bonds with oxygen and NO_2 is a covalent molecular
compound with a low melting point. Phosphorus prefers to form single bonds with
oxygen. P_4O_{10} is a larger covalent molecular compound than NO_2, with a higher melting
point.

19.106 In silicate and phosphate anions, both Si and P are surrounded by tetrahedra of O atoms,
which can link together to form chains and rings.

19.108 Earth's crust: O, Si, Al, Fe;   Human body: O, C, H, N

19.110 C, Si, Ge and Sn have allotropes with the diamond structure.
Sn and Pb have metallic allotropes.
C (nonmetal), Si (semimetal), Ge (semimetal), Sn (semimetal and metal), Pb (metal)

19.112 (a) $H_3PO_4(aq) + H_2O(l)\ \rightleftharpoons\ H_3O^+(aq) + H_2PO_4^-(aq)$
H_3PO_4 is a Brønsted-Lowry acid.
(b) $B(OH)_3(aq) + 2\ H_2O(l)\ \rightleftharpoons\ B(OH)_4^-(aq) + H_3O^+(aq)$
$B(OH)_3$ is a Lewis acid.

19.114 (a) In diamond each C is covalently bonded to four additional C atoms in a rigid three-
dimensional network solid. Graphite is a two-dimensional covalent network solid of
carbon sheets that can slide over each other. Both are high melting because melting
requires the breaking of C–C bonds.
(b) Chlorine does not form perhalic acids of the type H_5XO_6 because its smaller size
favors a tetrahedral structure over an octahedral one.

19.116 The angle required by P_4 is 60°. The strain would not be reduced by using sp^3 hybrid orbitals because their angle is ~109°.

Multi-Concept Problems

19.118 (a) $\cdot\ddot{N}=\ddot{O}:$

$$\left[:\ddot{O}-\ddot{O}\cdot\right]^{-} \longleftrightarrow \left[\cdot\ddot{O}-\ddot{O}:\right]^{-} \quad \left[:\ddot{O}=\ddot{N}-\ddot{O}-\ddot{O}:\right]^{-}$$

The O–N–O bond angle should be ~ 120°.

(b)

σ^{*}_{2p} —

π^{*}_{2p} ↑ __

σ_{2p} ↑↓

π_{2p} ↑↓ ↑↓

σ^{*}_{2s} ↑↓

σ_{2s} ↑↓   The bond order is 2½ with one unpaired electron.

19.120

$$2\,In^{+}(aq) + 2\,e^{-} \rightarrow 2\,In(s) \qquad E° = -0.14\ V$$
$$\underline{In^{+}(aq) \rightarrow In^{3+}(aq) + 2\,e^{-} \qquad E° = 0.44\ V}$$
$$3\,In^{+}(aq) \rightarrow In^{3+}(aq) + 2\,In(s) \qquad E° = 0.30\ V$$

$1\ J = 1\ V\cdot C$

$\Delta G° = -nFE° = -(2)(96{,}500\ C)(0.30\ V) = -5.8 \times 10^4\ J = -58\ kJ$

Because $\Delta G° < 0$, the disproportionation is spontaneous.

$$2\,Tl^{+}(aq) + 2\,e^{-} \rightarrow 2\,Tl(s) \qquad E° = -0.34\ V$$
$$\underline{Tl^{+}(aq) \rightarrow Tl^{3+}(aq) + 2\,e^{-} \qquad E° = -1.25\ V}$$
$$3\,Tl^{+}(aq) \rightarrow Tl^{3+}(aq) + 2\,Tl(s) \qquad E° = -1.59\ V$$

$\Delta G° = -nFE° = -(2)(96{,}500\ C)(-1.59\ V) = +3.07 \times 10^5\ J = +307\ kJ$

Because $\Delta G° > 0$, the disproportionation is nonspontaneous.

19.122 NH_4NO_3, 80.04 amu; $(NH_4)_2HPO_4$, 132.06 amu

(a) % N in $NH_4NO_3 = \dfrac{2 \times (14.007\ amu\ N)}{80.04\ amu\ NH_4NO_3} \times 100\% = 35.0\%\ N$

% N in $(NH_4)_2HPO_4 = \dfrac{2 \times (14.007\ amu\ N)}{132.06\ amu\ (NH_4)_2HPO_4} \times 100\% = 21.2\%\ N$

Let x equal the fraction of NH_4NO_3 and $(1-x)$ equal the fraction of $(NH_4)_2HPO_4$ in the mixture.

$0.3381 = x(0.350) + (1-x)(0.212) = 0.138x + 0.212$

$0.1261 = 0.138x \qquad x = \dfrac{0.1261}{0.138} = 0.9138$

$(1-x) = (1 - 0.9138) = 0.0862$

There is 8.62 % $(NH_4)_2HPO_4$ and 91.38% NH_4NO_3 in the mixture.

(b) sample mass = 0.965 g

mass $NH_4NO_3 = (0.965\ g)(0.9138) = 0.8818\ g$

mass $(NH_4)_2HPO_4 = 0.965\ g - 0.8818\ g = 0.0832\ g$

$$mol\ NH_4NO_3 = 0.8818\ g \times \frac{1\ mol\ NH_4NO_3}{80.04\ g\ NH_4NO_3} = 0.0110\ mol\ NH_4NO_3$$

$$mol\ (NH_4)_2HPO_4 = 0.0832\ g \times \frac{1\ mol\ (NH_4)_2HPO_4}{132.06\ g\ (NH_4)_2HPO_4} = 6.30 \times 10^{-4}\ mol\ (NH_4)_2HPO_4$$

$$[NH_4^+] = \frac{0.0110\ mol + 2(6.30 \times 10^{-4}\ mol)}{0.0500\ L} = 0.245\ M$$

$$[HPO_4^{2-}] = \frac{6.30 \times 10^{-4}\ mol}{0.0500\ L} = 0.0126\ M$$

The equilibrium constant for the transfer of a proton from the cation of a salt to the anion of the salt is equal to $\dfrac{K_a K_b}{K_w}$.

$K_a(NH_4^+) = 5.56 \times 10^{-10}$ and $K_b(HPO_4^{2-}) = 1.61 \times 10^{-7}$

$$K = \frac{K_a K_b}{K_w} = \frac{(5.56 \times 10^{-10})(1.61 \times 10^{-7})}{1.0 \times 10^{-14}} = 9.12 \times 10^{-3}$$

$$NH_4^+(aq) + HPO_4^{2-}(aq) \rightleftharpoons H_2PO_4^-(aq) + NH_3(aq)$$

	NH_4^+	HPO_4^{2-}	$H_2PO_4^-$	NH_3
initial (M)	0.245	0.0126	0	0
change (M)	$-x$	$-x$	$+x$	$+x$
equil (M)	$0.245 - x$	$0.0126 - x$	x	x

$$K = \frac{[H_2PO_4^-][NH_3]}{[NH_4^+][HPO_4^{2-}]} = 9.12 \times 10^{-3} = \frac{x^2}{(0.245 - x)(0.0126 - x)}$$

$0.991x^2 + 0.00235x - (2.82 \times 10^{-5}) = 0$

Use the quadratic formula to solve for x.

$$x = \frac{(-0.00235) \pm \sqrt{(0.00235)^2 - (4)(0.991)(-2.82 \times 10^{-5})}}{2(0.991)} = \frac{(-0.00235) \pm 0.01083}{2(0.991)}$$

$x = 0.00428$ and -0.00665

Of the two solutions for x, only the positive value of x has physical meaning because x is equal to both the $[H_2PO_4^-]$ and $[NH_3]$.

$$K_a(NH_4^+) = \frac{[H_3O^+][NH_3]}{[NH_4^+]} = 5.56 \times 10^{-10} = \frac{[H_3O^+](0.00428)}{(0.245 - 0.00428)}$$

$$[H_3O^+] = \frac{(5.56 \times 10^{-10})(0.245 - 0.00428)}{0.00428} = 3.127 \times 10^{-8}\ M$$

$pH = -\log[H_3O^+] = -\log(3.127 \times 10^{-8}) = 7.50$

19.124 $N_2O_4(g) \rightleftharpoons 2\ NO_2(g)$

$$P_{Total} = 753\ mm\ Hg \times \frac{1.00\ atm}{760\ mm\ Hg} = 0.991\ atm$$

$$P_{Total} = P_{N_2O_4} + P_{NO_2} = 0.991 \text{ atm}$$

$$P_{N_2O_4} = 0.991 \text{ atm} - P_{NO_2}$$

$$K_p = \frac{(P_{NO_2})^2}{P_{N_2O_4}} = 0.113$$

$$K_p = \frac{(P_{NO_2})^2}{(0.991 \text{ atm} - P_{NO_2})} = 0.113$$

$$(P_{NO_2})^2 + 0.113(P_{NO_2}) - 0.112 = 0$$

Use the quadratic formula to solve for P_{NO_2}.

$$P_{NO_2} = \frac{-(0.113) \pm \sqrt{(0.113)^2 - (4)(1)(-0.112)}}{2(1)} = \frac{-0.113 \pm 0.679}{2}$$

$$P_{NO_2} = -0.0396 \text{ and } 0.283$$

Of the two solutions for P_{NO_2}, only 0.283 has physical meaning because NO_2 can't have a negative partial pressure.

$$P_{NO_2} = 0.283 \text{ atm and } P_{N_2O_4} = 0.991 \text{ atm} - P_{NO_2} = 0.991 - 0.283 = 0.708 \text{ atm}$$

	$N_2O_4(g)$	$\rightarrow$	$2 NO_2(g)$
before reaction (atm)	0.708		0.283
change (atm)	−0.708		+2(0.708)
after reaction (atm)	0		0.283 + 2(0.708) = 1.70 atm

$PV = nRT$, $25°C = 298 \text{ K}$

$$n_{NO_2} = \frac{PV}{RT} = \frac{(1.70 \text{ atm})(0.5000 \text{ L})}{\left(0.082\ 06\ \dfrac{L \cdot atm}{K \cdot mol}\right)(298 \text{ K})} = 0.0348 \text{ mol } NO_2$$

(a)

	$2 NO_2(aq)$	$+ 2 H_2O(l)$	$\rightarrow$	$HNO_2(aq)$	$+ H_3O^+(aq)$	$+ NO_3^-(aq)$
before reaction (mol)	0.0348			0	~0	0
change (mol)	− 0.0348			+0.0348/2	+0.0348/2	+0.0348/2
after reaction (mol)	0			0.0174	0.0174	0.0174

(b) $[HNO_2] = [H_3O^+(aq)] = 0.0174 \text{ mol}/0.250 \text{ L} = 0.0696 \text{ M}$

	$HNO_2(aq)$	$+ H_2O(l)$	$\rightleftharpoons$	$H_3O^+(aq)$	$+ NO_2^-(aq)$
initial (M)	0.0696			0.0696	0
change (M)	−x			+x	+x
equil (M)	0.0696 − x			0.0696 + x	x

$$K_a = \frac{[H_3O^+][NO_2^-]}{HNO_2} = 4.5 \times 10^{-4} = \frac{(0.0696 + x)x}{(0.0696 - x)}$$

$$x^2 + 0.07005 - (3.132 \times 10^{-5}) = 0$$

Use the quadratic formula to solve for x.

$$x = \frac{-(0.07005) \pm \sqrt{(0.07005)^2 - (4)(1)(-3.132 \times 10^{-5})}}{2(1)} = \frac{-0.07005 \pm 0.07094}{2}$$

$x = -0.0705$ and 4.45×10^{-4}

Of the two solutions for x only 4.45×10^{-4} has physical meaning because -0.0705 leads to negative concentrations.

$[NO_2^-] = x = 4.45 \times 10^{-4}$ M $= 4.4 \times 10^{-4}$ M

$[H_3O^+] = 0.0696 + x = 0.0696 + 4.45 \times 10^{-4} = 0.0700$ M

$pH = -\log[H_3O^+] = -\log(0.0700) = 1.15$

(c) Total solution molarity $= [NO_3^-] + [NO_2^-] + [H_3O^+] + [HNO_2]$
$= 0.0696$ M $+ (4.45 \times 10^{-4}) + 0.0700$ M $+ 0.0692$ M $= 0.2092$ M

$$\Pi = MRT = (0.2092 \text{ M})\left(0.082\,06\,\frac{L \cdot atm}{K \cdot mol}\right)(298 \text{ K}) = 5.12 \text{ atm}$$

(d) mol $H_3O^+ = (0.0700 \text{ mol/L})(0.250 \text{ L}) = 0.0175$ mol H_3O^+

mol $HNO_2 = (0.0692 \text{ mol/L})(0.250 \text{ L}) = 0.0173$ mol HNO_2

Total mol of acid to neutralize $= 0.0175 + 0.0173 = 0.0348$ mol

$$\text{mass CaO} = 0.0348 \text{ mol acid} \times \frac{1 \text{ mol CaO}}{2 \text{ mol acid}} \times \frac{56.08 \text{ g CaO}}{1 \text{ mol CaO}} = 0.976 \text{ g CaO}$$

20   Transition Elements and Coordination Chemistry

20.1   (a) V, [Ar] $3d^3 4s^2$ (b) Co^{2+}, [Ar] $3d^7$
      (c) Mn^{4+} in MnO_2, [Ar] $3d^3$ (d) Cu^{2+} in $CuCl_4^{2-}$, [Ar] $3d^9$

20.2

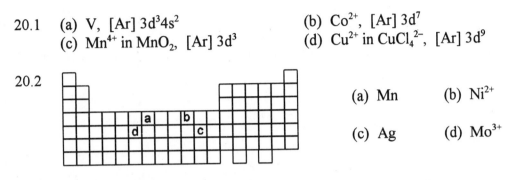

          (a) Mn   (b) Ni^{2+}

          (c) Ag   (d) Mo^{3+}

20.3 Z_{eff} increases from left to right across the first transition series.
      (a) The transition metal with the lowest Z_{eff} (Ti) should be the strongest reducing agent because it is easier for Ti to lose its valence electrons. The transition metal with the highest Z_{eff} (Zn) should be the weakest reducing agent because it is more difficult for Zn to lose its valence electrons.
      (b) The oxoanion with the highest Z_{eff} (FeO_4^{2-}) should be the strongest oxidizing agent because of the greater attraction for electrons. The oxoanion with the lowest Z_{eff} (VO_4^{3-}) should be the weakest oxidizing agent because of the lower attraction for electrons.

20.4 $[Cr(NH_3)_2(SCN)_4]^-$

20.5   In $Na_4[Fe(CN)_6]$ each sodium is in the +1 oxidation state (+4 total); each cyanide (CN^-) has a −1 charge (−6 total). The compound is neutral; therefore, the oxidation state of the iron is +2.

20.6   (a) tetraamminecopper(II) sulfate   (b) sodium tetrahydroxochromate(III)
      (c) triglycinatocobalt(III)   (d) pentaaquaisothiocyanatoiron(III) ion

20.7   (a) $[Zn(NH_3)_4](NO_3)_2$ (b) $Ni(CO)_4$ (c) $K[Pt(NH_3)Cl_3]$ (d) $[Au(CN)_2]^-$

20.8   (a) Two diastereoisomers are possible.

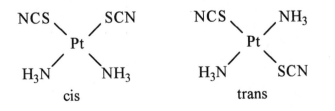

         cis   trans

      (b) No isomers are possible for a tetrahedral complex of the type MA_2B_2.

(c) Two diastereoisomers are possible.

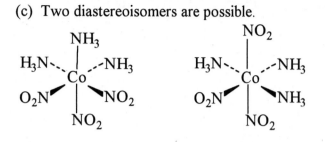

(d) No isomers are possible for a complex of this type.

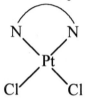

(e) Two diastereoisomers are possible.

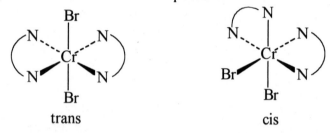

trans   cis

(f) No diastereoisomers are possible for a complex of this type.

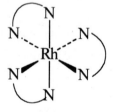

20.9   (1) and (2) are the same. (3) and (4) are the same. (1) and (2) are different from (3) and (4).

20.10 (a) chair, no   (b) foot, yes   (c) pencil, no
       (d) corkscrew, yes   (e) banana, no   (f) football, no

20.11 (a) (2) and (3) are chiral and (1) and (4) are achiral.
       (b) enantiomer of (2)   enantiomer of (3)

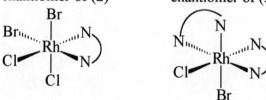

20.12 (a) $[Fe(C_2O_4)_3]^{3-}$ can exist as enantiomers.

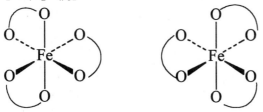

(b) $[Co(NH_3)_4en]^{3+}$ cannot exist as enantiomers.
(c) $[Co(NH_3)_2(en)_2]^{3+}$ can exist as enantiomers.

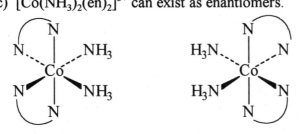

(d) $[Cr(H_2O)_4Cl_2]^+$ cannot exist as enantiomers.

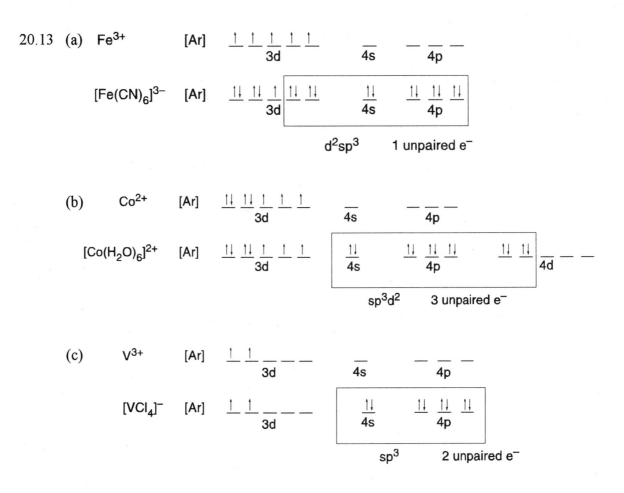

20.13 (a) Fe^{3+} [Ar] 3d 4s 4p

$[Fe(CN)_6]^{3-}$ [Ar] 3d 4s 4p

d^2sp^3 1 unpaired e^-

(b) Co^{2+} [Ar] 3d 4s 4p

$[Co(H_2O)_6]^{2+}$ [Ar] 3d 4s 4p 4d

sp^3d^2 3 unpaired e^-

(c) V^{3+} [Ar] 3d 4s 4p

$[VCl_4]^-$ [Ar] 3d 4s 4p

sp^3 2 unpaired e^-

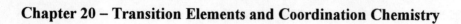

(d)   Pt²⁺   [Xe] 5d 6s 6p

[PtCl₄]²⁻ [Xe] 5d 6s 6p

dsp²   no unpaired e⁻

20.14

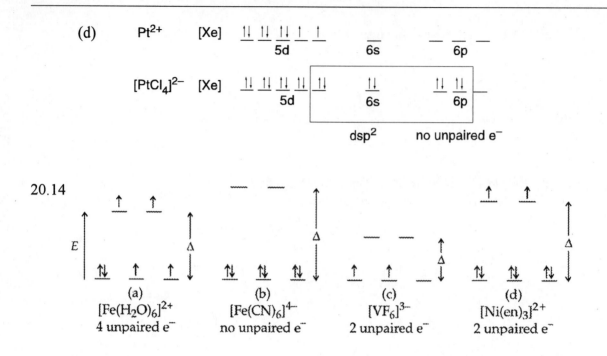

(a)	(b)	(c)	(d)
$[Fe(H_2O)_6]^{2+}$	$[Fe(CN)_6]^{4-}$	$[VF_6]^{3-}$	$[Ni(en)_3]^{2+}$
4 unpaired e⁻	no unpaired e⁻	2 unpaired e⁻	2 unpaired e⁻

20.15 Both $[NiCl_4]^{2-}$ and $[Ni(CN)_4]^{2-}$ contain Ni^{2+} with a [Ar] $3d^8$ electron configuration.

(a) $[NiCl_4]^{2-}$ (tetrahedral)

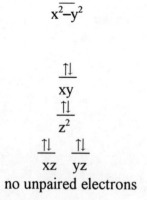

$\underset{xy}{\uparrow\downarrow}\ \underset{xz}{\uparrow}\ \underset{yz}{\uparrow}$

$\underset{z^2}{\uparrow\downarrow}\ \underset{x^2-y^2}{\uparrow\downarrow}$

2 unpaired electrons

(b) $[Ni(CN)_4]^{2-}$ (square planar)

$\overline{}$
x^2-y^2

$\underset{xy}{\uparrow\downarrow}$
$\underset{z^2}{\uparrow\downarrow}$
$\underset{xz}{\uparrow\downarrow}\ \underset{yz}{\uparrow\downarrow}$

no unpaired electrons

20.16 Although titanium has a large positive $E°$ for oxidation, the bulk metal is remarkably immune to corrosion because its surface becomes coated with a thin, protective oxide film.

20.17 Actual yield $= 1.00 \times 10^5$ tons Ti $\times \dfrac{2000 \text{ lbs}}{1 \text{ ton}} \times \dfrac{453.6 \text{ g}}{1 \text{ lb}} = 9.07 \times 10^{10}$ g Ti

Theoretical yield $= \dfrac{9.07 \times 10^{10} \text{ g Ti}}{0.935} = 9.70 \times 10^{10}$ g Ti

mol Ti $= 9.70 \times 10^{10}$ g Ti $\times \dfrac{1 \text{ mol Ti}}{47.88 \text{ g Ti}} = 2.03 \times 10^9$ mol Ti

mol $Cl_2 = 2.03 \times 10^9$ mol Ti $\times \dfrac{2 \text{ mol } Cl_2}{1 \text{ mol Ti}} = 4.06 \times 10^9$ mol Cl_2

$20°C = 293$ K
$PV = nRT$

$$V = \dfrac{nRT}{P} = \dfrac{(4.06 \times 10^9 \text{ mol})\left(0.082\ 06\ \dfrac{L \cdot atm}{K \cdot mol}\right)(293 \text{ K})}{\left(740 \text{ mm Hg} \times \dfrac{1.00 \text{ atm}}{760 \text{ mm Hg}}\right)} = 1.00 \times 10^{11} \text{ L of } Cl_2$$

Understanding Key Concepts

20.18

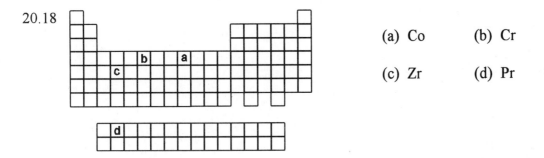

(a) Co   (b) Cr

(c) Zr   (d) Pr

20.20 (a) NH_2–CH_2–CH_2–NH_2 is a bidentate ligand. It can form a chelate ring using the atoms indicated in bold.
(b) CH_3–CH_2–CH_2–NH_2 is a monodentate ligand.
(c) NH_2–CH_2–CH_2–NH–CH_2–CO_2^- is a tridentate ligand. It can form chelate rings using the atoms indicated in bold.
(d) NH_2–CH_2–CH_2–NH_3^+ is a monodentate ligand. The first N can coordinate to a metal.

20.22. (a) (1) cis;   (2) trans;   (3) trans;   (4) cis
(b) (1) and (4) are the same.   (2) and (3) are the same.
(c) None of the isomers exist as enantiomers because their mirror images are identical.

20.24

Additional Problems
Electron Configurations and Properties of Transition Elements

20.26 (a) Cr, [Ar] $3d^5 4s^1$ (b) Zr, [Kr] $4d^2 5s^2$ (c) Co^{2+}, [Ar] $3d^7$
       (d) Fe^{3+}, [Ar] $3d^5$ (e) Mo^{3+}, [Kr] $4d^3$ (f) Cr(VI), [Ar] $3d^0$

20.28 (a) Cu^{2+},   [Ar] $3d^9$ ⇅ ⇅ ⇅ ⇅ ↑   1 unpaired e^-
                           3d

      (b) Ti^{2+},   [Ar] $3d^2$ ↑ ↑ _ _ _   2 unpaired e^-
                           3d

      (c) Zn^{2+},   [Ar] $3d^{10}$ ⇅ ⇅ ⇅ ⇅ ⇅   no unpaired e^-
                           3d

      (d) Cr^{3+},   [Ar] $3d^3$ ↑ ↑ ↑ _ _   3 unpaired e^-
                           3d

20.30 Ti is harder than K and Ca largely because the sharing of d, as well as s, electrons results in stronger metallic bonding.

20.32 (a) The decrease in radii with increasing atomic number is expected because the added d electrons only partially shield the added nuclear charge. As a result, Z_{eff} increases. With increasing Z_{eff}, the electrons are more strongly attracted to the nucleus, and atomic size decreases.
      (b) The densities of the transition metals are inversely related to their atomic radii.

20.34 The smaller than expected sizes of the third-transition series atoms are associated with what is called the lanthanide contraction, the general decrease in atomic radii of the f-block lanthanide elements. The lanthanide contraction is due to the increase in Z_{eff} as the 4f subshell is filled.

20.36   Sc $(631 + 1235) = 1866$ kJ/mol
        Ti $(658 + 1310) = 1968$ kJ/mol
        V $(650 + 1414) = 2064$ kJ/mol
        Cr $(653 + 1592) = 2225$ kJ/mol
        Mn $(717 + 1509) = 2226$ kJ/mol
        Fe $(759 + 1561) = 2320$ kJ/mol
        Co $(758 + 1646) = 2404$ kJ/mol
        Ni $(737 + 1753) = 2490$ kJ/mol
        Cu $(745 + 1958) = 2703$ kJ/mol
        Zn $(906 + 1733) = 2639$ kJ/mol

Across the first transition element series, Z_{eff} increases and there is an almost linear increase in the sum of the first two ionization energies. This is what is expected if the two electrons are removed from the 4s orbital. Higher than expected values for the sum of the first two ionization energies are observed for Cr and Cu because of their anomalous electron configurations (Cr $3d^5 4s^1$; Cu $3d^{10} 4s^1$). An increasing Z_{eff} affects 3d orbitals more than the 4s orbital and the second ionization energy for an electron from the 3d orbital is higher than expected.

20.38   (a) $Cr(s) + 2 H^+(aq) \rightarrow Cr^{2+}(aq) + H_2(g)$
        (b) $Zn(s) + 2 H^+(aq) \rightarrow Zn^{2+}(aq) + H_2(g)$
        (c) N.R.
        (d) $Fe(s) + 2 H^+(aq) \rightarrow Fe^{2+}(aq) + H_2(g)$

Oxidation States

20.40   (b) Mn   (d) Cu

20.42   Sc(III), Ti(IV), V(V), Cr(VI), Mn(VII), Fe(VI), Co(III), Ni(II), Cu(II), Zn(II)

20.44 Cu^{2+} is a stronger oxidizing agent than Cr^{2+} because of a higher Z_{eff}.

20.46 Cr^{2+} is more easily oxidized than Ni^{2+} because of a smaller Z_{eff}.

20.48 $Mn^{2+} < MnO_2 < MnO_4^-$ because of increasing oxidation state of Mn.

Chemistry of Selected Transition Elements

20.50   (a) $Cr_2O_3(s) + 2 Al(s) \rightarrow 2 Cr(s) + Al_2O_3(s)$
        (b) $Cu_2S(l) + O_2(g) \rightarrow 2 Cu(l) + SO_2(g)$

20.52   (a) $Cr_2O_7^{2-}$ (b) Cr^{3+} (c) Cr^{2+} (d) Fe^{2+} (e) Cu^{2+}

20.54 $Cr(OH)_2 < Cr(OH)_3 < CrO_2(OH)_2$
        Acid strength increases with polarity of the O–H bond, which increases in turn with the oxidation state of Cr.

20.56 (c) $Cr(OH)_3$

20.58 (a) Add excess $KOH(aq)$ and Fe^{3+} will precipitate as $Fe(OH)_3(s)$. $Na^+(aq)$ will remain in solution.
(b) Add excess $NaOH(aq)$ and Fe^{3+} will precipitate as $Fe(OH)_3(s)$. $Cr(OH)_4^-(aq)$ will remain in solution.
(c) Add excess $NH_3(aq)$ and Fe^{3+} will precipitate as $Fe(OH)_3(s)$. $Cu(NH_3)_4^{2+}(aq)$ will remain in solution.

20.60 (a) $Cr_2O_7^{2-}(aq) + 6\ Fe^{2+}(aq) + 14\ H^+(aq) \rightarrow 2\ Cr^{3+}(aq) + 6\ Fe^{3+}(aq) + 7\ H_2O(l)$
(b) $4\ Fe^{2+}(aq) + O_2(g) + 4\ H^+(aq) \rightarrow 4\ Fe^{3+}(aq) + 2\ H_2O(l)$
(c) $Cu_2O(s) + 2\ H^+(aq) \rightarrow Cu(s) + Cu^{2+}(aq) + H_2O(l)$
(d) $Fe(s) + 2\ H^+(aq) \rightarrow Fe^{2+}(aq) + H_2(g)$

20.62 (a) $2\ CrO_4^{2-}(aq) + 2\ H_3O^+(aq) \rightarrow Cr_2O_7^{2-}(aq) + 3\ H_2O(l)$
　　(yellow)　　　　　　　　　　(orange)
(b) $[Fe(H_2O)_6]^{3+}(aq) + SCN^-(aq) \rightarrow [Fe(H_2O)_5(SCN)]^{2+}(aq) + H_2O(l)$
　　　　　　　　　　　　　　　　(red)
(c) $3\ Cu(s) + 2\ NO_3^-(aq) + 8\ H^+(aq) \rightarrow 3\ Cu^{2+}(aq) + 2\ NO(g) + 4\ H_2O(l)$
　　　　　　　　　　　　　　(blue)
(d) $Cr(OH)_3(s) + OH^-(aq) \rightarrow Cr(OH)_4^-(aq)$
　　$2\ Cr(OH)_4^-(aq) + 3\ HO_2^-(aq) \rightarrow 2\ CrO_4^{2-}(aq) + 5\ H_2O(l) + OH^-(aq)$
　　　　　　　　　　　　(yellow)

Coordination Compounds; Ligands

20.64 Ni^{2+} accepts six pairs of electrons, two each from the three ethylenediamine ligands. Ni^{2+} is an electron pair acceptor, a Lewis acid. The two nitrogens in each ethylenediamine donate a pair of electrons to the Ni^{2+}. The ethylenediamine is an electron pair donor, a Lewis base. The formation of $[Ni(en)_3]^{2+}$ is a Lewis acid-base reaction.

20.66 (a) $[Ag(NH_3)_2]^+$　　　(b) $[Ni(CN)_4]^{2-}$　　　(c) $[Cr(H_2O)_6]^{3+}$

20.68 (a) $AgCl_2^-$
2 Cl^-
The oxidation state of the Ag is +1.

(b) $[Cr(H_2O)_5Cl]^{2+}$
$4\ H_2O$ (no charge)　　　1 Cl^-
The oxidation state of the Cr is +3.

(c) $[Co(NCS)_4]^{2-}$
4 NCS^-
The oxidation state of the Co is +2.

(d) $[ZrF_8]^{4-}$
8 F^-
The oxidation state of the Zr is +4.

(e) $Co(NH_3)_3(NO_2)_3$
3 NH_3 (no charge)   3 NO_2^-
The oxidation state of the Co is +3.

20.70 (a) $Ni(CO)_4$ (b) $[Ag(NH_3)_2]^+$ (c) $[Fe(CN)_6]^{3-}$ (d) $[Ni(CN)_4]^{2-}$

20.72

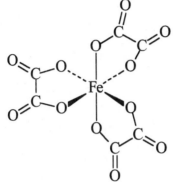

The iron is in the +3 oxidation state, and the coordination number is six. The geometry about the Fe is octahedral. The oxalate ligand is behaving as a bidentate chelating ligand. There are three chelate rings, one formed by each oxalate ligand.

20.74 (a) $Co(NH_3)_3(NO_2)_3$
3 NH_3 (no charge)   3 NO_2^-
The oxidation state of the Co is +3.

(b) $[Ag(NH_3)_2]NO_3$
2 NH_3 (no charge)   1 NO_3^-
The oxidation state of the Ag is +1

(c) $K_3[Cr(C_2O_4)_2Cl_2]$
3 K^+ 2 $C_2O_4^{2-}$ 2 Cl^-
The oxidation state of the Cr is +3.

(d) $Cs[CuCl_2]$
1 Cs^+ 2 Cl^-
The oxidation state of the Cu is +1

Naming Coordination Compounds

20.76 (a) tetrachloromanganate(II)   (b) hexaamminenickel(II)
(c) tricarbonatocobaltate(III)   (d) bis(ethylenediamine)dithiocyanatoplatinum(IV)

20.78 (a) cesium tetrachloroferrate(III)   (b) hexaaquavanadium(III) nitrate
(c) tetraamminedibromocobalt(III) bromide   (d) diglycinatocopper(II)

20.80   (a) $[Pt(NH_3)_4]Cl_2$ (b) $Na_3[Fe(CN)_6]$

        (c) $[Pt(en)_3](SO_4)_2$ (d) $Rh(NH_3)_3(SCN)_3$

Isomers

20.82

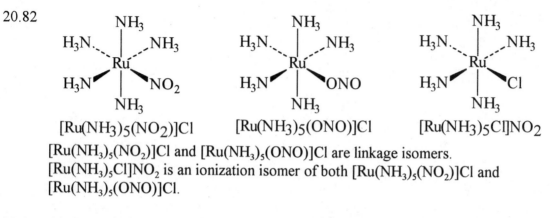

$[Ru(NH_3)_5(NO_2)]Cl$ $[Ru(NH_3)_5(ONO)]Cl$ $[Ru(NH_3)_5Cl]NO_2$

$[Ru(NH_3)_5(NO_2)]Cl$ and $[Ru(NH_3)_5(ONO)]Cl$ are linkage isomers.
$[Ru(NH_3)_5Cl]NO_2$ is an ionization isomer of both $[Ru(NH_3)_5(NO_2)]Cl$ and
$[Ru(NH_3)_5(ONO)]Cl$.

20.84   (a) $[Cr(NH_3)_2Cl_4]^-$ can exist as cis and trans diastereoisomers.

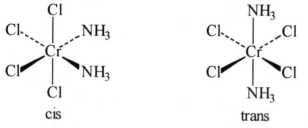

            cis   trans

      (b) $[Co(NH_3)_5Br]^{2+}$ cannot exist as diastereoisomers.

      (c) $[FeCl_2(NCS)_2]^{2-}$ (tetrahedral) cannot exist as diastereoisomers.

      (d) $[PtCl_2Br_2]^{2-}$ (square planar) can exist as cis and trans diastereoisomers.

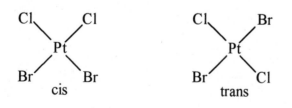

            cis   trans

20.86   (c) cis-$[Cr(en)_2(H_2O)_2]^{3+}$ (d) $[Cr(C_2O_4)_3]^{3-}$

20.88   (a) $Ru(NH_3)_4Cl_2$ can exist as cis and trans diastereoisomers.

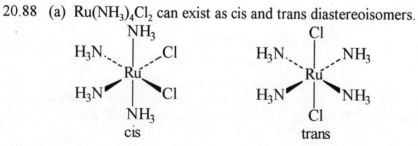

            cis   trans

(b) $[Pt(en)_3]^{4+}$ can exist as enantiomers.

(c) $[Pt(en)_2ClBr]^{2+}$ can exist as both diastereoisomers and enantiomers.

diastereoisomers

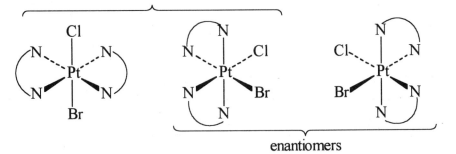

enantiomers

20.90 Plane-polarized light is light in which the electric vibrations of the light wave are restricted to a single plane. The following chromium complex can rotate the plane of plane-polarized light.

$[Cr(en)_3]^{3+}$

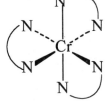

Color of Complexes; Valence Bond and Crystal Field Theories

20.92 The measure of the amount of light absorbed by a substance is called the absorbance, and a graph of absorbance versus wavelength is called an absorption spectrum.
If a complex absorbs at 455 nm, its color is orange (use the color wheel in Figure 20.26).

20.94 (a) $[Ti(H_2O)_6]^{3+}$

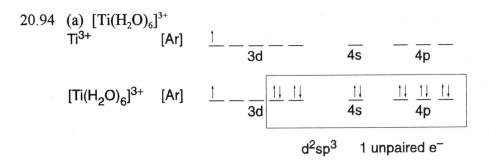

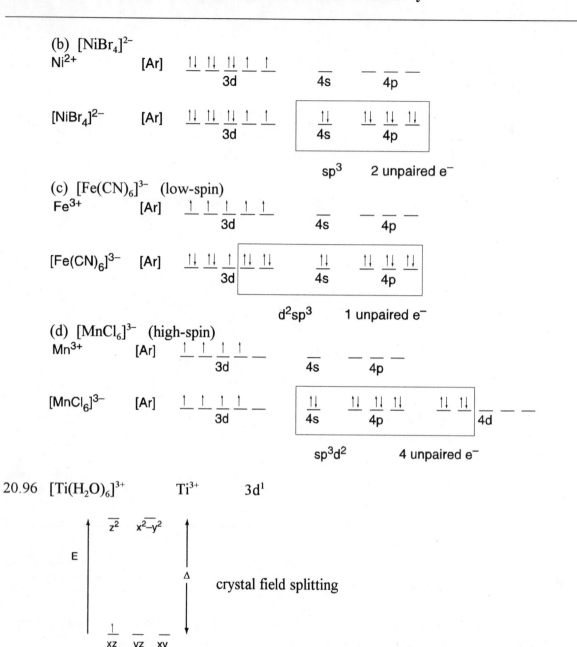

(b) $[NiBr_4]^{2-}$

Ni^{2+} [Ar] 3d 4s 4p

$[NiBr_4]^{2-}$ [Ar] 3d 4s 4p

sp^3 2 unpaired e$^-$

(c) $[Fe(CN)_6]^{3-}$ (low-spin)

Fe^{3+} [Ar] 3d 4s 4p

$[Fe(CN)_6]^{3-}$ [Ar] 3d 4s 4p

d^{2}sp^3 1 unpaired e$^-$

(d) $[MnCl_6]^{3-}$ (high-spin)

Mn^{3+} [Ar] 3d 4s 4p

$[MnCl_6]^{3-}$ [Ar] 3d 4s 4p 4d

sp^{3}d^2 4 unpaired e$^-$

20.96 $[Ti(H_2O)_6]^{3+}$ Ti^{3+} 3d^1

E

z^2 x^2–y^2

Δ

crystal field splitting

xz   yz   xy

$[Ti(H_2O)_6]^{3+}$ is colored because it can absorb light in the visible region, exciting the electron to the higher-energy set of orbitals

20.98 $\lambda = 544$ nm $= 544 \times 10^{-9}$ m

$$\Delta = \frac{hc}{\lambda} = \frac{(6.626 \times 10^{-34} \text{ J} \cdot \text{s})(3.00 \times 10^8 \text{ m/s})}{(544 \times 10^{-9} \text{ m})} = 3.65 \times 10^{-19} \text{ J}$$

$\Delta = (3.65 \times 10^{-19} \text{ J/ion})(6.022 \times 10^{23} \text{ ion/mol}) = 219{,}803$ J/mol $= 220$ kJ/mol

For $[Ti(H_2O)_6]^{3+}$, $\Delta = 240$ kJ/mol

Because $\Delta_{NCS^-} < \Delta_{H_2O}$ for the Ti complex, NCS$^-$ is a weaker-field ligand than H$_2$O. If

$[Ti(NCS)_6]^{3-}$ absorbs at 544 nm, its color should be red (use the color wheel in Figure 20.26).

20.100 (a) $[CrF_6]^{3-}$ (b) $[V(H_2O)_6]^{3+}$ (c) $[Fe(CN)_6]^{3-}$

$\underline{\quad}\ \underline{\quad}$ $\underline{\quad}\ \underline{\quad}$ $\underline{\quad}\ \underline{\quad}$

$\underline{\uparrow}\ \underline{\uparrow}\ \underline{\uparrow}$
3 unpaired e⁻

$\underline{\uparrow}\ \underline{\uparrow}\ \underline{\quad}$
2 unpaired e⁻

$\underline{\uparrow\downarrow}\ \underline{\uparrow\downarrow}\ \underline{\uparrow}$
1 unpaired e⁻

20.102 $Ni^{2+}(aq)$ $Zn^{2+}(aq)$

$\underline{\uparrow}\ \underline{\uparrow}$ $\underline{\uparrow\downarrow}\ \underline{\uparrow\downarrow}$

$\underline{\uparrow\downarrow}\ \underline{\uparrow\downarrow}\ \underline{\uparrow\downarrow}$ $\underline{\uparrow\downarrow}\ \underline{\uparrow\downarrow}\ \underline{\uparrow\downarrow}$

$Ni^{2+}(aq)$ is green because the Ni^{2+} ion can absorb light, which promotes electrons from the filled d orbitals to the higher energy half-filled d orbitals. $Zn^{2+}(aq)$ is colorless because the d orbitals are completely filled and no electrons can be promoted, so no light is absorbed.

20.104 Weak-field ligands produce a small Δ. Strong-field ligands produce a large Δ. For a metal complex with weak-field ligands, $\Delta < P$, where P is the pairing energy, and it is easier to place an electron in either d_{z^2} or $d_{x^2-y^2}$ than to pair up electrons; high-spin complexes result. For a metal complex with strong-field ligands, $\Delta > P$ and it is easier to pair up electrons than to place them in either d_{z^2} or $d_{x^2-y^2}$; low-spin complexes result.

20.106 $\underline{\quad}\ x^2-y^2$

$\underline{\uparrow\downarrow}\ xy$

$\underline{\uparrow\downarrow}\ z^2$

$\underline{\uparrow\downarrow}\ \underline{\uparrow\downarrow}\ xz, yz$

Square planar geometry is most common for metal ions with d^8 configurations because this configuration favors low-spin complexes in which all four lower energy d orbitals are filled and the higher energy $d_{x^2-y^2}$ orbital is vacant.

General Problems

20.108 (a) $[Mn(CN)_6]^{3-}$ $\qquad$ Mn^{3+} $\quad$ [Ar] $3d^4$
$\quad$ CN^- is a strong-field ligand. The Mn^{3+} complex is low-spin.

$\underline{-}$ $\underline{-}$

$\underline{\uparrow\downarrow}$ $\underline{\uparrow}$ $\underline{\uparrow}$ $\quad$ 2 unpaired e^-, paramagnetic

(b) $[Zn(NH_3)_4]^{2+}$ $\qquad$ Zn^{2+} $\quad$ [Ar] $3d^{10}$
$[Zn(NH_3)_4]^{2+}$ is tetrahedral.

$\underline{\uparrow\downarrow}$ $\underline{\uparrow\downarrow}$ $\underline{\uparrow\downarrow}$

$\underline{\uparrow\downarrow}$ $\underline{\uparrow\downarrow}$ $\quad$ no unpaired e^-, diamagnetic

(c) $[Fe(CN)_6]^{4-}$ $\qquad$ Fe^{2+} $\quad$ [Ar] $3d^6$
$\quad$ CN^- is a strong-field ligand. The Fe^{2+} complex is low-spin.

$\underline{-}$ $\underline{-}$

$\underline{\uparrow\downarrow}$ $\underline{\uparrow\downarrow}$ $\underline{\uparrow\downarrow}$ $\quad$ no unpaired e^-, diamagnetic

(d) $[FeF_6]^{4-}$ $\qquad$ Fe^{2+} $\quad$ [Ar] $3d^6$
F^- is a weak-field ligand. The Fe^{2+} complex is high-spin.

$\underline{\uparrow}$ $\underline{\uparrow}$

$\underline{\uparrow\downarrow}$ $\underline{\uparrow}$ $\underline{\uparrow}$ $\quad$ 4 unpaired e^-, paramagnetic

20.110 (a) $4 [Co^{3+}(aq) + e^- \rightarrow Co^{2+}(aq)]$
$\quad$ $\underline{2 H_2O(l) \rightarrow O_2(g) + 4 H^+(aq) + 4 e^-}$
$\quad$ $4 Co^{3+}(aq) + 2 H_2O(l) \rightarrow 4 Co^{2+}(aq) + O_2(g) + 4 H^+(aq)$

(b) $4 Cr^{2+}(aq) + O_2(g) + 4 H^+(aq) \rightarrow 4 Cr^{3+}(aq) + 2 H_2O(l)$

(c) $3 [Cu(s) \rightarrow Cu^{2+}(aq) + 2 e^-]$
$\quad$ $\underline{Cr_2O_7^{2-}(aq) + 14 H^+(aq) + 6 e^- \rightarrow 2 Cr^{3+}(aq) + 7 H_2O(l)}$
$\quad$ $3 Cu(s) + Cr_2O_7^{2-}(aq) + 14 H^+(aq) \rightarrow 3 Cu^{2+}(aq) + 2 Cr^{3+}(aq) + 7 H_2O(l)$

(d) $2 CrO_4^{2-}(aq) + 2 H^+(aq) \rightarrow Cr_2O_7^{2-}(aq) + H_2O(l)$

20.112 $\quad$ mol Fe^{2+} = (0.1000 L)(0.400 mol/L) = 0.0400 mol Fe^{2+}
$\quad$ mol $Cr_2O_7^{2-}$ = (0.1000 L)(0.100 mol/L) = 0.0100 mol $Cr_2O_7^{2-}$
$\quad$ $6 [Fe^{2+}(aq) \rightarrow Fe^{3+}(aq) + e^-]$
$\quad$ $\underline{Cr_2O_7^{2-}(aq) + 14 H^+(aq) + 6 e^- \rightarrow 2 Cr^{3+}(aq) + 7 H_2O(l)}$
$\quad$ $6 Fe^{2+}(aq) + Cr_2O_7^{2-}(aq) + 14 H^+(aq) \rightarrow 6 Fe^{3+}(aq) + 2 Cr^{3+}(aq) + 7 H_2O(l)$

	$\underline{Fe^{2+}}$	$\underline{Cr_2O_7^{2-}}$	$\underline{Fe^{3+}}$	$\underline{Cr^{3+}}$
initial (mol)	0.0400	0.0100	0	0
change (mol)	−0.0400	$-\dfrac{0.0400}{6}$	+0.0400	$+\dfrac{0.0400}{3}$
after (mol)	0	0.00333	0.0400	0.0133

$$[Fe^{3+}] = \frac{0.0400 \text{ mol}}{0.2000 \text{ L}} = 0.200 \text{ M} \qquad [Cr^{3+}] = \frac{0.0133 \text{ mol}}{0.200 \text{ L}} = 0.0665 \text{ M}$$

$$[Cr_2O_7^{2-}] = \frac{0.00333 \text{ mol}}{0.200 \text{ L}} = 0.0166 \text{ M}$$

20.114 (a) sodium aquabromodioxalatoplatinate(IV)
　　　　(b) hexaamminechromium(III) trioxalatocobaltate(III)
　　　　(c) hexaamminecobalt(III) trioxalatochromate(III)
　　　　(d) diamminebis(ethylenediamine)rhodium(III) sulfate

20.116　Cl^- is a weak-field ligand, whereas CN^- is a strong-field ligand. Δ for $[Fe(CN)_6]^{3-}$ is larger than the pairing energy P; Δ for $[FeCl_6]^{3-}$ is smaller than P. Fe^{3+} has a $3d^5$ electron configuration.

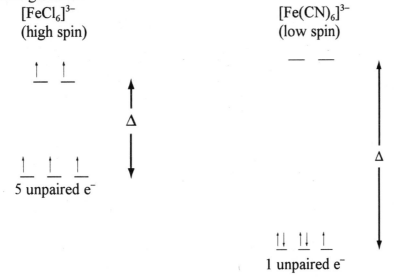

$[FeCl_6]^{3-}$
(high spin)

$[Fe(CN)_6]^{3-}$
(low spin)

5 unpaired e^-

1 unpaired e^-

Because of the difference in Δ, $[FeCl_6]^{3-}$ is high-spin with five unpaired electrons, whereas $[Fe(CN)_6]^{3-}$ is low-spin with only one unpaired electron.

20.118　A choice between high-spin and low-spin electron configurations arises only for complexes of metal ions with four to seven d electrons, the so-called d^4-d^7 complexes. For d^1-d^3 and d^8-d^{10} complexes, only one ground-state electron configuration is possible. In d^1-d^3 complexes, all the electrons occupy the lower-energy d orbitals, independent of the value of Δ. In d^8-d^{10} complexes, the lower-energy set of d orbitals is filled with three pairs of electrons, while the higher-energy set contains two, three, or four electrons, again independent of the value of Δ.

$$\underline{-}\ \underline{-}\qquad\qquad \underline{-}\ \underline{-}\qquad\qquad \underline{-}\ \underline{-}$$

$$\underline{\uparrow}\ \underline{-}\ \underline{-}\qquad \underline{\uparrow}\ \underline{\uparrow}\ \underline{-}\qquad \underline{\uparrow}\ \underline{\uparrow}\ \underline{\uparrow}$$
$$d^1\qquad\qquad\quad d^2\qquad\qquad\quad d^3$$

$$\underline{\uparrow}\ \underline{-}\qquad\qquad \underline{-}\ \underline{-}\qquad\qquad \underline{\uparrow}\ \underline{\uparrow}\qquad\qquad \underline{-}\ \underline{-}$$

$$\underline{\uparrow}\ \underline{\uparrow}\ \underline{\uparrow}$$
d^4 high-spin $\quad\underline{\uparrow\downarrow}\ \underline{\uparrow}\ \underline{\uparrow}$ $\qquad\underline{\uparrow}\ \underline{\uparrow}\ \underline{\uparrow}$
$\qquad\qquad\qquad d^4$ low-spin $\qquad\ d^5$ high-spin $\quad\underline{\uparrow\downarrow}\ \underline{\uparrow\downarrow}\ \underline{\uparrow}$
$\qquad\qquad\qquad\qquad\qquad\qquad\qquad\qquad\qquad\qquad d^5$ low-spin

---

$$\underline{\uparrow}\ \underline{\uparrow}\qquad\qquad \underline{-}\ \underline{-}\qquad\qquad \underline{\uparrow}\ \underline{\uparrow}\qquad\qquad \underline{\uparrow}\ \underline{-}$$

$$\underline{\uparrow\downarrow}\ \underline{\uparrow}\ \underline{\uparrow}$$
d^6 high-spin $\quad\underline{\uparrow\downarrow}\ \underline{\uparrow\downarrow}\ \underline{\uparrow\downarrow}$ $\qquad\underline{\uparrow\downarrow}\ \underline{\uparrow\downarrow}\ \underline{\uparrow}$
$\qquad\qquad\qquad d^6$ low-spin $\qquad\ d^7$ high-spin $\quad\underline{\uparrow\downarrow}\ \underline{\uparrow\downarrow}\ \underline{\uparrow\downarrow}$
$\qquad\qquad\qquad\qquad\qquad\qquad\qquad\qquad\qquad\qquad d^7$ low-spin

---

$$\underline{\uparrow}\ \underline{\uparrow}\qquad\qquad \underline{\uparrow\downarrow}\ \underline{\uparrow}\qquad\qquad \underline{\uparrow\downarrow}\ \underline{\uparrow\downarrow}$$

$$\underline{\uparrow\downarrow}\ \underline{\uparrow\downarrow}\ \underline{\uparrow\downarrow}\qquad \underline{\uparrow\downarrow}\ \underline{\uparrow\downarrow}\ \underline{\uparrow\downarrow}\qquad \underline{\uparrow\downarrow}\ \underline{\uparrow\downarrow}\ \underline{\uparrow\downarrow}$$
$$d^8\qquad\qquad\quad d^9\qquad\qquad\quad d^{10}$$

20.120 $[CoCl_4]^{2-}$ is tetrahedral. $[Co(H_2O)_6]^{2+}$ is octahedral. Because $\Delta_{tet} < \Delta_{oct}$, these complexes have different colors. $[CoCl_4]^{2-}$ has absorption bands at longer wavelengths.

20.122 Co(gly)$_3$

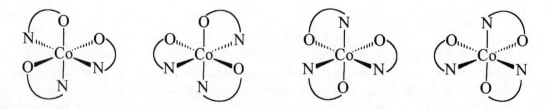

20.124

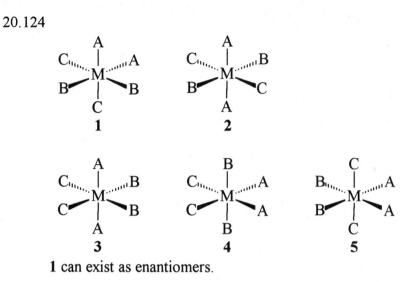

1 can exist as enantiomers.

20.126

	weak-field ligands	strong-field ligands	
Ti^{2+} [Ar] $3d^2$	BM = $\sqrt{2(2+2)}$ = 2.83	BM = $\sqrt{2(2+2)}$ = 2.83	BM cannot distinguish between high-spin and low-spin electron configurations
V^{2+} [Ar] $3d^3$	BM = $\sqrt{3(3+2)}$ = 3.87	BM = $\sqrt{3(3+2)}$ = 3.87	BM cannot distinguish between high-spin and low-spin electron configurations
Cr^{2+} [Ar] $3d^4$	BM = $\sqrt{4(4+2)}$ = 4.90	BM = $\sqrt{2(2+2)}$ = 2.83	BM can distinguish between high-spin and low-spin electron configurations

Mn^{2+} [Ar] $3d^5$	↑ ↑ ↑ ↑ ↑ BM $= \sqrt{5(5+2)} = 5.92$	— — ↑↓ ↑↓ ↑ BM $= \sqrt{1(1+2)} = 1.73$	BM can distinguish between high-spin and low-spin electron configurations
Fe^{2+} [Ar] $3d^6$	↑ ↑ ↑↓ ↑ ↑ BM $= \sqrt{4(4+2)} = 4.90$	— — ↑↓ ↑↓ ↑↓ BM $= 0$	BM can distinguish between high-spin and low-spin electron configurations
Co^{2+} [Ar] $3d^7$	↑ ↑ ↑↓ ↑↓ ↑ BM $= \sqrt{3(3+2)} = 3.87$	↑ — ↑↓ ↑↓ ↑↓ BM $= \sqrt{1(1+2)} = 1.73$	BM can distinguish between high-spin and low-spin electron configurations
Ni^{2+} [Ar] $3d^8$	↑ ↑ ↑↓ ↑↓ ↑↓ BM $= \sqrt{2(2+2)} = 2.83$	↑ ↑ ↑↓ ↑↓ ↑↓ BM $= \sqrt{2(2+2)} = 2.83$	BM cannot distinguish between high-spin and low-spin electron configurations
Cu^{2+} [Ar] $3d^9$	↑↓ ↑ ↑↓ ↑↓ ↑↓ BM $= \sqrt{1(1+2)} = 1.73$	↑↓ ↑ ↑↓ ↑↓ ↑↓ BM $= \sqrt{1(1+2)} = 1.73$	BM cannot distinguish between high-spin and low-spin electron configurations
Zn^{2+} [Ar] $3d^{10}$	↑↓ ↑↓ ↑↓ ↑↓ ↑↓ BM $= 0$	↑↓ ↑↓ ↑↓ ↑↓ ↑↓ BM $= 0$	BM cannot distinguish between high-spin and low-spin electron configurations

20.128 (a)

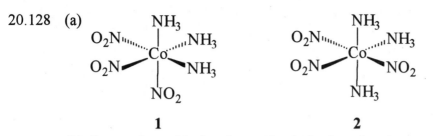

1   **2**

(b) Isomer **2** would give rise to the desired product because it has two trans NO_2 groups.

20.130 (a) $(NH_4)[Cr(H_2O)_6](SO_4)_2$, ammonium hexaaquachromium(III) sulfate
Cr^{3+} __ __

$\underline{\uparrow}$ $\underline{\uparrow}$ $\underline{\uparrow}$
3 unpaired e^-

(b) $Mo(CO)_6$, hexacarbonylmolybdenum(0)
Mo^0 __ __

$\underline{\uparrow\downarrow}$ $\underline{\uparrow\downarrow}$ $\underline{\uparrow\downarrow}$
low-spin, no unpaired e^-

(c) $[Ni(NH_3)_4(H_2O)_2](NO_3)_2$, tetraamminediaquanickel(II) nitrate
Ni^{2+} $\underline{\uparrow}$ $\underline{\uparrow}$

$\underline{\uparrow\downarrow}$ $\underline{\uparrow\downarrow}$ $\underline{\uparrow\downarrow}$
2 unpaired e^-

(d) $K_4[Os(CN)_6]$, potassium hexacyanoosmate(II)
Os^{2+} __ __

$\underline{\uparrow\downarrow}$ $\underline{\uparrow\downarrow}$ $\underline{\uparrow\downarrow}$
low-spin, no unpaired e^-

(e) $[Pt(NH_3)_4](ClO_4)_2$, tetraammineplatinum(II) perchlorate
Pt^{2+} $\underline{\uparrow\downarrow}$

$\underline{\uparrow\downarrow}$

$\underline{\uparrow\downarrow}$ $\underline{\uparrow\downarrow}$
low spin, no unpaired e^-

(f) $Na_2[Fe(CO)_4]$, sodium tetracarbonylferrate(–II)

Fe^{2-} ⇅ ⇅ ⇅

⇅ ⇅

no unpaired e^-

20.132 For transition metal complexes, observed colors and absorbed colors are generally complementary. Using the color wheel (Figure 20.26), the absorbed colors below are complementary colors to those observed.

	Observed Color	Absorbed Color	Approximate λ (nm)
$Cr(acac)_3$	red	green	530
$[Cr(H_2O)_6]^{3+}$	violet	yellow	580
$[CrCl_2(H_2O)_4]^+$	green	red	700
$[Cr(urea)_6]^{3+}$	green	red	700
$[Cr(NH_3)_6]^{3+}$	yellow	violet	420
$Cr(acetate)_3(H_2O)_3$	blue-violet	orange-yellow	600

The magnitude of Δ is comparable to the energy of the absorbed light from the low energy red end to the high energy violet end (ROYGBIV). The red of $[CrCl_2(H_2O)_4]^+$ is lower energy than the yellow of $[Cr(H_2O)_6]^{3+}$, so $Cl^- < H_2O$. Because $[CrCl_2(H_2O)_4]^+$ and $[Cr(urea)_6]^{3+}$ are both red, Δ for 6 urea's is approximately equal to Δ for 2 Cl^-'s and 4 H_2O's. Therefore, urea is between Cl^- and H_2O.

The spectrochemical series is: $Cl^- <$ urea $<$ acetate $< H_2O <$ acac $< NH_3$

Multi-Concept Problems

20.134 (1) $Ni(H_2O)_6^{2+}(aq) + 6\ NH_3(aq) \rightleftharpoons Ni(NH_3)_6^{2+}(aq) + 6\ H_2O(l)$ $K_f = 2.0 \times 10^8$

(2) $Ni(H_2O)_6^{2+}(aq) + 3\ en(aq) \rightleftharpoons Ni(en)_3^{2+}(aq) + 6\ H_2O(l)$ $K_f = 4 \times 10^{17}$

(a) Reaction (2) should have the larger entropy change because three bidentate en ligands displace six water molecules.

(b) $\Delta G° = \Delta H° - T\Delta S°$

Because $\Delta H°_1$ and $\Delta H°_2$ are almost the same, the difference in $\Delta G°$ is determined by the difference in $\Delta S°$. Because $\Delta S°_2$ is larger than $\Delta S°_1$, $\Delta G°_2$ is more negative than $\Delta G°_1$ which is consistent with the greater stability of $Ni(en)_3^{2+}$.

(c) $\Delta H° - T\Delta S° = \Delta G° = -RT \ln K_f$

$\Delta H°_1 - T\Delta S°_1 - (\Delta H°_2 - T\Delta S°_2) = -RT \ln K_f(1) - [-RT \ln K_f(2)]$

$$T\Delta S°_2 - T\Delta S°_1 = RT \ln K_f(2) - RT \ln K_f(1) = RT \ln \frac{K_f(2)}{K_f(1)}$$

$$\Delta S°_2 - \Delta S°_1 = R \ln \frac{K_f(2)}{K_f(1)} = [8.314\ J/(K \cdot mol)] \ln \frac{4 \times 10^{17}}{2.0 \times 10^8}$$

$\Delta S°_2 - \Delta S°_1 = 178\ J/(K \cdot mol)$ or $180\ J/(K \cdot mol)$

20.136 (a) $Cr(s) + 2 H^+(aq) \rightarrow Cr^{2+}(aq) + H_2(g)$

(b) $mol\ Cr = 2.60\ g\ Cr \times \dfrac{1\ mol\ Cr}{52.00\ g\ Cr} = 0.0500\ mol\ Cr$

$mol\ H_2SO_4 = (0.05000\ L)(1.200\ mol/L) = 0.06000\ mol\ H_2SO_4$

The stoichiometry between Cr and H_2SO_4 is one to one, therefore Cr is the limiting reagent because of the smaller number of moles.

$mol\ H_2 = 0.0500\ mol\ Cr \times \dfrac{1\ mol\ H_2}{1\ mol\ Cr} = 0.0500\ mol\ H_2$

$25°C = 298\ K$

$PV = nRT$

$V = \dfrac{nRT}{P} = \dfrac{(0.0500\ mol)\left(0.082\ 06\ \dfrac{L \cdot atm}{K \cdot mol}\right)(298\ K)}{\left(735\ mm\ Hg \times \dfrac{1.00\ atm}{760\ mm\ Hg}\right)} = 1.26\ L\ of\ H_2$

(c) $0.06000\ mol\ H_2SO_4$ can provide $0.1200\ mol\ H^+$. $0.0500\ mol\ Cr$ reacts with $2 \times (0.0500\ mol\ H^+) = 0.100\ mol\ H^+$. This leaves $0.0200\ mol\ H^+$ and $0.0600\ mol\ SO_4^{2-}$, which will give, after neutralization, $0.0200\ mol\ HSO_4^-$ and $0.0400\ mol\ SO_4^{2-}$.

$[HSO_4^-] = 0.0200\ mol/0.05000\ L = 0.400\ M$

$[SO_4^{2-}] = 0.0400\ mol/0.05000\ L = 0.800\ M$

The pH of this solution can be determined from the following equilibrium:

	$HSO_4^-(aq)$	+	$H_2O(l)$	⇌	$H_3O^+(aq)$	+	$SO_4^{2-}(aq)$
initial (M)	0.400				0		0.800
change (M)	−x				+x		+x
equil (M)	0.400 − x				x		0.800 + x

$K_{a2} = \dfrac{[H_3O^+][SO_4^{2-}]}{[HSO_4^-]} = 1.2 \times 10^{-2} = \dfrac{(x)(0.800 + x)}{0.400 - x}$

$x^2 + 0.812x - 0.0048 = 0$

Use the quadratic formula to solve for x.

$x = \dfrac{-(0.812) \pm \sqrt{(0.812)^2 - 4(1)(-0.0048)}}{2(1)} = \dfrac{-0.812 \pm 0.8237}{2}$

$x = 0.00585\ and\ -0.818$

Of the two solutions for x, only the positive value of x has physical meaning, because x is the $[H_3O^+]$.

$[H_3O^+] = x = 0.00585\ M$

$pH = -\log[H_3O^+] = -\log(0.00585) = 2.23$

(d) Crystal field d-orbital energy level diagram

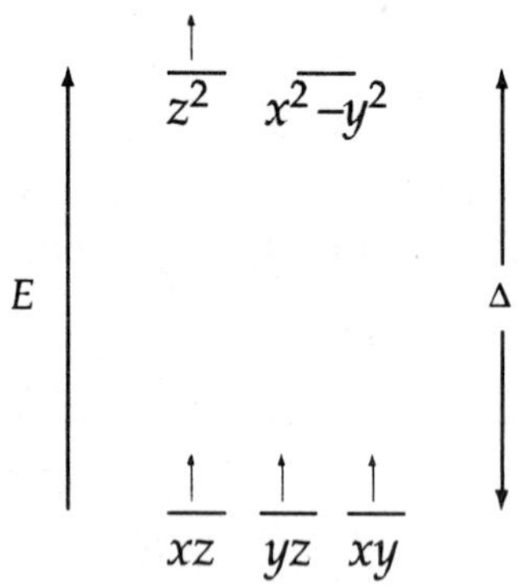

Valence bond orbital diagram

$Cr(H_2O)_6^{2+}$ [Ar] ↑ ↑ ↑ ↑ __ (3d) | ↑↓ (4s) ↑↓ ↑↓ ↑↓ (4p) ↑↓ ↑↓ | __ __ __ (4d)

sp^3d^2 4 unpaired e^-

(e) The addition of excess KCN converts $Cr(H_2O)_6^{2+}(aq)$ to $Cr(CN)_6^{4-}(aq)$. CN^- is a strong field ligand and increases Δ changing the chromium complex from high spin, with 4 unpaired electrons, to low spin, with only 2 unpaired electrons.

20.138 (a) Assume a 100.0 g sample of the chromium compound.

$$19.52 \text{ g Cr} \times \frac{1 \text{ mol Cr}}{51.996 \text{ g Cr}} = 0.3754 \text{ mol Cr}$$

$$39.91 \text{ g Cl} \times \frac{1 \text{ mol Cl}}{35.453 \text{ g Cl}} = 1.126 \text{ mol Cl}$$

$$40.57 \text{ g } H_2O \times \frac{1 \text{ mol } H_2O}{18.015 \text{ g } H_2O} = 2.252 \text{ mol } H_2O$$

$Cr_{0.3754}Cl_{1.126}(H_2O)_{2.252}$, divide each subscript by the smallest, 0.3754.

$Cr_{0.3754\,/\,0.3754}Cl_{1.126\,/\,0.3754}(H_2O)_{2.252\,/\,0.3754}$

$CrCl_3(H_2O)_6$

(b) $Cr(H_2O)_6Cl_3$, 266.45 amu; AgCl, 143.32 amu

For **A**: mol Cr complex = mol Cr = $0.225 \text{ g Cr complex} \times \frac{1 \text{ mol Cr complex}}{266.45 \text{ g Cr complex}} =$

8.44×10^{-4} mol Cr

$$\text{mol Cl} = \text{mol AgCl} = 0.363 \text{ g AgCl} \times \frac{1 \text{ mol AgCl}}{143.32 \text{ g AgCl}} = 2.53 \times 10^{-3} \text{ mol Cl}$$

$$\frac{\text{mol Cl}}{\text{mol Cr}} = \frac{2.53 \times 10^{-3} \text{ mol Cl}}{8.44 \times 10^{-4} \text{ mol Cr}} = 3 \text{ Cl/Cr}$$

For **B**: mol Cr complex = mol Cr = 0.263 g Cr complex x $\dfrac{1\ mol\ Cr\ complex}{266.45\ g\ Cr\ complex}$ =

9.87×10^{-4} mol Cr

mol Cl = mol AgCl = 0.283 g AgCl x $\dfrac{1\ mol\ AgCl}{143.32\ g\ AgCl}$ = 1.97×10^{-3} mol Cl

$\dfrac{mol\ Cl}{mol\ Cr} = \dfrac{1.97 \times 10^{-3}\ mol\ Cl}{9.87 \times 10^{-4}\ mol\ Cr}$ = 2 Cl/Cr

For **C**: mol Cr complex = mol Cr = 0.358 g Cr complex x $\dfrac{1\ mol\ Cr\ complex}{266.45\ g\ Cr\ complex}$ =

1.34×10^{-3} mol Cr

mol Cl = mol AgCl = 0.193 g AgCl x $\dfrac{1\ mol\ AgCl}{143.32\ g\ AgCl}$ = 1.34×10^{-3} mol Cl

$\dfrac{mol\ Cl}{mol\ Cr} = \dfrac{1.34 \times 10^{-3}\ mol\ Cl}{1.34 \times 10^{-3}\ mol\ Cr}$ = 1 Cl/Cr

Because only the free Cl$^-$ ions (those not bonded to the Cr^{3+}) give an immediate precipitate of AgCl, the probable structural formulas are

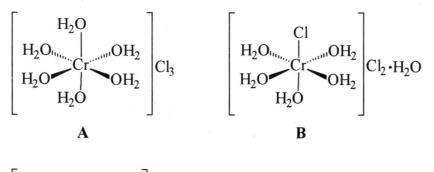

A   **B**

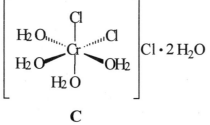

C

Structure **C** can exist as either cis or trans diastereoisomers.

(c) H_2O is a stronger field ligand than Cl$^-$. Compound **A** is likely to be violet absorbing in the yellow. Compounds **B** and **C** have weaker field ligands and would appear blue or green absorbing in the orange or red, respectively.

(d) $\Delta T = K_f \cdot m \cdot i$
For **A**, i = 4; for **B**, i = 3; and for **C**, i = 2.
For **A**, $\Delta T = K_f \cdot m \cdot i = (1.86°C/m)(0.25\ m)(4) = 1.86°C$

freezing point = 0ºC − ΔT = 0ºC − 1.86ºC = −1.86ºC

For **B**, $\Delta T = K_f \cdot m \cdot i = (1.86ºC/m)(0.25\ m)(3) = 1.39ºC$

freezing point = 0ºC − ΔT = 0ºC − 1.39ºC = −1.39ºC

For **C**, $\Delta T = K_f \cdot m \cdot i = (1.86ºC/m)(0.25\ m)(2) = 0.93ºC$

freezing point = 0ºC − ΔT = 0ºC − 0.93ºC = −0.93ºC

Metals and Solid-State Materials

21.1 (a) $Cr_2O_3(s) + 2\,Al(s) \rightarrow 2\,Cr(s) + Al_2O_3(s)$
(b) $Cu_2S(s) + O_2(g) \rightarrow 2\,Cu(s) + SO_2(g)$
(c) $PbO(s) + C(s) \rightarrow Pb(s) + CO(g)$
(d) $2\,K^+(l) + 2\,Cl^-(l) \xrightarrow{\text{electrolysis}} 2\,K(l) + Cl_2(g)$

21.2 The electron configuration for Hg is $[Xe]\,4f^{14}\,5d^{10}\,6s^2$. Assuming the 5d and 6s bands overlap, the composite band can accomodate 12 valence electrons per metal atom. Weak bonding and a low melting point are expected for Hg because both the bonding and antibonding MOs are occupied.

21.3 (a) The composite s-d band can accomodate 12 valence electrons per metal atom.
Hf $[Xe]\,6s^2\,4f^{14}\,5d^2$, 4 valence electrons (4 bonding, 0 antibonding)
The s-d band is 1/4 full, so Hf is picture (1).
Pt $[Xe]\,6s^2\,4f^{14}\,5d^8$, 10 valence electrons (6 bonding, 4 antibonding)
The s-d band is 5/6 full, so Pt is picture (2).
Re $[Xe]\,6s^2\,4f^{14}\,5d^5$, 7 valence electrons (6 bonding, 1 antibonding)
The s-d band is 7/12 full, so Re is picture (3).
(b) Re has an excess of 5 bonding electrons and it has the highest melting point and is the hardest of the three.
(c) Pt has an excess of only 2 bonding electrons and it has the lowest melting point and is the softest of the three.

21.4 Ge doped with As is an n-type semiconductor because As has an additional valence electron. The extra electrons are in the conduction band. The number of electrons in the conduction band of the doped Ge is much higher than for pure Ge, and the conductivity of the doped semiconductor is higher.

21.5 (a) (1), silicon; (2), white tin; (3), diamond; (4), silicon doped with aluminum
(b) (3) < (1) < (4) < (2)
Diamond (3) is an insulator with a large band gap. Silicon (1) is a semiconductor with a band gap smaller than diamond. The conduction band is partially occupied with a few electrons and the valence band is partially empty. Silicon doped with aluminum (4) is a p-type semiconductor that has fewer electrons than needed for bonding and has vacancies (positive holes) in the valence band. White tin (2) has a partially filled s-p composite band and is a metallic conductor.

21.6 8 Cu at corners   8 x 1/8 = 1 Cu
8 Cu on edges   8 x 1/4 = 2 Cu
                           Total = 3 Cu

12 O on edges	12 x 1/4 = 3 O
8 O on faces	8 x 1/2 = <u>4 O</u>
	Total = 7 O

21.7 $Si(OCH_3)_4 + 4 H_2O \rightarrow Si(OH)_4 + 4 HOCH_3$

21.8 $Ba[OCH(CH_3)_2]_2 + Ti[OCH(CH_3)_2]_4 + 6 H_2O \rightarrow BaTi(OH)_6(s) + 6 HOCH(CH_3)_2$

$BaTi(OH)_6(s) \xrightarrow{heat} BaTiO_3(s) + 3 H_2O(g)$

21.9   (a) cobalt/tungsten carbide is a ceramic-metal composite.
(b) silicon carbide/zirconia is a ceramic-ceramic composite.
(c) boron nitride/epoxy is a ceramic-polymer composite.
(d) boron carbide/titanium is a ceramic-metal composite.

21.10   Diamond is a ceramic because it is an inorganic, nonmetallic, nonmolecular solid. It shares many properties with other ceramics including hardness and high melting point.

21.11

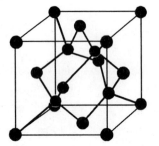

In diamond each carbon atom is covalently bonded to four other carbons. This network of strong bonds makes the diamond structure very rigid which is why diamond is so hard.

Understanding Key Concepts

21.12   A – metal oxide;   B – metal sulfide;   C – metal carbonate;   D – free metal

21.14   (a) (1) and (4) are semiconductors; (2) is a metal; (3) is an insulator
(b) (3) < (1) < (4) < (2). The conductivity increases with decreasing band gap.
(c) (1) and (4) increases; (2) decreases; (3) not much change.

21.16

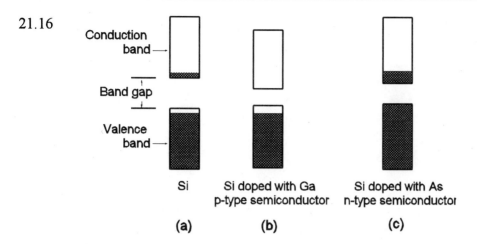

(d) The electrical conductivity of the doped silicon in both cases is higher than for pure silicon. Si doped with Ga is a p-type semiconductor with many more positive holes in the conduction band than in pure Si. This results in a higher conductivity. Si doped with As is an n-type semiconductor with many more electrons in the conduction band than in pure Si. This also results in a higher conductivity.

Additional Problems
Sources of the Metallic Elements

21.18 TiO_2, MnO_2, and Fe_2O_3

21.20   (a) Cu is found in nature as a sulfide.   (b) Zr is found in nature as an oxide.
         (c) Pd is found in nature uncombined.   (d) Bi is found in nature as a sulfide.

21.22   The less electronegative early transition metals tend to form ionic compounds by losing electrons to highly electronegative nonmetals such as oxygen. The more electronegative late transition metals tend to form compounds with more covalent character by bonding to the less electronegative nonmetals such as sulfur.

21.24   (a) Fe_2O_3, hematite   (b) PbS, galena
         (c) TiO_2, rutile   (d) $CuFeS_2$, chalcopyrite

Metallurgy

21.26   The flotation process exploits the differences in the ability of water and oil to wet the surfaces of the mineral and the gangue. The gangue, which contains ionic silicates, is moistened by the polar water molecules and sinks to the bottom of the tank. The mineral particles, which contain the less polar metal sulfide, are coated by the oil and become attached to the soapy air bubbles created by the detergent. The metal sulfide particles are carried to the surface in the soapy froth, which is skimmed off at the top of the tank. This process would not work well for a metal oxide because it is too polar and will be wet by the water and sink with the gangue.

21.28   Because $E° < 0$ for Zn^{2+}, the reduction of Zn^{2+} is not favored.
Because $E° > 0$ for Hg^{2+}, the reduction of Hg^{2+} is favored.
The roasting of CdS should yield CdO because, like Zn^{2+}, $E° < 0$ for the reduction of Cd^{2+}.

21.30   (a) $V_2O_5(s) + 5\ Ca(s) \rightarrow 2\ V(s) + 5\ CaO(s)$
(b) $2\ PbS(s) + 3\ O_2(g) \rightarrow 2\ PbO(s) + 2\ SO_2(g)$
(c) $MoO_3(s) + 3\ H_2(g) \rightarrow Mo(s) + 3\ H_2O(g)$
(d) $3\ MnO_2(s) + 4\ Al(s) \rightarrow 3\ Mn(s) + 2\ Al_2O_3(s)$

(e) $MgCl_2(l) \xrightarrow{\text{electrolysis}} Mg(l) + Cl_2(g)$

21.32 $2\ ZnS(s) + 3\ O_2(g) \rightarrow 2\ ZnO(s) + 2\ SO_2(g)$
$\Delta H° = [2\ \Delta H°_f(ZnO) + 2\ \Delta H°_f(SO_2)] - [2\ \Delta H°_f(ZnS)]$
$\Delta H° = [(2\ mol)(-348.3\ kJ/mol) + (2\ mol)(-296.8\ kJ/mol)]$
$\qquad\qquad - (2\ mol)(-206.0\ kJ/mol) = -878.2\ kJ$
$\Delta G° = [2\ \Delta G°_f(ZnO) + 2\ \Delta G°_f(SO_2)] - [2\ \Delta G°_f(ZnS)]$
$\Delta G° = [(2\ mol)(-318.3\ kJ/mol) + (2\ mol)(-300.2\ kJ/mol)]$
$\qquad\qquad - (2\ mol)(-201.3\ kJ/mol) = -834.4\ kJ$

$\Delta H°$ and $\Delta G°$ are different because of the entropy change associated with the reaction. The minus sign for $(\Delta H° - \Delta G°)$ indicates that the entropy is negative, which is consistent with a decrease in the number of moles of gas from 3 mol to 2 mol.

21.34 $FeCr_2O_4(s) + 4\ C(s) \rightarrow Fe(s) + 2\ Cr(s) + 4\ CO(g)$
$\qquad\qquad\qquad\qquad$ ferrochrome

(a) $FeCr_2O_4$, 223.84 amu; Cr, 52.00 amu;   236 kg = 236 x 10^3 g

$$\text{mass Cr} = 236 \times 10^3\ g \times \frac{1\ mol\ FeCr_2O_4}{223.84\ g} \times \frac{2\ mol\ Cr}{1\ mol\ FeCr_2O_4} \times \frac{52.00\ g\ Cr}{1\ mol\ Cr} \times \frac{1.00\ kg}{1000\ g} = 110\ kg\ Cr$$

(b) $\text{mol CO} = 236 \times 10^3\ g \times \dfrac{1\ mol\ FeCr_2O_4}{223.84\ g} \times \dfrac{4\ mol\ CO}{1\ mol\ FeCr_2O_4} = 4217.3\ mol\ CO$

$$PV = nRT;\quad V = \frac{nRT}{P} = \frac{(4217.3\ mol)\left(0.08206\ \dfrac{L \cdot atm}{K \cdot mol}\right)(298\ K)}{\left(740\ mm\ Hg \times \dfrac{1.00\ atm}{760\ mm\ Hg}\right)} = 1.06 \times 10^5\ L\ CO$$

21.36 $Ni^{2+}(aq) + 2\ e^- \rightarrow Ni(s)$; 1 A = 1 C/s

$$\text{mass Ni} = 52.5\ \frac{C}{s} \times 8\ h \times \frac{3600\ s}{1\ h} \times \frac{1\ mol\ e^-}{96,500\ C} \times \frac{1\ mol\ Ni}{2\ mol\ e^-} \times \frac{58.69\ g\ Ni}{1\ mol\ Ni} \times \frac{1.00\ kg}{1000\ g}$$

mass Ni = 0.460 kg Ni

Iron and Steel

21.38 $Fe_2O_3(s) + 3\ CO(g) \rightarrow 2\ Fe(l) + 3\ CO_2(g)$
Fe_2O_3 is the oxidizing agent. CO is the reducing agent.

21.40   Slag is a byproduct of iron production, consisting mainly of $CaSiO_3$. It is produced from the gangue in iron ore.

21.42   Molten iron from a blast furnace is exposed to a jet of pure oxygen gas for about 20 minutes. The impurities are oxidized to yield a molten slag that can be poured off.

$P_4(l)\ +\ 5\ O_2(g) \rightarrow\ P_4O_{10}(l)$
$6\ CaO(s)\ +\ P_4O_{10}(l) \rightarrow\ 2\ Ca_3(PO_4)_2(l)$ (slag)

$2\ Mn(l)\ +\ O_2(g) \rightarrow\ 2\ MnO(s)$
$MnO(s)\ +\ SiO_2(s) \rightarrow\ MnSiO_3(l)$ (slag)

21.44 $SiO_2(s)\ +\ 2\ C(s) \rightarrow\ Si(s)\ +\ 2\ CO(g)$
$Si(s)\ +\ O_2(g) \rightarrow\ SiO_2(s)$
$CaO(s)\ +\ SiO_2(s) \rightarrow\ CaSiO_3(l)$ (slag)

Bonding in Metals

21.46   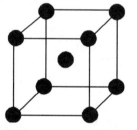   Each K has a single valence electron and has eight nearest neighbor K atoms. The valence electrons can't be localized in an electron-pair bond between any particular pair of K atoms.

21.48   Malleability and ductility of metals follow from the fact that the delocalized bonding extends in all directions. When a metallic crystal is deformed, no localized bonds are broken. Instead, the electron sea simply adjusts to the new distribution of cations, and the energy of the deformed structure is similar to that of the original. Thus, the energy required to deform a metal is relatively small.

21.50   The energy required to deform a transition metal like W is greater than that for Cs because W has more valence electrons and hence more electrostatic "glue".

21.52   The difference in energy between successive MOs in a metal decreases as the number of metal atoms increases so that the MOs merge into an almost continuous band of energy levels. Consequently, MO theory for metals is often called band theory.

21.54   The energy levels within a band occur in degenerate pairs; one set of energy levels applies to electrons moving to the right, and the other set applies to electrons moving to the left. In the absence of an electrical potential, the two sets of levels are equally populated. As a result there is no net electric current. In the presence of an electrical potential those electrons moving to the right are accelerated, those moving to the left are slowed down, and some change direction. Thus, the two sets of energy levels are now unequally populated. The number of electrons moving to the right is now greater than the number moving to the left, and so there is a net electric current.

21.56   (a)   (b)

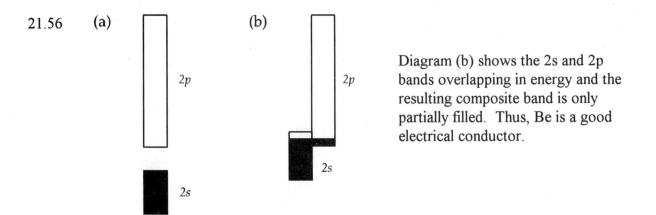

Diagram (b) shows the 2s and 2p bands overlapping in energy and the resulting composite band is only partially filled. Thus, Be is a good electrical conductor.

21.58   Transition metals have a d band that can overlap the s band to give a composite band consisting of six MOs per metal atom. Half of the MOs are bonding and half are antibonding, and thus one expects maximum bonding for metals that have six valence electrons per metal atom. Accordingly, the melting points of the transition metals go through a maximum at or near group 6B.

Semiconductors

21.60   A semiconductor is a material that has an electrical conductivity intermediate between that of a metal and that of an insulator. Si, Ge, and Sn (gray) are semiconductors.

21.62   (a)   (b)

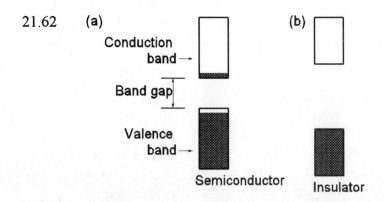

The MOs of a semiconductor are similar to those of an insulator, but the band gap in a semiconductor is smaller. As a result, a few electrons have enough energy to jump the gap and occupy the higher-energy, conduction band. The conduction band is thus partially filled, and the valence band is partially empty. When an electrical potential is applied to a semiconductor, it conducts a small amount of current because the potential can accelerate the electrons in the partially filled bands.

21.64   As the band gap increases, the number of electrons able to jump the gap and occupy the higher-energy conduction band decreases, and thus the conductivity decreases.

21.66   An n-type semiconductor is a semiconductor doped with a substance with more valence electrons than the semiconductor itself. Si doped with P is an example.

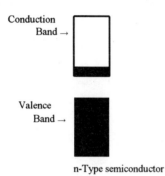

Conduction Band →

Valence Band →

n-Type semiconductor

21.68   In the MO picture, the extra electrons occupy the conduction band. The number of electrons in the conduction band of the doped Ge is much greater than for pure Ge, and the conductivity of the doped semiconductor is correspondingly higher.

21.70   (a) p-type (In is electron deficient with respect to Si)
         (b) n-type (Sb is electron rich with respect to Ge)
         (c) n-type (As is electron rich with respect to gray Sn)

21.72 Al_2O_3 < Ge < Ge doped with In < Fe < Cu

Superconductors

21.74   (1) A superconductor is able to levitate a magnet.
         (2) In a superconductor, once an electric current is started, it flows indefinitely without loss of energy. A superconductor has no electrical resistance.

21.76   The fullerides are three-dimensional superconductors, whereas the copper oxide ceramics are two-dimensional superconductors.

Ceramics and Composites

21.78   Ceramics are inorganic, nonmetallic, nonmolecular solids, including both crystalline and amorphous materials. Ceramics have higher melting points, and they are stiffer, harder, and more resistant to wear and corrosion than are metals.

21.80   Ceramics have higher melting points, and they are stiffer, harder, and more wear resistant than metals because they have stronger bonding. They maintain much of their strength at high temperatures, where metals either melt or corrode because of oxidation.

21.82   The brittleness of ceramics is due to strong chemical bonding. In silicon nitride each Si atom is bonded to four N atoms and each N atom is bonded to three Si atoms. The strong, highly directional covalent bonds prevent the planes of atoms from sliding over one another when the solid is subjected to a stress. As a result, the solid can't deform to relieve the stress. It maintains its shape up to a point, but then the bonds give way suddenly and the material fails catastrophically when the stress exceeds a certain threshold value. By contrast, metals are able to deform under stress because their planes of metal cations can slide easily in the electron sea.

21.84   Ceramic processing is the series of steps that leads from raw material to the finished ceramic object.

21.86   A sol is a colloidal dispersion of tiny particles. A gel is a more rigid gelatin-like material consisting of larger particles.

21.88 $Zr[OCH(CH_3)_2]_4 + 4\ H_2O \rightarrow Zr(OH)_4 + 4\ HOCH(CH_3)_2$

21.90 $(HO)_3Si-O-H + H-O-Si(OH)_3 \rightarrow (HO)_3Si-O-Si(OH)_3 + H_2O$
Further reactions of this sort give a three-dimensional network of Si–O–Si bridges. On heating, SiO_2 is obtained.

21.92 $3\ SiCl_4(g) + 4\ NH_3(g) \rightarrow Si_3N_4(s) + 12\ HCl(g)$

21.94   Graphite/epoxy composites are good materials for making tennis rackets and golf clubs because of their high strength-to-weight ratios.

General Problems

21.96   The chemical composition of the alkaline earth minerals is that of metal sulfates and sulfites, MSO_4 and MSO_3.

21.98   Band theory better explains how the number of valence electrons affects properties such as melting point and hardness.

21.100   V $[Ar]\ 3d^3\ 4s^2$ Zn $[Ar]\ 3d^{10}\ 4s^2$
Transition metals have a d band that can overlap the s band to give a composite band consisting of six MOs per metal atom. Half of the MOs are bonding and half are antibonding. Strong bonding and a high enthalpy of vaporization are expected for V because almost all of the bonding MOs are occupied and all of the antibonding MOs are empty. Weak bonding and a low enthalpy of vaporization are expected for Zn because both the bonding and the antibonding MOs are occupied.

21.102   With a band gap of 130 kJ/mol, GaAs is a semiconductor. Because Ge lies between Ga and As in the periodic table, GaAs is isoelectronic with Ge.

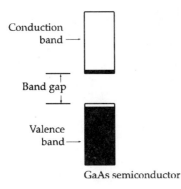

GaAs semiconductor

21.104 $YBa_2Cu_3O_7$, 666.20 amu; $Cu(OCH_2CH_3)_2$, 153.67 amu
$Y(OCH_2CH_3)_3$, 224.09 amu; $Ba(OCH_2CH_3)_2$, 227.45 amu

$$\text{mol } Cu(OCH_2CH_3)_2 = 75.4 \text{ g} \times \frac{1 \text{ mol}}{153.67 \text{ g}} = 0.4907 \text{ mol } Cu(OCH_2CH_3)_2$$

mass $Y(OCH_2CH_3)_3$ = 0.4907 mol $Cu(OCH_2CH_3)_2$ x

$$\frac{1 \text{ mol } Y(OCH_2CH_3)_3}{3 \text{ mol } Cu(OCH_2CH_3)_2} \times \frac{224.09 \text{ g } Y(OCH_2CH_3)_3}{1 \text{ mol } Y(OCH_2CH_3)_3} = 36.7 \text{ g } Y(OCH_2CH_3)_3$$

mass $Ba(OCH_2CH_3)_2$ = 0.4907 mol $Cu(OCH_2CH_3)_2$ x

$$\frac{2 \text{ mol } Ba(OCH_2CH_3)_2}{3 \text{ mol } Cu(OCH_2CH_3)_2} \times \frac{227.45 \text{ g } Ba(OCH_2CH_3)_2}{1 \text{ mol } Ba(OCH_2CH_3)_2} = 74.4 \text{ g } Ba(OCH_2CH_3)_2$$

mass $YBa_2Cu_3O_7$ = 0.4907 mol $Cu(OCH_2CH_3)_2$ x

$$\frac{1 \text{ mol } YBa_2Cu_3O_7}{3 \text{ mol } Cu(OCH_2CH_3)_2} \times \frac{666.20 \text{ g } YBa_2Cu_3O_7}{1 \text{ mol } YBa_2Cu_3O_7} = 109 \text{ g } YBa_2Cu_3O_7$$

21.106   (a) $6 Al(OCH_2CH_3)_3 + 2 Si(OCH_2CH_3)_4 + 26 H_2O \rightarrow$
$6 Al(OH)_3(s) + 2 Si(OH)_4(s) + 26 HOCH_2CH_3$
sol

(b) H_2O is eliminated from the sol through a series of reactions linking the sol particles together through a three-dimensional network of O bridges to form the gel.
$(HO)_2Al–O–H + H–O–Si(OH)_3 \rightarrow (HO)_2Al–O–Si(OH)_3 + H_2O$
(c) The remaining H_2O and solvent are removed from the gel by heating to produce the ceramic, $3 Al_2O_3 \cdot 2 SiO_2$.

21.108   (a)

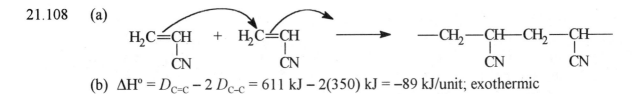

(b) $\Delta H° = D_{C=C} - 2 D_{C-C} = 611 \text{ kJ} - 2(350) \text{ kJ} = -89 \text{ kJ/unit}$; exothermic

21.110   (a)

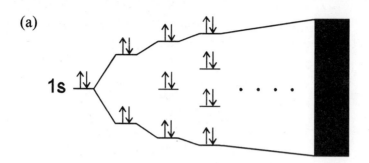

This material is an insulator because all MOs are filled, preventing the movement of electrons.

(b)

Neutral hydrogen atoms have only 1 valence electron, compared with 2 in H^-. Partially empty antibonding MOs will allow the movement of electrons, so the doped material will be a conductor.

(c) The missing electrons in the doped material create "holes" which are positive charge carriers. This type of doped material is a p-type conductor.

Multi-Concept Problems

21.112   (a) 431 pm = 431 x 10^{-12} m
There are 4 oxygen atoms in the face-centered cubic unit cell.
mass of unit cell = (5.75 g/cm³)(431 x 10^{-12} m)³(100 cm/1 m)³ = 4.604 x 10^{-22} g

mass of Fe in unit cell = (4.604 x 10^{-22} g) – 4 O atoms x $\dfrac{15.9994 \text{ g O}}{6.022 \times 10^{23} \text{ O atoms}}$

= 3.541 x 10^{-22} g Fe

number of Fe atoms in unit cell = 3.541 x 10^{-22} g Fe x $\dfrac{6.022 \times 10^{23} \text{ Fe atoms}}{55.847 \text{ g Fe}}$

= 3.818 Fe atoms

For Fe_xO, x = $\dfrac{3.818 \text{ Fe atoms}}{4 \text{ O atoms}}$ = 0.954

(b) The average oxidation state of Fe = $\dfrac{+2}{0.954}$ = 2.095

(c) Let X equal the fraction of Fe^{3+} and Y equal the fraction of Fe^{2+} in wustite.
So, X + Y = 1 and 3X + 2Y = 2.095
Y = 1 – X

$3X + 2(1 - X) = 2.095$

$3X + 2 - 2X = 2.095$

$X + 2 = 2.095$

$X = 2.095 - 2 = 0.095$

9.5% of the Fe in wustite is Fe^{3+}.

(d) d = 431 pm

$$d = \frac{n\lambda}{2\sin\theta} = \frac{(3)(70.93\,\text{pm})}{2\sin\theta} = 431 \text{ pm}$$

$$\sin\theta = \frac{(3)(70.93\,\text{pm})}{(2)(431\,\text{pm})} = 0.247 \text{ and } \theta = 14.3\,^\circ$$

(e) The presence of Fe^{3+} in the semiconductor leads to missing electrons that create "holes" which are positive charge carriers. This type of doped material is a p-type semiconductor.

21.114 $Ni(s) + 4\,CO(g) \rightleftharpoons Ni(CO)_4(g)$

$\Delta H^\circ = -160.8$ kJ; $\Delta S^\circ = -410$ J/K $= -410 \times 10^{-3}$ kJ/K

(a) 150°C = 423 K

$\Delta G^\circ = \Delta H^\circ - T\Delta S^\circ = -160.8$ kJ $-(423\text{ K})(-410 \times 10^{-3}$ kJ/K$) = +12.6$ kJ

$\Delta G^\circ = -RT \ln K$

$$\ln K = \frac{-\Delta G^\circ}{RT} = \frac{-12.6 \text{ kJ/mol}}{[8.314 \times 10^{-3} \text{ kJ/(K}\cdot\text{mol)}](423\text{ K})} = -3.58$$

$K = K_p = e^{-3.58} = 0.028$

(b) 230°C = 503 K

$\Delta G^\circ = \Delta H^\circ - T\Delta S^\circ = -160.8$ kJ $- (503\text{ K})(-410 \times 10^{-3}$ kJ/K$) = +45.4$ kJ

$\Delta G^\circ = -RT \ln K$

$$\ln K = \frac{-\Delta G^\circ}{RT} = \frac{-45.4 \text{ kJ/mol}}{[8.314 \times 10^{-3} \text{ kJ/(K}\cdot\text{mol)}](503\text{ K})} = -10.86$$

$K = K_p = e^{-10.86} = 1.9 \times 10^{-5}$

(c) ΔS° is large and negative because as the reaction proceeds in the forward direction, the number of moles of gas decrease from four to one.

Because ΔS° is negative, $-T\Delta S^\circ$ is positive, and as T increases, ΔG° becomes more positive because $\Delta G^\circ = \Delta H^\circ - T\Delta S^\circ$.

(d) The reaction is exothermic because ΔH° is negative.

$Ni(s) + 4\,CO(g) \rightleftharpoons Ni(CO)_4(g) + \text{heat}$

Heat is added as the temperature is raised and the reaction proceeds in the reverse direction to relieve this stress, as predicted by Le Châtelier's principle. As the reverse reaction proceeds, the partial pressure of CO increases and the partial pressure of $Ni(CO)_4$ decreases. K_p decreases as calculated because $K_p = \dfrac{P_{Ni(CO)_4}}{(P_{CO})^4}$.

21.116   (a) $(NH_4)_2Zn(CrO_4)_2(s) \rightarrow ZnCr_2O_4(s) + N_2(g) + 4\,H_2O(g)$

(b) $mol\ (NH_4)_2Zn(CrO_4)_2 = 10.36\ g\ (NH_4)_2Zn(CrO_4)_2 \times \dfrac{1\ mol\ (NH_4)_2Zn(CrO_4)_2}{333.45\ g\ (NH_4)_2Zn(CrO_4)_2}$

$$= 0.03107\ mol\ (NH_4)_2Zn(CrO_4)_2$$

$$mass\ ZnCr_2O_4 = 0.03107\ mol\ (NH_4)_2Zn(CrO_4)_2 \times \dfrac{1\ mol\ ZnCr_2O_4}{1\ mol\ (NH_4)_2Zn(CrO_4)_2}$$

$$\times\ \dfrac{233.38\ g\ ZnCr_2O_4}{1\ mol\ ZnCr_2O_4} = 7.251\ g\ ZnCr_2O_4$$

(c) $mol\ N_2 + mol\ H_2O = 0.03107\ mol\ (NH_4)_2Zn(CrO_4)_2 \times \dfrac{5\ mol\ gas}{1\ mol\ (NH_4)_2Zn(CrO_4)_2}$

$$= 0.1554\ mol\ gaseous\ by\text{-}products$$

$292^\circ C = 292 + 273 = 565\ K; \qquad PV = nRT$

$$V = \dfrac{nRT}{P} = \dfrac{(0.1554\ mol)\left(0.082\ 06\ \dfrac{L \cdot atm}{K \cdot mol}\right)(565\ K)}{\left(745\ mm\ Hg \times \dfrac{1.00\ atm}{760\ mm\ Hg}\right)} = 7.35\ L$$

(d) A face-centered cubic unit cell has four octahedral holes and eight tetrahedral holes. This unit cell contains one Zn^{2+} ion in a tetrahedral hole and two Cr^{3+} ions in octahedral holes, therefore 1/8 of the tetrahedral holes and 1/2 of the octahedral holes are filled.

(e)

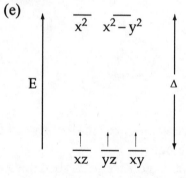

Octahedral Cr^{3+} has three unpaired electrons in the lower-energy d orbitals (xy, xz, yz). Cr^{3+} can absorb visible light to promote one of these d electrons to one of the higher-energy d orbitals making this compound colored. All of the d orbitals in Zn^{2+} are filled and no d electrons can be promoted, consequently the Zn^{2+} ion does not contribute to the color.

21.118

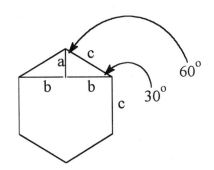

$c = 141.5$ pm $= 141.5 \times 10^{-12}$ m
$\cos(60) = a/c$ and $\sin(60) = b/c$
$a = \cos(60) \cdot c = (0.5)(141.5 \times 10^{-12}$ m$) = 7.075 \times 10^{-11}$ m
$b = \sin(60) \cdot c = (0.866)(141.5 \times 10^{-12}$ m$) = 1.225 \times 10^{-10}$ m
diameter $= 1.08$ nm $= 1.08 \times 10^{-9}$ m $= 1080 \times 10^{-12}$ m $= 1080$ pm

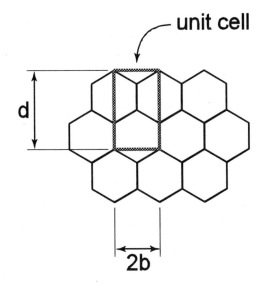

$d = 2a + 2c = 2(7.075 \times 10^{-11}$ m$) + 2(141.5 \times 10^{-12}$ m$) = 4.245 \times 10^{-10}$ m
$d = 424.5 \times 10^{-12}$ m $= 424.5$ pm
$2b = 2(1.225 \times 10^{-10}$ m$) = 2.450 \times 10^{-10}$ m $= 245.0 \times 10^{-12}$ m $= 245.0$ pm
Area of unit cell $= (424.5$ pm$)(245.0$ pm$) = 1.04 \times 10^5$ pm^2/cell
No. of C atoms/cell $= (4)(1/4) + (2)(1/2) + 2 = 4$ C atoms/cell
Surface area of nanotube $= \pi d l = \pi(1080$ pm$)(1.0 \times 10^9$ pm$) = 3.39 \times 10^{12}$ pm^2

No. of C atoms $= (3.39 \times 10^{12}$ pm$^2)\left(\dfrac{1 \text{ cell}}{1.04 \times 10^5 \text{ pm}^2} \right)\left(\dfrac{4 \text{ C atoms}}{\text{cell}} \right) = 1.3 \times 10^8$ C atoms

22 Nuclear Chemistry

22.1 (a) In beta emission, the mass number is unchanged, and the atomic number increases by one. $\quad ^{106}_{44}Ru \rightarrow \ ^{0}_{-1}e \ + \ ^{106}_{45}Rh$

(b) In alpha emission, the mass number decreases by four, and the atomic number decreases by two. $\quad ^{189}_{83}Bi \rightarrow \ ^{4}_{2}He \ + \ ^{185}_{81}Tl$

(c) In electron capture, the mass number is unchanged, and the atomic number decreases by one. $\quad ^{204}_{84}Po \ + \ ^{0}_{-1}e \rightarrow \ ^{204}_{83}Bi$

22.2 The mass number decreases by four, and the atomic number decreases by two. This is characteristic of alpha emission. $\quad ^{214}_{90}Th \rightarrow \ ^{210}_{88}Ra \ + \ ^{4}_{2}He$

22.3 $\quad t_{1/2} = \dfrac{0.693}{k} = \dfrac{0.693}{1.08 \times 10^{-2} \ h^{-1}} = 64.2 \ h$

22.4 $\quad k = \dfrac{0.693}{t_{1/2}} = \dfrac{0.693}{5730 \ y} = 1.21 \times 10^{-4} \ y^{-1}$

22.5 $\quad \ln\left(\dfrac{N}{N_0}\right) = -0.693\left(\dfrac{t}{t_{1/2}}\right) = -0.693\left(\dfrac{16,230 \ y}{5730 \ y}\right) = -1.963$

$\dfrac{N}{N_0} = e^{-1.963} = 0.140; \quad \dfrac{N}{100\%} = 0.140; \quad N = 14.0\%$

22.6 $\quad \ln\left(\dfrac{N}{N_0}\right) = (-0.693)\left(\dfrac{t}{t_{1/2}}\right); \quad \dfrac{N}{N_0} = \dfrac{\text{Decay rate at time t}}{\text{Decay rate at t} = 0}$

$\ln\left(\dfrac{10,860}{16,800}\right) = (-0.693)\left(\dfrac{28.0 \ d}{t_{1/2}}\right); \quad -0.436 = (-0.693)\left(\dfrac{28.0 \ d}{t_{1/2}}\right)$

$t_{1/2} = \dfrac{(-0.693)(28.0 \ d)}{(-0.436)} = 44.5 \ d$

22.7 $\ln\left(\dfrac{N}{N_0}\right) = (-0.693)\left(\dfrac{t}{t_{1/2}}\right)$

$\ln\left(\dfrac{10}{100}\right) = (-0.693)\left(\dfrac{10\,d}{t_{1/2}}\right)$; $\qquad -2.303 = (-0.693)\left(\dfrac{10\,d}{t_{1/2}}\right)$

$t_{1/2} = \dfrac{(-0.693)(10\,d)}{(-2.303)} = 3.0\,d$

22.8 $^{59}_{29}\text{Cu} \rightarrow \,^{0}_{1}\text{e} + \,^{59}_{28}\text{Ni}$

$16\,^{59}\text{Cu} \rightarrow 8\,^{59}\text{Cu} \rightarrow 4\,^{59}\text{Cu} \rightarrow 2\,^{59}\text{Cu}$; three half-lives have passed.

22.9 ^{199}Au has a higher neutron/proton ratio and decays by beta emission. ^{173}Au has a lower neutron/proton ratio and decays by alpha emission.

22.10 The shorter arrow pointing right is for beta emission. The longer arrow pointing left is for alpha emission.

$A = \,^{153+97}_{97}X = \,^{250}_{97}\text{Bk}$

$B = \,^{152+98}_{98}X = \,^{250}_{98}\text{Cf}$

$C = \,^{150+96}_{96}X = \,^{246}_{96}\text{Cm}$

$D = \,^{148+94}_{94}X = \,^{242}_{94}\text{Pu}$

$E = \,^{146+92}_{92}X = \,^{238}_{92}\text{U}$

22.11 For $^{16}_{8}\text{O}$:

First, calculate the total mass of the nucleons (8 n + 8 p)
Mass of 8 neutrons = (8)(1.008 66 amu) = 8.069 28 amu
<u>Mass of 8 protons = (8)(1.007 28 amu) = 8.058 24 amu</u>
Mass of 8 n + 8 p $\qquad\qquad$ = 16.127 52 amu

Next, calculate the mass of a ^{16}O nucleus by subtracting the mass of 8 electrons from the mass of a ^{16}O atom.
Mass of ^{16}O atom $\qquad\qquad\qquad\qquad$ = 15.994 92 amu
<u>–Mass of 8 electrons = –(8)(5.486 x 10^{-4} amu)</u> $\quad$ = –0.004 39 amu
Mass of ^{16}O nucleus $\qquad\qquad\qquad\qquad$ = 15.990 53 amu
Then subtract the mass of the ^{16}O nucleus from the mass of the nucleons to find the mass defect:
Mass defect = mass of nucleons – mass of nucleus
$\qquad\qquad$ = (16.127 52 amu) – (15.990 53 amu) = 0.136 99 amu
Mass defect in g/mol:
(0.136 99 amu)(1.660 54 x 10^{-24} g/amu)(6.022 x 10^{23} mol^{-1}) = 0.136 99 g/mol
Now, use the Einstein equation to convert the mass defect into the binding energy.
$\Delta E = \Delta mc^2 = (0.136\,99\text{ g/mol})(10^{-3}\text{ kg/g})(3.00 \times 10^8\text{ m/s})^2$
$\Delta E = 1.233 \times 10^{13}$ J/mol = 1.233×10^{10} kJ/mol

$$\Delta E = \frac{1.233 \times 10^{13} \text{ J/mol}}{6.022 \times 10^{23} \text{ nuclei/mol}} \times \frac{1 \text{ MeV}}{1.60 \times 10^{-13} \text{ J}} \times \frac{1 \text{ nucleus}}{16 \text{ nucleons}} = 8.00 \frac{\text{MeV}}{\text{nucleon}}$$

22.12 $\Delta E = -852 \text{ kJ/mol} = -852 \times 10^3 \text{ J/mol}$; $1 \text{ J} = 1 \text{ kg} \cdot \text{m}^2/\text{s}^2$

$$\Delta E = \Delta mc^2; \quad \Delta m = \frac{\Delta E}{c^2} = \frac{\left(-852 \times 10^3 \dfrac{\text{kg} \cdot \text{m}^2}{\text{s}^2 \cdot \text{mol}}\right)}{\left(3.00 \times 10^8 \text{ m/s}\right)^2} = -9.47 \times 10^{-12} \text{ kg/mol}$$

$$\Delta m = -9.47 \times 10^{-12} \frac{\text{kg}}{\text{mol}} \times \frac{1000 \text{ g}}{1 \text{ kg}} = -9.47 \times 10^{-9} \text{ g/mol}$$

22.13 $${}_{0}^{1}\text{n} + {}_{92}^{235}\text{U} \rightarrow {}_{52}^{137}\text{Te} + {}_{40}^{97}\text{Zr} + 2 \, {}_{0}^{1}\text{n}$$

mass ${}_{92}^{235}\text{U}$	235.0439	amu
mass ${}_{0}^{1}\text{n}$	1.008 66	amu
$-$mass ${}_{52}^{137}\text{Te}$	-136.9254	amu
$-$mass ${}_{40}^{97}\text{Zr}$	-96.9110	amu
$-$mass $2 \, {}_{0}^{1}\text{n}$	$-(2)(1.008\ 66)$	amu
mass change	0.1988	amu

$(0.1988 \text{ amu})(1.660\ 54 \times 10^{-24} \text{ g/amu})(6.022 \times 10^{23} \text{ mol}^{-1}) = 0.1988 \text{ g/mol}$
$\Delta E = \Delta mc^2 = (0.1988 \text{ g/mol})(10^{-3} \text{ kg/g})(3.00 \times 10^8 \text{ m/s})^2$
$\Delta E = 1.79 \times 10^{13} \text{ J/mol} = 1.79 \times 10^{10} \text{ kJ/mol}$

22.14 $${}_{1}^{1}\text{H} + {}_{1}^{2}\text{H} \rightarrow {}_{2}^{3}\text{He}$$

mass ${}^1\text{H}$	1.007 83 amu
mass ${}^2\text{H}$	2.014 10 amu
$-$mass ${}^3\text{He}$	$-3.016\ 03$ amu
mass change	0.005 90 amu

$(0.005\ 90 \text{ amu})(1.660\ 54 \times 10^{-24} \text{ g/amu})(6.022 \times 10^{23} \text{ mol}^{-1}) = 0.005\ 90 \text{ g/mol}$
$\Delta E = \Delta mc^2 = (0.005\ 90 \text{ g/mol})(10^{-3} \text{ kg/g})(3.00 \times 10^8 \text{ m/s})^2$
$\Delta E = 5.31 \times 10^{11} \text{ J/mol} = 5.31 \times 10^8 \text{ kJ/mol}$

22.15 $${}_{18}^{40}\text{Ar} + {}_{1}^{1}\text{p} \rightarrow {}_{19}^{40}\text{K} + {}_{0}^{1}\text{n}$$

22.16 $${}_{92}^{238}\text{U} + {}_{1}^{2}\text{H} \rightarrow {}_{93}^{238}\text{Np} + 2 \, {}_{0}^{1}\text{n}$$

22.17 $\ln\left(\dfrac{N}{N_0}\right) = (-0.693)\left(\dfrac{t}{t_{1/2}}\right)$; $\dfrac{N}{N_0} = \dfrac{\text{Decay rate at time t}}{\text{Decay rate at time t} = 0}$

$\ln\left(\dfrac{2.4}{15.3}\right) = (-0.693)\left(\dfrac{t}{5730\ \text{y}}\right)$; $t = 1.53 \times 10^4\ \text{y}$

22.18   Elements heavier than iron arise from nuclear reactions occuring as a result of supernova explosions.

Understanding Key Concepts

22.20   The isotope contains 8 neutrons and 6 protons. The isotope symbol is $^{14}_{6}\text{C}$.

$^{14}_{6}\text{C}$ would decay by beta emission because the n/p ratio is high.

22.22   The shorter arrow pointing right is for beta emission. The longer arrow pointing left is for alpha emission.

$A = {}^{147+94}_{\quad 94}X = {}^{241}_{94}\text{Pu}$

$B = {}^{146+95}_{\quad 95}X = {}^{241}_{95}\text{Am}$

$C = {}^{144+93}_{\quad 93}X = {}^{237}_{93}\text{Np}$

$D = {}^{142+91}_{\quad 91}X = {}^{233}_{91}\text{Pa}$

$E = {}^{141+92}_{\quad 92}X = {}^{233}_{92}\text{U}$

Additional Problems
Nuclear Reactions and Radioactivity

22.24   Positron emission is the conversion of a proton in the nucleus into a neutron plus an ejected positron.
Electron capture is the process in which a proton in the nucleus captures an inner-shell electron and is thereby converted into a neutron.

22.26   Alpha particles move relatively slowly and can be stopped by the skin. However, inside the body, alpha particles give up their energy to the immediately surrounding tissue. Gamma rays move at the speed of light and are very penetrating. Therefore they are equally hazardous internally and externally.

22.28   There is no radioactive "neutralization" reaction like there is an acid-base neutralization reaction.

22.30   (a) $^{126}_{50}\text{Sn} \rightarrow {}^{0}_{-1}\text{e} + {}^{126}_{51}\text{Sb}$ (b) $^{210}_{88}\text{Ra} \rightarrow {}^{4}_{2}\text{He} + {}^{206}_{86}\text{Rn}$

(c) $^{77}_{37}\text{Rb} \rightarrow {}^{0}_{1}\text{e} + {}^{77}_{36}\text{Kr}$ (d) $^{76}_{36}\text{Kr} + {}^{0}_{-1}\text{e} \rightarrow {}^{76}_{35}\text{Br}$

22.32   (a) $^{188}_{80}\text{Hg} \rightarrow \,^{188}_{79}\text{Au} + \,^{0}_{1}\text{e}$ (b) $^{218}_{85}\text{At} \rightarrow \,^{214}_{83}\text{Bi} + \,^{4}_{2}\text{He}$

        (c) $^{234}_{90}\text{Th} \rightarrow \,^{234}_{91}\text{Pa} + \,^{0}_{-1}\text{e}$

22.34   (a) $^{162}_{75}\text{Re} \rightarrow \,^{158}_{73}\text{Ta} + \,^{4}_{2}\text{He}$ (b) $^{138}_{62}\text{Sm} + \,^{0}_{-1}\text{e} \rightarrow \,^{138}_{61}\text{Pm}$

        (c) $^{188}_{74}\text{W} \rightarrow \,^{188}_{75}\text{Re} + \,^{0}_{-1}\text{e}$ (d) $^{165}_{73}\text{Ta} \rightarrow \,^{165}_{72}\text{Hf} + \,^{0}_{1}\text{e}$

22.36 ^{160}W is neutron poor and decays by alpha emission. ^{185}W is neutron rich and decays by beta emission.

22.38 $^{241}_{95}\text{Am} \rightarrow \,^{237}_{93}\text{Np} + \,^{4}_{2}\text{He}$

        $^{237}_{93}\text{Np} \rightarrow \,^{233}_{91}\text{Pa} + \,^{4}_{2}\text{He}$

        $^{233}_{91}\text{Pa} \rightarrow \,^{233}_{92}\text{U} + \,^{0}_{-1}\text{e}$

        $^{233}_{92}\text{U} \rightarrow \,^{229}_{90}\text{Th} + \,^{4}_{2}\text{He}$

        $^{229}_{90}\text{Th} \rightarrow \,^{225}_{88}\text{Ra} + \,^{4}_{2}\text{He}$

        $^{225}_{88}\text{Ra} \rightarrow \,^{225}_{89}\text{Ac} + \,^{0}_{-1}\text{e}$

        $^{225}_{89}\text{Ac} \rightarrow \,^{221}_{87}\text{Fr} + \,^{4}_{2}\text{He}$

        $^{221}_{87}\text{Fr} \rightarrow \,^{217}_{85}\text{At} + \,^{4}_{2}\text{He}$

        $^{217}_{85}\text{At} \rightarrow \,^{213}_{83}\text{Bi} + \,^{4}_{2}\text{He}$

        $^{213}_{83}\text{Bi} \rightarrow \,^{213}_{84}\text{Po} + \,^{0}_{-1}\text{e}$

        $^{213}_{84}\text{Po} \rightarrow \,^{209}_{82}\text{Pb} + \,^{4}_{2}\text{He}$

        $^{209}_{82}\text{Pb} \rightarrow \,^{209}_{83}\text{Bi} + \,^{0}_{-1}\text{e}$

22.40   Each alpha emission decreases the mass number by four and the atomic number by two. Each beta emission increases the atomic number by one.

        $^{232}_{90}\text{Th} \rightarrow \,^{208}_{82}\text{Pb}$

$$\text{Number of } \alpha \text{ emissions} = \frac{\text{Th mass number} - \text{Pb mass number}}{4}$$

$$= \frac{232 - 208}{4} = 6 \ \alpha \text{ emissions}$$

        The atomic number decreases by 12 as a result of 6 alpha emissions. The resulting atomic number is $(90 - 12) = 78$.

        Number of β emissions = Pb atomic number $- 78 = 82 - 78 = 4 \ \beta$ emissions

Radioactive Decay Rates

22.42   If the half-life of ^{59}Fe is 44.5 d, it takes 44.5 days for half of the original amount of ^{59}Fe to decay.

22.44 $k = \dfrac{0.693}{t_{1/2}} = \dfrac{0.693}{2.805 \text{ d}} = 0.247 \text{ d}^{-1}$

22.46 $t_{1/2} = \dfrac{0.693}{k} = \dfrac{0.693}{0.228 \text{ d}^{-1}} = 3.04 \text{ d}$

22.48   After 65 d: $\ln\left(\dfrac{N}{N_0}\right) = -0.693\left(\dfrac{t}{t_{1/2}}\right) = -0.693\left(\dfrac{\left[\dfrac{65 \text{ d}}{365 \text{ d/y}}\right]}{432.2 \text{ y}}\right) = -0.000\,285\,5$

$\dfrac{N}{N_0} = e^{-0.0002855} = 0.9997;$ $\dfrac{N}{100\%} = 0.9997;$ $N = 99.97\%$

After 65 y: $\ln\left(\dfrac{N}{N_0}\right) = -0.693\left(\dfrac{t}{t_{1/2}}\right) = -0.693\left(\dfrac{65 \text{ y}}{432.2 \text{ y}}\right) = -0.1042$

$\dfrac{N}{N_0} = e^{-0.1042} = 0.9010;$ $\dfrac{N}{100\%} = 0.9010;$ $N = 90.10\%$

After 650 y: $\ln\left(\dfrac{N}{N_0}\right) = -0.693\left(\dfrac{t}{t_{1/2}}\right) = -0.693\left(\dfrac{650 \text{ y}}{432.2 \text{ y}}\right) = -1.042$

$\dfrac{N}{N_0} = e^{-1.042} = 0.3527;$ $\dfrac{N}{100\%} = 0.3527;$ $N = 35.27\%$

22.50 $\ln\left(\dfrac{N}{N_0}\right) = (-0.693)\left(\dfrac{t}{t_{1/2}}\right);$ $\ln(0.43) = (-0.693)\left(\dfrac{t}{5730 \text{ y}}\right);$ $t = 6980 \text{ y}$

22.52 $t_{1/2} = \dfrac{0.693}{k} = \dfrac{0.693}{7.89 \times 10^{-3} \text{ d}^{-1}} = 87.83 \text{ d}$

$\ln\left(\dfrac{N}{N_0}\right) = (-0.693)\left(\dfrac{t}{t_{1/2}}\right) = (-0.693)\left(\dfrac{185 \text{ d}}{87.83 \text{ d}}\right) = -1.4597$

$\dfrac{N}{N_0} = e^{-1.4597} = 0.2323;$ $\dfrac{N}{100\%} = 0.2323;$ $N = 23.2\%$

22.54 $t_{1/2} = (102 \text{ y})(365 \text{ d/y})(24 \text{ h/d})(3600 \text{ s/h}) = 3.2167 \times 10^9 \text{ s}$

$k = \dfrac{0.693}{t_{1/2}} = \dfrac{0.693}{3.2167 \times 10^9 \text{ s}} = 2.1544 \times 10^{-10} \text{ s}^{-1}$

$N = (1.0 \times 10^{-9} \text{ g})\left(\dfrac{1 \text{ mol Po}}{209 \text{ g Po}}\right)(6.022 \times 10^{23} \text{ atoms/mol}) = 2.881 \times 10^{12} \text{ atoms}$

Decay rate = kN = $(2.1544 \times 10^{-10}~\text{s}^{-1})(2.881 \times 10^{12}~\text{atoms})$ = $6.21 \times 10^2~\text{s}^{-1}$
621 α particles are emitted in 1.0 s.

22.56   Decay rate = kN

$$N = (1.0 \times 10^{-3}~\text{g})\left(\frac{1~\text{mol}~^{79}\text{Se}}{79~\text{g}}\right)(6.022 \times 10^{23}~\text{atoms/mol}) = 7.6 \times 10^{18}~\text{atoms}$$

$$k = \frac{\text{Decay rate}}{N} = \frac{1.5 \times 10^5/\text{s}}{7.6 \times 10^{18}} = 2.0 \times 10^{-14}~\text{s}^{-1}$$

$$t_{1/2} = \frac{0.693}{k} = \frac{0.693}{2.0 \times 10^{-14}~\text{s}^{-1}} = 3.5 \times 10^{13}~\text{s}$$

$$t_{1/2} = (3.5 \times 10^{13}~\text{s})\left(\frac{1~\text{h}}{3600~\text{s}}\right)\left(\frac{1~\text{d}}{24~\text{h}}\right)\left(\frac{1~\text{y}}{365~\text{d}}\right) = 1.1 \times 10^6~\text{y}$$

22.58 $\ln\left(\dfrac{N}{N_0}\right) = (-0.693)\left(\dfrac{t}{t_{1/2}}\right)$; $\dfrac{N}{N_0} = \dfrac{\text{Decay rate at time } t}{\text{Decay rate at time } t = 0}$

$\ln\left(\dfrac{6990}{8540}\right) = (-0.693)\left(\dfrac{10.0~\text{d}}{t_{1/2}}\right)$; $t_{1/2} = 34.6~\text{d}$

Energy Changes During Nuclear Reactions

22.60   The loss in mass that occurs when protons and neutrons combine to form a nucleus is called the mass defect. The lost mass is converted into the binding energy that is used to hold the nucleons together.

22.62 $E = (1.50~\text{MeV})\left(\dfrac{1.60 \times 10^{-13}~\text{J}}{1~\text{MeV}}\right) = 2.40 \times 10^{-13}~\text{J}$

$\lambda = \dfrac{hc}{E} = \dfrac{(6.626 \times 10^{-34}~\text{J} \cdot \text{s})(3.00 \times 10^8~\text{m/s})}{2.40 \times 10^{-13}~\text{J}} = 8.28 \times 10^{-13}~\text{m} = 0.000~828~\text{nm}$

22.64   (a) For $^{52}_{26}\text{Fe}$:
First, calculate the total mass of the nucleons (26 n + 26 p)
Mass of 26 neutrons = (26)(1.008 66 amu) = 26.225 16 amu
Mass of 26 protons = (26)(1.007 28 amu) = 26.189 28 amu
Mass of 26 n + 26 p   = 52.414 44 amu
Next, calculate the mass of a ^{52}Fe nucleus by subtracting the mass of 26 electrons from the mass of a ^{52}Fe atom.
 Mass of ^{52}Fe atom   = 51.948 11 amu
–Mass of 26 electrons = –(26)(5.486 × 10⁻⁴ amu)   = –0.014 26 amu
Mass of ^{52}Fe nucleus   = 51.933 85 amu
Then subtract the mass of the ^{52}Fe nucleus from the mass of the nucleons to find the mass defect:

Mass defect = mass of nucleons − mass of nucleus
$$= (52.414\ 44\ \text{amu}) - (51.933\ 85\ \text{amu}) = 0.480\ 59\ \text{amu}$$
Mass defect in g/mol:
$$(0.480\ 59\ \text{amu})(1.660\ 54 \times 10^{-24}\ \text{g/amu})(6.022 \times 10^{23}\ \text{mol}^{-1}) = 0.480\ 59\ \text{g/mol}$$

(b) For $^{92}_{42}\text{Mo}$:

First, calculate the total mass of the nucleons (50 n + 42 p)
Mass of 50 neutrons = (50)(1.008 66 amu) = 50.433 00 amu
<u>Mass of 42 protons = (42)(1.007 28 amu) = 42.305 76 amu</u>
Mass of 50 n + 42 p $\qquad\qquad\qquad$ = 92.738 76 amu
Next, calculate the mass of a ^{92}Mo nucleus by subtracting the mass of 42 electrons from the mass of a ^{92}Mo atom.
Mass of ^{92}Mo atom $\qquad\qquad\qquad\qquad$ = 91.906 81 amu
<u>−Mass of 42 electrons = −(42)(5.486 × 10⁻⁴ amu)</u> $\quad$ <u>= −0.023 04 amu</u>
Mass of ^{92}Mo nucleus $\qquad\qquad\qquad\qquad$ = 91.883 77 amu
Then subtract the mass of the ^{92}Mo nucleus from the mass of the nucleons to find the mass defect:
Mass defect = mass of nucleons − mass of nucleus
$$= (92.738\ 76\ \text{amu}) - (91.883\ 77\ \text{amu}) = 0.854\ 99\ \text{amu}$$
Mass defect in g/mol:
$$(0.854\ 99\ \text{amu})(1.660\ 54 \times 10^{-24}\ \text{g/amu})(6.022 \times 10^{23}\ \text{mol}^{-1}) = 0.854\ 99\ \text{g/mol}$$

22.66 $\quad$ (a) For $^{58}_{28}\text{Ni}$:

First, calculate the total mass of the nucleons (30 n + 28 p)
Mass of 30 neutrons = (30)(1.008 66 amu) = 30.259 80 amu
<u>Mass of 28 protons = (28)(1.007 28 amu) = 28.203 84 amu</u>
Mass of 30 n + 28 p $\qquad\qquad\qquad$ = 58.463 64 amu
Next, calculate the mass of a ^{58}Ni nucleus by subtracting the mass of 28 electrons from the mass of a ^{58}Ni atom.
Mass of ^{58}Ni atom $\qquad\qquad\qquad\qquad$ = 57.935 35 amu
<u>−Mass of 28 electrons = −(28)(5.486 × 10⁻⁴ amu)</u> $\quad$ <u>= −0.015 36 amu</u>
Mass of ^{58}Ni nucleus $\qquad\qquad\qquad\qquad$ = 57.919 99 amu
Then subtract the mass of the ^{58}Ni nucleus from the mass of the nucleons to find the mass defect:
Mass defect = mass of nucleons − mass of nucleus
$$= (58.463\ 64\ \text{amu}) - (57.919\ 99\ \text{amu}) = 0.543\ 65\ \text{amu}$$
Mass defect in g/mol:
$$(0.543\ 65\ \text{amu})(1.660\ 54 \times 10^{-24}\ \text{g/amu})(6.022 \times 10^{23}\ \text{mol}^{-1}) = 0.543\ 65\ \text{g/mol}$$
Now, use the Einstein equation to convert the mass defect into the binding energy.
$$\Delta E = \Delta mc^2 = (0.543\ 65\ \text{g/mol})(10^{-3}\ \text{kg/g})(3.00 \times 10^8\ \text{m/s})^2$$
$$\Delta E = 4.893 \times 10^{13}\ \text{J/mol} = 4.893 \times 10^{10}\ \text{kJ/mol}$$
$$\Delta E = \frac{4.893 \times 10^{13}\ \text{J/mol}}{6.022 \times 10^{23}\ \text{nuclei/mol}} \times \frac{1\ \text{MeV}}{1.60 \times 10^{-13}\ \text{J}} \times \frac{1\ \text{nucleus}}{58\ \text{nucleons}} = 8.76\ \frac{\text{MeV}}{\text{nucleon}}$$

(b) For $^{84}_{36}$Kr:

First, calculate the total mass of the nucleons (48 n + 36 p)

Mass of 48 neutrons = (48)(1.008 66 amu) = 48.415 68 amu

Mass of 36 protons = (36)(1.007 28 amu) = 36.262 08 amu

Mass of 48 n + 36 p = 84.677 76 amu

Next, calculate the mass of a ^{84}Kr nucleus by subtracting the mass of 36 electrons from the mass of a ^{84}Kr atom.

Mass of ^{84}Kr atom = 83.911 51 amu

−Mass of 36 electrons = −(36)(5.486 x 10^{-4} amu) = −0.019 75 amu

Mass of ^{84}Kr nucleus = 83.891 76 amu

Then subtract the mass of the ^{84}Kr nucleus from the mass of the nucleons to find the mass defect:

Mass defect = mass of nucleons − mass of nucleus

= (84.677 76 amu) − (83.891 76 amu) = 0.786 00 amu

Mass defect in g/mol:

(0.786 00 amu)(1.660 54 x 10^{-24} g/mol)(6.022 x 10^{23} mol^{-1}) = 0.786 00 g/mol

Now, use the Einstein equation to convert the mass defect into the binding energy.

$\Delta E = \Delta mc^2$ = (0.786 00 g/mol)(10^{-3} kg/g)(3.00 x 10^8 m/s)^2

ΔE = 7.074 x 10^{13} J/mol = 7.074 x 10^{10} kJ/mol

$$\Delta E = \frac{7.074 \times 10^{13} \text{ J/mol}}{6.022 \times 10^{23} \text{ nuclei/mol}} \times \frac{1 \text{ MeV}}{1.60 \times 10^{-13} \text{ J}} \times \frac{1 \text{ nucleus}}{84 \text{ nucleons}} = 8.74 \ \frac{\text{MeV}}{\text{nucleon}}$$

22.68 $^{174}_{77}$Ir $\rightarrow$ $^{170}_{75}$Re + $^4_2$He

mass $^{174}_{77}$Ir   173.966 66 amu

−mass $^{170}_{75}$Re   −169.958 04 amu

−mass $^4_2$He   −4.002 60 amu

───────────────────────────────────

mass change   0.006 02 amu

(0.006 02 amu)(1.660 54 x 10^{-24} g/amu)(6.022 x 10^{23} mol^{-1}) = 0.006 02 g/mol

$\Delta E = \Delta mc^2$ = (0.006 02 g/mol)(10^{-3} kg/g)(3.00 x 10^8 m/s)^2

ΔE = 5.42 x 10^{11} J/mol = 5.42 x 10^8 kJ/mol

22.70 $\Delta m = \dfrac{\Delta E}{c^2} = \dfrac{92.2 \times 10^3 \text{ J}}{(3.00 \times 10^8 \text{ m/s})^2} = \dfrac{92.2 \times 10^3 \text{ kg·m}^2/\text{s}^2}{(3.00 \times 10^8 \text{ m/s})^2} = 1.02 \times 10^{-12}$ kg

Δm = 1.02 x 10^{-9} g

22.72   Mass of positron and electron

= 2(9.109 x 10^{-31} kg)(6.022 x 10^{23} mol^{-1}) = 1.097 x 10^{-6} kg/mol

$\Delta E = \Delta mc^2$ = (1.097 x 10^{-6} kg/mol)(3.00 x 10^8 m/s)^2

ΔE = 9.87 x 10^{10} J/mol = 9.87 x 10^7 kJ/mol

Chapter 22 – Nuclear Chemistry

Nuclear Transmutation

22.74 (a) $^{109}_{47}\text{Ag} + {}^{4}_{2}\text{He} \rightarrow {}^{113}_{49}\text{In}$ (b) $^{10}_{5}\text{B} + {}^{4}_{2}\text{He} \rightarrow {}^{13}_{7}\text{N} + {}^{1}_{0}\text{n}$

22.76 $^{209}_{83}\text{Bi} + {}^{58}_{26}\text{Fe} \rightarrow {}^{266}_{109}\text{Mt} + {}^{1}_{0}\text{n}$

22.78 $^{238}_{92}\text{U} + {}^{12}_{6}\text{C} \rightarrow {}^{246}_{98}\text{Cf} + 4\,{}^{1}_{0}\text{n}$

General Problems

22.80 $^{232}_{90}\text{Th} \rightarrow {}^{208}_{82}\text{Pb} + 6\,{}^{4}_{2}\text{He} + 4\,{}^{0}_{-1}\text{e}$

Reactant: $^{232}_{90}\text{Th}$ nucleus $= {}^{232}_{90}\text{Th}$ atom $- 90\ e^-$

Product: $^{208}_{82}\text{Pb}$ nucleus $+ (6)({}^{4}_{2}\text{He}$ nucleus$) + 4\ e^-$

$= ({}^{208}_{82}\text{Pb}$ atom $- 82\ e^-) + (6)({}^{4}_{2}\text{He atom} - 2\ e^-) + 4\ e^-$

$= {}^{208}_{82}\text{Pb}$ atom $+ (6)({}^{4}_{2}\text{He atom}) - 90\ e^-$

Change: $({}^{232}_{90}\text{Th}$ atom $- 90\ e^-) - [{}^{208}_{82}\text{Pb}$ atom $+ (6)({}^{4}_{2}\text{He atom}) - 90\ e^-]$

$= {}^{232}_{90}\text{Th}$ atom $- [{}^{208}_{82}\text{Pb}$ atom $+ (6)({}^{4}_{2}\text{He atom})]$ (electrons cancel)

Mass change $= 232.038\ 054$ amu $- [207.976\ 627$ amu $+ (6)(4.002\ 603$ amu$)]$
$= 0.045\ 809$ amu

$(0.045\ 809\ \text{amu})(1.660\ 54 \times 10^{-24}\ \text{g/amu})(6.022 \times 10^{23}\ \text{mol}^{-1}) = 0.045\ 809\ \text{g/mol}$
$\Delta E = \Delta mc^2 = (0.045\ 809\ \text{g/mol})(10^{-3}\ \text{kg/g})(3.00 \times 10^8\ \text{m/s})^2$
$\Delta E = 4.12 \times 10^{12}\ \text{J/mol} = 4.12 \times 10^9\ \text{kJ/mol}$

22.82 $^{241}_{94}\text{Pu} \rightarrow {}^{233}_{92}\text{U} + 2\,{}^{4}_{2}\text{He} + 2\,{}^{0}_{-1}\text{e}$

Reactant: $^{241}_{94}\text{Pu}$ nucleus $= {}^{241}_{94}\text{Pu}$ atom $- 94\ e^-$

Product: $^{233}_{92}\text{U}$ nucleus $+ (2)({}^{4}_{2}\text{He nucleus}) + 2\ e^-$

$= ({}^{233}_{92}\text{U atom} - 92\ e^-) + (2)({}^{4}_{2}\text{He atom} - 2\ e^-) + 2\ e^-$

$= {}^{233}_{92}\text{U atom} + (2)({}^{4}_{2}\text{He atom}) - 94\ e^-$

Change: $({}^{241}_{94}\text{Pu}$ atom $- 94\ e^-) - [{}^{233}_{92}\text{U atom} + (2)({}^{4}_{2}\text{He atom}) - 94\ e^-]$

$= {}^{241}_{94}\text{Pu}$ atom $- [{}^{233}_{92}\text{U atom} + (2)({}^{4}_{2}\text{He atom})]$

Mass change $= 241.056\ 845$ amu $- [233.039\ 628$ amu $+ (2)(4.002\ 603$ amu$)]$
$= 0.012\ 011$ amu

$(0.012\ 011\ \text{amu})(1.660\ 54 \times 10^{-24}\ \text{g/amu})(6.022 \times 10^{23}\ \text{mol}^{-1}) = 0.012\ 011\ \text{g/mol}$

$\Delta E = \Delta mc^2 = (0.012\ 011\ \text{g/mol})(10^{-3}\ \text{kg/g})(3.00 \times 10^8\ \text{m/s})^2$

$\Delta E = 1.08 \times 10^{12}\ \text{J/mol} = 1.08 \times 10^9\ \text{kJ/mol}$

22.84 $\ln\left(\dfrac{N}{N_0}\right) = (-0.693)\left(\dfrac{t}{t_{1/2}}\right)$; $\dfrac{N}{N_0} = \dfrac{\text{Decay rate at time t}}{\text{Decay rate at time t} = 0}$

$\ln\left(\dfrac{100 - 99.99}{100}\right) = (-0.693)\left(\dfrac{t}{1.53\ \text{s}}\right)$; $t = 20.3\ \text{s}$

22.86   (a) For $^{50}_{24}\text{Cr}$:

First, calculate the total mass of the nucleons (26 n + 24 p)

Mass of 26 neutrons = (26)(1.008 66 amu) = 26.225 16 amu

Mass of 24 protons = (24)(1.007 28 amu) = 24.174 72 amu

Mass of 26 n + 24 p   = 50.399 88 amu

Next, calculate the mass of a ^{50}Cr nucleus by subtracting the mass of 24 electrons from the mass of a ^{50}Cr atom.

Mass of ^{50}Cr atom   = 49.946 05 amu

−Mass of 24 electrons = −(24)(5.486 × 10^{-4} amu)   = −0.013 17 amu

Mass of ^{50}Cr nucleus   = 49.932 88 amu

Then subtract the mass of the ^{50}Cr nucleus from the mass of the nucleons to find the mass defect:

Mass defect = mass of nucleons − mass of nucleus

           = (50.399 88 amu) − (49.932 88 amu) = 0.467 00 amu

Mass defect in g/mol:

$(0.467\ 00\ \text{amu})(1.660\ 54 \times 10^{-24}\ \text{g/amu})(6.022 \times 10^{23}\ \text{mol}^{-1}) = 0.467\ 00\ \text{g/mol}$

Now, use the Einstein equation to convert the mass defect into the binding energy.

$\Delta E = \Delta mc^2 = (0.467\ 00\ \text{g/mol})(10^{-3}\ \text{kg/g})(3.00 \times 10^8\ \text{m/s})^2$

$\Delta E = 4.203 \times 10^{13}\ \text{J/mol} = 4.203 \times 10^{10}\ \text{kJ/mol}$

$\Delta E = \dfrac{4.203 \times 10^{13}\ \text{J/mol}}{6.022 \times 10^{23}\ \text{nuclei/mol}} \times \dfrac{1\ \text{MeV}}{1.60 \times 10^{-13}\ \text{J}} \times \dfrac{1\ \text{nucleus}}{50\ \text{nucleons}} = 8.72\ \dfrac{\text{MeV}}{\text{nucleon}}$

(b) For $^{64}_{30}\text{Zn}$:

First, calculate the total mass of the nucleons (34 n + 30 p)

Mass of 34 neutrons = (34)(1.008 66 amu) = 34.294 44 amu

Mass of 30 protons = (30)(1.007 28 amu) = 30.218 40 amu

Mass of 34 n + 30 p   = 64.512 84 amu

Next, calculate the mass of a ^{64}Zn nucleus by subtracting the mass of 30 electrons from the mass of a ^{64}Zn atom.

Mass of ^{64}Zn atom   = 63.929 15 amu

−Mass of 30 electrons = −(30)(5.486 × 10^{-4} amu)   = −0.016 46 amu

Mass of ^{64}Zn nucleus   = 63.912 69 amu

Then subtract the mass of the ^{64}Zn nucleus from the mass of the nucleons to find the mass defect:

Mass defect = mass of nucleons – mass of nucleus

$$= (64.512\ 84\ \text{amu}) - (63.912\ 69\ \text{amu}) = 0.600\ 15\ \text{amu}$$

Mass defect in g/mol:

$(0.600\ 15\ \text{amu})(1.660\ 54 \times 10^{-24}\ \text{g/amu})(6.022 \times 10^{23}\ \text{mol}^{-1}) = 0.600\ 15\ \text{g/mol}$

Now, use the Einstein equation to convert the mass defect into the binding energy.

$\Delta E = \Delta mc^2 = (0.600\ 15\ \text{g/mol})(10^{-3}\ \text{kg/g})(3.00 \times 10^8\ \text{m/s})^2$

$\Delta E = 5.401 \times 10^{13}\ \text{J/mol} = 5.401 \times 10^{10}\ \text{kJ/mol}$

$$\Delta E = \frac{5.401 \times 10^{13}\ \text{J/mol}}{6.022 \times 10^{23}\ \text{nuclei/mol}} \times \frac{1\ \text{MeV}}{1.60 \times 10^{-13}\ \text{J}} \times \frac{1\ \text{nucleus}}{64\ \text{nucleons}} = 8.76\ \frac{\text{MeV}}{\text{nucleon}}$$

The ^{64}Zn is more stable.

22.88 $^{2}_{1}\text{H} + {}^{3}_{2}\text{He} \rightarrow {}^{4}_{2}\text{He} + {}^{1}_{1}\text{H}$

mass $^{2}_{1}\text{H}$	2.0141 amu
mass $^{3}_{2}\text{He}$	3.0160 amu
–mass $^{4}_{2}\text{He}$	–4.0026 amu
–mass $^{1}_{1}\text{H}$	–1.0078 amu
mass change	0.0197 amu

$(0.0197\ \text{amu})(1.660\ 54 \times 10^{-24}\ \text{g/amu})(6.022 \times 10^{23}\ \text{mol}^{-1}) = 0.0197\ \text{g/mol}$

$\Delta E = \Delta mc^2 = (0.0197\ \text{g/mol})(10^{-3}\ \text{kg/g})(3.00 \times 10^8\ \text{m/s})^2$

$\Delta E = 1.77 \times 10^{12}\ \text{J/mol} = 1.77 \times 10^9\ \text{kJ/mol}$

22.90 $^{238}_{92}\text{U} + {}^{1}_{0}\text{n} \rightarrow {}^{239}_{94}\text{Pu} + 2\ {}^{0}_{-1}\text{e}$

22.92 $^{10}\text{B} + {}^{1}\text{n} \rightarrow {}^{4}\text{He} + {}^{7}\text{Li} + \gamma$

mass ^{10}B	10.012 937 amu
mass ^{1}n	1.008 665 amu
–mass ^{4}He	– 4.002 603 amu
–mass ^{7}Li	–7.016 004 amu
mass change	0.002 995 amu

$(0.002\ 995\ \text{amu})(1.660\ 54 \times 10^{-24}\ \text{g/amu}) = 4.973 \times 10^{-27}\ \text{g}$

$\Delta E = \Delta mc^2 = (4.973 \times 10^{-27}\ \text{g})(10^{-3}\ \text{kg/g})(3.00 \times 10^8\ \text{m/s})^2 = 4.476 \times 10^{-13}\ \text{J}$

$$\text{Kinetic energy} = 2.31\ \text{MeV} \times \frac{1.60 \times 10^{-13}\ \text{J}}{1\ \text{MeV}} = 3.696 \times 10^{-13}\ \text{J}$$

γ photon energy $= \Delta E - \text{KE} = 4.476 \times 10^{-13}\ \text{J} - 3.696 \times 10^{-13}\ \text{J} = 7.80 \times 10^{-14}\ \text{J}$

22.94   (a) $^{100}_{43}\text{Tc} \rightarrow {}^{0}_{1}\text{e} + {}^{100}_{42}\text{Mo}$ (positron emission)

$^{100}_{43}\text{Tc} + {}^{0}_{-1}\text{e} \rightarrow {}^{100}_{42}\text{Mo}$ (electron capture)

(b) Positron emission

Reactant: $^{100}_{43}\text{Tc}$ nucleus = $^{100}_{43}\text{Tc}$ atom – 43 e⁻

Product: $^{100}_{42}\text{Mo}$ nucleus + e⁺ = $^{100}_{42}\text{Mo}$ atom – 42 e⁻ + 1 e⁺

Change: $(^{100}_{43}\text{Tc}$ atom – 43 e⁻$) - (^{100}_{42}\text{Mo}$ atom – 42 e⁻ + 1 e⁺$)$

= $^{100}_{43}\text{Tc}$ atom – $^{100}_{42}\text{Mo}$ atom – 2 e⁻

Mass change = 99.907 657 amu – 99.907 48 amu – (2)(0.000 5486 amu)
         = –0.000 92 amu

(–0.000 92 amu)(1.660 54 x 10⁻²⁴ g/amu)(6.022 x 10²³ mol⁻¹) = –0.000 92 g/mol
$\Delta E = \Delta mc^2$ = (–0.000 92 g/mol)(10⁻³ kg/g)(3.00 x 10⁸ m/s)²
ΔE = –8.3 x 10¹⁰ J/mol = –8.3 x 10⁷ kJ/mol

Electron Capture

Reactant: $^{100}_{43}\text{Tc}$ nucleus + e⁻ = $^{100}_{43}\text{Tc}$ atom – 42 e⁻

Product: $^{100}_{42}\text{Mo}$ nucleus = $^{100}_{42}\text{Mo}$ atom – 42 e⁻

Change: $(^{100}_{43}\text{Tc}$ atom – 42 e⁻$) - (^{100}_{42}\text{Mo}$ atom – 42 e⁻$)$

= $^{100}_{43}\text{Tc}$ atom – $^{100}_{42}\text{Mo}$ atom   (electrons cancel)

Mass change = 99.907 657 amu – 99.907 48 amu = 0.000 177 amu

(0.000 177 amu)(1.660 54 x 10⁻²⁴ g/amu)(6.022 x 10²³ mol⁻¹) = 0.000 177 g/mol
$\Delta E = \Delta mc^2$ = (0.000 177 g/mol)(10⁻³ kg/g)(3.00 x 10⁸ m/s)²
ΔE = 1.6 x 10¹⁰ J/mol = 1.6 x 10⁷ kJ/mol

Only electron capture is observed because there is a mass decrease and a release of energy.

Multi-Concept Problems

22.96 $BaCO_3$, 197.34 amu

$$1.000 \text{ g BaCO}_3 \times \frac{1 \text{ mol BaCO}_3}{197.34 \text{ g BaCO}_3} \times \frac{1 \text{ mol C}}{1 \text{ mol BaCO}_3} \times \frac{12.011 \text{ g C}}{1 \text{ mol C}} = 0.060 \text{ 86 g C}$$

4.0 x 10⁻³ Bq = 4.0 x 10⁻³ disintegrations/s
(4.0 x 10⁻³ Bq = 4.0 x 10⁻³ disintegrations/s)(60 s/min) = 0.24 disintegrations/min

$$\text{sample radioactivity} = \frac{0.24 \text{ disintegrations/min}}{0.060\ 86 \text{ g C}} = 3.94 \text{ disintegrations/min per gram of C}$$

$$\ln\left(\frac{N}{N_0}\right) = (-0.693)\left(\frac{t}{t_{1/2}}\right); \qquad \frac{N}{N_0} = \frac{\text{Decay rate at time } t}{\text{Decay rate at time } t = 0}$$

$$\ln\left(\frac{3.94}{15.3}\right) = (-0.693)\left(\frac{t}{5730 \text{ y}}\right); \qquad t = 11{,}000 \text{ y}$$

22.98   First find the activity of the ^{51}Cr after 17.0 days.

$$\ln\left(\frac{N}{N_0}\right) = (-0.693)\left(\frac{t}{t_{1/2}}\right); \qquad \frac{N}{N_0} = \frac{\text{Decay rate at time } t}{\text{Decay rate at time } t = 0}$$

$$\ln\left(\frac{N}{4.10}\right) = (-0.693)\left(\frac{17.0 \text{ d}}{27.7 \text{ d}}\right)$$

$\ln N - \ln(4.10) = -0.4253$

$\ln N = -0.4253 + \ln(4.10) = 0.9857$

$N = e^{0.9857} = 2.68 \ \mu\text{Ci/mL}$

$(20.0 \text{ mL})(2.68 \ \mu\text{Ci/mL}) = (\text{total blood volume})(0.009\ 35 \ \mu\text{Ci/mL})$

total blood volume $= 5732 \text{ mL} = 5.73 \text{ L}$

Organic Chemistry

23.1

H H H H H H H
H—C—C—C—C—C—C—C—H
H H H H H H H

23.2

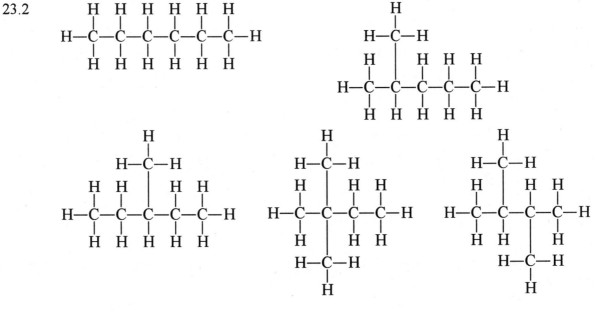

23.3

$CH_3CH_2CH_2CH_2CH_3$ $CH_3CH_2\overset{\displaystyle CH_3}{\underset{}{C}}HCH_3$ $CH_3\overset{\displaystyle CH_3}{\underset{\displaystyle CH_3}{C}}CH_3$

23.4 C_7H_{16}

$CH_3CH_2CH_2\overset{\displaystyle CH_3}{\underset{\displaystyle CH_3}{C}}CH_3$

23.5   Structures (a) and (c) are identical. They both contain a chain of six carbons with two –CH_3 branches at the fourth carbon and one –CH_3 branch at the second carbon. Structure (b) is different, having a chain of seven carbons.

23.6   The two structures are identical. The compound is

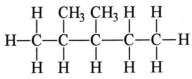

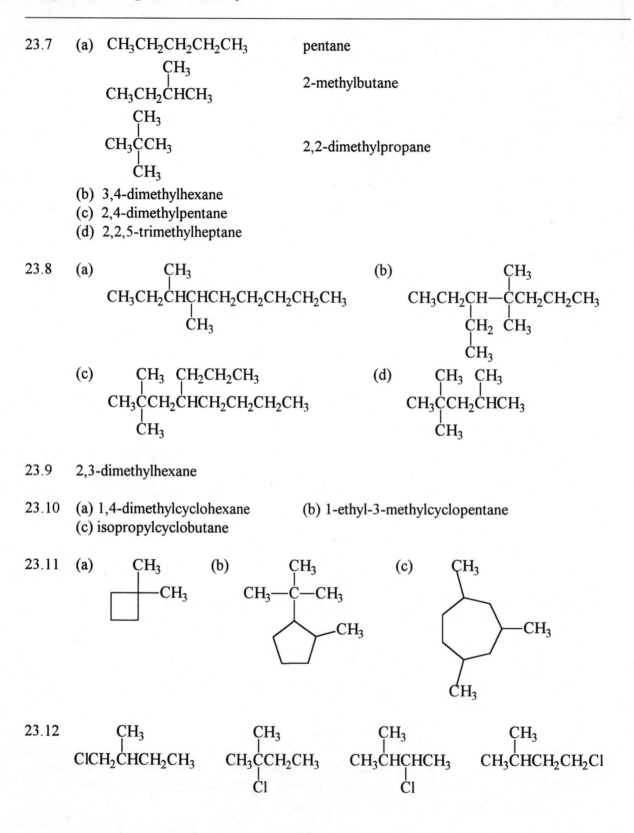

23.7   (a)   CH₃CH₂CH₂CH₂CH₃   pentane

                  CH₃
                  |
          CH₃CH₂CHCH₃   2-methylbutane

                CH₃
                |
          CH₃CCH₃   2,2-dimethylpropane
                |
                CH₃

    (b) 3,4-dimethylhexane
    (c) 2,4-dimethylpentane
    (d) 2,2,5-trimethylheptane

23.8   (a)

23.9   2,3-dimethylhexane

23.10 (a) 1,4-dimethylcyclohexane   (b) 1-ethyl-3-methylcyclopentane
     (c) isopropylcyclobutane

23.11 (a)   (b)   (c)

23.12

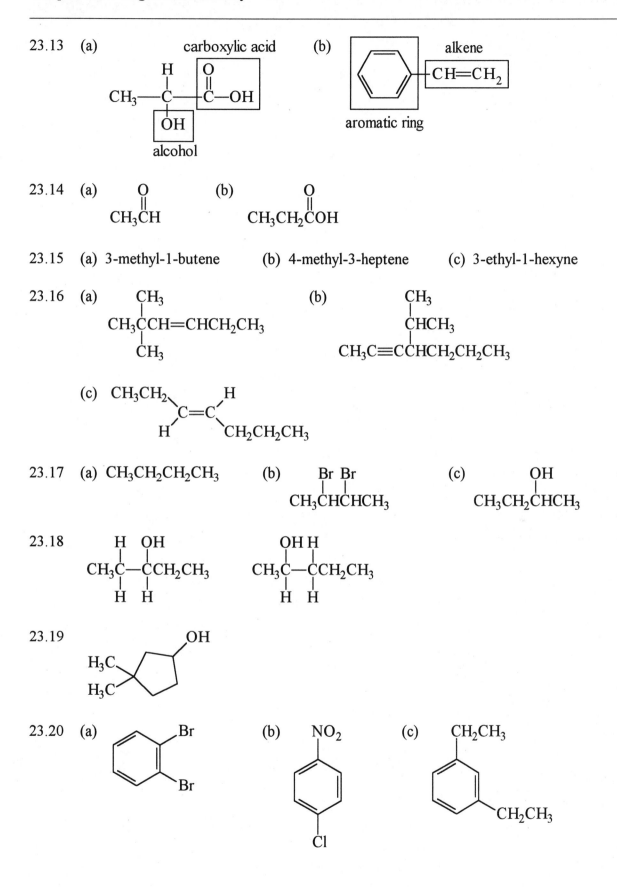

23.13 (a)

carboxylic acid

(b)

alkene

$CH_3-\overset{\overset{H}{|}}{\underset{\underset{OH}{|}}{C}}-\boxed{\overset{\overset{O}{\|}}{C}-OH}$

$\boxed{CH=CH_2}$

aromatic ring

alcohol

23.14 (a) $\overset{\overset{O}{\|}}{CH_3CH}$

(b) $\overset{\overset{O}{\|}}{CH_3CH_2COH}$

23.15 (a) 3-methyl-1-butene   (b) 4-methyl-3-heptene   (c) 3-ethyl-1-hexyne

23.16 (a) $CH_3\overset{\overset{CH_3}{|}}{\underset{\underset{CH_3}{|}}{C}}CH=CHCH_2CH_3$

(b) $CH_3C\equiv C\overset{\overset{CH_3}{\overset{|}{CHCH_3}}}{\underset{}{C}}HCH_2CH_2CH_3$

(c) $\overset{CH_3CH_2}{\underset{H}{}}C=C\overset{H}{\underset{CH_2CH_2CH_3}{}}$

23.17 (a) $CH_3CH_2CH_2CH_3$ (b) $CH_3\overset{\overset{Br}{|}}{C}H\overset{\overset{Br}{|}}{C}HCH_3$ (c) $CH_3CH_2\overset{\overset{OH}{|}}{C}HCH_3$

23.18 $CH_3\overset{\overset{H}{|}}{\underset{\underset{H}{|}}{C}}-\overset{\overset{OH}{|}}{\underset{\underset{H}{|}}{C}}CH_2CH_3$ $CH_3\overset{\overset{OH}{|}}{\underset{\underset{H}{|}}{C}}-\overset{\overset{H}{|}}{\underset{\underset{H}{|}}{C}}CH_2CH_3$

23.19

23.20 (a)   Br   Br

(b) NO_2 Cl

(c) CH_2CH_3 CH_2CH_3

409

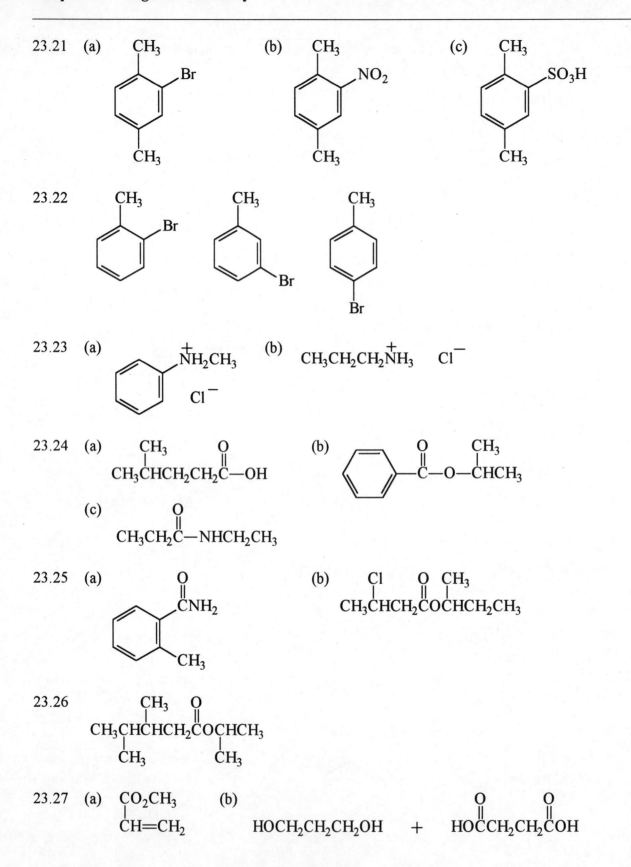

23.28

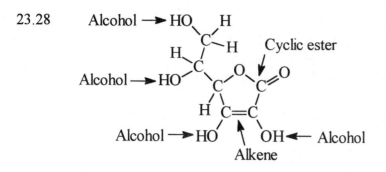

Understanding Key Concepts

23.30 (a)

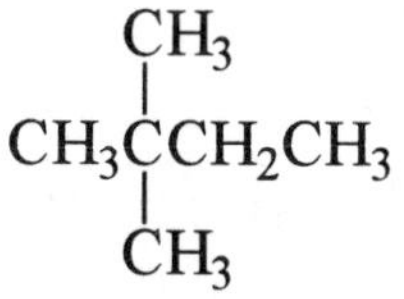

$$CH_3\underset{\underset{CH_3}{|}}{\overset{\overset{CH_3}{|}}{C}}CH_2CH_3$$

(b)

$$CH_3\overset{\overset{CH_3}{|}}{C}H\underset{\underset{OH}{|}}{C}HCH_3$$

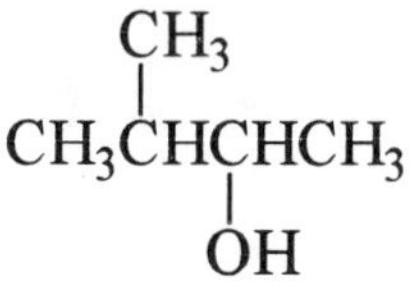

23.32 (a) alkene, ketone, ether   (b) alkene, amine, carboxylic acid

23.34 $CH_2{=}CCl_2$

23.36 There are many possibilities. Here are two:

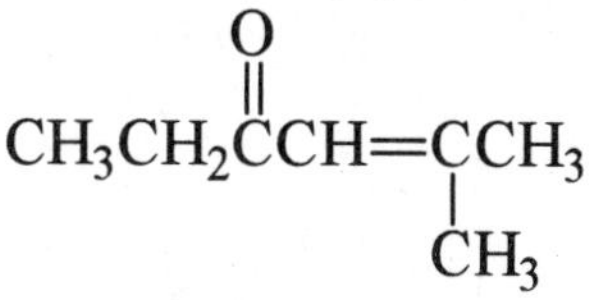

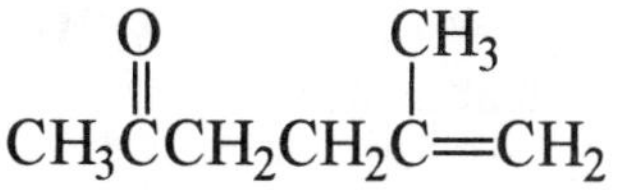

$$CH_3CH_2\overset{\overset{O}{\|}}{C}CH{=}\underset{\underset{CH_3}{|}}{C}CH_3 \qquad CH_3\overset{\overset{O}{\|}}{C}CH_2CH_2\overset{\overset{CH_3}{|}}{C}{=}CH_2$$

Additional Problems
Functional Groups and Isomers

23.38 A functional group is a part of a larger molecule and is composed of an atom or group of atoms that has a characteristic chemical behavior. They are important because their chemistry controls the chemistry in molecules that contain them.

23.40 (a)

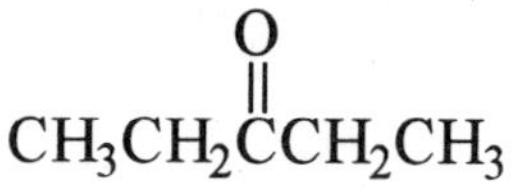

$$CH_3CH_2\overset{\overset{O}{\|}}{C}CH_2CH_3$$

(b)

$$CH_3CH_2CH_2\overset{\overset{O}{\|}}{C}OCH_2CH_3$$

(c)

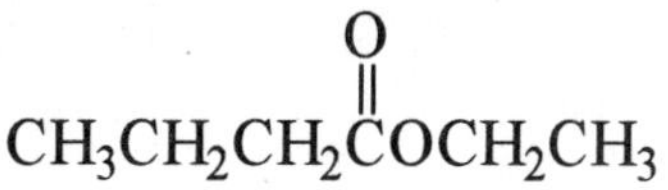

$$NH_2CH_2\overset{\overset{O}{\|}}{C}OH$$

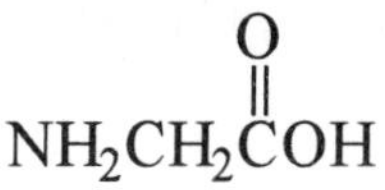

23.42

$$CH_3CH_2CH_2OH \qquad CH_3\overset{\overset{OH}{|}}{C}HCH_3 \qquad CH_3CH_2OCH_3$$

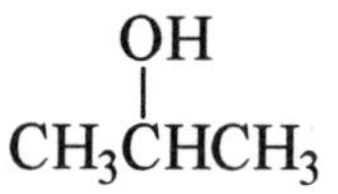

23.44 (a) alkene and aldehyde   (b) aromatic ring, alcohol, and ketone

Alkanes

23.46 In a straight-chain alkane, all the carbons are connected in a row. In a branched-chain alkane, there are branching connections of carbons along the carbon chain.

23.48 In forming alkanes, carbon uses sp^3 hybrid orbitals.

23.50 C_3H_9 contains one more H than needed for an alkane.

23.52 (a) 4-ethyl-3-methyloctane   (b) 4-isopropyl-2-methylheptane
     (c) 2,2,6-trimethylheptane   (d) 4-ethyl-4-methyloctane

23.54

(a)
$$\begin{array}{c}CH_2CH_3\\ |\\ CH_3CH_2CHCH_2CH_2CH_3\end{array}$$

(b)
$$\begin{array}{c}CH_3\ \ CH_3\\ |\ \ \ \ |\\ CH_3C\!\!-\!\!CHCH_2CH_3\\ |\\ CH_3\end{array}$$

(c)
$$\begin{array}{c}CH_2CH_3\\ |\\ CH_3CH_2C\!\!-\!\!CHCH_2CH_2CH_3\\ |\ \ \ \ |\\ CH_3\ CH_3\end{array}$$

(d)
$$\begin{array}{c}CH_3\\ |\\ CH_3\ \ \ \ \ \ CHCH_3\\ |\ \ \ \ \ \ \ \ |\\ CH_3CHCH_2CH_2CHCH_2CH_2CH_3\end{array}$$

23.56 (a) 1,1-dimethylcyclopentane   (b) 1-isopropyl-2-methylcyclohexane
     (c) 1,2,4-trimethylcyclooctane

23.58 The structures are shown in Problem 23.2.
hexane, 2-methylpentane, 3-methylpentane, 2,2-dimethylbutane, and 2,3-dimethylbutane

23.60 (a)

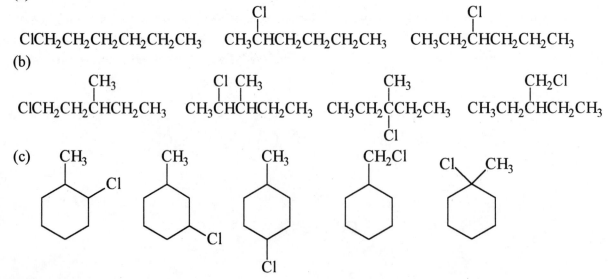

Alkenes, Alkynes, and Aromatic Compounds

23.62   (a) sp^2 (b) sp   (c) sp^2

23.64   Today the term "aromatic" refers to the class of compounds containing a six-membered
ring with three double bonds, not to the fragrance of a compound.

23.66   (a) $CH_3CH{=}CHCH_2CH_3$ (b) $HC{\equiv}CCH_2CH_3$ (c)

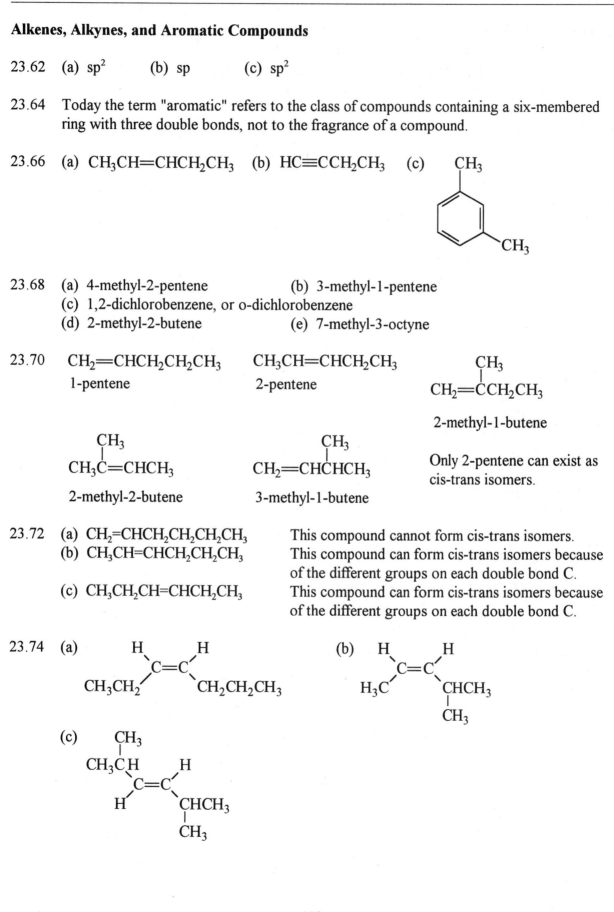

23.68   (a) 4-methyl-2-pentene   (b) 3-methyl-1-pentene
(c) 1,2-dichlorobenzene, or o-dichlorobenzene
(d) 2-methyl-2-butene   (e) 7-methyl-3-octyne

23.70 $CH_2{=}CHCH_2CH_2CH_3$ $CH_3CH{=}CHCH_2CH_3$
1-pentene   2-pentene

2-methyl-1-butene

$CH_3\overset{\underset{|}{CH_3}}{C}{=}CHCH_3$ $CH_2{=}CH\overset{\underset{|}{CH_3}}{C}HCH_3$ Only 2-pentene can exist as
cis-trans isomers.

2-methyl-2-butene   3-methyl-1-butene

23.72   (a) $CH_2{=}CHCH_2CH_2CH_2CH_3$ This compound cannot form cis-trans isomers.
(b) $CH_3CH{=}CHCH_2CH_2CH_3$ This compound can form cis-trans isomers because
of the different groups on each double bond C.

(c) $CH_3CH_2CH{=}CHCH_2CH_3$ This compound can form cis-trans isomers because
of the different groups on each double bond C.

23.74   (a)   (b)

(c)

23.76 Cis-trans isomers are possible for substituted alkenes because of the lack of rotation about the carbon-carbon double bond. Alkanes and alkynes cannot form cis-trans isomers because alkanes have free rotation about carbon-carbon single bonds and alkynes are linear about the carbon-carbon triple bond.

23.78 (a)

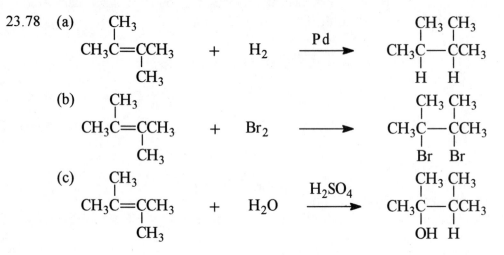

(b)

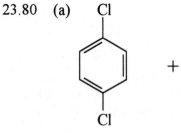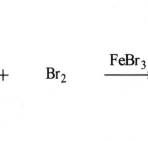

(c)

23.80 (a)

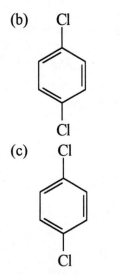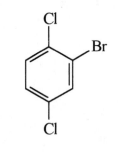

(b)

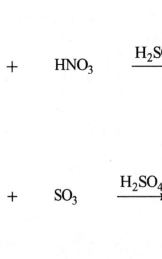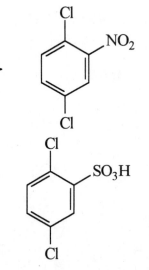

(c)

(d)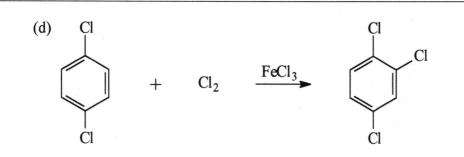

Alcohols, Amines, and Carbonyl Compounds

23.82 (a)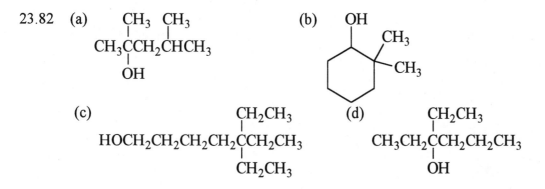

23.84 Quinine, a base will dissolve in aqueous acid, but menthol is insoluble.

23.86 An aldehyde has a terminal carbonyl group. A ketone has the carbonyl group located between two carbon atoms.

23.88 The industrial preparation of ketones and aldehydes involves the oxidation of the related alcohol.

23.90 (a) ketone   (b) aldehyde   (c) ketone   (d) amide   (e) ester

23.92

$$C_6H_5CO_2H(aq) + H_2O(l) \rightleftarrows H_3O^+(aq) + C_6H_5CO_2^-(aq)$$

	$C_6H_5CO_2H$	H_3O^+	$C_6H_5CO_2^-$
initial (M)	1.0	~0	0
change (M)	−x	+x	+x
equil (M)	1.0 − x	x	x

$$K_a = \frac{[H_3O^+][C_6H_5CO_2^-]}{[C_6H_5CO_2H]} = 6.5 \times 10^{-5} = \frac{x^2}{1.0-x} \approx \frac{x^2}{1.0}$$

$$x = [H_3O^+] = [C_6H_5CO_2H]_{diss} = 0.0081 \text{ M}$$

$$\% \text{ dissociation} = \frac{[C_6H_5CO_2H]_{diss}}{[C_6H_5CO_2H]_{initial}} \times 100\% = \frac{0.0081 \text{ M}}{1.0 \text{ M}} \times 100\% = 0.81\%$$

23.94 (a) methyl 4-methylpentanoate   (b) 4,4-dimethylpentanoic acid
(c) 2-methylpentanamide

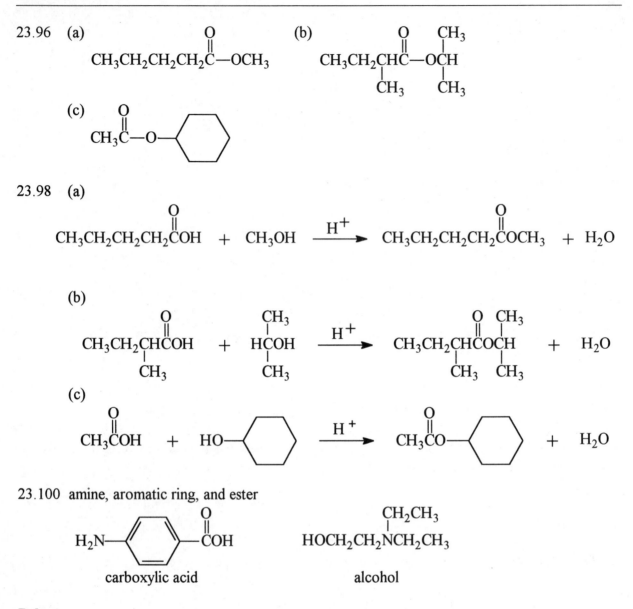

23.96   (a)

CH₃CH₂CH₂CH₂C—OCH₃

(b)

CH₃CH₂CHC—OCH
     | |
    CH₃   CH₃

(c)

CH₃C—O—⬡

23.98   (a)

CH₃CH₂CH₂CH₂COH + CH₃OH —H⁺→ CH₃CH₂CH₂CH₂COCH₃ + H₂O

(b)

CH₃CH₂CHCOH + HCOH —H⁺→ CH₃CH₂CHCOCH + H₂O
       | | | |
      CH₃   CH₃   CH₃ CH₃

(c)

CH₃COH   +   HO—⬡   —H⁺→   CH₃CO—⬡   +   H₂O

23.100 amine, aromatic ring, and ester

H₂N—⬡—COH

carboxylic acid

          CH₂CH₃
           |
HOCH₂CH₂NCH₂CH₃

alcohol

Polymers

23.102 Polymers are large molecules formed by the repetitive bonding together of many smaller molecules, called monomers.

23.104

    ⎛   Cl   Cl   Cl   Cl   ⎞
    | | | | | |
  —⎜CH₂CHCH₂CHCH₂CHCH₂CH⎟—
    ⎝   ⎠ₙ

23.106 (a) CH₂=CH   (b) CH₂=CH   (c) CH₂=CCl₂
               | |
               CN   CH₃

416

23.108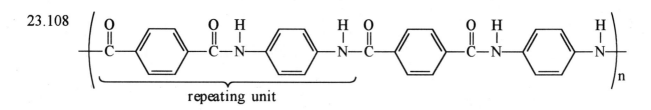
repeating unit

General Problems

23.110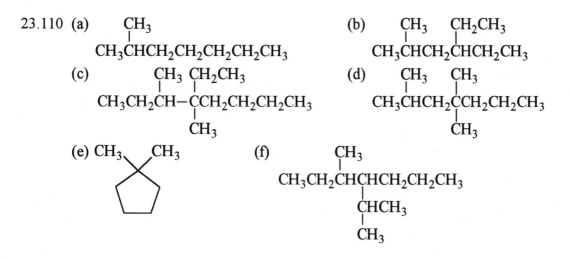

(a)
$$CH_3$$
$$CH_3CHCH_2CH_2CH_2CH_2CH_3$$

(b)
$$CH_3 \quad CH_2CH_3$$
$$CH_3CHCH_2CHCH_2CH_3$$

(c)
$$CH_3 \quad CH_2CH_3$$
$$CH_3CH_2CH-CCH_2CH_2CH_2CH_3$$
$$CH_3$$

(d)
$$CH_3 \quad CH_3$$
$$CH_3CHCH_2CCH_2CH_2CH_3$$
$$CH_3$$

(e)
$$CH_3 \quad CH_3$$
[cyclopentane ring with two CH₃ groups]

(f)
$$CH_3$$
$$CH_3CH_2CHCHCH_2CH_2CH_3$$
$$CHCH_3$$
$$CH_3$$

23.112 Cyclohexene will react with Br_2 and decolorize it. Cyclohexane will not react.

23.114 (a) 
$$CH_3$$
$$CH_3CH_2CH_2CH_2CHCH_3$$

(b) 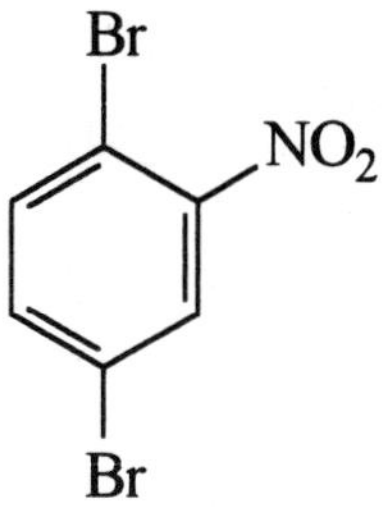
Br
NO₂
Br

(c) 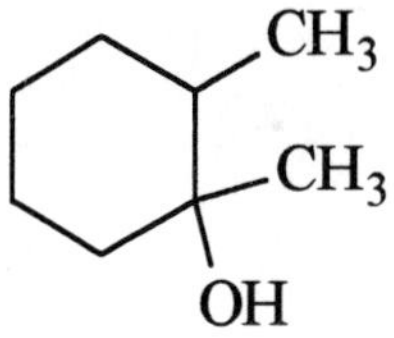
$$CH_3$$
$$CH_3$$
OH
[cyclohexane ring]

Multi-Concept Problems

23.116 (a) Calculate the empirical formula. Assume a 100.0 g sample of fumaric acid.

$$41.4 \text{ g C} \times \frac{1 \text{ mol C}}{12.01 \text{ g C}} = 3.45 \text{ mol C}; \qquad 3.5 \text{ g H} \times \frac{1 \text{ mol H}}{1.008 \text{ g H}} = 3.47 \text{ mol H}$$

$$55.1 \text{ g O} \times \frac{1 \text{ mol O}}{16.00 \text{ g O}} = 3.44 \text{ mol O}$$

Because the mol amounts for the three elements are essentially the same, the empirical formula is CHO (29 amu).

(b) Calculate the molar mass from the osmotic pressure.

417

$$\Pi = MRT; \quad M = \frac{\Pi}{RT} = \frac{\left(240.3 \text{ mm Hg} \times \dfrac{1.00 \text{ atm}}{760 \text{ mm Hg}}\right)}{\left(0.082\,06 \dfrac{L \cdot atm}{K \cdot mol}\right)(298 \text{ K})} = 0.0129 \text{ M}$$

$(0.0129 \text{ mol/L})(0.1000 \text{ L}) = 1.29 \times 10^{-3}$ mol fumaric acid

fumaric acid molar mass $= \dfrac{0.1500 \text{ g}}{1.29 \times 10^{-3} \text{ mol}} = 116$ g/mol

molecular mass $= 116$ amu

(c) Determine the molecular formula. $\dfrac{\text{molar mass}}{\text{empirical formula mass}} = \dfrac{116}{29} = 4$

molecular formula $= C_{(1 \times 4)}H_{(1 \times 4)}O_{(1 \times 4)} = C_4H_4O_4$

From the titration, the number of carboxylic acid groups can be determined.

mol $C_4H_4O_4 = 0.573$ g $\times \dfrac{1 \text{ mol } C_4H_4O_4}{116 \text{ g}} = 0.004\,94$ mol $C_4H_4O_4$

mol NaOH used $= (0.0941 \text{ L})(0.105 \text{ mol/L}) = 0.0099$ mol NaOH

$\dfrac{\text{mol NaOH}}{\text{mol } C_4H_4O_4} = \dfrac{0.0099 \text{ mol}}{0.004\,94 \text{ mol}} = 2$

Because 2 mol of NaOH are required to titrate 1 mol $C_4H_4O_4$, $C_4H_4O_4$ is a diprotic acid. Because $C_4H_4O_4$ gives an addition product with HCl and a reduction product with H_2, it contains a double bond.

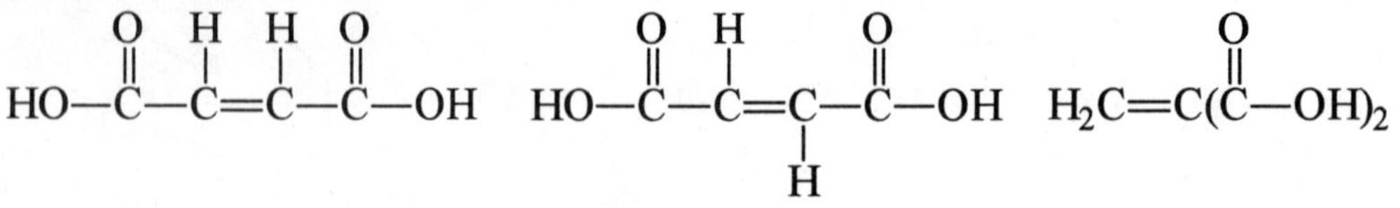

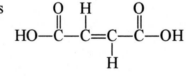

(d) The correct structure is

23.118 (a) 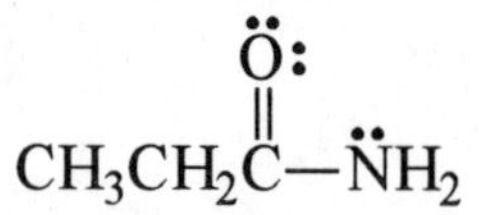   propanamide

(b)

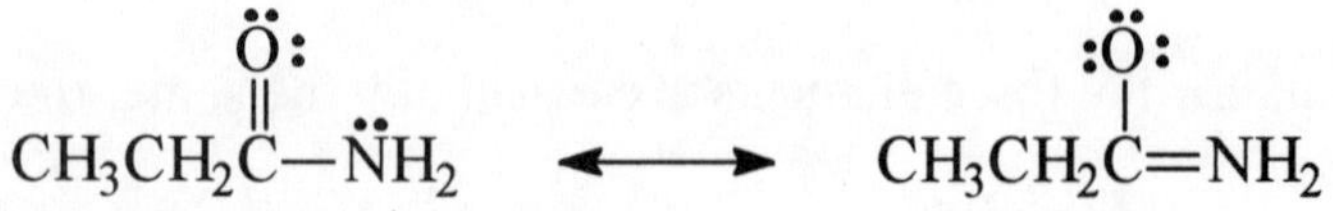

(c)

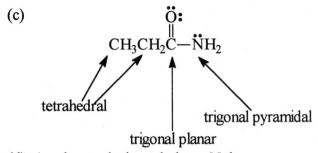

tetrahedral

trigonal pyramidal

trigonal planar

(d) An observed trigonal planar N does not agree with the VSEPR prediction. The second resonance structure is consistent with a trigonal planar N.

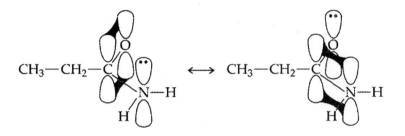

24   **Biochemistry**

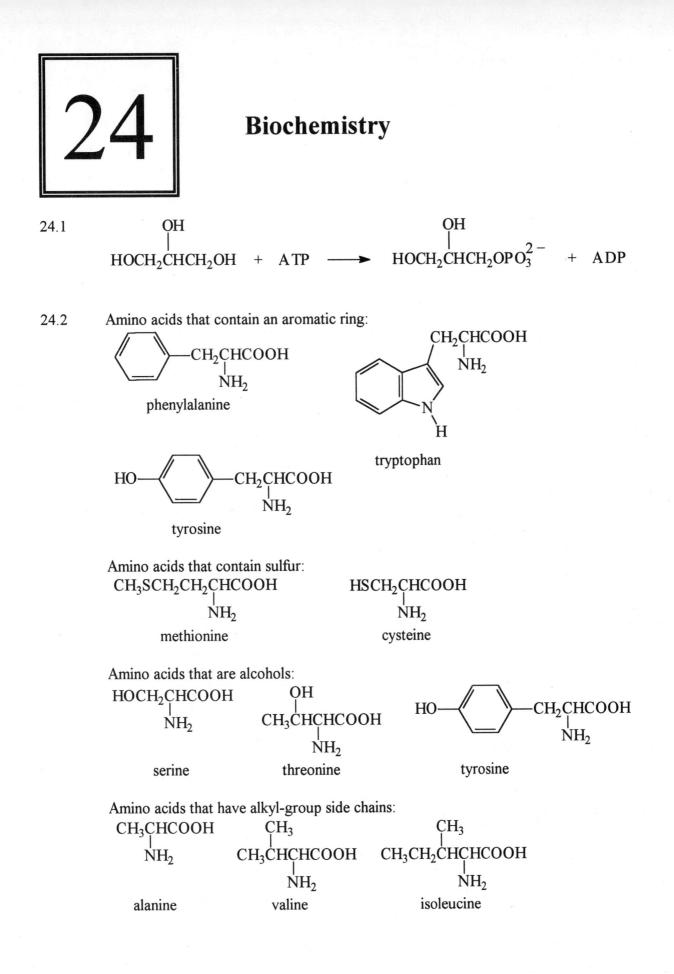

24.1

$$HOCH_2\overset{\displaystyle OH}{\underset{\displaystyle |}{C}}HCH_2OH \ + \ ATP \ \longrightarrow \ HOCH_2\overset{\displaystyle OH}{\underset{\displaystyle |}{C}}HCH_2OPO_3^{2-} \ + \ ADP$$

24.2   Amino acids that contain an aromatic ring:

phenylalanine

tryptophan

tyrosine

Amino acids that contain sulfur:

$$CH_3SCH_2CH_2\overset{\displaystyle |}{C}HCOOH \qquad HSCH_2\overset{\displaystyle |}{C}HCOOH$$
$$\underset{\displaystyle NH_2}{} \qquad \qquad \underset{\displaystyle NH_2}{}$$

methionine   cysteine

Amino acids that are alcohols:

serine   threonine   tyrosine

Amino acids that have alkyl-group side chains:

alanine   valine   isoleucine

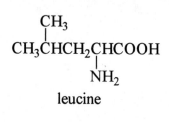

leucine

24.3

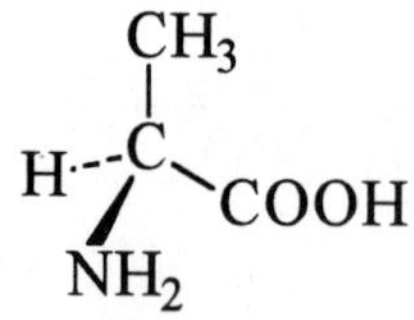

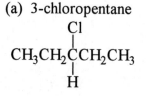

24.4   The amino acid is threonine with a neutral polar side chain.

24.5   (a) a glove and   (c) a screw are chiral

24.6   2-aminopropane

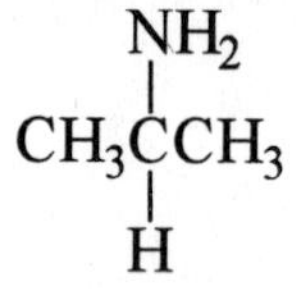

$$\underset{\overset{|}{H}}{\overset{\overset{NH_2}{|}}{CH_3CCH_3}}$$

No carbon in 2-aminopropane has four different groups attached to it so the molecule is achiral.

2-aminobutane

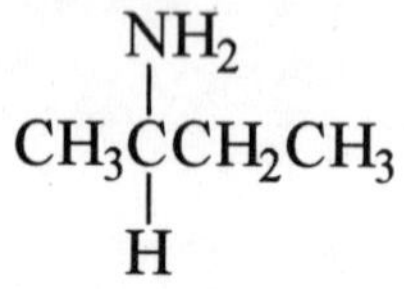

$$\underset{\overset{|}{H}}{\overset{\overset{NH_2}{|}}{CH_3CCH_2CH_3}}$$

The second carbon in 2-aminobutane has four different groups attached to it so the molecule is chiral.

24.7   (a) 3-chloropentane

$$\underset{\overset{|}{H}}{\overset{\overset{Cl}{|}}{CH_3CH_2CCH_2CH_3}}$$

No carbon in 3-chloropentane has four different groups attached to it so the molecule is achiral.

(b) 2-chloropentane

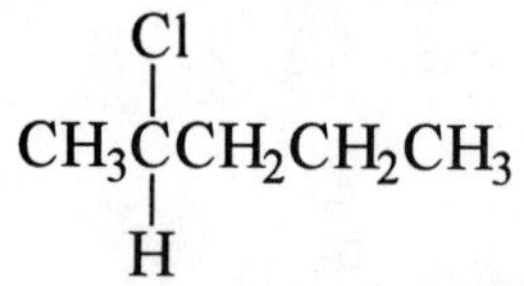

$$\underset{\overset{|}{H}}{\overset{\overset{Cl}{|}}{CH_3CCH_2CH_2CH_3}}$$

The second carbon in 2-chloropentane has four different groups attached to it so the molecule is chiral.

(c)

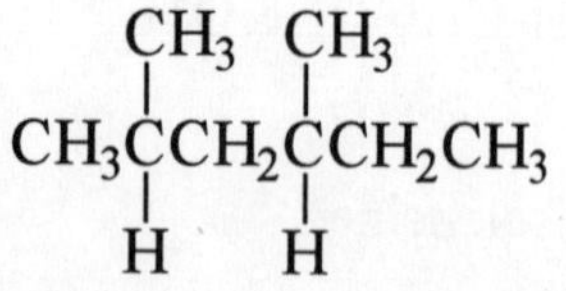

$$\underset{\overset{|}{H}\ \ \ \overset{|}{H}}{\overset{\overset{CH_3}{|}\ \ \ \overset{CH_3}{|}}{CH_3CCH_2CCH_2CH_3}}$$

The fourth carbon in the straight chain has four different groups attached to it so the molecule is chiral.

24.8   isoleucine and threonine

24.9   (b) and (c) are identical. (a) is an enantiomer.

24.10   Val-Cys   Cys-Val

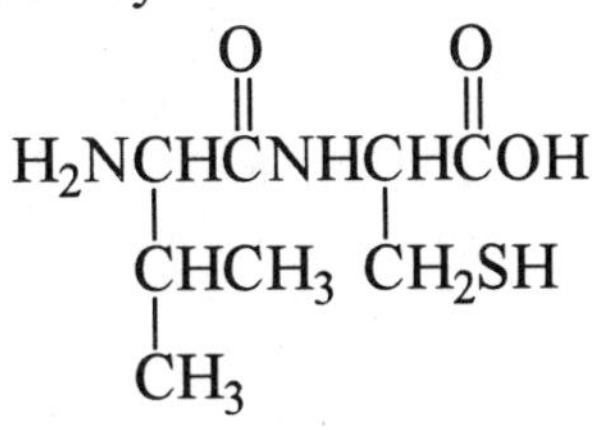

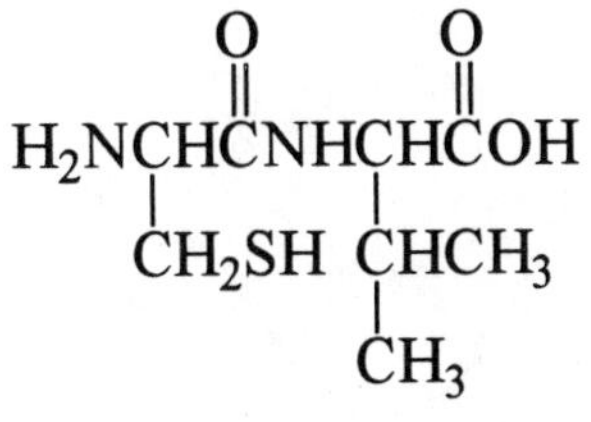

$$\underset{\underset{\underset{CH_3}{|}}{\underset{CHCH_3}{|}}{H_2NCHCNHCHCOH}}{\overset{\overset{O}{\parallel}}{}\overset{\overset{O}{\parallel}}{}}$$

Val-Cys:
$$H_2N\underset{\underset{\underset{CH_3}{|}}{\underset{CHCH_3}{|}}}{CH}\overset{O}{\underset{\parallel}{C}}NH\underset{CH_2SH}{CH}\overset{O}{\underset{\parallel}{C}}OH$$

Cys-Val:
$$H_2N\underset{\underset{\underset{CH_3}{|}}{\underset{CHCH_3}{|}}}{CH}\overset{O}{\underset{\parallel}{C}}NH\underset{CH_2SH}{CH}\overset{O}{\underset{\parallel}{C}}OH$$

24.11   Val-Tyr-Gly; Val-Gly-Tyr; Tyr-Gly-Val; Tyr-Val-Gly; Gly-Tyr-Val; Gly-Val-Tyr

24.12   (a) aldopentose   (b) ketotriose   (c) aldotetrose

24.13

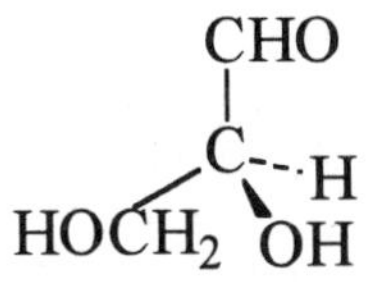

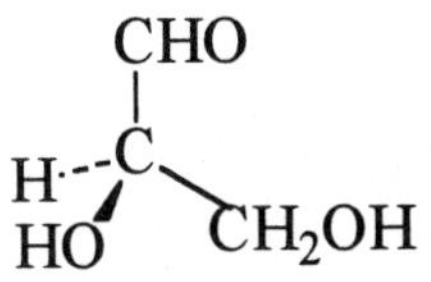

$$\underset{HOCH_2\quad OH}{\overset{CHO}{\overset{|}{\underset{}{C}}}\cdots H}$$
$$\underset{HO\quad CH_2OH}{H\cdots\overset{CHO}{\overset{|}{C}}}$$

24.14   In general, a compound with n chiral carbon atoms has a maximum of 2^n possible forms. Because ribose has three chiral carbon atoms, the maximum number of ribose forms $= 2^3 = 8$.

24.15   Aldohexose; 4 chiral carbons

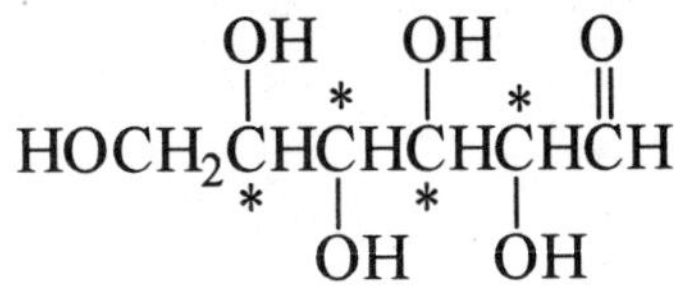

$$\underset{\underset{OH\quad OH}{\underset{*\;|\;\;\;*\;|}{}}}{HOCH_2\overset{\overset{OH\quad OH\quad O}{|\;\;*\;|\;\;*\;\parallel}}{CHCHCHCHCH}}$$

24.16

$$\underset{\underset{\underset{CH_2OC(CH_2)_7CH=CH(CH_2)_7CH_3}{|}}{\underset{CHOC(CH_2)_7CH=CH(CH_2)_7CH_3}{\overset{O}{\parallel}}}}{CH_2O\overset{O}{\overset{\parallel}{C}}(CH_2)_7CH=CH(CH_2)_7CH_3}$$

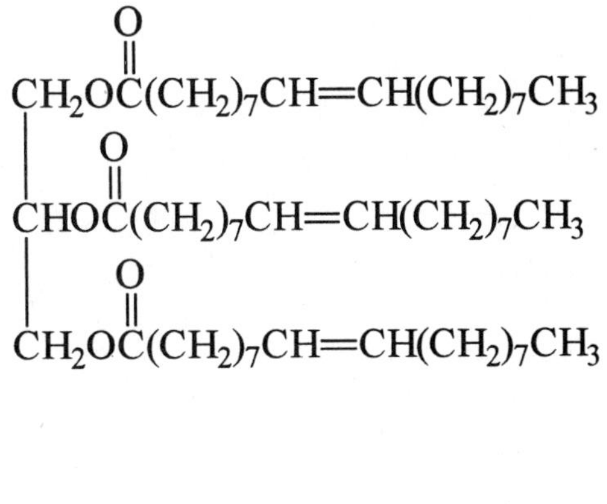

24.17   DNA dinucleotide A – G.

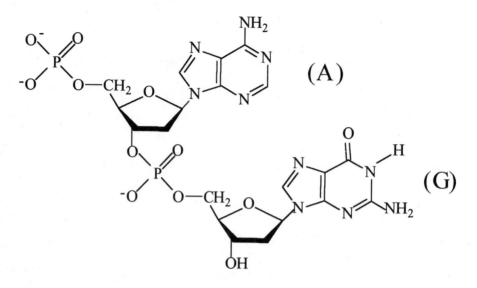

(A)

(G)

24.18   RNA dinucleotide U – A.

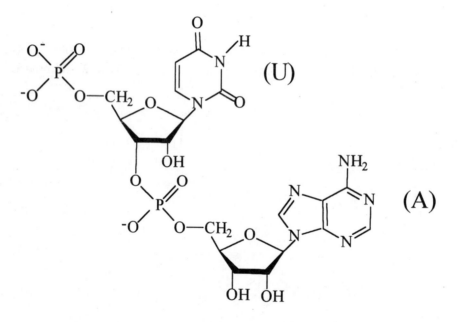

(U)

(A)

24.19   (a) adenine (DNA, RNA)   (b) thymine (DNA)

24.20   Original: G–G–C–C–C–G–T–A–A–T
         Complement: C–C–G–G–G–C–A–T–T–A

24.21

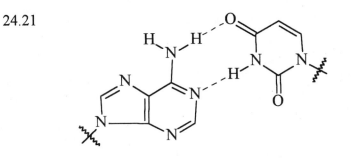

24.22   C–G–T–G–A–T–T–A–C–A (DNA)
            G–C–A–C–U–A–A–U–G–U (RNA)

24.23   U–G–C–A–U–C–G–A–G–U (RNA)
            A–C–G–T–A–G–C–T–C–A (DNA)

24.24   Human genes contain short, repeating sequences of noncoding DNA, called short tandem repeat (STR) loci. The base sequences in these STR loci are slightly different for every individual.

Understanding Key Concepts

24.26   (a) serine   (b) methionine

24.28   β form

24.30   Ser-Val

Additional Problems
Amino Acids, Peptides, and Proteins

24.32   The "α" in α-amino acids means that the amino group in each is connected to the carbon atom alpha to (next to) the carboxylic acid group.

24.34   (a) serine   (b) threonine   (c) proline   (d) phenylalanine   (e) cysteine

24.36   (a) $H_2NCH_2\overset{\overset{\displaystyle O}{\|}}{C}OCH_3$ (b) $HO\overset{\overset{\displaystyle O}{\|}}{C}CH_2\overset{+}{N}H_3$ Cl^-

24.38   Val-Ser-Phe-Met-Thr-Ala

24.40   (a) The primary structure of a protein is the sequence in which amino acids are linked together.
           (b) The secondary structure of a protein is the orientation of segments of a protein chain into a regular pattern.

(c) The tertiary structure of a protein is the way in which a protein chain folds into a specific three-dimensional shape.

24.42   A protein's tertiary structure is stabilized by amino acid side-chain interactions, which include hydrophobic interactions, covalent disulfide bridges, electrostatic salt bridges, and hydrogen bonds.

24.44   Cysteine is an important amino acid for defining the tertiary structure of proteins because the nearby cysteine residues can link together forming a disulfide bridge.

24.46   Met-Ile-Lys, Met-Lys-Ile, Ile-Met-Lys, Ile-Lys-Met, Lys-Met-Ile, Lys-Ile-Met

24.48

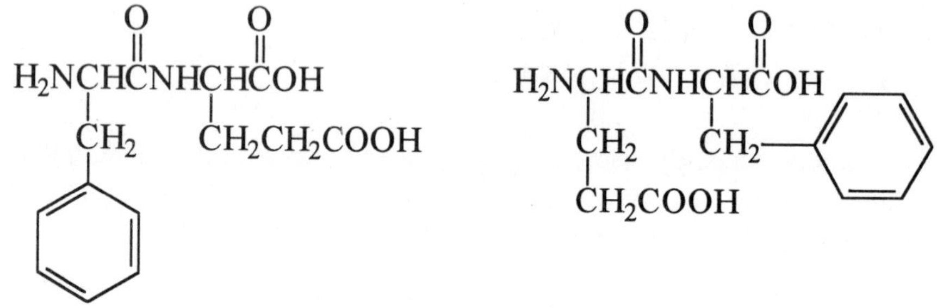

Molecular Handedness

24.50   (a) a shoe and   (c) a light bulb are chiral.

24.52   (a) 2,4-dimethylheptane

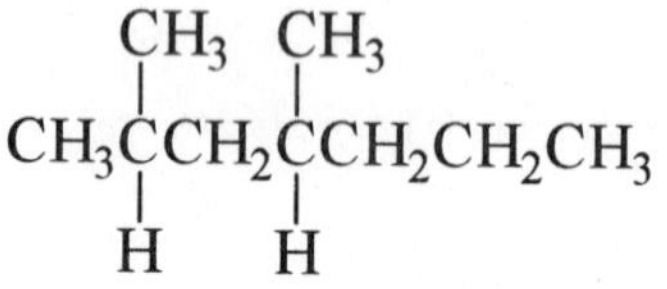

The fourth carbon in the straight chain has four different groups attached to it so the molecule is chiral.

(b) 5–ethyl–3,3–dimethylheptane

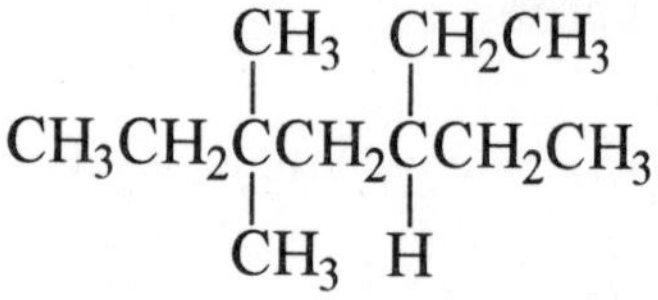

No carbon in 5–ethyl–3,3–dimethylheptane has four different groups attached to it so the molecule is achiral.

24.54 $CH_3CH_2CH_2CH_2CH_2OH$

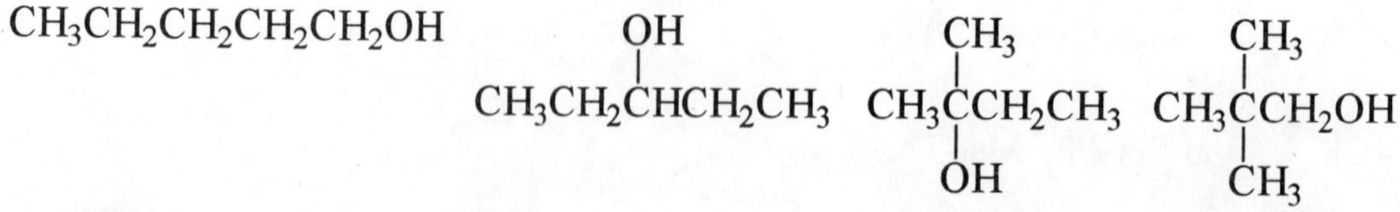

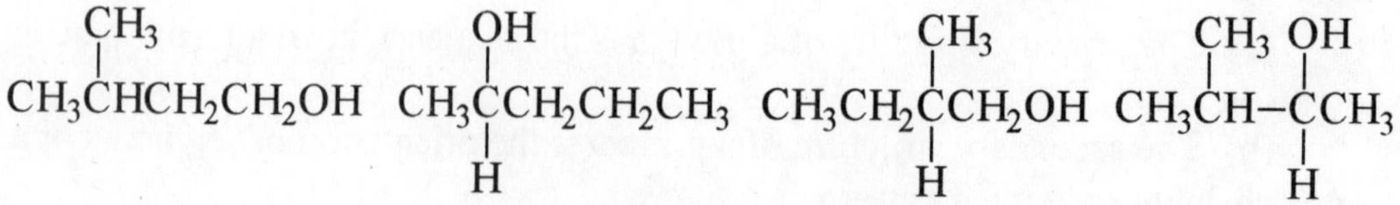

426

Carbohydrates

24.56   An aldose contains the aldehyde functional group while a ketose contains the ketone functional group.

24.58   Structurally, starch differs from cellulose in that it contains α- rather than β-glucose units.
Cellulose is the fibrous substance used by plants as a structural material in grasses, leaves, and stems. It consists of several thousand β-glucose molecules joined together by 1,4 links to form an immense polysaccharide.
Starch is a polymer made up of α-glucose units. Unlike cellulose, starch is easily digested. The starch in beans, rice, and potatoes is an essential part of the human diet.

24.60

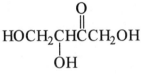

24.62

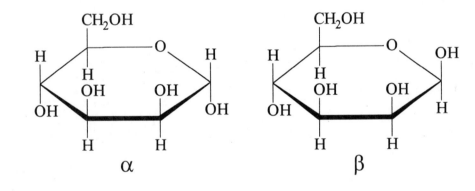

24.64

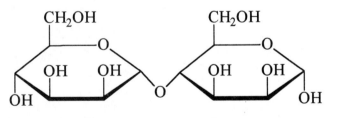

Lipids

24.66   Long-chain carboxylic acids are called fatty acids. Fatty acids are usually unbranched and have an even number of carbon atoms in the range of 12-22.

24.68

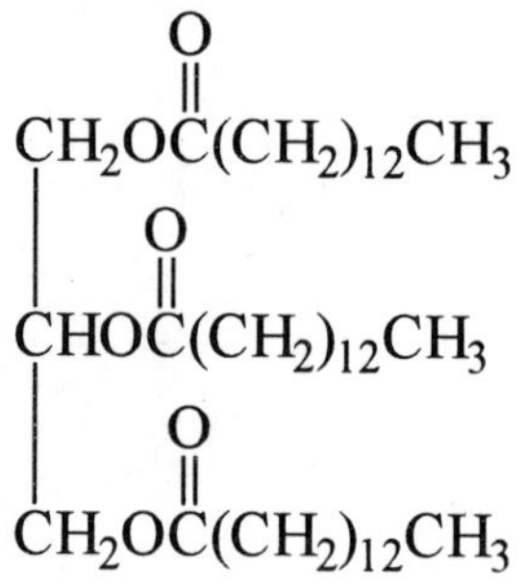

$$CH_2O\overset{\overset{O}{\|}}{C}(CH_2)_{12}CH_3$$
$$CHO\overset{\overset{O}{\|}}{C}(CH_2)_{12}CH_3$$
$$CH_2O\overset{\overset{O}{\|}}{C}(CH_2)_{12}CH_3$$

24.70

$$CH_2O\overset{\overset{O}{\|}}{C}(CH_2)_{16}CH_3$$
$$CHO\overset{\overset{O}{\|}}{C}(CH_2)_{16}CH_3$$
$$CH_2O\overset{\overset{O}{\|}}{C}(CH_2)_{14}CH_3$$

$$CH_2O\overset{\overset{O}{\|}}{C}(CH_2)_{16}CH_3$$
$$CHO\overset{\overset{O}{\|}}{C}(CH_2)_{14}CH_3$$
$$CH_2O\overset{\overset{O}{\|}}{C}(CH_2)_{16}CH_3$$

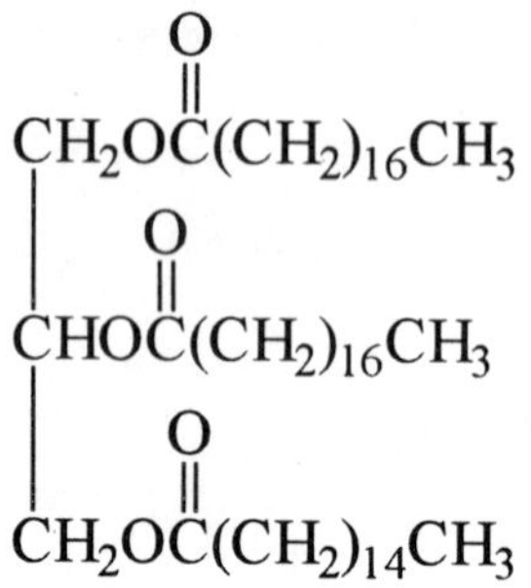

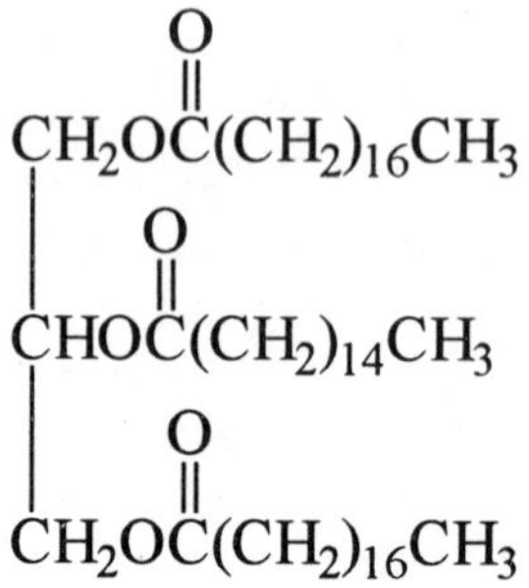

The two fat molecules differ from each other depending on where the palmitic acid chain is located. In the first fat molecule, palmitic acid is on an end and in the second it is in the middle.

24.72

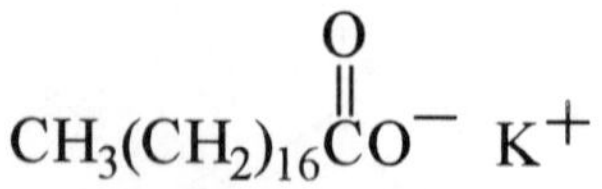

$$CH_3(CH_2)_{16}\overset{\overset{O}{\|}}{C}O^-\ K^+$$
potassium stearate

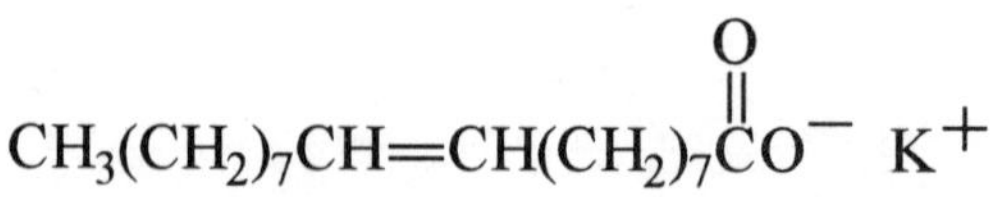

$$CH_3(CH_2)_7CH{=}CH(CH_2)_7\overset{\overset{O}{\|}}{C}O^-\ K^+$$
potassium oleate

$$CH_3CH_2CH{=}CHCH_2CH{=}CHCH_2CH{=}CH(CH_2)_7\overset{\overset{O}{\|}}{C}O^-\ K^+$$
potassium linolenate

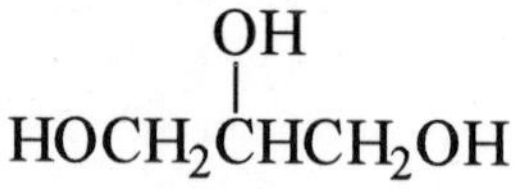

$$HOCH_2\overset{\overset{OH}{|}}{C}HCH_2OH$$
glycerol

24.74 (a)

$$CH_3(CH_2)_7\overset{\overset{Br}{|}}{C}H\overset{\overset{Br}{|}}{C}H(CH_2)_7COOH$$

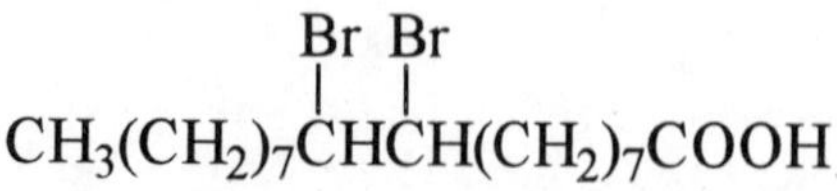

(b) $CH_3(CH_2)_{16}COOH$

(c)

$$CH_3(CH_2)_7CH{=}CH(CH_2)_7\overset{\overset{O}{\|}}{C}OCH_3$$

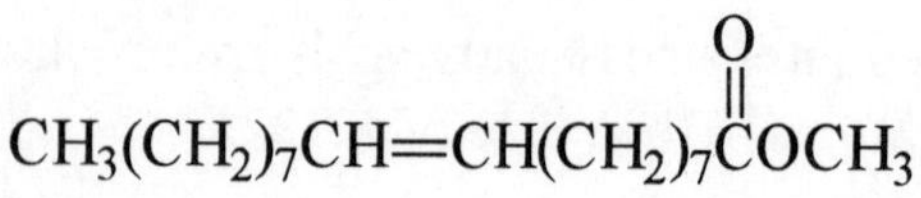

Nucleic Acids

24.76   Just as proteins are polymers made of amino acid units, nucleic acids are polymers made up of nucleotide units linked together to form a long chain. Each nucleotide contains a phosphate group, an aldopentose sugar, and an amine base.

24.78   Most DNA of higher organisms is found in the nucleus of cells.

24.80   A chromosome is a threadlike strand of DNA in the cell nucleus. A gene is a segment of a DNA chain that contains the instructions necessary to make a specific protein.

24.82

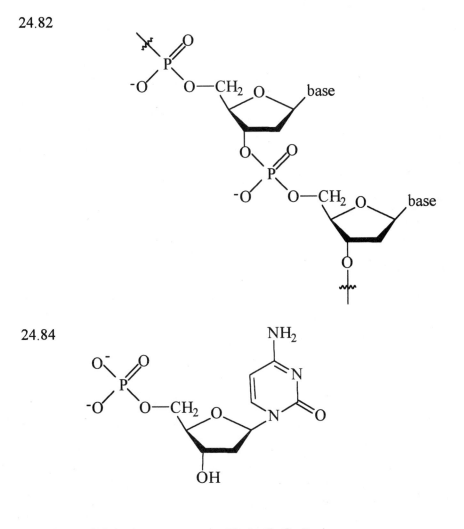

24.84

24.86   Original: T–A–C–C–G–A
         Complement: A–T–G–G–C–T

24.88   It takes three nucleotides to code for a specific amino acid. In insulin the 21 amino acid chain would require (3 x 21) = 63 nucleotides to code for it, and the 30 amino acid chain would require (3 x 30) = 90 nucleotides to code for it.

General Problems

24.90

$$CH_3(CH_2)_{18}\overset{\overset{\displaystyle O}{\|}}{C}O(CH_2)_{31}CH_3$$

24.92 $\Delta G° = +2870$ kJ/mol, because photosynthesis is the reverse reaction.

24.94   (a)

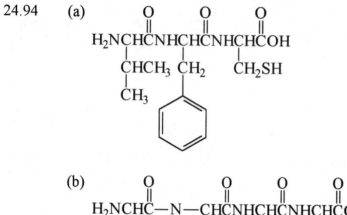

(b)

24.96

$$CH_3(CH_2)_{16}\overset{\overset{\displaystyle O}{\|}}{C}O(CH_2)_{21}CH_3$$

24.98   Original: A–G–T–T–C–A–T–C–G
         Complement: T–C–A–A–G–T–A–G–C

Multi-Concept Problem

24.100   (a) grams $I_2 = (0.0250 \text{ L})(0.200 \text{ mol/L})\left(\dfrac{253.81 \text{ g } I_2}{1 \text{ mol } I_2}\right) = 1.27 \text{ g } I_2$

         (b) mol $Na_2S_2O_3 = (0.08199 \text{ L})(0.100 \text{ mol/L}) = 8.20 \times 10^{-3}$ mol

         grams excess $I_2 = 8.20 \times 10^{-3}$ mol $Na_2S_2O_3 \times \dfrac{1 \text{ mol } I_2}{2 \text{ mol } Na_2S_2O_3} \times \dfrac{253.81 \text{ g } I_2}{1 \text{ mol } I_2} = 1.04 \text{ g } I_2$

         grams I_2 reacted $= 1.27 \text{ g} - 1.04 \text{ g} = 0.23 \text{ g } I_2$

         (c) iodine number $= \dfrac{0.23 \text{ g } I_2}{0.500 \text{ g milkfat}} \times 100 = 46$

         (d) mol milkfat $= 0.500 \text{ g milkfat} \times \dfrac{1 \text{ mol milkfat}}{800 \text{ g milkfat}} = 6.25 \times 10^{-4}$ mol

$$\text{mol } I_2 \text{ reacted} = 0.23 \text{ g } I_2 \times \frac{1 \text{ mol } I_2}{253.81 \text{ g } I_2} = 9.06 \times 10^{-4} \text{ mol}$$

$$\text{number of double bonds per molecule} = \frac{9.06 \times 10^{-4} \text{ mol } I_2}{6.25 \times 10^{-4} \text{ mol milkfat}} = 1.4 \text{ double bonds}$$

eMedia Solutions

Chemistry: Matter and Measurement

1.112 **(a)** metals
(b) halogens

1.114 **(a)** **(a)** No. For multiplication and division, the answer cannot have more significant figures than the original numbers.
(b) Yes. For addition and subtraction, the answer cannot have more numbers to the right of the decimal point than the original numbers, but it can have more significant figures. For example, in the equation: $72.2 + 86.9 = 159.1$ each original number has 3 significant figures, while their sum contains 4.

Atom, Molecule, and Ions

2.114 Most of the alpha particles pass unscattered through the gold foil. A few, however, are scattered through large angles with some even bouncing back in the direction they'd come. These observations resulted in the development of a new view of the atom. Rutherford proposed that the gold atoms consisted of a small, dense nucleus surrounded by mostly empty space. The alpha particles passed freely through the empty space but were scattered through larger angles when they happened to hit the tiny nucleus.

2.116

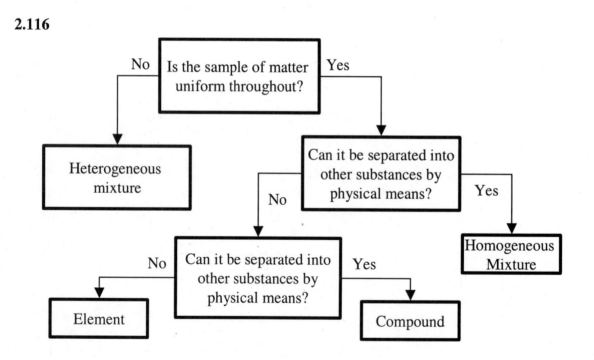

2.118 The 3 cations that form soluble compounds are ammonium, NH_4^+, lithium, Li^+, and sodium, Na^+. The two anions that form soluble compounds are nitrate, NO_3^- and perchlorate, ClO_4^-.

3

Formulas, Equations, and Moles

3.132 *Steps required to determine the mass of NaOH:*

(a) Balance the equation.

$$H_2SO_4 + 2\,NaOH \rightarrow 2H_2O + Na_2SO_4$$

(b) Determine number of moles of H_2SO_4 in 20.5 grams.

$$20.5\,g\,H_2SO_4 \times \frac{1\,mole\,H_2SO_4}{98.08\,g\,H_2SO_4} = 0.209\,mol\,H_2SO_4$$

(c) Use coefficients in the balanced equation to determine number of moles of NaOH.

$$0.209\,mol\,H_2SO_4 \times \frac{2\,mol\,NaOH}{1\,mol\,H_2SO_4} = 0.418\,mol\,NaOH$$

(d) Using molar mass determine number of grams of NaOH required.

$$0.418\,mol\,NaOH \times \frac{40.00g}{1\,mol\,NaOH} = 16.7g\,NaOH$$

Mass of water produced when 17.4 g of Na$_2$SO$_4$ is generated:

(a) Determine molar mass of Na_2SO_4.

$$(2 \times \frac{22.99\,g\,Na}{mole}) + (\frac{32.07\,g\,S}{mole}) + (4 \times \frac{16.00\,g\,O}{mole}) = 142.05\,\frac{g\,Na_2SO_4}{mole}$$

(b) Determine number of moles of Na_2SO_4 in 17.4 grams.

$$\frac{17.4g\,Na_2SO_4}{142.05g\,mol^{-1}} = 0.122\,mol\,Na_2SO_4$$

(c) Use coefficients in the balanced equation to determine number of moles of H_2O.

$$0.122\,mol\,Na_2SO_4 \times \frac{2\,mol\,H_2O}{1\,mol\,Na_2SO_4} = 0.245\,mol\,H_2O$$

(d) Determine mass of water.

$$0.245\,mol\,H_2O \times \frac{18.00g\,H_2O}{mol} = 4.41g\,H_2O$$

3.134

(a) $25.0g\ H_2SO_4 \times \dfrac{1\,mole\ H_2SO_4}{98.07g\ H_2SO_4} = 0.255\ moles\ H_2SO_4$

$0.255\ moles\ H_2SO_4 \times \dfrac{1\,Liter}{0.073\ moles\ H_2SO_4} = 3.49\ Liter$

(b) $40.0g\ KCl \times \dfrac{1\,mole\ KCl}{74.551g\ KCl} = 0.536\ moles\ KCl$

$0.536\ moles\ KCl \times \dfrac{Liter}{0.101\ moles} = 5.31\ Liters$

(d) 22 g. It requires less HCl. Since the molecular weight of NaCl is greater than the molecular weight of HCl, you need more mass to equal the same number of moles.

3.136

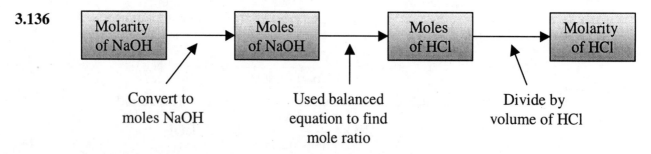

To determine the volume of NaOH solution needed to achieve the equivalence point, we do the following:

(3.18 M HCl)x (20.00 L)= 63.6 moles HCl

$63.6\ moles\ HCl \times \dfrac{1\,mole\ NaOH}{1\,mole\ HCl} = 63.6\ moles\ NaOH$

$\dfrac{63.6\ moles\ NaOH}{0.134\ M\ NaOH} = 475\,L\ \ NaOH\ solution$

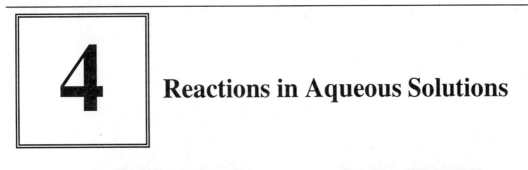

Reactions in Aqueous Solutions

4.114

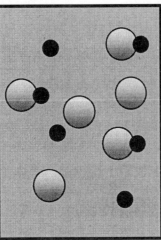 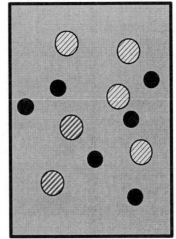

Acetic acid (weak electrolyte)   Hydrochloric acid (strong electrolyte)

4.116 FeCl₃(aq) + 3 NaOH(aq) → Fe(OH)₃(s) + NaCl(aq)

$FeCl_3(aq) + 3 NaOH(aq) \rightarrow Fe(OH)_3(s) + NaCl(aq)$

$Fe^{3+}(aq) + 3 Cl^-(aq) + 3 Na^+(aq) + 3OH(aq) \rightarrow Fe(OH)_3(s) + Na^+(aq) + Cl^-(aq)$

$Fe^{3+}(aq) + 3 OH^-(aq) \rightarrow Fe(OH)_3(s)$

$2 FeCl_3(aq) + 3 Na_2S(aq) \rightarrow Fe_2S_3(s) + 6 NaCl(aq)$

$2 Fe^{3+}(aq) + 6 Cl^-(aq) + 6 Na^+(aq) + 3S^{2-}(aq) \rightarrow Fe_2S_3(s) + 6Na^+(aq) + 6Cl^-(aq)$

$2 Fe^{3+}(aq) + 3S^{2-}(aq) \rightarrow Fe_2S_3(s)$

$2 FeCl_3(aq) + 3 K_2CO_3(aq) \rightarrow Fe_2(CO_3)_3(s) + 6 KCl(aq)$

$2 Fe^{3+}(aq) + 6 Cl^-(aq) + 6 K^+(aq) + 3 CO_3^{2-}(aq) \rightarrow Fe_2(CO_3)_3(s) + 6 K^+(aq) + 6 Cl^-(aq)$

$2 Fe^{3+}(aq) + 3 CO_3^{2-}(aq) \rightarrow Fe_2(CO_3)_3(s)$

$FeCl_3(aq) + Na_3PO_4(aq) \rightarrow FePO_4(s) + 3 NaCl(aq)$

$Fe^{3+}(aq) + 3 Cl^-(aq) + 3 Na^+(aq) + PO_4^{3-}(aq) \rightarrow FePO_4(s) + 3 Na^+(aq) + 3 Cl^-(aq)$

$Fe^{3+}(aq) + PO_4^{3-}(aq) \rightarrow FePO_4(s)$

4.118 $AgNO_3(aq) + KCl(aq) \rightarrow AgCl(s) + KNO_3(aq)$

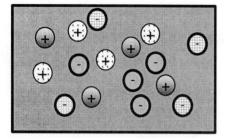   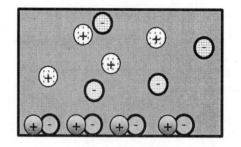

Net ionic equation:
$$Ag^+(aq) + Cl^-(aq) \rightarrow AgCl(s)$$

$$NaCl(aq) + NH_4(NO_3)(aq) \rightarrow Na(NO_3)(aq) + NH_4Cl(aq)$$

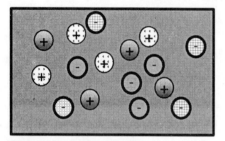   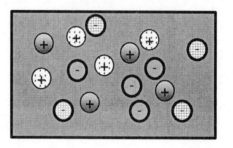

There is no net reaction.

Periodicity and Atomic Structure

5.116 (a)   red light = 627 – 752 nm
yellow light = 521 – 585 nm
blue light = 395 – 477 nm

 (b)   red light frequency = 4.78×10^{14} Hz to 3.99×10^{14} Hz
   photon energy = 3.17×10^{-19} J – 2.64×10^{-19} J
yellow light frequency = 5.76×10^{14} Hz to 5.13×10^{14} Hz
   photon energy = 3.82×10^{-19} J – 3.40×10^{-19} J
blue light frequency = 7.59×10^{14} Hz to 6.29×10^{14} Hz
   photon energy = 5.03×10^{-19} J – 4.17×10^{-19} J

 (c)   Frequency decreases as wavelength increases.

 (d)   As wavelength increases, the energy of the photon decreases.

5.118 (a)   Nodes are places where the electron density is zero. The principle quantum number determines this characteristic.

 (b)   Ne and Ar have more protons in the nucleu that draw the 1s electrons in close. He has fewer protons and cannot hold its electrons as tightly.

 (c)   The inner shells of electrons shield neon's nucleus, thus allowing the outermost shell of electrons to exist farther from the nucleus.

5.120   Atomic radii decrease as you move across a period. This is because the atoms are getting heavier and the positively charged nucleus can pull its electrons tighter to its core.
Atomic radii increase as you move down a group since electrons are filling shells farther away from the nucleus.

6 Ionic Bonds and Some Main-Group Chemistry

6.116 (a) non-metal
   (b) metal
   (c) The highest energy level.

6.118 The ionic radius is larger than the atomic radius in group 17. The ions have gained electrons, thereby both decreasing the effective nuclear charge and increasing the electron-electron repulsion.

6.120 Na^+, O^{2-}, Se^{2-}, F^-, Ca^{2+}
   Br^-, Al^{3+}, N^{2-}, Rb^+, P^{3-}
   Poorly shielded electrons are held tight to the nucleus. They don't have many electrons shielding them from the nucleus and so are difficult to remove. Strongly shielded electrons are far removed from the positively charged nucleus. These electrons are easy to remove.

Covalent Bonds and Molecular Structure

7.126 The strongest dipole is found when the molecule is nonsymmetrical and has components of differing electronegativity. The larger the difference in electronegativity, the larger the dipole. For example, AlH_2F has a strong dipole pointing toward the fluorine.

7.128 The formal charge equals: (number of valence electrons) – (number of bonds + (2 x non-bonded electron pairs))
For oxygen, formal charge: 6-(2+4) = 0
For carbon, formal charge: 4-(4+0) = 0
Thus, a Lewis structure, which has carbon double-bonded to each oxygen, is the most stable.

7.130

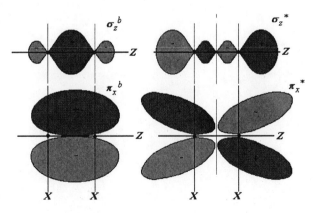

Thermochemistry: Chemical Energy

8.122 As an example of how to do the calculations, we'll do Benzoic acid.
   (a) The temperature change is $2.87°C$.
   (b) The heat gained by water= $(4.184 \text{ J/g}°C)(1000 \text{ g water})(2.87°C)= 1.20 \times 10^4$ J
   (c) $(420 \text{ J/}°C)(2.87°C)=1205$ J
   (d) $\Delta E=1.2 \times 10^4$ J + 1205 J= 1.45×10^7 J

$$\frac{1.45 \times 10^7 \text{ J}}{100 \text{ mg}} \times \frac{1000 \text{ mg}}{\text{g}} \times \frac{\text{kJ}}{1000 \text{ J}} \times \frac{123.06 \text{ g Benzoic Acid}}{\text{mol}} = 1.78 \times 10^9 \text{ kJ/mol}$$

   (e) We find ΔE because the reaction takes place at constant volume but not constant pressure.

8.124 For the reaction, ΔH is negative (exothermic), ΔS is positive and ΔG is negative. No. The reaction starts with a switch heating the NaN_3. Since the reaction is exothermic, it proceeds rapidly and becomes self-perpetuating. In addition, it is the hot $N_2(g)$ that inflates the airbag. An endothermic reaction would not produce hot $N_2(g)$, thus reducing the inflation.

Gases: Their Properties and Behavior

9.120 When the pressure is doubled, the volume is reduced by half. This illustrates Boyle's Law. When the amount of gas is doubled, the volume also doubles. This illustrates Avogadro's Law.

9.122 **(a)** 2.4 atm.
(b) The number of He molecules stays the same.
(c) The partial pressure of He stays the same. The partial pressure of He is the pressure it would exert if it were alone in the container.
(d) Total pressure= $p_{He} + p_{N2}$

9.124 The average kinetic energies for helium and neon:

$$u = \sqrt{\frac{3\,RT}{M}} = \sqrt{\frac{3 \times 8.314 \dfrac{J}{K\ mol} \times 298K}{4.00 \times 10^{-3} \dfrac{kg}{mol}}} = 1.36 \times 10^{3}\ m/s\ \text{Helium}$$

$$u = \sqrt{\frac{3\,RT}{M}} = \sqrt{\frac{3 \times 8.314 \dfrac{J}{K\ mol} \times 298K}{20.18 \times 10^{-3} \dfrac{kg}{mol}}} = 6.07 \times 10^{3}\ m/s\ \text{Neon}$$

So the ratio of He/Ne=2.24. Helium moves 2.24 times faster than neon on average.

Gas pressure is a measure of the number and forcefulness of collisions between gas particles and the walls of their container. When the kinetic energy of a gas increases, the atoms move faster in their container, increasing the number of collisions with the walls. Thus we see that the pressure increases at higher temperatures.

Liquids, Solids and Phase Changes

10.122   The B-F bond has a strong dipole moment toward the fluorine. However, BF_3 has trigonal planer geometry and each dipole cancels each other out.

10.124 **(a)**   The energy required for evaporation does not change.
  (b)   At higher temperatures, there are a greater number of molecules with the minimum amount of kinetic energy needed to escape the liquid.
  (c)   At higher temperatures, more of the molecules can exist in the vapor phase, leading to an increased rate of evaporation and increased pressure.

10.126 Ionic solids are composed of ions. The ions are ordered in a regular three-dimensional array and are held together by ionic bonds. Molecular solids, like ice, contain molecules held together by intermolecular forces such as hydrogen bonding. Covalent network solids have atoms linked together by covalent bonds into a giant three-dimensional array. Metallic solids also consist of large arrays of atoms, but their crystals have metallic properties such as electrical conductivity.

Solutions and Their Properties

11.136 The units based on mass are mass percent, parts per million and parts per billion. They only differ by the number the ratio is multiplied by- mass percent uses 100, ppm uses 10^6 and ppb uses 10^9. Molarity, molality and mole fraction are determined directly from the moles of solute. With mole fraction, we look at the # of moles of solute/ total # of moles; molarity is the ratio of moles of solute/ volume of solution in L and molality is the ratio of moles solute/ kg of solution.

11.138 **(a)** Ammonium nitrate does not dissociate as readily in solution as either NaCl or KOH. Thus the number of ions present in solution is less than with either of the more ionic compounds. Since boiling-point elevation and freezing point depression are determined by the number of ions in solution, we see a lesser effect.

(b) Ethylene glycol increases the entropy of solution, thereby lowering the freezing point.

Chemical Kinetics

12.132 At 100 seconds:

$[N_2O_5]$= 0.017 M
$[NO_2]$= 0.006 M
$[O_2]$= 0.0015 M

At 500 seconds:

$[N_2O_5]$= 0.009 M
$[NO_2]$=0.024 M
$[O_2]$= 0.006 M

The rate of appearance of NO_2 is always four times the rate of appearance of O_2. The rate of appearance for each compound is determined by the net ionic equation.

Starting with equal concentrations of N_2O_5 and O_2 would not necessarily change the relative rates of appearance. The reaction is only dependent upon the concentration of N_2O_5.

12.134 At 0°C, $k = \dfrac{\text{rate}}{[A]} = \dfrac{0.00729 \text{ M/s}}{[0.4 \text{ M}]} = 0.018 \text{ s}^{-1}$

The graph shows the half life is 40.0 s. Doubling the initial concentration does not affect the half life at all. It remains 40.0 s.

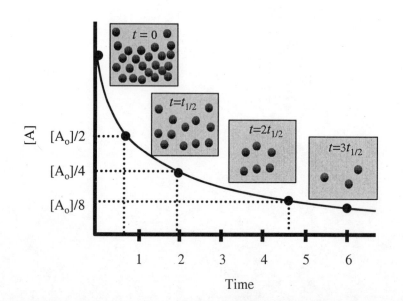

12.136

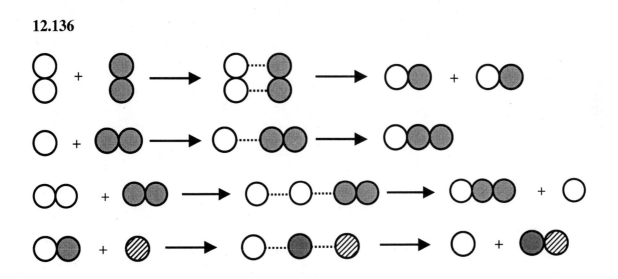

Chemical Equilibrium

13.130 (a) $K_c = \dfrac{[NO_2]^2}{[N_2O_4]}$

(b) The equilibrium lies to the left. If the equilibrium lay to the right, the $[NO_2]$ would be greater than the $[N_2O_4]$.

(c) Yes. Since the reaction is reversible, the $[NO_2]$ can combine to form $[N_2O_4]$.

(d) $K_c = \dfrac{[N_2O_4]}{[NO_2]^2}$

13.132 (a) $K_p = 1.83$. For this reaction $K_c = K_p$ because the only components in the reaction are gases and solids.

(b) $CaCO_3$ and CaO are not included in the equilibrium constant expressions because they are solids.

(c) There are no gases so a K_p expression would not be used.

Hydrogen, Oxygen, and Water

14.114 $H_2(g) + O_2(g) \rightarrow 2\,H_2O(l)$

Yes, this reaction is an oxidation-reduction reaction. Oxygen is reduced and so is the oxidizing agent. Hydrogen is oxidized and so is the reducing agent.

14.116 $CO_2(g) + H_2O(l) \rightarrow H^+(aq) + HCO_3^-(aq)$

Anhydride means "without hydrogen" it is the product that results from dehydration of an acid. The acid anhydride of sulfuric acid is SO_3 and the acid anhydride of nitric acid is N_2O_5.

$SO_3 + H_2O \rightarrow H_2SO_4$

$N_2O_5 + H_2O \rightarrow 2\,HNO_3$

(Note: NO_2 is not the anhydride of HNO_3 because no change in oxidation number can take place when an acid anhydride reacts with water to form the acid. The oxidation number of nitrogen is +4 in NO_2 and +5 in HNO_3. Thus the acid anhydride of HNO_3 is N_2O_5 where N has an oxidation number of +5.)

14.118 $2\,Na(s) + 2\,H_2O(l) \rightarrow H_2(g) + 2\,Na^+(aq) + 2\,OH^-(aq)$

This is an oxidation-reduction reaction with water acting as the oxidizing agent (and being reduced) and sodium acting as the reducing agent (and being oxidized).

Aqueous Equilibria: Acids and Bases

15.140 (a) $[H_3O^+] = 1 \times 10^{-14}/[OH^-]$

(b) $[OH^-]$ concentration decreases when acid is added, because the H^+ bonds with OH^- to form water.

15.142 $pH = 11.83 = -\log[H^+]$
$[H^+] = 1.48 \times 10^{-12}$ M

% dissociation at 0.1 M $= (1.48 \times 10^{-12}$ M $/0.1$ M$) \times 100$
$= 1.48 \times 10^{-9}$ %

% dissociation at 0.001 M $= 1.48 \times 10^{-7}$ %
% dissociation at 1.0×10^{-5} M $= 1.48 \times 10^{-5}$ %
The lower the pH, the greater the percent dissociation. For acids, the higher the pH, the greater the dissociation.

15.144

$$LiNO_2 + H_2O \rightarrow HNO_2 + LiOH$$
base   conj. Acid

$$NH_4Cl + H_2O \rightarrow NH_3 + H_3O^+ + Cl^-$$
Acid   conj. Base

$$Ca(NO_3)_2 + 2\,H_2O \rightarrow Ca(OH)_2 + 2\,HNO_3$$
Base   conj. Acid

$$MgSO_4 + H_2O \rightarrow HSO_4^- + MgOH$$
Base   conj. Acid

$$AgClO_3 + H_2O \rightarrow AgOH + HClO_3$$
Acid   conj. Base

$$CsCN + H_2O \rightarrow HCN + CsOH$$
Base   conj. acid

Applications of Aqueous Equilibria

16.150 Using HOAc as an example:
When you add a strong base, we see this reaction
$$HOAc + OH^- \rightarrow H_2O + OAc^-$$

Adding a strong acid, gives us
$$H^+ + OAc^- \rightarrow HOAc$$
To increase the pH of our buffer, we would add more base.

16.152 **(a)** pH= pK_a + log [B]/[A]
pK_a= 3.755
K_a= 1.760 X 10^{-4}

(b) The volume of water comes into the calculation with the concentrations of B and A. Since the Henderson-Hasselbach equation utilizes the ratio of [B]/[A], the volume components cancel each other out.

(c) HClO (pK_a= 7.304) would be the best buffer choice. A buffer solution is most effective when the pH is close in value to the acids' pK_a.

16.154 **(a)** The precipitation with H_2S of metal cations occurs only when the solution is strongly acidic. If H_2S is added first, no group II ions will precipitate. When acid is then added (HCl), both the group I and the group II ions will precipitate. This gives misleading results, suggesting that the precipitate is a chloride when in reality it could also be a sulfide.
(b) Addition of NH_3 makes the solution basic, causing the metal sulfide solubility equilibrium to shift to the left. Thus, MnS, FeS, CoS, NiS, ZnS and the insoluble hydroxides $Al(OH)_3$ and $Cr(OH)_3$ will precipitate out.
(c) Mg is identified by adding $(NH_4)_2HPO_4$. White $Mg(NH_4)PO$ will precipitate out. Sodium and potassium are identified by flame tests. Sodium burns yellow while potassium burns violet.

Thermodynamics: Entropy, Free Energy, and Equilibrium

17.124 The gas moves to fill up the empty space. Yes, this is a spontaneous process. When the gas expands to fill the additional spaced, it increases in disorder (increases entropy). It is not possible to spontaneously have the gas return to the one side of the vessel. This would decrease the entropy of the gas and so would require an outside force acting on the gas to make it go back.

17.126 **(i)** Molecules are warming in the solid phase. The intermolecular forces become strained as molecules have more kinetic energy.
(ii) The compound is melting. Intermolecular forces break and a liquid is created. The molecules can now move freely with respect to each other.
(iii) The liquid warms. Molecules have more and more movement in solution.
(iv) Vaporization begins. Molecules escape the solution and enter the vapor phase.
(v) The gas heats up, resulting in more and more movement amongst the molecules.

17.128 **(a)** ΔG is negative, ΔH is negative, ΔS is positive
(b) No. The reaction will always be spontaneous, because $-\Delta H - (T)(+\Delta S)$ will always have a negative value.

Electrochemistry

18.130   This will not generate voltage because the redox reaction takes place at the metal-solution interface and involves direct transfer of $2e^-$ from Zn atoms to Cu^{2+} ions. The enthalpy of this reaction is lost to the surroundings as heat. If the reaction were carried out using an electrochemical cell, then some of the chemical energy could be converted to electrical energy.

18.132   **(a)** $Cu^{2+}(aq) + Zn(s) \rightarrow Cu(s) + Zn^{2+}(aq)$
(b) If you add Cu^{2+} to the solution, it increases the tendency of the reaction to take place, thus increasing the cell potential. According to Le Chatelier's principle, adding reactant will cause the reaction to shift toward products.
(c) If you add more Zn^{2+} to the solution, it reduces the tendency of the reaction to happen and reduces the cell voltage. Adding product shifts the equilibrium back toward the reactants.

18.134   **(i)** $Al^{3+}(aq) + 3e^- \rightarrow Al(s)$
(ii) $Ag^+(aq) + 1e^- \rightarrow Ag(s)$
(iii) $Au^+(aq) + 1e^- \rightarrow Au(s)$
(iv) $Cu^{2+}(aq) + 2e^- \rightarrow Cu(s)$
The activity confirms that the metal ion charge, current level and duration of electrolysis all determine the moles of metal deposited. Silver and gold will deposit in the greatest number of moles because of their +1 charge.

The Main Group Elements

19.126 **(a)** Boron only has 6e- and as a result, is very reactive. The boron halides all have an empty p orbital available that can accept electrons from a Lewis base.

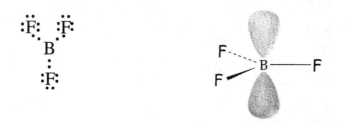

(b) The outer B-H bonds each contain 2 electrons, but the B-H bonds on the inside of the molecule only contain 1and a1/2 electrons each.

19.128 **(a)** $2 NO_2(g) \longleftrightarrow N_2O_4(g)$
(b) The dimer formation is exothermic, so at high temperatures, NO_2 predominates, and at low temperatures, N_2O_4 predominates.

Transition Elements and Coordination Chemistry

20.140 (i) Boiling point increases for transition metals and peaks in the middle of the periodic table. This is because the transition metals can share d as well as s electrons with each other, resulting in stronger bonding between atoms and a higher boiling point.

(ii) Atomic radius decreases as you move across and up the periodic table. The smaller atoms can hold onto their electrons more efficiently and occupy a smaller region of space. As you move to the right of the periodic table, the atoms have a greater positive charge in the nucleus and can bring their electrons in tighter.

(iii) Ionization energy increases as you move across and up the periodic table. Ionization energy is a measure of how tightly an atom holds onto its electrons and we see higher ionization energies with small atoms that have their valence electrons in the lower energy shells. Toward the right of the periodic table, we have larger positive charges present in the nucleus, allowing the atoms to hold tightly onto their electrons.

20.142 (a) cis-[Cr(en)$_2$Cl$_2$]$^+$

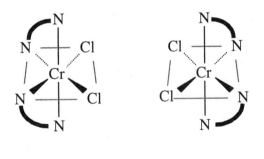

(b) cis-[Cr(C$_2$O$_4$)$_2$(H$_2$O$_2$)]$^-$

(c) $[Ni(en)_3]^{2+}$

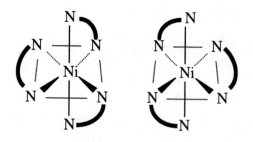

20.144 $[Cr(H_2O)_5]^{3+}$ appears violet and $[Cr(CN)_6]^3+$ appears yellow. H_2O is a weak field ligand and will result in a high-spin electronic configuration. CN^- is a strong-field ligand and will result in electrons in a low-spin configuration. The two compounds will therefore absorb light differently.

Metals and Solid-State Materials

21.120 $Fe_2O_3(s) + 2Al(s) \rightarrow Al_2O_3(s) + 2Fe(s)$

$KClO_4$, sugar and H_2SO_4 are catalysts that start the reaction.

21.122 (i) In an insulator, the bonding and antibonding orbitals are separated by a large band gap. There are no vacant MOs to accept an excited electron so the compound has low electrical conductivity.

(ii) An undoped semiconductor has a band gap that is smaller than what is seen with an insulator. A few electrons can jump the gap and occupy the higher-energy conduction band. A small amount of current can be transmitted because the electrons can be accelerated in the partially filled bands.

(iii) An n-type semiconductor has an impurity like phosphorus added to the semiconductor. Phosphorus has an extra electron available that is not required for bonding. The extra electrons occupy the conduction band and the conductivity is correspondingly higher.

(iv) In a p-type semiconductor, the semiconductor is doped with boron. Boron only has 6 electrons surrounding it so there are not enough electrons to fill all the bonding molecular orbitals. The vacancies in the filled bands therefore allow for greater conductivity.

21.124 Ceramics have higher melting points than metals and are stiffer, harder and more resistant to wear and corrosion. They also retain their strength at high temperatures. The disadvantage of ceramics is that they are brittle compared to metal. The ceramic can't deform under stress because its bonds are strong and highly directional. When placed under stress, the bonds will break suddenly and the ceramic will shatter, while metal is malleable.

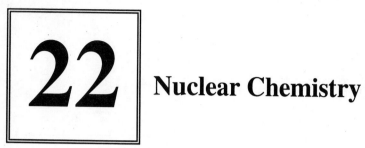

Nuclear Chemistry

22.100 (a) It takes 5 half-lives for 96.875% of the sample to decompose.

(b) Since the half life is 10 minutes, twenty-five percent of the original sample will still remain after twenty minutes.

(c) We use the integrated rate law:

$$\ln\left(\frac{N}{N_o}\right) = (-\ln 2)\left(\frac{t}{t_{1/2}}\right)$$

so, $\quad \ln\dfrac{(0.211\,\text{g})}{(13.5\,\text{g})} = (-0.693)\dfrac{(t)}{10\,\text{min}}$

t=60 minutes

22.102 We use the integrated rate law to solve for ^{232}Th half-life:

$$\ln\frac{15.64\,\text{kg}}{20\,\text{kg}} = (-\ln 2)\left(\frac{5,000,000,000\ \text{years}}{t_{1/2}}\right)$$

$t_{1/2}= 1.41 \times 10^{10}$ years

Now, we can use the calculated half-life and solve for N:

$$\ln\frac{N}{20\,\text{kg}} = (-0.693)\left(\frac{5.05 \times 10^9\ \text{years}}{1.41 \times 10^{10}\ \text{years}}\right)$$

N= 15.61 kg

So $\dfrac{15.61\,\text{kg}}{20\,\text{kg}} \times 100 = 78.0\%$ would remain after 5.05×10^9 years.

22.104 Section one is a closed system and has water around the control rods. This water is exposed to uranium and becomes contaminated by radioactivity. The control rods heat the water as the reaction proceeds. Section two is a closed system and contains water that becomes heated from contact with the hot pipes of section one. The water converts to steam, which drives a turbine, generating electricity. Section three is cool water pulled from a large outside source which condenses the steam in section two back to liquid. The heated water of section three is then dumped back into the lake or ocean.

Organic Chemistry

23.120 (a) Hydroxyl groups participate better in hydrogen bonding than amines. Boiling points for the alcohols are always higher than boiling points for the corresponding amines.
(b) The oxygen has two lone pairs of electrons available to form a hydrogen bond with another alcohol. The nitrogen on amines only has one lone pair of electrons and thus doesn't participate as strongly.

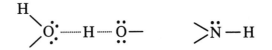

23.122

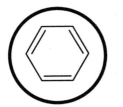

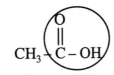

The $CH_3-CH_2-\boxed{NH_2}$ functional $CH_3-\overset{\overset{\displaystyle O}{\|}}{C}-NH_2$ groups that can participate in hydrogen bonding are alcohols, carboxylic acids and acetamides because the OH and NH groups have a hydrogen and a lone pair of electrons available.

23.124 The structural formula for nylon.

$$Cl-\overset{\overset{\displaystyle O}{\|}}{C}-(CH_2)_8-\overset{\overset{\displaystyle O}{\|}}{C}-Cl \ + \ H_2N-(CH_2)_6-NH_2 \longrightarrow \left[HN-(CH_2)_6-NH-\overset{\overset{\displaystyle O}{\|}}{C}-(CH_2)_8-\overset{\overset{\displaystyle O}{\|}}{C}\right]_n \ + \ HCl$$

Biochemistry

24.102 (a) A carbon must be bonded to four different groups to be chiral.
(b) There is only one chiral carbon, C3, bonded to H, methyl, ethyl and isobutyl groups.

$$CH_3 - CH - CH_2 - CH - CH_2 - CH_3$$

with CH_3 branches on the 2nd and 4th carbons.

24.104 The four functional groups are the methyl, hydrogen, amine and carboxylic acid.

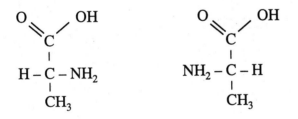

24.106

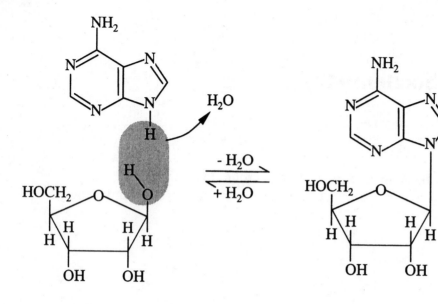

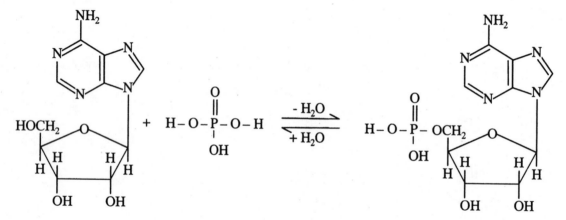